PERSPECTIVES IN CONTROL ENGINEERING

IEEE Press
445 Hoes Lane, P.O. Box 1331
Piscataway, NJ 08855-1331

IEEE Press Editorial Board
Robert J. Herrick, *Editor in Chief*

M. Akay	M. Eden	M. Padgett
J. B. Anderson	M. E. El-Hawary	W. D. Reeve
P. M. Anderson	R. F. Hoyt	M. S. Newman
J. E. Brewer	S. V. Kartalopoulos	G. Zobrist
	D. Kirk	

Kenneth Moore, *Director of IEEE Press*
Catherine Faduska, *Senior Acquisitions Editor*
John Griffin, *Acquisitions Editor*
Robert Bedford, *Assistant Acquisitions Editor*
Anthony VenGraitis, *Project Editor*
Marilyn Catis, *Marketing Manager*

IEEE Control Systems Society, *Sponsor*
CSS Liaison to IEEE Press, Bruce M. Krogh

Cover design: William T. Donnelly, *WT Design*

Technical Reviewers
Bruce H. Krogh, *Carnegie Mellon University, Pittsburgh, PA*
Dr. Michael K. Masten, *Texas Instruments, Plano, TX*
Harris N. McClamroch, *University of Michigan, Ann Arbor, MI*
Siva S. Banda, *Wright Patterson AFB, OH*

Books of Related Interest from the IEEE Press

THE CONTROL HANDBOOK
Edited by William S. Levine
A CRC Handbook published in cooperation with IEEE Press
1995   Hardcover   1,568 pp   IEEE Order No. PC5649   ISBN 0-8493-8570-9

INTELLIGENT CONTROL SYSTEMS: Theory and Applications
Edited by Madan M. Gupta and Naresh K. Sinha
1996   Hardcover   865 pp   IEEE Order No. PC4176   ISBN 0-7803-1063-2

UNDERSTANDING ELECTRO-MECHANICAL ENGINEERING: An Introduction to Mechatronics
Lawerence J. Kamm
1996   Softcover   416 pp   IEEE Order No. PP3806   ISBN 0-7803-1031-4

ROBUST VISION FOR VISION-BASED CONTROL OF MOTION
Edited by Markus Vincze and Gregory D. Hager
2000   Hardcover   272 pp   IEEE Order No. PC5403   ISBN 0-7803-5378-1

PERSPECTIVES IN CONTROL ENGINEERING

Technologies, Applications, and New Directions

Tariq Samad
Honeywell Technology Center
Minneapolis, MN

IEEE Control Systems Society, *Sponsor*

IEEE PRESS

The Institute of Electrical and Electronics Engineers, Inc., New York

This book and other books may be purchased at a discount from the publisher when ordered in bulk quantities. Contact:

IEEE Press Marketing
Attn: Special Sales
445 Hoes Lane
P.O. Box 1331
Piscataway, NJ 0885-1331
Fax: +1 732 981 9334

For more information about IEEE Press products, visit the
IEEE Online Catalog & Store at http://www.ieee.org/iceestore.

©2001 by the Institute of Electrical and Electronics Engineers, Inc.
3 Park Avenue, 17th Floor, New York, NY 10016-5997

All rights reserved. No part of this book may be reproduced in any form, nor may it be stored in a retrieval system or transmitted in any form without written permission from the publisher.

Printed in the United States of America.

10 9 8 7 6 5 4 3 2 1

ISBN 0-7803-5356-0
IEEE Order No. PC5798

Library of Congress Cataloging-in-Publication Data

Samad, Tariq.
  Perspectives in control engineering: technologies, applications, and new directions Tariq Samad.
    p. cm.
  "IEEE Control Systems Society, sponsor."
  Includes bibliographical references and index.
  ISBN 0-7803-5356-0
  1. Automatic control. 2. Control theory. I. IEEE Control Systems Society. II. Title.

TJ213 .S1145 2000   00-038854
629.8–dc21   CIP

CONTENTS

Introduction xvii

Acknowledgments xxiii

PART I   CONTROL TECHNOLOGIES

CHAPTER 1   REAL-TIME COMPUTING AND CONTROL   1
Scott Bortoff

Editor's Summary   1
1.1 Introduction   1
    1.1.1 Background   2
1.2 Timing Is Everything   4
1.3 Low-Level Real-Time Programming   6
    1.3.1 Fixed-Priority Scheduling Theory   8
    1.3.2 Data Dependence   10
1.4 Real-Time Operating Systems and Programming Languages   11
    1.4.1 Real-Time Operating Systems at Run-Time   13
1.5 Hardware Issues   17
    1.5.1 Desktop PCs   17
    1.5.2 Single-Board Computers   17
    1.5.3 Digital Signal Processors   18
    1.5.4 Programmable Logic Controllers   18
1.6 Conclusion   19
    References   19

CHAPTER 2   DISCRETE-EVENT SYSTEMS AND THEIR OPTIMIZATION   20
Edwin K. P. Chong

Editor's Summary   20
2.1 Introduction   20
2.2 Discrete-Event Systems   22
    2.2.1 What Is a System?   22
    2.2.2 What Is a Discrete-Event System?   23
    2.2.3 Why the Need for Discrete-Event Systems?   24
2.3 Some Discrete-Event System Models   25
    2.3.1 State Trajectory of a Discrete-Event System   25
    2.3.2 State Model of a Discrete-Event System   26
        2.3.2.1 State Machines   26
        2.3.2.2 Nondeterministic State Machines   27
        2.3.2.3 Markov Chains   28

              2.3.2.4   State Machines and Discrete-
                        Event Systems   29
      2.3.3   State Models with Event Clocks   29
              2.3.3.1   Event Clocks   30
              2.3.3.2   Discrete-Event Simulations   32
              2.3.3.3   Markov and Semi-Markov
                        Processes   32
  2.4   Optimization of Discrete-Event Systems   33
      2.4.1   What Is Optimization?   33
      2.4.2   Gradient Algorithms for Optimization   34
      2.4.3   Gradient Estimation   35
      2.4.4   Online Optimization   37
              2.4.4.1   Basic Idea   37
              2.4.4.2   Example Application   37
  2.5   Further Reading   40
        Acknowledgments   40
        References   40

CHAPTER 3   COMPUTER-AUTOMATED CONTROL SYSTEM DESIGN   42
Georg Grübel

  Editor's Summary   42
  3.1   Introduction   42
  3.2   Control Design Life Cycle to Be Supported by CACSD   45
  3.3   Design Modeling and Synthesis Algorithms   47
      3.3.1   Physical System Modeling   48
      3.3.2   Synthesis Algorithms and Controller Modeling   51
      3.3.3   Performance Evaluation Setup   53
  3.4   Quality Modeling for Design Analysis and Decision Making   54
      3.4.1   Quality Functions   56
      3.4.2   Feasible Design and Competing Requirements   58
      3.4.3   Visualization for Comparative Design Exploration   59
  3.5   Automatic Tuning and Declarative Compromising   60
      3.5.1   Automated Tuning by Multi-Objective Parameter Optimization   62
      3.5.2   Declarative Compromising   63
      3.5.3   Robust Control Laws by Multimodel Compromising   66
  3.6   Further CACSD Technology   66
        Acknowledgments   69
        References   69

CHAPTER 4   SYSTEM MODELING   71
Pradeep Misra

  Editor's Summary   71
  4.1   Introduction   71
      4.1.1   Historical Perspective   72
      4.1.2   Modeling and Control   73
      4.1.3   Classification   73
  4.2   Static Models   75
      4.2.1   Linear Models   75
      4.2.2   Nonlinear Models   77
  4.3   Dynamical Models   79

	4.3.1	Lumped Parameter Models 80
	4.3.2	System Identification 82
		4.3.2.1 Transfer Function Models 83
		4.3.2.2 State Space Models 84
	4.3.3	Model Reduction 86
		4.3.3.1 Modal Truncation 86
		4.3.3.2 Singular Perturbation 86
		4.3.3.3 Balanced Reduction 87

4.4 Nonlinear Dynamical Systems 88
    4.4.1 Common Effects of Nonlinearities 89
    4.4.2 Linearization 92
        4.4.2.1 Local Linearized Approximation 92
        4.4.2.2 Describing Function Approximation 93
        4.4.2.3 Feedback Linearization 95

4.5 Models of Distributed Parameter Systems 95
    4.5.1 Classification of PDEs 96
    4.5.2 Finite Difference Models of PDEs 97
        4.5.2.1 Explicit Models (Forward Differences) 98
        4.5.2.2 Implicit Models (Backward Differences) 99

4.6 Macromodels: Scope and Future 99
4.7 Remarks 101
Acknowledgment 102
References 102

CHAPTER 5   INTELLIGENT CONTROL: AN OVERVIEW OF TECHNIQUES   104

Kevin M. Passino

Editor's Summary 104
5.1 Introduction 104
5.2 Intelligent Control Techniques 105
    5.2.1 Fuzzy Control 105
        5.2.1.1 Fuzzy Control Design 106
        5.2.1.2 Ship Example 109
        5.2.1.3 Design Concerns 109
    5.2.2 Neural Networks 110
        5.2.2.1 Multilayer Perceptrons 110
        5.2.2.2 Training Neural Networks 113
        5.2.2.3 Design Concerns 115
    5.2.3 Genetic Algorithms 116
        5.2.3.1 The Population of Individuals 116
        5.2.3.2 Genetic Operators 117
        5.2.3.3 Design Concerns 118
    5.2.4 Expert and Planning Systems 119
        5.2.4.1 Expert Control 119
        5.2.4.2 Planning Systems for Control 120
    5.2.5 Intelligent and Autonomous Control 121
5.3 Applications 122
    5.3.1 Heuristic Construction of Nonlinear Controllers 122
        5.3.1.1 Model-Free Control? 122
        5.3.1.2 Example: Vibration Damping in a Flexible-Link Robot 123
    5.3.2 Data-Based Nonlinear Estimation 124

                        5.3.2.1   Estimator Construction
                                  Methodology   124
                        5.3.2.2   Example: Automotive Engine Failure
                                  Estimation   125
               5.3.3   Intelligent Adaptive Control Strategies   126
                        5.3.3.1   Fuzzy, Neural, and Genetic Adaptive
                                  Control   126
                        5.3.3.2   Example: Adaptive Fuzzy Control for
                                  Ship Steering   128
      5.4   Concluding Remarks: Outlook on Intelligent Control   130
               Acknowledgments   132
               References   132

CHAPTER 6   NEURAL, FUZZY, AND APPROXIMATION-BASED CONTROL   134
Jay A. Farrell and Marios M. Polycarpou

      Editor's Summary   134
      6.1   Introduction   134
              6.1.1   Components of Approximation-Based Control   135
                       6.1.1.1   Control Architecture   135
                       6.1.1.2   Approximator   136
                       6.1.1.3   Stable Training Algorithm   137
              6.1.2   Problem Statement   138
              6.1.3   Discussion   139
      6.2   Control Architectures   140
              6.2.1   Indirect Methods   141
              6.2.2   Direct Methods   142
      6.3   Approximator Properties   142
              6.3.1   Universal Approximator   143
              6.3.2   Parameter (Non)Linearity   145
              6.3.3   Best Approximator Property   147
              6.3.4   Generalization   147
              6.3.5   Extent of Influence Function Support   149
                       6.3.5.1   Approximators with Local Influence
                                 Functions   149
                       6.3.5.2   Lattice-Based Approximators   151
                       6.3.5.3   Curse of Dimensionality   151
              6.3.6   Approximator Transparency   151
      6.4   Parameter Estimation: Online Approximation   152
              6.4.1   Parametric Models   152
              6.4.2   Gradient Algorithms   155
              6.4.3   Least-Squares Algorithms   157
              6.4.4   Lyapunov-Based Algorithms   158
              6.4.5   Robust Learning Algorithms   160
      6.5   Conclusions   162
              References   163

CHAPTER 7   SUPERVISORY HYBRID CONTROL SYSTEMS   165
Michael D. Lemmon

      Editor's Summary   165
      7.1   Introduction   165
      7.2   Examples of Supervisory Hybrid Systems   166
              7.2.1   Switched Dynamical Systems   167
              7.2.2   Asynchronous Sequential Circuits   169
      7.3   Hybrid Automaton   170

Contents

        7.3.1   Definition of the Hybrid Automaton   171
        7.3.2   Robotic System Example: Revisited   174
   7.4 Hybrid Specifications   177
   7.5 Hybrid System Analysis   179
   7.6 Hybrid Control System Synthesis   183
   7.7 Summary   185
        Acknowledgments   186
        References   186

CHAPTER 8   VARIABLE STRUCTURE AND SLIDING-MODE CONTROL   189
Fumio Hamano and Younchan Kim

   Editor's Summary   189
   8.1 Introduction   189
   8.2 Basic Idea of Sliding-Mode Control   193
        8.2.1   Tracking Problem and Tracking Error Dynamics   193
        8.2.2   Choosing a Sliding Surface (or Line)   194
        8.2.3   Control Law to Confine the State on the Sliding Surface   194
        8.2.4   Control Law for Reaching the Sliding Surface (and Staying on It)   195
        8.2.5   Robust Sliding-Mode Control   195
        8.2.6   Generalized Lyapunov Function   196
        8.2.7   Preventing Chattering by Continuous Approximation   197
   8.3 Sliding-Mode Control: General Case   198
        8.3.1   Problem Formulation   199
        8.3.2   Sliding Surface   200
        8.3.3   Robust Sliding-Mode Control   201
        8.3.4   Continuous Approximation to Avoid Chattering   203
        8.3.5   Example: Single Degree of Freedom Robot   203
   8.4 Sliding-Mode-Like Control For Sampled Data Control Systems   206
   8.5 Concluding Remarks   216
        References   216

CHAPTER 9   CONTROL SYSTEMS FOR "COMPLEXITY MANAGEMENT"   218
Tariq Samad

   Editor's Summary   218
   9.1 Introduction   218
        9.1.1   Control Systems: Domain Knowledge and Solution Technologies   219
   9.2 Control and Automation Tomorrow: Toward Complexity Management   221
   9.3 Objectives for Control and Automation   221
        9.3.1   Human and Environmental Safety   222
        9.3.2   Regulatory Compliance   223
        9.3.3   Time and Cost to Market   223
        9.3.4   Increased Autonomy   224
        9.3.5   Other Criteria: Yield, Capacity, Efficiency, and More   224
   9.4 Emerging Control Technologies for Complex Systems   225

		9.4.1	Randomized Algorithms 225	

9.4.1 Randomized Algorithms 225
9.4.2 Biologically Motivated Control 226
9.4.3 Complex Adaptive Systems 227
9.4.4 Distributed Parameter Systems 228
9.5 New Application Opportunities for Control and Automation 228
    9.5.1 Large-Scale and Enterprisewide Optimization 228
    9.5.2 Integration of Business and Physical Systems 229
    9.5.3 Autonomous Vehicles 229
    9.5.4 Data Mining and Intelligent Data Analysis 231
    9.5.5 Control Systems and the World Wide Web 231
9.6 Schools of Complexity Management 231
    9.6.1 Human and Environmental Safety: Forfeiture and Risk Assessment 232
    9.6.2 Efficiency in Design: System Engineering and Virtuality 233
    9.6.3 Nature and Biology: Evolution, Emergence, and Power Laws 234
    9.6.4 Societal Connections 235
9.7 Conclusions 236
References 237

PART II   CONTROL APPLICATIONS

CHAPTER 10   CONTROL OF MULTIVEHICLE AEROSPACE SYSTEMS   239
Jorge Tierno, Joseph Jackson, and Steven Green

Editor's Summary 239
10.1 Introduction 239
10.2 Future Controls Applications and Challenges in ATM 241
    10.2.1 Preliminaries: Airspace and Air Traffic Management 241
    10.2.2 Air Traffic Capacity Management in the Presence of Disturbances 242
        10.2.2.1 Initial Conditions and Framework 243
        10.2.2.2 Control Variables 244
        10.2.2.3 State Variables 244
        10.2.2.4 Disturbances 245
        10.2.2.5 Control System 245
    10.2.3 Enabling User Preferences in a Safety-Constrained ATM System 246
        10.2.3.1 Development of Distributed Separation Assurance Procedures 247
        10.2.3.2 ATM Considerations 247
        10.2.3.3 Flight Deck Considerations 248
        10.2.3.4 Airline Operating Center (AOC) Considerations 248
    10.2.4 "Executing to Plan" in Constrained Airspace: Terminal Area Operations 249
10.3 Example 2: Uninhabited (Combat) Air Vehicles 249
    10.3.1 Inter-Fleet and Central-Command-to-Fleet Communications 250
    10.3.2 Safety Analysis and Conflict Resolution 251
    10.3.3 Autonomy 252
10.4 Example 3: Formation Flying and Satellite Clusters 253
    10.4.1 Multi-Agent Systems and Decentralized

Contents xi

       Distributed Control 254
        10.4.1.1 Emergent Behavior 255
        10.4.1.2 Flocking 255
        10.4.1.3 Market-Oriented Programming 255
     10.4.2 Distributed Processing 255
  10.5 Conclusions 256
     Acknowledgments 257
     References 257

CHAPTER 11 AFFORDABLE FLIGHT CONTROL FOR AIRCRAFT AND MISSILES 259
Kevin A. Wise

  Editor's Summary 259
  11.1 Introduction 259
  11.2 Aircraft and Missile Dynamics and Linear Models 260
  11.3 Simulation Tools 268
  11.4 Flight Control System Design 269
     11.4.1 Aircraft Control Law Design Using Dynamic Inversion 270
     11.4.2 Missile Control Law Design Using Linear Quadratic Optimal Control 273
     11.4.3 Zero Shaping to Improve Control System Design 279
  11.5 Analysis Tools 281
     11.5.1 Linear Analysis Models 281
     11.5.2 Performance Analysis 283
     11.5.3 Robustness Analysis 283
  11.6 Digital Implementation, Reusable Software, and Autocode 286
  11.7 Flight Control Challenges in the Twenty-First Century: Unmanned Aircraft 287
     References 290

CHAPTER 12 INDUSTRIAL PROCESS CONTROL 291
Michael A. Johnson and Michael J. Grimble

  Editor's Summary 291
  12.1 Introduction 291
  12.2 Industrial Process Control Technology: State of the Art 292
     12.2.1 The Information Technology Infrastructure for Process Control 293
     12.2.2 Process Control Applications Software 294
       12.2.2.1 Control Application Suite 1 294
       12.2.2.2 Control Application Suite 2 295
       12.2.2.3 Control Application Suite 3 295
     12.2.3 Data Communications and Standards 296
     12.2.4 Summary Conclusions 296
  12.3 Organizing Process Control Applications/Production Processes 297
     12.3.1 The Industrial Operations Hierarchy: Strategy Issues 297
     12.3.2 The Industrial Operations Hierarchy: Information Issues 299
  12.4 Performance Monitoring 300
     12.4.1 Statistical Process Control 301
     12.4.2 Performance Quality Indices 302

　　　　　12.4.3　Benchmarking Process Control　　305
　　　　　12.4.4　Summary Conclusions　　306
　　12.5　Industrial Three Term Control　　306
　　　　　12.5.1　The Sustained Oscillation Procedure　　307
　　　　　　　　　12.5.1.1　Procedure 1: Method of Sustained Oscillation　　308
　　　　　12.5.2　Why Autotune?　　309
　　　　　　　　　12.5.2.1　Problems with Ziegler–Nichols PID Tuning　　309
　　　　　　　　　12.5.2.2　A Technology Changeover in the 1980s　　310
　　　　　　　　　12.5.2.3　Process Controller Technology Today　　310
　　　　　12.5.3　The Relay Experiment　　310
　　　　　　　　　12.5.3.1　Nonparametric Identification by Relay Experiment　　311
　　　　　　　　　12.5.3.2　PID Control　　311
　　　　　　　　　12.5.3.3　Procedure 2: The Relay Experiment　　312
　　　　　12.5.4　Recent Directions for Industrial PID　　312
　　12.6　Adaptation and Robustness　　313
　　　　　12.6.1　Adaptation　　313
　　　　　12.6.2　Robustness　　313
　　12.7　Aspects of Global System Optimization　　314
　　　　　12.7.1　The Supervisory System Command Structure　　315
　　　　　　　　　12.7.1.1　Low-Level Control Strategies　　315
　　　　　　　　　12.7.1.2　Dynamic Setpoint Maneuvers　　315
　　　　　　　　　12.7.1.3　Setpoint Optimization and Load Management Strategy　　316
　　　　　12.7.2　Model-Based Predictive Control　　316
　　　　　　　　　12.7.2.1　The Basics of Model-Based Predictive Control　　317
　　　　　　　　　12.7.2.2　A Process Model　　317
　　　　　　　　　12.7.2.3　A Predictive Model Equation　　317
　　　　　　　　　12.7.2.4　A Process Cost Function　　317
　　　　　　　　　12.7.2.5　A Receding Horizon Control Philosophy　　318
　　　　　　　　　12.7.2.6　Some MPC Tuning Parameters　　318
　　　　　　　　　12.7.2.7　The Two Key Advantages of MPC　　318
　　　　　　　　　12.7.2.8　MPC Architectures　　319
　　　　　　　　　12.7.2.9　Finally, the Industrial Varieties of MPC　　319
　　12.8　Conclusions　　319
　　　　　Acknowledgements　　321
　　　　　References　　321

CHAPTER 13　POWER SYSTEM CONTROL AND ESTIMATION IN A COMPETITIVE ENVIRONMENT　　324
Christopher L. DeMarco

　　Editor's Summary　　324
　　13.1　Introduction: Electric Power System Structure and Forces for Change　　324
　　13.2　Power System Dynamics and the Historical Structure of Grid Control　　327
　　　　　13.2.1　Control Objectives in Power Systems　　327
　　　　　13.2.2　Synchronous Generator Dynamics: A Brief Tutorial　　328
　　　　　13.2.3　Grid Frequency Regulation　　331

Contents   xiii

                 13.2.4 Stability-Enhancing Controls in Power Systems   334
     13.3 Institutional Changes Impacting Control Techniques   337
                 13.3.1 Power Grid Control Structures: If They're Not Broken, Why Fix Them?   338
     13.4 New Technologies Impacting Restructuring and Control in a Competitive Environment   339
                 13.4.1 The Impact of Efficient Gas Turbines   339
                 13.4.2 The Role of New Information and Measurement Technologies   340
                 13.4.3 Control Opportunities for Flexible AC Transmission Systems   342
     13.5 A Perspective on Future Directions for Power System Control Development and Research   343
            References   346

CHAPTER 14   INTELLIGENT TRANSPORTATION SYSTEMS: ROADWAY APPLICATIONS   348
Ümit Özgüner

     Editor's Summary   348
     14.1 Introduction   348
     14.2 Traffic-Related Issues   351
                 14.2.1 Signalization   351
                 14.2.2 Networks of Intersections   352
                 14.2.3 Routing   353
                 14.2.4 Control of Traffic on Highways   354
                           14.2.4.1 Convoys, Platoons, et al.   354
                           14.2.4.2 Ramp Control and Merging   354
                           14.2.4.3 Automated Highway Systems   355
                 14.2.5 Some Practical Concerns   355
     14.3 Intelligent Vehicles   357
                 14.3.1 Pre-IV Autonomy: Cruise Control and ABS   357
                           14.3.1.1 Preliminary Needs: Drive-by-Wire Vehicles   357
                 14.3.2 Car Following and Advanced Cruise Control   358
                 14.3.3 Lane Tracking   361
                           14.3.3.1 Vehicle Model   362
                           14.3.3.2 A Nonlinear Lane-Keeping Controller   363
                 14.3.4 A Lateral Lane Change Controller   365
                 14.3.5 Hybrid Systems and Scenario Resolution   365
     14.4 Conclusions   366
                 14.4.1 Related Problems   366
                           14.4.1.1 Precision Movement   366
                           14.4.1.2 Coupled Systems   366
                           14.4.1.3 Autonomy versus Full Information Exchange   366
                           14.4.1.4 Fault Tolerance/Safety   368
                 14.4.2 And Technology Keeps Marching On...   368
            References   369

CHAPTER 15   AUTOMOTIVE POWERTRAIN CONTROLLER DEVELOPMENT USING CACSD   370
K. Butts, J. Cook, C. Davey, J. Friedman, P. Menter, S. Raman, N. Sivashankar, P. Smith, and S. Toeppa

     Editor's Summary   370

15.1 Introduction 370
    15.1.1 The Role of the Powertrain Control System 371
    15.1.2 The Powertrain Controller Development Organization 372
15.2 The Systems Engineering Process 373
    15.2.1 The Powertrain Controller Development Process 374
15.3 Computer-Aided Control System Design for Powertrain Controller Development 376
    15.3.1 Software Requirements Capture 377
    15.3.2 Software Application Architecture Design 377
    15.3.3 Control Feature Design and Validation 379
    15.3.4 Software Application Validation 382
    15.3.5 Control Feature Software Design 382
    15.3.6 Control Feature Software Implementation 383
    15.3.7 Control Feature Structural Verification 383
    15.3.8 Control Feature Functional Verification 384
    15.3.9 Software Application Structural Verification 385
    15.3.10 Software Application Functional Verification 386
    15.3.11 Software/Module Integration Verification 386
    15.3.12 User Documentation 387
    15.3.13 Configuration Management 388
    15.3.14 Software Engineering Project Management 388
15.4 Conclusion 390
References 391

CHAPTER 16 BUILDING CONTROL AND AUTOMATION SYSTEMS 393
Albert T. P. So

Editor's Summary 393
16.1 Introduction 393
16.2 Existing Building Control Technologies 395
    16.2.1 Applications of PID Loops 396
    16.2.2 Programmable Logic Control 398
    16.2.3 Direct Digital Controls 399
16.3 Information Technology for Building Systems Control 399
    16.3.1 Control Networks 400
    16.3.2 Protocols 402
16.4 Building Automation Systems (BASs) 404
    16.4.1 Hardware Structure 404
    16.4.2 Software Features 406
16.5 Advanced Building Controls Technologies 407
    16.5.1 Applications of Expert Systems 407
    16.5.2 Neural Network-Based Control 408
    16.5.3 Fuzzy Logic-Based Control 410
    16.5.4 Computer Vision-Based Control 412
16.6 Difficulties with Building Systems Control 413
16.7 Conclusion 414
References 415

CHAPTER 17 CONTROLLING CIVIL INFRASTRUCTURES 417
B. F. Spencer Jr. and Michael K. Sain

Editor's Summary 417
17.1 Introduction 417
17.2 Hybrid Control Systems 420
    17.2.1 Hybrid Mass Damper 420
    17.2.2 Hybrid Base Isolation 429

Contents   xv

    17.3   Semiactive Control Systems   430
           17.3.1   Variable-Orifice Dampers   430
           17.3.2   Variable-Friction Dampers   431
           17.3.3   Controllable Tuned Liquid Dampers   431
           17.3.4   Controllable Fluid Dampers   432
           17.3.5   Semiactive Impact Dampers   434
    17.4   Semiactive Control of Civil Engineering Structures   435
           17.4.1   Scale-Model Studies   435
           17.4.2   Full-Scale Seismic MR Damper   436
    17.5   Conclusions   439
          Acknowledgements   440
          References   440

CHAPTER 18   ROBOT CONTROL   442
Bruno Siciliano

    Editor's Summary   442
    18.1   A Historical Perspective   442
    18.2   Kinematic Control   443
    18.3   Dynamic Control   446
    18.4   Force Control   451
    18.5   Visual Servoing   457
    18.6   The Future   459
          References   460

CHAPTER 19   CONTROL OF COMMUNICATION NETWORKS   462
R. Srikant

    Editor's Summary   462
    19.1   Introduction   462
    19.2   Network Control and Management   464
           19.2.1   Admission Control for Real-Time Sources   464
           19.2.2   Congestion Control for Best-Effort Sources   465
           19.2.3   Routing   466
           19.2.4   Scheduling   467
    19.3   QoS, Admission Control, and Calculus of Variations   468
           19.3.1   Large Deviations of the Empirical Mean of a Sequence of Random Variables   468
           19.3.2   Large Deviations of a Random Process from Its Fluid Limit   469
           19.3.3   Estimating Probabilities of Rare Events in Queues   471
           19.3.4   Examples   474
    19.4   Congestion Control   476
           19.4.1   Model   477
           19.4.2   Implementation Issues   480
           19.4.3   Simulations   483
    19.5   Conclusions   486
          Acknowledgements   487
          References   488

INDEX   491

ABOUT THE EDITOR   503

INTRODUCTION

Automation systems that affect the physical world must ultimately exploit concepts that control engineering and science have always been at the forefront of developing—concepts such as feedback, dynamical systems, optimization, modeling, and estimation. It is thus no wonder that controls has, in the past, been a linchpin of our modern technological world. Achievements as numerous and diverse as space missions, petroleum refining, climate-controlled homes and buildings, commercial and military airplanes, innumerable chemical products, reliable electric power, and many, many others have been rendered possible because of control technology. Controls is one of a handful of disciplines that can truly claim to be a common enabler across such a spectrum of applications.

Today, governments, societies, and corporations are attempting to close the gap on ever-larger-scale systems and ever-more-complex problems. In response to the dictates of human and environmental safety, national defense, corporate cost-reduction and profitability, and other factors, a new generation of automation and control systems is being envisioned and developed.

Many of today's technologically motivated trends augur well for control, but significant extensions in the existing controls technology base are required. For example, while advances in single-loop and low-level control will always be of interest, the real opportunities for impact are increasingly at higher levels of systems. The past successes of control can be attributed in part to the effectiveness with which control technologists have uncovered new applications of their theories, algorithms, and heuristics. Similar diligence is still mandatory and, in addition, we are being challenged to understand a new and larger class of problems, and to develop new tools and techniques for their solutions.

Some degree of redefinition of controls as a discipline is needed to ensure both that the expertise of control engineers and scientists continues to be viewed as essential to the evolutions and revolutions in automation, and that safe, efficient, and performative solutions are ultimately developed. The redefinition is not a rewriting but a broadening—the foundations of modern control have served society and industry in exemplary ways over the past half-century or so; our challenge is to extend these foundations and the intellectual edifices we construct from them. One of the themes of this book is that control technologists are meeting these challenges across a broad spectrum of techniques and application arenas.

The revolution in information technologies that we have witnessed over the last decade or two is, of course, part of the picture. The dramatic advances in processors, memory, communications, displays, and other hardware and software "infrastructure" have profoundly changed how we live and work. These advances are also being recognized as facilitating new levels of functionality and intelligence in control systems. In particular, we now have the computational infrastructure available to perform the complicated calculations implicit in so many control theoretic developments—developments that, for lack of processing and memory resources, have been gathering dust on the bookshelf.

Some readers may contrast the optimism expressed above with the lack of recognition of control in the broader technical community. Ask a randomly selected person what the key technologies are for the future, and it is unlikely that you will hear control engineering mentioned. Instead, computing, networks, and perhaps robotics and autonomous vehicles will be on the list. This lack of recognition is not late-breaking news to control engineers and scientists—most of us have become somewhat inured by now to questions such as "So what does control have to do with anything anyway?" In fact, however, there are several reasons for the underappreciation of control, notably the following:

- Control is a "hidden" component in all automated systems, invisible to end users—the general public in particular.
- The breadth and multifaceted nature of controls is such that relatively few of its practitioners and developers themselves appreciate its entire scope.
- Controls is among the most mathematically rigorous of the engineering disciplines: the intimidation factor limits broad-based appreciation.
- As an established and historically successful discipline, controls has tended to eschew hyperbole and self-promotion, perhaps to a fault.

None of these reasons has any bearing on the reality of the current impact of control technology or its future relevance. We can be excited about the substance of our discipline even as we acknowledge its rhetorical shortcomings!

ABOUT THIS BOOK

This edited volume brings together a set of chapters authored by experts in a number of specialized subfields of control technology. Our objective is to provide a broad review of the state of the art in control science and engineering, with particular emphasis on new research and application directions. The "take home message" is that controls is a vibrant, exciting, diverse field. Its new initiatives are likely to ensure its central role in technological solutions for the increasingly complex challenges facing society and industry in a new millennium.

This book is targeted to control engineers of all stripes, from industrial practitioners to academic researchers. This is a broad audience, and the book attempts to appeal to this diversity by not narrowly constraining the style, tone, or technical depth of individual chapters. Some chapters are technical tutorials; others focus on discussions of today's state of the art; some provide experimental results; several emphasize future visions; and so on. We hope that the heterogeneity will create cross-cutting appeal; our goal is that every control engineer will find parts of this book of significant interest both intellectually and professionally. Most chapters are written at a level appropriate for an undergraduate-degree control engineer, but a few may require an introductory graduate-level mathematical background.

Although this book covers most of the key technical specializations of control, it is not intended to be an encyclopedic compilation. Readability and a reasonable length were important considerations in planning the contents. As discussed below, the subjects covered reflect the emphases of the IEEE Control Systems Society (CSS). In any case, controls is too dynamic a discipline to expect one snapshot such as a printed book to be truly comprehensive, even in principle.

Book Outline

The technical contributions to this volume are structured into two parts. The first set of chapters is devoted to control "technologies"; the second focuses on traditional and novel application domains for control systems. To help unify and integrate the different chapters, each is preceded by an Editor's Summary and, where appropriate, includes at the end a list of related chapters. The book includes affiliation and contact information for the chapter contributors and a comprehensive index.

Part I. The nine chapters in the first part are concerned with the following technology-oriented topics:

- Chapter 1: Scott Bortoff describes the challenges and solution approaches for implementing control algorithms on real-time digital computing platforms. The complications discussed include sampling rate variations, variable processing delays, task scheduling, interprocess communication, and sensor and actuator failure.
- Chapter 2: Edwin Chong reviews discrete-event systems, contrasting them with the continuous-time systems (and sampled equivalents) which control science and engineering have traditionally focused on. Techniques for optimizing discrete-event systems for applications in communications networks, manufacturing systems, and other fields are presented.
- Chapter 3: George Grübel offers a general introduction to the topic of computer-aided control system design (CACSD). The chapter emphasizes the importance of system modeling, performance specifications, and the iterative process of controller development. As an example, flight control system design is considered.
- Chapter 4: Pradeep Misra gives a broad overview of the basic concepts of modeling and simulation as they relate to control systems. He discusses a number of modeling methodologies and topics, including system identification, model reduction, linearization, and distributed parameter systems.
- Chapter 5: Kevin Passino reviews a number of intelligent control techniques. The central concepts of fuzzy logic, neural networks, genetic algorithms, and planning systems are outlined and illustrated with examples from ship maneuvering, robotics, and automotive diagnostics. Remarks on autonomous and adaptive control are also included.
- Chapter 6: Jay Farrell and Marios Polycarpou give the reader a technical introduction to "nonlinear approximators," specifically neural networks and fuzzy models. The authors describe algorithms for estimating values of approximator

parameters, and they introduce the concepts of generalization, approximator transparency, and linear versus nonlinear parametrizations.
- Chapter 7: Hybrid dynamical systems—systems that contain both continuous-time and discrete-event dynamics—are the subject of this chapter by Michael Lemmon. The focus here is on hybrid systems where the discrete-event component models supervisory commands. Deadlock avoidance for a two-arm robotic platform serves as the motivating example.
- Chapter 8: Fumio Hamano and Younchan Kim also treat hybrid systems. Their subject is variable structure control—control schemes in which the control structure can be modified dynamically. One particular variable structure control technique, sliding mode control, is developed in some detail.
- Chapter 9: To conclude the first part of the book, Tariq Samad discusses the increasing complexity of automation and control systems and attempts to relate it to developments in control technology and control applications.

Part II. Each of the ten chapters in the second part of this volume discusses the application of control to an important application domain.

- Chapter 10: Jorge Tierno, Joseph Jackson, and Steven Green describe issues related to the control of multiple aerospace vehicles. They consider three application areas, all of which are driving new research: "free flight" commercial air transportation, autonomous formation flight of uninhabited vehicles, and precise positioning of satellite clusters.
- Chapter 11: Kevin Wise discusses the development of flight control laws in military high-performance aircraft and missiles, with an emphasis on cost efficiency and exploitation of CACSD tools. Dynamic models for control design are presented, and the role of simulation and analysis software is emphasized.
- Chapter 12: Michael Johnson and Michael Grimble outline the hierarchical organization of process control systems and review several control technologies for the process industries. These include performance monitoring, PID controller tuning, adaptive and robust control, model predictive control, and plantwide optimization.
- Chapter 13: Christopher DeMarco reviews the historical operation of electric power networks and the control challenges arising from the deregulation and competition that are now driving the evolution of the electric power industry. New technological developments, from flexible AC transmission devices to agent-based optimization, are proposed to meet these challenges.
- Chapter 14: An area of increasing engineering interest in general is intelligent transportation systems. Such systems pose a number of challenging control problems, and in this chapter Ümit Özgüner discusses several of these—with specific attention to traffic control and intelligent road vehicles.
- Chapter 15: As a sophisticated example of CACSD, Ken Butts and colleagues discuss in detail a process for team-based development of automotive powertrain controllers. The process is based on systems engineering principles. Key aspects include validation and verification, feedback mechanisms, and analysis and design support.

- Chapter 16: Albert So starts his chapter with a review of the history of building control and automation systems. He notes the revolution in building automation caused by the personal computer platform, and he discusses local area networks for building management systems. Finally, applications of intelligent control in this domain are reviewed.
- Chapter 17: A relatively new area for control is the control of civil structures such as buildings, bridges, and towers. In this chapter, Michael Sain and Bill Spencer note some relevant actuation technologies and present several examples of operational structural control systems. Semiactive control actuators are discussed at some length.
- Chapter 18: Bruno Siciliano presents traditional techniques and recent developments in robot control. The chapter sketches the evolution from kinematic control to dynamic control to force control, which permits precise tasks in elastic or compliant environments to be accomplished. Vision-based robot control is highlighted as the next frontier.
- Chapter 19: The last chapter also highlights a new, and promising avenue for control applications: communication networks. R. Srikant points out the control problems involved: admission control, congestion control, packet routing, and scheduling of node bandwidth. He also discusses specific considerations for asynchronous transfer mode networks and the Internet.

CSS Technical Activities Board—Providing Resources for Control Engineers

This book is an initiative of the Technical Activities Board of the IEEE Control Systems Society. The majority of the chapter contributors are leaders of CSS TAB, and in most cases are chairs of technical committees on specialized control topics. The difficult process of identifying which topics to include in a book of limited size was facilitated by the structure of TAB: the topics largely correspond to the technical committees.

CSS TAB provides resources and collaboration opportunities for control engineers and scientists, whether students, industrial practitioners, or academic researchers. Up-to-date information about the Board and about available resources can be accessed through the Web site of the Control Systems Society http://www.ieeecss.org. The CSS home page can also be accessed through the central IEEE Web site at http://ieee.org/organizations/tab/cur_soc_hps.html. Furthermore, information on joining the Society and on membership benefits can also be obtained through this Web site.

The flagship periodical of IEEE CSS, *Control Systems Magazine*, regularly carries feature articles on emerging technologies and application domains for control. Many of these articles are written at introductory or tutorial levels. (Earlier versions of a couple of the chapters in this book appeared in *CSM*.) Readers who find this book of interest may find *CSM* a useful vehicle for keeping abreast of future developments in control technology.

Tariq Samad
Honeywell Technology Center
Minneapolis, MN

ACKNOWLEDGMENTS

I am grateful to the contributors to this volume for graciously tolerating several rounds of review and revision. The Publications Activities Board of the IEEE Control Systems Society, chaired by Bruce Krogh, supported this project and solicited several reviewers whose feedback on an earlier draft of the manuscript resulted in a significantly improved final product. Finally, it has been a pleasure working with the IEEE Press staff on this project.

Tariq Samad
Honeywell Technology Center
Minneapolis, MN

LIST OF CONTRIBUTORS

Ken Butts
Ford Research Laboratory
Powertrain Control Systems Department
MD 2036 SRL
2101 Village Road
Dearborn, MI 48121 USA

Edwin K. P. Chong
Purdue University
School of Electrical and Computer
Engineering
1285 Electrical Engineering Building
West Lafayette, IN 47907-1285 USA

Christopher DeMarco
Department of Electrical and Computer
Engineering
University of Wisconsin–Madison
1415 Engineering Drive
Madison, WI 53706 USA

Jay A. Farrell
Department of Electrical Engineering
Marlan and Rosemary Bourns College of
Engineering
University of California, Riverside
Riverside, CA 92521 USA

George Grübel
formerly with:
Institute of Robotics and System Dynamics
DLR – German Aerospace Center
Oberpfaffenhofen
D-82234 Wessling, GERMANY

Fumio Hamano
California State University, Long Beach
Department of Electrical Engineering
1250 Bellflower Blvd.
Long Beach, CA 90840 USA

Michael A. Johnson
Industrial Control Centre
University of Strathclyde
George Street
Glasgow G1 1QE
Scotland, UK

Younchan Kim
California State University, Long Beach
Department of Electrical Engineering
1250 Bellflower Blvd.
Long Beach, CA 90840 USA

Michael D. Lemmon
Department of Electrical Engineering
University of Notre Dame
Notre Dame, IN 46556 USA

Pradeep Misra
Wright State University
Electrical Engineering Department
3640 Col. Glenn
Dayton, OH 45435 USA

Ümit Özgüner
Ohio State University
Department of Electrical Engineering
2015 Neil Avenue
Columbus, OH 43210 USA

List of Contributors

Kevin M. Passino
Department of Electrical Engineering
The Ohio State University
2015 Neil Ave.
Columbus, OH 43210 USA

Marlos M. Polycarpou
University of Cincinnati
Department of Electrical and Computer
Engineering and Computer Science
Cincinnati, OH 45221-0030 USA

Bruno Siciliano
IEEE Robitics and Automation Society Vice-President for Publications
PRISMA Lab. Dipartimento di Informatica e Sistemistica
Università degli Studi di Napoli Federico II
Via Claudio 21, 80125 Napoli, ITALY

Michael K. Sain
University of Notre Dame
Department of Electrical Engineering
275 Fitzpatrick Hall
Notre Dame, IN 46556 USA

R. Srikant
Coordinated Science Lab and Department of General Engineering
University of Illinois
1308 W. Main Street
Urbana, IL 61801 USA

Tariq Samad
Honeywell Technology Center
3660 Technology Drive
Minneapolis, MN 55418 USA

Albert T. P. So
Department of Building & Construction,
City University of Hong Kong
Tat Chee Avenue, Kowloon,
HONG KONG

B. F. Spencer, Jr.
University of Notre Dame
Department of Civil Engineering & Geological Sciences
156 Fitzpatrick Hall
Notre Dame, IN 46556 USA

Jorge Tierno
Honeywell Technology Center
3660 Technology Dr.
Minneapolis, MN 55418 USA

Kevin A. Wise
Boeing Technical Fellow
The Boeing Company
P.O. Box 516
St. Louis, MO 63166 USA

Chapter 1

REAL-TIME COMPUTING AND CONTROL

Scott Bortoff

Editor's Summary

To engineers involved in designing, developing, or operating control systems for practical applications, research in control may seem an exercise in mathematical abstractions. As an engineering discipline, however, the connection with the physical world is intrinsic to control. Interfacing with sensors and actuators, implementing advanced algorithms on real-time platforms, dealing with sampling time issues, and other such pragmatic matters may seem to be taken for granted in much of advanced control, but in fact there is an extensive body of research that is concerned with these very topics.

All advanced control algorithms today are hosted on digital computing systems, and any discussion of real-time control applications must address the specific issues and challenges associated with digital implementation. The benefits of digital realization are numerous: software-based computing allows more sophisticated control laws; updates and maintenance are rendered easier in many cases; control systems can be made more compact; and control can more readily be integrated with ancillary functions such as information display, system health management, and data recording.

But the digital world brings complications too. For example, the continuous variables of a physical system must now be discretely sampled. Variations in the sampling rate are generally assumed to be negligible, but this is not always the case and a significant adverse impact on control quality can result. Similarly, variable delays arise in the processing of sensory data. In today's processors, even the same sequence of arithmetic operations can take more or less time to execute depending on the state of the processor and operand values.

As control programs become more complex, so do their real-time implementations. Most control system computing platforms are used for several tasks in addition to executing the base control law. Real-time computing and control thus also involves the solution of difficult scheduling problems. Different tasks can have different priorities, desired execution frequencies, and execution times; there may be dependences between them that require interprocess communication and accessing of shared resources; and failures of sensors and actuators and other abnormal situations cannot be allowed to compromise safety.

Scott Bortoff is an associate professor in the Department of Electrical and Computer Engineering at the University of Toronto, and a former chair of the IEEE-CSS Technical Committee on Real-Time Computing, Control, and Signal Processing.

1.1 INTRODUCTION

It is safe to say that most modern control systems are implemented digitally. From the fly-by-wire systems that control modern commercial and military aircraft to the digital proportional-integral-derivative (PID) boxes that regulate everything from temperature to pH in a process control setting, there is a clear preference for digital realizations. This is true even though the plant might be a continuous-time system, the control law might be designed using continuous-time mathematics, and the controller itself could otherwise be realized with inexpensive analog components.

The reasons for this trend are well-known: Software is easier to modify than analog hardware; both dynamic range and signal-to-noise ratio of digital signals can be made larger (especially given today's high-precision digital-to-analog (D/A) converters [1]), larger time constants can be implemented with software, complex nonlinear and adaptive control laws can be realized only by using a computer, overall system reliability is increased, overall control system weight can be reduced, and so on. In addition, a computer in the loop can add functionality to the system as a whole, much of which might be considered outside the traditional domain of control systems. For example, a processor in the loop can log data, a critical feature to the process control industry. With off-the-shelf networking technology, the controller can interface with the Internet, giving control engineers remote access to the system. Built-in testing and fault detection, a necessity for avionics systems, can also be coded into the controller.

In this chapter, we focus on the methods and tools used to implement digital controllers. Our main focus is on software, an area that has undergone a tremendous transformation in the last decade or so. Indeed, all areas of real-time systems, including real-time operating systems, programming languages, scheduling theory, and formal methods, are very active areas of research within both computer science and electrical engineering, and there are now a number of monographs on the subject, for example, [2]–[6].

After providing some background in Section 1.1, we present an example in Section 1.2 that illustrates why real-time aspects should be of interest to the control engineer. We then turn our attention to the methods used to implement digital controllers, beginning in Section 1.3 with a low-level approach that is best when the control law is simple, for example, single-input, single-output PID. Of course, computers in-the-loop are often used for more than just control. The real-time control system might be designed to realize a conventional control law *and also* to provide a timely response, within a specified deadline, to other asynchronous events. In this case, designing a single program to execute as a single task becomes unwieldy. Therefore, as the number of real-time specifications and tasks increases, the software is best designed to run not as a single process but as as multiple, cooperating, communicating processes. In Section 1.3.1. we present an introduction to scheduling theory, which is used to assign priorities to processes running under a priority-based preemptive operating system. Our focus then turns to higher level approaches to real-time control system implementation, namely, real-time operating systems and programming languages, in Section 1.4. We close the chapter in Section 1.5 with a brief look at hardware, including single-board computers and Programmable Logic Controllers (PLCs).

1.1.1 Background

Most feedback control systems can be represented by the generic feedback loop shown in Figure 1.1. Here, the *Plant* includes the actuators, the system to be controlled, and the sensor dynamics. The *Controller* is comprised of the sampler, which may be an analog-to-digital (A/D) converter, a shaft encoder, or some other instrument that converts each measured plant signal $y_i(t)$, $1 \leq i \leq p$, into a discrete-time measurement $\widehat{y}_i(t)$; a processor, which computes the control law based on these measurements; and a

Section 1.1 Introduction

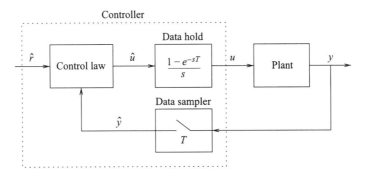

Figure 1.1 The general form of a digital control system.

sample-and-hold, which converts the discrete-time control signals $\widehat{u}_i$, $1 \le i \le m$, back into continuous-time signals u_i, $1 \le i \le m$, for example, a digital-to-analog converter. (Throughout this chapter, the symbol $\;\widehat{}\;$ is used to denote a sampled, discrete-time quantity.) The controller is usually designed to sample $y_i(t)$ periodically, at uniformly spaced instances of time $t = kT$, where $k > 0$ is an integer index and T is the sample period. The control law itself is usually realized as a set of discrete-time state equations

$$\widehat{u}(kT) = h\Big(\widehat{x}(kT), \widehat{y}(kT), \widehat{r}(kT)\Big), \tag{1.1}$$

$$\widehat{x}(kT + T) = f\Big(\widehat{x}(kT), \widehat{y}(kT), \widehat{r}(kT)\Big), \tag{1.2}$$

where the vectors $\widehat{x} = [\widehat{x}_1, \ldots, \widehat{x}_n]^T$, $\widehat{u} = [\widehat{u}_1, \ldots, \widehat{u}_m]^T$, and $\widehat{y} = [\widehat{y}_1, \ldots, \widehat{y}_p]^T$ are the controller state, controller output, and sampled plant output, respectively, $\widehat{r} \in \mathcal{R}^r$ is a vector of reference input signals, and f and h are (perhaps nonlinear) maps of appropriate dimension.

Example 1.1.1

Consider the servo control system diagrammed in Figure 1.2, where the plant has transfer function $P(s) = \frac{10}{s(s+1)}$. Suppose that in addition to several transient response specifications, the servo angle y must also track a ramp input applied at r with zero steady-state error. To meet these specifications, assume that a digital PID control structure is used, as shown in Figure 1.2. The three gains, K_p, K_d, and K_i, are designed to satisfy the specifications under two assumptions: (1) $y(t)$ is sampled uniformly in time, at $t = kT$, for some $T > 0$; and (2) the processor's computational delay is fixed and known to the designer. Often, the computational delay is assumed to be zero because the time required to compute the control law is much less than T. In this case, the controller is being designed under the assumption that the output $u(kT)$ is available simultaneously with the sampled input $y(kT)$. In any case, the state equations (1.1)–(1.2) for the digital PID controller shown in Figure 1.2 are

$$\widehat{u}(kT) = \frac{K_d}{T}\widehat{x}_1(kT) + \left(K_i T - \frac{K_d}{T}\right)\widehat{x}_2(kT) + \left(K_p + K_i T + \frac{K_d}{T}\right)\widehat{e}(kT) \tag{1.3}$$

$$\widehat{x}_1(kT + T) = \widehat{x}_2(kT) \tag{1.4}$$

$$\widehat{x}_2(kT + T) = \widehat{x}_2(kT) + \widehat{e}(kT) \tag{1.5}$$

where $\widehat{e}(kT) = \Big(\widehat{r}(kT) - \widehat{y}(kT)\Big)$ is the tracking error. The states are usually initialized at the origin, so $\widehat{x}_1(0) = \widehat{x}_2(0) = 0$.

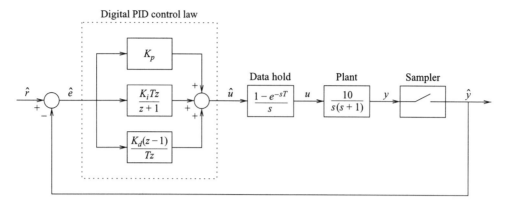

Figure 1.2 Servomotor with digital PID controller.

Coding (1.1)–(1.2) on any processor is a usually an elementary task in any programming language. The real challenge is timing: The processor must be programmed to sample y uniformly in time, at T-second intervals, and the controller output $\hat{u}$ must be computed and applied to the plant at a time consistent with the design assumptions. In particular, if the computational delay is assumed to be zero, then $u(kT)$ must be applied to the plant as soon as possible after $y(kT)$ is sampled. This is because the control law gains (or more generally, the functions f and h) are typically quite sensitive to the sample interval T; a small change in T can result in a large change in closed-loop pole locations, for example. Moreover, the performance of the overall system is usually very sensitive to timing *jitter*, which is a variance in the time that y is sampled and/or the time that u is applied to the plant.

1.2 TIMING IS EVERYTHING

In an actual digital implementation, delays are present in both the signal conversion hardware and the processor. If the delay is of fixed duration, then (1.1)–(1.2) can often be modified to compensate the delay. The simplest case occurs when the delay is a multiple of T, that is, qT for some positive integer q. In this case, q additional shift operators ($1/z^q$, where z is the shift operator) can be put into the loop, between the sampler and the data hold, and the control law can be designed to meet performance specifications despite the delay. Of course, the additional shift operators are not part of the control law—their presence in the loop is to model the total controller delay.

Unfortunately, the delays in a digital control system are not always of fixed duration. Depending on the hardware used, an analog-to-digital converter may take a varying amount of time to complete the data conversion. Or if a timer is used to generate a processor interrupt and an interrupt service routine (ISR) is then used to trigger the data-conversion hardware, then a delay will be associated with servicing the interrupt, called the *interrupt latency*. This is seldom a fixed duration of time because the processor could be servicing an interrupt of higher priority, whose ISR must finish before the data-conversion ISR can run. In addition, processors, especially those with complex instruction set architectures (CISC) such as the Pentium, can take a varying number of central processing unit (CPU) cycles—a varying amount of time—to complete the calculation of (1.1)–(1.2). This is due to instruction branching, main memory caches (which reduce the *average* amount of time to fetch instructions and operands but increase the *variance*), and the fact that the number of CPU cycles required to execute some instructions, such as floating point multiplication, is dependent on the operands themselves. All of these varying delays result in timing jitter, so that $y(kT)$ is not sampled at time kT but some time later, and $u(kT)$ is not applied to the plant at exactly time kT, but some time later.

Example 1.2.1

Continuing the digital PID control system introduced in Example 1.1.1, let us illustrate the effect of timing jitter. First, assume that the PID gains, listed in Table 1.1, have been *designed* under the ideal assumptions of zero computational delay and a fixed, uniform sample time of $T = 50$ ms. In this case, if we apply a ramp input $\hat{r}(kT) = kT$, then the internal model principle tells us that the tracking error will converge to zero as $k \to \infty$.

How is performance affected if this same control law is implemented with hardware that samples $y(t)$ not at time $t = kT$ but rather at time $t = kT + D$, where D is a random variable, uniformly distributed in the interval [0.005, 0.010]? Thus, the actual measurement occurs 5 to 10 ms after time kT. What if we assume further that the time required to compute $\hat{u}(k)$ is not zero but is a random variable C, uniformly distributed in the interval [0.015, 0.030], meaning the processor takes 15 to 30 ms to complete the calculation of $\hat{u}(kT)$? In this case, $\hat{u}(k)$ could be applied to the plant as late as 40 ms after time kT, or, more importantly, as late as 30 ms after y is measured.

Figure 1.3 shows the effects of this timing jitter on the tracking error for the servomotor system by comparing it with the ideal delay-free case, when the ramp input $\hat{r}(k) = kT$ is applied. (All initial conditions are zero.) As expected, in the ideal case, $\hat{e}$ converges to zero exponentially. However, when computational and measurement delay are included in the simulation, $\hat{e}$ no longer converges to zero. In fact, it suffers from a "noisy" steady-state error. What causes this? During the interval of time $[kT, kT + C + D]$, the previously computed control, $u(kT - T)$, is still being applied to the plant. The design has not taken this into account.

Judging from this example, it is dangerous for the control system designer to be ignorant of real-time implementation issues. Both delay and timing jitter, present in any real-world application, can adversely affect closed-loop performance. At the very least, the designer should be aware of their effects. Better still, the designer should incorporate the computational delays and worst-case timing jitter estimates into the control law specifications. Including the time-varying effects of timing jitter into controller design methodologies is very much an open area of research. Nonetheless, the designer can always perform simulations that include the effects of delay and timing jitter, as we have here. Finally, the designer should be actively involved at the implementation stage, in case the control law must be modified should the timing specifications change. In short, the control system designer's efforts do not end with Eqs. (1.1)–(1.2).

TABLE 1.1 PID Controller Parameters

Parameter	Value
T	50 ms
K_p	1.0
K_i	0.5
K_d	0.5

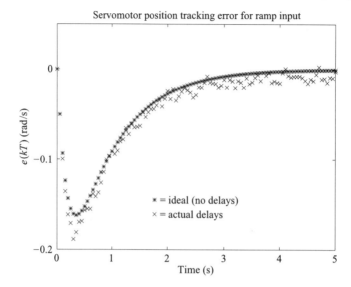

Figure 1.3 Ramp response of a servomotor with digital PID control. The ∗ symbols denote ideal response, when the rotor position is sampled every 50 ms, and there is no computational delay. The × symbols denote the response when there is both measurement and computational delay in the loop.

1.3 LOW-LEVEL REAL-TIME PROGRAMMING

There are several ways to approach the task of programming a processor to implement a digital controller. At one end of the spectrum, assembly language or a sequential language such as C can be used without the benefit of an operating system. This is probably the easiest approach if the processor is to be used only to realize a single control law, and the sensor and actuator interfaces are relatively simple. At the other end of the spectrum lie real-time programming languages such as Ada and commercial real-time operating systems such as QNX and Lynx. These are most useful in more complex situations, when the processor has other tasks to accomplish in addition to realizing a digital control law, and these tasks have hard real-time constraints. Between these extremes lies a myriad of real-time kernels, executives, and "soft" real-time operating systems such as Windows NT. Depending on application, these can simplify the often challenging task of real-time programming.

Let us first suppose that the designer must realize (1.1)–(1.2) using a single processor for a relatively simple system, such as the digital PID servo system. In this case, a combination of C and assembler without any operating system support is probably the best choice. The program is simply designed to use any available timer to generate a processor interrupt every T seconds. (The timer is simply a counter that decrements from an initial value at a fixed frequency. When it reaches zero, it issues an interrupt to the processor, resets to its initial value, and begins counting toward zero again.) Physically, the timer may be part of the processor hardware itself, as is the case with most digital signal processors, or it may be part of external hardware such as an A/D card. The control law is then programmed as the ISR, which first converts $y(kT)$ to $\hat{y}(kT)$ (or, more precisely, triggers the conversion hardware and waits for its comple-

Section 1.3 Low-Level Real-Time Programming 7

tion), then computes $\widehat{u}(kT)$ using (1.1), converts $\widehat{u}(kT)$ to $u(kT)$, and finally computes $\widehat{x}(kT + T)$ using (1.2). The software is written to initialize the timer and the ISR, and then to enter into an infinite loop to wait for the interrupt. In this loop, it can execute any other tasks, such as built-in testing or fault diagnosis. When the interrupt is generated, the processor immediately executes the ISR. Of course, the interrupt latency, delays in signal conversion, and time to execute the interrupt service routine all should be taken into account in the design of the control law. If the processor has no other jobs to execute, these delays are usually measured in microseconds on a modern microprocessor, so their relevance is implementation-dependent.

Now let us suppose that the processor is to accomplish several other tasks in addition to just realizing a simple controller. For example, suppose that a single processor is to be used for the following three tasks:

1. Realize the digital PID controller for the servo system diagrammed in Figure 1.2.
2. Interface with an ultrasonic transducer, which measures a distance in space. This task is really two separate tasks. First, the processor must periodically send a signal to the sensor, which then emits an ultrasonic pulse. At the same time, the processor must start a timer. When the returned echo is received by the sensor, it raises an interrupt line to the processor. The processor must respond by stopping the timer and reading its value, which is subsequently converted to a distance estimate.
3. Pass data to and from a second processor via an RS-232 serial port.

The designer who chooses to write a single low-level program for all three tasks will soon realize that it will contain four ISRs: one for the digital PID control law, which is driven by a timer; one to trigger the ultrasonic sensor, which is also driven by a timer; one to stop the ultrasonic sensor counter and read its value, which is driven by the sensor itself; and one to attend to the serial port buffer, which is driven by the RS-232 port hardware. If we assume that there is no dependence among these three tasks, then any of the interrupts could occur at any time. In particular, two could occur simultaneously, or one could occur while the processor is executing the ISR for another. Thus, the designer must assign *priorities* to each interrupt.

Most modern processors have special hardware that orders the interrupts and their ISRs according to a priority assignment. Should interrupt X occur while the ISR for interrupt Y is running, then ISR Y will be interrupted only if X has a higher priority than Y. In this case, ISR Y would be stopped, and ISR X would run to completion: ISR X has *preempted* ISR Y. When finished, the processor will return to ISR Y. Otherwise, if X has a lower or equal priority than Y, then ISR Y is allowed to run to completion, at which point ISR X begins processing. By convention, the integer 1 is usually associated with the "highest" priority task, while higher integer values imply lower priorities.

Returning to our example, we see that each of the four tasks has severe timing constraints. One way software can be designed for these specifications is to code each of the four tasks as an ISR and then program the hardware to generate the appropriate interrupts. But each ISR would have to be assigned a priority. How should the designer assign these priorities? For a relatively simple example such as this, an exhaustive search of all possible assignments is possible. We can simply check all 24 possible priority assignments and determine if the timing constraints are met for each assignment. However, since the number of possible priority assignments is equal to the factorial of the number of tasks, a more formal method is needed. Fortunately, scheduling theory provides several sufficiently mature tools to aid the designer in this task.

1.3.1 Fixed-Priority Scheduling Theory

In any real-time system, process scheduling is the programmer's responsibility. In the previous example, the processes to be scheduled were four ISRs. More generally, each might be an independent process running under a fully preemptive, priority-based, real-time operating system. (Here, *fully preemptive* means that a higher-priority process will always be able to interrupt a lower priority process, and the higher-priority process will then run until completion.) In either case, process scheduling means assigning a priority P_i, where $P_i \in \{1, 2, \ldots, N\}$, to each process, denoted τ_i, $1 \le i \le N$. There are several methods to assign priorities, and indeed scheduling theory for real-time systems continues to be a vigorous area of research. Fortunately, several so-called fixed-priority algorithms, where the process priority P_i is assigned to the process before run-time and remains fixed during run-time, are adequate for most control applications.

Let us first assume we have a set of N periodic processes, denoted $\{\tau_1, \ldots, \tau_N\}$. Each might be an ISR triggered by a timer, for example. For each process, assume the following data are known to the designer:

- The computation time C_i, which is the worst-case (least upper bound) time required to complete process τ_i, assuming no other processes can run;
- A *deadline* D_i, which is the maximum allowable time between the *release time* of a process τ_i (e.g., the time that an interrupt for a particular ISR occurs) and when the process must complete; and
- A period T_i, which is the time between releases of process τ_i.

Note that we can relax the periodic assumption and include sporadic processes in the analysis, if we assume that the time T_i is the minimum arrival time for the process. In our previous example, the ultrasonic sensor ISR that stops the timer and reads its value is a sporadic process. The interrupt is generated when the echo returns to the sensor, not by a timer. But this interrupt has a minimum arrival time that is equal to the product of the speed of sound and twice the minimum distance that can be sensed.

The *rate monotonic* (*RM*) algorithm is probably the most popular method of priority assignment. Priorities P_i are assigned inversely related to the process period T_i. In particular, the process with the shortest period is assigned the highest priority (1), while the process with the longest period is assigned the lowest priority (N). In its simplest form, the process deadlines D_i are assumed to be equal to the period T_i. In this case, several simple necessary and/or sufficient conditions can be checked to see if the schedule is feasible, that is, if all processes will meet their deadlines. For example, if

$$\sum_{i=1}^{N} \frac{C_i}{T_i} \le N(2^{1/N} - 1),$$

where the left-hand side is the total processor utilization, then the process set is feasible. If the deadlines D_i are less than T_i, then this formula no longer holds, but other necessary and sufficient conditions are easily checked [4].

Another similar algorithm is the so-called *Deadline Monotonic* (*DM*) scheduling [7]. Here, fixed priorities P_i are assigned inversely to the process deadlines D_i, instead of the periods. (This should not be confused with the *earliest deadline first* algorithm, which is a *dynamic*-priority scheduling algorithm that executes the process with the earliest deadline.) This approach has several advantages over RM, including a greater emphasis on process deadline. This can result in a reduction in timing jitter (the variation in the total time required to complete a process) when compared to RM, which can be critical to a control application.

After assigning priorities, we must check to see if each process will meet its deadline. Both RM and DM schedules, and most other fixed-priority algorithms, are analyzed the same way. For each τ_i, define the *response time* R_i as the worst-case amount of time required for the process τ_i to complete. Since process τ_j can interrupt process τ_i if $P_j < P_i$, the response time for τ_j must be added to C_i when computing R_i. That is, we must add to the computation time C_i the so-called maximum *interference* that τ_i receives from all tasks of higher priority. Once all the response times are computed, we simply check if R_i is less than D_i for $1 \leq i \leq N$, and if so, then the process set is *schedulable*, meaning all processes are guaranteed to meet their deadlines.

It can be shown that the response time R_i satisfies the equation

$$R_i = C_i + \sum_{i \in H} \left\lceil \frac{R_i}{T_j} \right\rceil C_j, \qquad (1.6)$$

where H is the set of processes with priority strictly greater than P_i, and where $\lceil \cdot \rceil$ is the ceiling operator, defined for all real numbers s as $\lceil s \rceil = \bar{s}$, where $\bar{s}$ is the smallest integer that is greater than or equal to s. Equation (1.6) is not difficult to understand: The term $\lceil \frac{R_i}{T_j} \rceil$ is simply the (integer) number of times that a higher-priority process τ_j can interrupt task τ_i before τ_i has completed. Multiplying this term by C_j and summing over all tasks of higher priority than task τ_i gives the maximum (worst-case) amount of time spent servicing all the higher priority tasks. Adding C_i gives the response time.

Note that Eq. (1.6) is nonlinear because of the ceiling operator. However, it is easily solved by converting it into the following recurrence relation [7]

$$R_i^{k+1} = C_i + \sum_{j \in H} \left\lceil \frac{R_i^k}{T_j} \right\rceil C_j, \qquad (1.7)$$

for $k \geq 1$, where $R_i^1 = C_i$. If $R_i^{k+1} = R_i^k$ for any $k > 1$, then $R_i = R_i^k$. In this case, the task τ_i is guaranteed to meet its deadline if and only if $R_i < D_i$.

Example 1.3.1

Consider three processes τ_1, τ_2, and τ_3, with computation times C_i, periods T_i, and deadlines D_i as given in Table 1.2. The RM schedule would assign the highest priority to τ_1 and the lowest to τ_3, that is, $P_1 = 1$, $P_2 = 2$, and $P_3 = 3$. Using Eq. (1.7) to compute the response times, we find that $R_1 = 1$ because τ_1 has the highest priority. Since $R_1 < D_1 = 5$, τ_1 is guaranteed to meet its deadline. Computing R_2 recursively using (1.7), we have

$$R_2^1 = 2$$

$$R_2^2 = 2 + \left\lceil \frac{2}{5} \right\rceil \cdot 1 = 3$$

$$R_2^3 = 2 + \left\lceil \frac{3}{5} \right\rceil \cdot 1 = 3,$$

so $R_2 = 3 < 10 = D_2$, and so τ_2 also meets its deadline. Computing R_3 the same way, we have

$$R_3^1 = 3$$

$$R_3^2 = 3 + \left\lceil \frac{3}{5} \right\rceil \cdot 1 + \left\lceil \frac{3}{10} \right\rceil \cdot 2 = 6$$

$$R_3^3 = 3 + \left\lceil \frac{6}{5} \right\rceil \cdot 1 + \left\lceil \frac{6}{10} \right\rceil \cdot 2 = 7$$

$$R_3^4 = 3 + \left\lceil \frac{7}{5} \right\rceil \cdot 1 + \left\lceil \frac{7}{10} \right\rceil \cdot 2 = 7.$$

Since $R_3 = 7 > 4 = D_3$, τ_3 is not guaranteed to meet its deadline, and the rate monotonic schedule fails the schedulability test.

On the other hand, if we use a Deadline Monotonic schedule, then the priorities are reassigned as $P_1 = 2$, $P_2 = 3$, and $P_3 = 1$. A very similar calculation shows that

$$R_3 = 3 < 4 = D_3$$
$$R_1 = 4 < 5 = D_1$$
$$R_2 = 7 < 10 = D_2$$

and all three processes are guaranteed to meet their deadlines.

1.3.2 Data Dependence

Thus far, we have assumed that the processes $\tau_1, \ldots, \tau_N$ are independent. Of course, in most real-time systems, processes communicate among themselves, share resources, and so on. For example, it is very common for two processes to communicate by using *shared memory*; that is, certain variables can be stored in memory that can be accessed by two separate processes.

TABLE 1.2 Computation Times, Periods, and Deadlines for an Example Schedule

Process	C	T	D
τ_1	1	5	5
τ_2	2	10	10
τ_3	3	20	4

Example 1.3.2

Suppose three processes, τ_i, $1 \leq i \leq 3$, are running on a single processor, and τ_1 and τ_3 share some memory, denoted X, that stores a vector x of real numbers. The process τ_3 implements a user interface that displays x on a screen, while τ_1 is a device driver that interfaces with data-conversion hardware, reading the data x from a hardware register and storing it in X. Assume τ_1 is at a higher priority than τ_3. Now, τ_1 might interrupt (preempt) τ_3 while τ_3 is accessing X. Thus, when τ_1 finishes putting new data into X and τ_3 begins to run again, then the (old) data that τ_3 was reading will be inconsistent with the (new) data that τ_1 put into X.

The sections of code that access a shared resource such as shared memory are said to be *critical sections*. Computer scientists have known for decades that such critical sections must be protected, meaning only one process at a time may access the shared resource. For example, if a process and an ISR share a memory location that stores an integer, then neither one can interrupt (preempt) the other while the integer is being written to that memory location, an operation that might take several processor instructions to complete. Otherwise, the data stored will be corrupted.

Now, if each "process" is really an ISR, then the simplest way to protect a shared resource is to disable all interrupts before entering the critical section and then reenable all of them after leaving the critical section. In this way, once an ISR begins the critical section, it is guaranteed to complete its execution without interruption. This is a very common way for ISRs to communicate. In Example 1.3.2, τ_3 would disable interrupts before reading from X, thus ensuring that it will not be interrupted by τ_1 until it is finished. Note, however, that disabling interrupts for a period of time increases the interrupt latency, since a higher-priority interrupt (τ_1) will not be processed until a lower-priority process (τ_3) has reenabled the interrupts. Thus, every effort should be made to minimize the number of CPU cycles over which interrupts remain disabled.

Real-time programming at this level becomes awkward as the number of processor tasks increases. A single program becomes difficult to test because the usual tools such as debuggers are often useless: A real-time system is not correct unless both its logic and its timing are correct. The designer might be able to verify the logical correctness of the code with a debugger, test vectors, and so forth, but proving that the timing constraints are also satisfied under all conditions can be very difficult. Moreover, if the designer must add another task after the code is complete, then the entire design might have to be redone because the timing of the whole system has changed. Finally, worst-case interrupt latency might become intolerable due to a large number of critical sections of code, each protected by disabling and reenabling interrupts. When enough of these effects conspire to make a low-level approach too difficult, either a real-time operating system or a high-level real-time programming language is probably in order.

1.4 REAL-TIME OPERATING SYSTEMS AND PROGRAMMING LANGUAGES

An operating system is a program, or a set of programs, that manages the hardware and software resources of a computer. At the heart of any operating system is its *kernel*. The kernel is the program that schedules application programs, manages main memory, handles communication among application programs, and usually contains drivers for devices such as disks, user interfaces, and network interfaces. The kernel insulates each application program from the details of opening files, communicating with other programs, and so forth.

Most preemptive, priority-based, real-time operating systems are built around relatively small kernels that provide a minimum of functionality. For example, the QNX kernel [8] provides only two basic functions:

- Process scheduling, which manages the state of each process and determines when each process will run; and
- Interprocess communication, which allows processes to send messages to each other.

All other functions that might normally be associated with the kernel, such as device drivers and file system support, are provided as separate processes. This architecture offers several advantages over large, feature-rich kernels (such as Linux). First, because the kernel is small, the worst-case interrupt latency time (the time elapsed after an interrupt occurs but before its ISR runs) can be shorter and simpler to determine. Latency time can be both long and difficult to measure for a large kernel for the following reason [9]. Suppose a low-priority user process executes a system call to the kernel just before a high-priority interrupt occurs. When the interrupt does occur, that system call might not be preemptable *because the kernel itself is not preemptable*. Thus, the interrupt might not be serviced until control is returned to the low-priority process. To make matters worse, the nonpreemptable system call could in turn issue a second and then a third nonpreemptable system call. Thus, with a large number of available, nonpreemptable system calls, the worst case interrupt latency could be not only long, but also quite difficult to determine. Moreover, this worst-case might occur only very rarely and would therefore be an overly conservative estimate of latency time.

This is the case with many popular operating systems, such as Windows NT [10]. Because the NT kernel is not fully preemptable, it is not possible to determine a worst-case bound on the amount of time that will elapse between an event, such as a software interrupt, and the operating system's response. This makes such operating systems unsuitable for so-called hard real-time applications, where the response time must be precise and guaranteed. For example, imagine a process that generates gating signals in a pulse-width modulation (PWM) motor driver. The duration of the pulse is the control input. Suppose the process uses a system timer to control the duration of the pulses. The timer is initialized at a number proportional to the desired pulse duration and counts down, generating a software interrupt when it reaches zero. The process then responds to the interrupt by changing the state of a bit on a digital output card via a device driver for that card. If this process is running under Windows NT, then the amount of time that will elapse between the timer interrupt and the hardware bit flip will vary, leading to a "noisy" control signal. The variance is not entirely deterministic. (In particular, it is impossible to compute an upper bound on this time, such that the operating system will always satisfy the bound.) Moreover, as the PWM switching frequency increases, this varying delay will remain the same (at best), meaning the signal-to-noise ratio will decrease. Thus, if the PWM switching frequency is sufficiently high, then such a software-only solution using Windows NT will probably fail to meet a specification on signal-to-noise ratio. On the other hand, for many control systems the response time of an operating system like NT is "fast enough." Indeed, NT is designed to minimize response times. In such a case, the latency and its variance, that is, the timing jitter, would be relatively small and performance would not be noticeably affected. Returning to the PWM example, if the pulses have a duration of between one and two seconds, then a few milliseconds of timing jitter will probably not adversely affect performance.

When the sampling frequency is relatively high and a hard bound on response time is necessary, then a truly real-time operating system becomes necessary. By limiting the functionality of the kernel to the bare essentials and by designing the kernel to be fully preemptable, meaning kernel system calls can be interrupted, a real-time kernel will provide not only a lower latency time, but also a more predictable latency time. Note that this does not necessarily limit the functionality of the operating system as a whole. Device drivers, file system support, user interfaces, and so on, can be added as processes whose priority can be assigned by the programmer.

The primary advantage of using a real-time priority-based operating system over writing a single low-level program is that each task can be coded as a separate process. There are two prevailing philosophies here: Either a programming language such as C, extended by real-time libraries, can be used to develop each separate program, or alternatively, a real-time language such as Ada can be used to generate cooperating processes. In both scenarios, the result is really the same: A number of cooperating processes are created that run together, for all practical purposes simultaneously on a single processor.

Developing and testing each process separately offer the same advantages that breaking down a single, large program into separate subroutines does. Management is simplified since coding can be done by several developers. Testing is often simplified because the logical correctness of each process can be determined independently from other processes, and the timing correctness can be determined by the proper application of scheduling theory. Determining the worst-case run-times (C_i), necessary to assign priorities (P_i), can be done for each process independently. Finally, adding functionality is simply a matter of writing new code for new processes and perhaps redefining priorities—existing processes do not require modification. Thus, for sufficiently complex real-time programming tasks, fully preemptive, priority-based operating systems and real-time programming languages are a major asset.

1.4.1 Real-Time Operating Systems at Run-Time

In a preemptive, priority-based system, priority is assigned to a process by the designer, not the operating system (OS), assuming static priorities are being used. It is important to understand how a typical real-time OS schedules these processes. Roughly speaking, each process can assume one of two states: ready or blocked. The operating system maintains a list of all processes and allows the process that has the highest priority and is also in the ready state to execute. The OS will continue to monitor the status of all other processes. The highest-priority process continues to execute until either another higher-priority process becomes ready, because of an external signal, for example, or the process itself becomes blocked, for example, it completes its calculations.

In the QNX operating system, for example, processes either are in the ready state, or they assume one of a number of different blocked states, each of which is related to interprocess communication. A simplified version of the situation is diagrammed in Figure 1.4. Again, the highest-priority ready process runs, until either another higher-priority process becomes ready or it issues either a `send()` or `receive()` call, both of which are kernel calls. (Other processes might enter the ready state because of a hardware interrupt or a timer interrupt, for example.) When this occurs, the process becomes blocked, ceases to run, and the kernel executes the next process in the ready state with the highest priority. Should two or more processes be ready and at the same priority, then they can be scheduled round-robin, or first-in first-out (FIFO). This client-server architecture is very well-suited to control applications, as the following example illustrates.

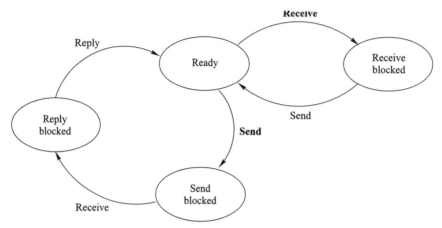

Figure 1.4 The QNX process states are based on a client-server model. A process begins in the ready state. If it issues a **send** message to another process, it becomes send-blocked until it receives a receive message from that process. It then becomes reply-blocked, until it receives the reply signal from that process, when it again becomes ready. Otherwise, it can issue a **receive** message and become receive-blocked, until it receives a send message from another process. Messages issued by the process are shown in **bold**, while those issued by other processes are in normal typeface. Taken from [8].

Example 1.4.1

Let us return to the three processes introduced in Example 1.3.2. Again, process τ_1 is a simple device driver. When it runs, it reads the data x from several A/D registers, copies it into the shared memory X, and then blocks. Suppose now that process τ_2 is a control law, which uses x to compute a control using an expression such as (1.1)–(1.2). As such, τ_2 is timer-driven, meaning it will sleep (block) until a timer expires, at which point it becomes ready. Finally, τ_3 is the user interface, which displays x on the computer screen. Suppose priorities are assigned as $P_1 = 1$, $P_2 = 3$, and $P_3 = 5$, and all other processes running have lower priority ($P_i > 5$ for $i > 3$).

At time t_0, when all three processes are started, τ_1 (the device driver) will run because it has the highest priority. After properly initializing the A/D hardware, it issues a `receive($\tau_2$)` call and becomes receive-blocked until τ_2 issues a `send($\tau_1$)`. Being the highest-priority process now in the ready state, τ_2 (the controller) now runs. It initializes its timer, and it goes to sleep (blocks) until the timer expires. When this occurs, at every time kT, it first issues a `send($\tau_1$)`, causing τ_2 to reply-block. (It would normally receive-block, but τ_1 has already issued a `receive($\tau_1$)`, so τ_2 reply-blocks.) Now τ_1 moves to the ready state, and being the highest priority process, it runs. It reads the A/D, moves the data into shared memory, and issues a `reply($\tau_2$)`, which makes τ_2 ready. Finally, it issues a `receive($\tau_2$)` to become receive-blocked again. At this point, τ_2 will run again,

Section 1.4 Real-Time Operating Systems and Programming Languages

copy the data from shared memory, compute the control law, and output it. It then goes back to sleep (blocks), awaiting the next timer signal. Finally, τ_3 (the user interface) can run because by now both τ_1 and τ_2 are blocked. It will continue to run until the timer driving τ_2 expires at time $kT + T$, τ_2 becomes ready, and the cycle begins anew. Note that the processes τ_1 and τ_2 can never access the shared memory simultaneously because of the priority assignment. Also note that additional processes could run at lower priorities, but they would not affect τ_1, τ_2, and τ_3. Finally, note how the logical correctness of all three processes can be tested (debugged) independently of one another. The situation is diagrammed in Figure 1.5.

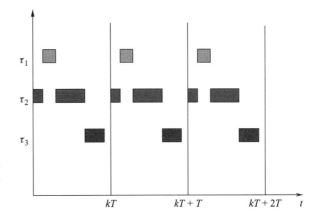

Figure 1.5 Three QNX process states that are used to implement a control law in Example 1.4.1. The shaded bars indicate when each process executes.

When a low-level interrupt-driven approach is used, mutual exclusion of shared resources is assured by disabling interrupts during critical sections of code. With a real-time operating system, this is no longer necessary. (In fact, it might not be possible!) To protect critical sections of code, well-known software constructs such as *semaphores* can be used, or priorities can be assigned which guarantee that critical sections cannot be interrupted, as in Example 1.4.1. A semaphore acts as a lock, preventing all other processes from accessing a shared resource, until the process that entered its critical section finishes.

Example 1.4.2

Let us revisit the situation in Examples 1.3.2 and 1.4.1 a final time, and suppose the user-interface process τ_3 decreases (locks) a semaphore before entering its critical section of code that accesses the shared memory X. If τ_3 is preempted by τ_2 while in the critical section, τ_2 would first check the semaphore and find that X is being used by another process. It would then block, at which point τ_3 would complete its access to X. This ensures correct logical behavior: Process τ_2 must wait until process τ_3 completes its access and unlocks the critical section.

When τ_2 blocks, the operating system will run the process that has the highest priority that is ready, which will be τ_3. This will complete its critical section and unlock the shared resource. At this point, τ_2 becomes ready and preempts τ_3, as it should. But suppose that a fourth process, τ_4, with a priority between those of τ_2 and τ_3, that is, $P_4 = 4$, becomes ready while τ_3 is executing its critical section. Suppose τ_4 does not use X, so it is not affected by the locked semaphore. Thus, it will preempt τ_3. Since all other processes are effectively blocked, τ_4 will run to completion. In this case, the priority of τ_4 is in effect higher than τ_2, which is probably not what the designer had intended. The situation is illustrated in Figure 1.6.

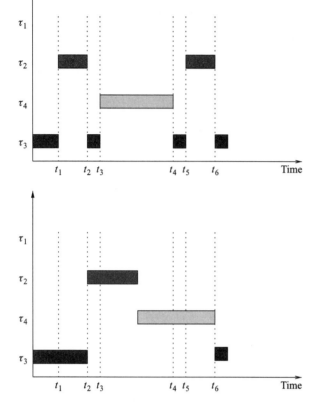

Figure 1.6 Protecting critical sections with a semaphore, for example, can lead to priority inversion (above). Process τ_3, with the lowest priority, enters a critical section when, at t_1, τ_2 becomes ready. It runs until t_2, at which point it attempts to enter the same critical section and becomes blocked. At this point, τ_3 is the highest priority ready process, and it runs. But it has not completed the critical section when, at t_3, it is preempted by τ_4, which has a priority between that of τ_2 and τ_3. τ_4 will run to completion, while τ_2 remains blocked. τ_3 finally completes its critical section at t_5. The figure (below) shows the same four processes where a priority ceiling algorithm has promoted the priority of τ_3 to that of τ_1 while it executes in its critical section. Notice the reduced number of context switches.

Example 1.4.2 illustrates an undesirable phenomenon known as *priority inversion*, in which a higher-priority task (τ_2) is effectively preempted by a lower-priority process (τ_4). This can be overcome by the use of a *priority ceiling protocol*. The idea is to promote the priority of any process that accesses a shared resource to exactly the priority of the highest-priority process that also shares that particular resource. Returning to our example, the priority of τ_3 would be promoted to $P_1 = 1$ while it is accessing x. If τ_2 becomes ready during this period, it waits until τ_3 is finished (because its priority is equal, not less than the newly promoted τ_3). When τ_3 finishes its critical section, its priority is returned to $P_3 = 5$, and τ_2 will preempt. The priority ceiling protocol is also illustrated in Figure 1.6.

The priority ceiling protocol has several desirable properties, including preventing priority inversion, preventing deadlock, and minimizing the number of context switches [7]. Importantly, the schedulability test (1.6) can be modified accordingly: We simply have to add to R_i the worst-case *blocking time*, which we denote B_i,

$$R_i = C_i + B_i + \sum_{i \in H} \left\lceil \frac{R_i}{T_j} \right\rceil C_j. \tag{1.8}$$

The priority ceiling protocol guarantees that τ_i will be blocked by a lower priority process at most *once* while it is running. (The process that is blocking τ_i will run at least the priority of τ_i while accessing the shared resource.) Thus, B_i is just the largest C_i of all tasks sharing the resource. A more complete discussion is beyond our scope, and we refer the reader to any one of a number of textbooks on the subject, for example [4].

There are alternatives to the use of preemptive, priority-based, multitasking operating systems that we have presented here. For example, so called *cyclic executives* execute processes on a fixed schedule and have certain advantages, such as minimizing timing jitter. But the trend in industry seems to be toward multitasking, real-time operating systems, in which programs are written using well-known languages such as C, extended with libraries of real-time functions, or perhaps Ada. Given the widespread familiarity of both C and multitasking operating systems such as Linux, this trend will doubtless continue.

1.5 HARDWARE ISSUES

Our emphasis to this point has been on the real-time software aspects of controller implementation using a general processor. The chapter would be incomplete without some discussion of the hardware requirements. In the following subsections, we outline a few of the more popular off-the-shelf hardware platforms used to realize control systems.

1.5.1 Desktop PCs

The commodity pricing of desktop PCs, along with their continuously improving performance, storage capacity, and ability to network, has made them increasingly popular platforms for real-time control. Of course, it is often necessary to enclose the delicate electronics into an appropriate industrial-strength case. But, under the hood, there is little difference between an industrially hardened PC and its desktop version. Input-output can be provided by using appropriate expansion cards. The primary advantage of using PCs is, of course, low cost, relatively high performance, and a large variety of available software.

The question is what operating system to use. As discussed earlier, Windows NT is not a "hard" real-time operating system: A controller that is running as a process on such a machine will suffer from some timing jitter. Depending on the other executing processes, this may or may not be an issue for the particular control problem at hand. As an alternative to developing a controller in a language such as C, the designer can turn to number of Windows NT applications that provide graphically programmable controllers. Labview from National Instruments is a popular example. In a matter of minutes, a designer can program a PID controller using its graphical programming language. However, if more demanding real-time performance is required, a "hard" real-time OS such as the Unix-derivatives QNX or Lynx should be used.

1.5.2 Single-Board Computers

Desktop PCs can be too large and bulky for embedded use. When size is an issue, a so-called single-board computer can be used. Functionally, these are complete PCs on a single printed circuit board, perhaps lacking the keyboard and monitor, and manufactured with smaller footprints than would be used for a desktop PC. Typically, they are equipped with popular busses such as the PCI bus and the relatively new PC-104 bus. The latter accommodates very small expansion cards for input-output. These are stacked parallel to the "motherboard," so that the entire unit fills a very compact volume. Most support solid-state storage such as flash memory that can replace disk drives, making the PC more rugged. Several operating systems are now available which can be loaded from flash memory instead of disk drives.

1.5.3 Digital Signal Processors

Several manufacturers market Digital Signal Processors (DSPs) and complete stand-alone DSP boards for control system implementation. DSPs excel at the numerical aspects of control, such as floating point multiplication and indexing arithmetic. Moreover, most DSPs do not have memory caches and use a RISC architecture, meaning instructions take a fixed number of CPU cycles to complete. Thus the time required to execute a set of instructions is easy to predict. When the timing aspects of a controller implementation are critical, both in terms of speed and timing jitter, a DSP will offer the best solution.

1.5.4 Programmable Logic Controllers

Programmable Logic Controllers (PLCs) are special-purpose industrial computers that are used extensively in industry to implement control systems [12, 13]. Historically, PLCs were invented as a programmable replacement for large hardwired control panels that were used to control machines on the factory floor in the automotive industry. Typically, these panels connected switches to relays, valves, lights, sensors, and so on, all wired together to make a complex factory system work. Thus, early PLCs acted as simple on/off control devices, taking input from perhaps a large number of switches and sensors, and providing output to electromechanical relays.

Traditionally, PLCs have been programmed using a graphical language called a *ladder diagram*. A ladder diagram is essentially a schematic wiring diagram that shows the logical behavior of an electrical circuit. It is familiar to electricians and technicians who might work on a factory floor. Each "rung" of a ladder represents an electrical circuit, which might include a switch, relay, light, and so forth. "Rungs" can be added to the PLC program just as a new line of C code can be added to a C program.

PLCs have evolved into much more than just programmable switches. Elements such as timers, logic statements, and arithmetic operations can be used to make decisions based on timing and logic. They can include a rich set of input/output modules, including A/D converters, shaft encoder modules, and even vision systems. Modules particular to control include PID subsystems and ethernet communication modules. Thus, a custom control system can be put together using a PLC and a set of appropriate modules and software.

PLCs are generally used in an environment that has a large number of inputs and outputs, and where logical decisions based on these signals must be made. For example, the high-level control of an industrial robotic manipulator will often be done using a PLC. Such a controller would provide the reference set-points to the low-level joint control system. At the same time, it might monitor a safety system, stopping the robot should a light-beam break, indicating that a person has entered an unsafe area. The PLC might also interface with other factory-floor PLCs that control the systems that feed parts to the manipulator. More complete descriptions can be found in [12, 13].

1.6 CONCLUSION

Real-time programming is usually thought to be beyond the scope of conventional control theory and practice. This is rather ironic, given that so much advanced theory can only find application through advances in real-time technology. In this chapter, we have illustrated the danger of ignoring real-time issues, and we have introduced some modern tools that are extremely useful for complex real-time control system design. In the future, tools such as real-time operating systems, programming languages, and processor architectures will make possible control systems with increased functionality and complexity. Control engineers should not only reach out and embrace this technology, but they should play a role in its development. After all, the job does not end with a difference equation.

> **Related Chapters**
>
> - Some related types of real-time control issues also arise in communication networks—see Ch. 19.
> - The use of programmable logic controllers for real-time control programming is briefly discussed in Ch. 16.
> - In Ch. 13, a challenging class of real-time control problems that arise in power systems, due to couplings between geographically separate generators, is discussed.

REFERENCES

[1] S. R. Norsworthy, R. Schreier, and G. C. Temes (eds.), *Delta-Sigma Data Converters*. Piscataway, NJ: IEEE Press, 1997.

[2] R. J. A. Buhr and D. L. Bailey, *An Introduction to Real-Time Systems*. Upper Saddle River, NJ: Prentice Hall, 1999.

[3] M. Joseph (ed.), *Real-Time Systems: Specification, Verification and Analysis*. New York: Prentice Hall, 1996.

[4] C. M. Krishna and K. G. Shin, *Real-Time Systems*. New York: McGraw-Hill, 1997.

[5] J. Wikander and B. Svensson (eds.), *Real-Time Systems in Mechatronic Applications*. Boston, MA: Kluwer, 1998.

[6] G. Olsson and G. Piani, *Computer Systems for Automation and Control*. New York: Prentice Hall, 1992.

[7] N. C. Audsley, A. Burns, and A. J. Wellings, "Deadline monotonic scheduling theory and application." *Control Engineering Practice*, Vol. 1, no. 1, pp. 71–78, February 1993.

[8] *QNX OS System Architecture*. Kanata, Ontario, Canada: QNX Software Systems, Ltd., 1993.

[9] H. Rzehak, "Real-time Unix: What performance can we expect?" *Control Engineering Practice*, Vol. 1, no. 1, pp. 65–70, February 1993.

[10] M. Ragen, "Real-time systems with Microsoft Windows NT." Available at www.microsoft.com/embedded/winnt.htm.

[11] J. L. Peterson and A. Silberschatz, *Operating System Concepts*. Reading, MA: Addison-Wesley, 1987.

[12] J. Stenerson, *Fundamentals of Programmable Logic Controllers, Sensors, and Communications*. Englewood Cliffs, NJ: Prentice Hall, 1993.

[13] T. E. Kissell, *Understanding and Using Programmable Controllers*. Englewood Cliffs, NJ: Prentice-Hall, 1986.

Chapter 2 | DISCRETE-EVENT SYSTEMS AND THEIR OPTIMIZATION

Edwin K. P. Chong

Editor's Summary

Classical control technology has by and large focused on continuous-time systems—including their digitized and sampled equivalents. The unequivocal success in this arena has resulted in a broadening of interests and in explorations of the application of control concepts to other problems, even those that are not readily amenable to the techniques of traditional control. This chapter focuses on one such topic: discrete-event systems.

Whereas control technology is mostly concerned with systems with internal dynamics that can be mediated by continuous-valued inputs, discrete-event systems (DES) exhibit dynamics that evolve in accordance with external events—the state of the system changes only when an event occurs. Many problems in communication networks, manufacturing systems, transportation and traffic, and numerous other domains can be seen as DES applications. In the first case, for example, events of interest can be the arrival of packets of information at a node in the network. (Connections between DES and communication networks are further elaborated in Chapter 19.)

Modeling approaches for DES include state machines and automata, Markov chains, and timed models using event clocks. Simple examples are shown for each, drawn from computer systems with on, off, and failed states and single-server queues. Some of these models are also discussed in Chapter 7 which deals with systems that combine discrete-event and continuous-time dynamics.

This chapter also discusses the topic of optimization of DESs: how control parameters of a DES (for example, the mean service time for jobs in a queue) can be selected to optimize some performance measure. To use gradient-based optimization methods for discrete-event systems, gradient information must be estimated; since the systems and their representations are not continuous, gradients cannot be analytically calculated. Stochastic approximation algorithms are an effective option and can even allow DES optimization to be performed on-line, while the system is operating.

Edwin Chong is an associate professor in the School of Electrical and Computer Engineering at Purdue University, West Lafayette, and the chair of the IEEE-CSS Technical Committee on Discrete Event Systems.

2.1 INTRODUCTION

The twentieth century was dominated by the development of highly complex man-made systems that performed complicated tasks. From mobile telephone networks to satellite space stations, the development of such systems was accompanied by an ever-increasing demand for even more sophisticated systems. As we usher in a new century that promises the development of technology currently not even imaginable, the need for a systematic and mathematical approach to the analysis, design, and control of complicated large-scale systems is becoming increasingly important.

Such a need has long been recognized by researchers in a multitude of technological areas. In operations research, for example, researchers have been interested for a long time in systematic methods to deal with large-scale systems. However, only in relatively recent years have control engineers taken up this challenge. A result of this undertaking has been the birth and development of the area of study known as *discrete-event systems*.[1]

A discrete-event system is a dynamic system that evolves in accordance with the occurrence of events. An extensive literature on discrete-event systems has appeared in the last 15 years, and their study continues to be an area of ongoing research. Targeting application areas such as telecommunication networks and manufacturing systems, the area has attracted an interdisciplinary pool of researchers, from systems and control, theoretical computer science, operations research, and artificial intelligence. Within the domain of control engineering, the study of discrete-event systems attempts to address the following questions:

- To what extent can ideas from classical systems and control theory be used in discrete-event systems, and how?
- How do we specify and solve decision and control problems in discrete-event systems?
- What models are appropriate for performance analysis and optimization of discrete-event systems?
- How do we systematically and optimally design a discrete-event system to satisfy given design specifications?

The study of discrete-event systems has reached a stage where an undergraduate-level textbook on the subject is available, as well as several books on specialized topics in the area. (We provide some references at the end of the chapter.) Moreover, many academic institutions around the world have begun to offer courses on discrete-event systems, reflecting the increased recognition of the importance of the area. Nonetheless, to date, the state of the art does not yet fully address all of the questions listed above, and much remains to be done in the area.

In this chapter, we provide an overview of discrete-event systems at a level that should be accessible to engineers with no more than undergraduate training in systems and control. The treatment starts out at an elementary level and builds up to a discussion of optimization techniques that are in fact too advanced to be discussed in detail in a chapter like this. Our goal is simply to whet the reader's appetite, providing a glimpse of what promises to be an important topic for control engineers in years to come.

In the next section, we provide a simple definition of a discrete-event system. We begin in Section 2.2.1 with a basic discussion of systems, including input-output systems and states. Then, in Section 2.2.2, we introduce the idea of a discrete-event system, contrasting it with classical models based on differential and difference equations. In

---

[1] Often, the abbreviation DES is used. Some use the term *discrete-event dynamic system*, with the abbreviation DEDS.

Section 2.2.3, we argue that discrete-event systems arise naturally as models when considering many of today's complex engineering systems.

Section 2.3 is devoted to a discussion of some basic ideas in modeling discrete-event systems. We discuss state trajectories (Section 2.3.1) and state machine models (Section 2.3.2), as well as some extensions of these ideas. To explore discrete-event system models that consider event occurrence times, in Section 2.3.3 we describe a particular model using event clocks. Here, we also discuss discrete-event simulations, as well as Markov and semi-Markov processes. These discussions, though brief, assume some knowledge of probability and stochastic processes.

In Section 2.4, we discuss the problem of optimization in discrete-event systems, focusing on a particular approach involving gradients. First, in Section 2.4.1, we define the components of an optimization problem. Then, in Section 2.4.2 we discuss one approach to solving optimization problems based on gradient algorithms. To apply this approach to the optimization of discrete-event systems, in Section 2.4.3 we argue that gradient estimation is a key ingredient, and we introduce some ideas along these lines. Finally, in Section 2.4.4, we describe the use of gradient estimators and stochastic approximation algorithms for the on-line optimization of discrete-event systems. We illustrate the application of this approach via an example of a capacity allocation problem.

Finally, in Section 2.5, we provide directions for further reading. In keeping with the intended level of our exposition, we have attempted to restrict our reference list to books and overview articles. These references provide good starting points for investigation, and most contain further references to other useful sources, including research articles.

2.2 DISCRETE-EVENT SYSTEMS

2.2.1 What Is a System?

The heart of control engineering is the ability to analyze real-world "systems" using analytical tools (pen-and-paper or computer/software). Typically, this analysis is done by reasoning with a *model* of the system, the language of such reasoning being mathematics. Throughout this chapter, we use the term *system* to mean a mathematical model, constructed for the purpose of analysis. This use of the term is typical in formal methods of control engineering.

The prototypical system in control engineering is the *input-output system*. Here, we associate with the model an explicit *input* and *output*. Typically, the input is "controllable" in the sense that we can manipulate it to achieve some desired effect. The output is usually the manifestation of the resulting effect, such as a sensor measurement of some physical quantity.

It is also common in control engineering to use *state models*. Here, we associate with the model an explicit description of the *state* of the system. The state is usually an entity that determines or characterizes the "internal condition" of the system. For example, the state of a hot-water heater may be defined by the temperature and volume of water it contains.

Also relevant to our discussion is the notion of a *dynamic system*. Such a system "evolves" in the sense that its state changes in response to external factors. Such factors may include changing values of the input, or simply just the passing of time. In the latter case, we say that the dynamic system "evolves in accordance with time." For example, the water temperature of a hot-water heater may increase in response to a change of the temperature setting (input). If it has a leak, the volume may decrease with time even when there is no change in input.

Over the centuries, scientists (physicists, chemists, etc.) have provided engineers with system models based on physical laws. However these laws were developed, they agree with how the world operates—experimental measurements of the behavior of the world agree with what is predicted by these laws. Therefore, such physical laws appropriately provide the basis for system models, and many engineering marvels have been created as a result.

The prevailing mathematical principle underlying physical laws is the *differential equation*. From Maxwell to Schrödinger, scientists have used differential equations as the primary vehicle to describe how things work. Specifically, given certain variables that describe the state of a system, such as volume and temperature, a differential equation tells us how these variables change with time (and sometimes even how they vary with location in space). Not surprisingly, over the years differential equations have provided the primary basis for models in control engineering.

As the use of computers became mainstream in control systems, engineers began to use a slightly different form of model, called the *difference equation*. Functionally, the difference equation serves the same purpose as the differential equation in that it dictates how a system evolves with time. The distinguishing factor is that in a difference equation, time takes on values only at discrete points (e.g., every tick of a second). For this reason, difference equations give rise to what are called *discrete-time* models. Such models may be used to describe the value of the state of a system as sampled at discrete instants of time, such as is necessitated by the use of an A/D (analog-to-digital) converter.

2.2.2 What Is a Discrete-Event System?

A discrete-event system is a dynamic system that evolves in accordance with occurrences of *events*. It is instructive to distinguish discrete-event systems from models based on differential and difference equations. First, we note that while differential and difference equation models evolve with time, a discrete-event system evolves with the occurrence of events. The occurrence of such events may also be associated with instants of time, but this association is unnecessary in certain applications. (We will provide examples later.) A discrete-event system model may also involve a state. By definition, the state of a discrete-event system changes only when an event occurs. Note that the adjective *discrete* in "discrete-event system" is used to emphasize the notion that an event makes the system state change abruptly and not continuously. The reader should not confuse a discrete-*event* system with a discrete-*time* system.

Discrete-event system models are useful in a wide variety of situations and provide a flexible means to describe many systems of interest. An event in a discrete-event system can represent virtually anything that occurs abruptly. For example, an event may correspond to the arrival of a packet at a node in a communication network, the completion of a job by a central processing unit (CPU), or the failure of a machine in a manufacturing system. Discrete-event system models can be used in virtually all man-made systems, examples of which are endless—manufacturing systems, communication networks, computer systems, logistics, vehicular traffic, economic systems, stock markets, and so on. Discrete-event systems also provide suitable models for some natural phenomena, such as particle interactions in certain materials.

2.2.3 Why the Need for Discrete-Event Systems?

Discrete-event system models are useful when dealing with dynamic systems that are not fully captured by classical models, such as differential or difference equations. Such systems increasingly are dominating the modeling arena for two main reasons.

First, the complexity of many man-made systems necessitates viewing them at different levels of abstraction. It is often impossible to have a single picture of an entire system; engineering systems are just too complicated nowadays. The standard practice is to define various levels of abstraction, each of which can be handled in a manageable way. Usually, in such multilevel abstractions, a low-level view provides a detailed description of the dynamics of the various components in the system, whereas a high-level view is used to characterize the interactions between the components. Although the low-level dynamics are often adequately captured by classical differential equation models, the high-level view involves dynamics of a different nature, such as logical decision making, discrete control actions, abstract task descriptions, and switching between modes. Therefore, at sufficiently high levels of abstraction, the use of events to capture dynamics is natural, perhaps even inevitable.

Example

Control issues in manufacturing systems typically are approached in a hierarchical fashion (see Table 2.1). At the lowest level, we take into account individual machines in the manufacturing system, such as conveyor motors or robotic manipulators. Here, the time scale of the dynamics (i.e., how fast the state of the system is changing) is relatively short, perhaps on the order of milliseconds.

At a higher level, we consider the cells in the manufacturing system. Each cell consists of several machines, but we are not concerned with the details of individual machines. We focus only on the aggregate behavior of the machines as a cell. The time scale at which the dynamics evolve is longer here than in the low-level view, somewhere in the domain of minutes or hours.

At the highest level shown in Table 2.1, we are concerned with the overall behavior of the factory—the so-called big picture of the system. Here, even individual cells are not of interest, only how they interact together to affect the overall factory. Time scales at this level are relatively long, perhaps involving days or even months.

TABLE 2.1 Control Hierarchy in Manufacturing Systems

Level	Time Scale	Example Issues	Example Technologies
Machine	Short	Robot arm trajectory Relay ladder logic Servomotor control	PID, PLC, Fuzzy
Cell	Medium	Routing Scheduling	Simulation
Factory	Long	Material planning Inventory control Tooling problem	Just-in-time, Kanban

Section 2.3 Some Discrete-Event System Models 25

A second reason for considering discrete-event system models is that many decision and control problems involving complex systems are inherently discrete in nature. These include resource allocation, scheduling, policy selection, synchronization, routing, and admission control in communication networks. Although such issues give rise to challenging problems for control engineers, they cannot readily be tackled using only classical models such as differential equations.

2.3 SOME DISCRETE-EVENT SYSTEM MODELS

2.3.1 State Trajectory of a Discrete-Event System

Figure 2.1 shows the state of a discrete-event system model as a function of time (the *state trajectory*). Notice that the state of the system changes only when an event occurs. The labels α, β, and so on, signify the associated events. In between the instants of occurrence of events, the state remains fixed. Note that the vertical axis in Figure 2.1 does not necessarily represent "numerical" values, such as 1, 2, or 3 but may be arbitrary qualitative states, such as on or off. It is often natural in discrete-event systems to restrict the possible states of the system to some "discrete" (countable) set.

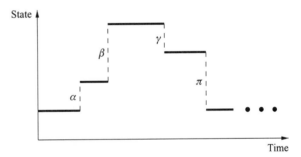

Figure 2.1 State trajectory of a discrete event system.

Example (Simple model of computer)

Consider a "high-level" model of a computer with three states: on, off, and down. When the state is off and we push the power switch (the event *push-switch*, denoted π, occurs), the state changes to on. Similarly, when the state is on and the event π occurs, the state changes to off. If the event *fault* (ϕ) occurs when the state is on, the state changes to down. Once in the state down, the state remains there until the event *repair* (ρ) occurs, in which case the state changes to off. Figure 2.2 illustrates a possible state trajectory of the discrete-event system model of the computer.

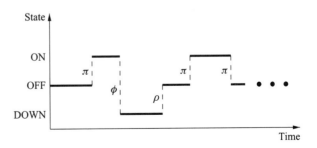

Figure 2.2 State trajectory of a simple model of a computer.

Note that in this example, the number of states in the system is finite (three). In general, it is possible to have an infinite number of states, as in the following example.

Example (Queue)

Consider the familiar scenario where customers arrive at a server that can serve only one customer at a time. Customers who have arrived but have not yet completed service must wait in line (they form a queue). These customers include those who are being served and those who have not yet begun service. As soon as the server completes service of a customer, that customer immediately leaves the system.

The model described here is called a *single-server queue*. The events in the system are customer *arrival* (α) and *service completion* (σ). The possible states of the system can be modeled simply as the number of customers in the queue (i.e., customers who have arrived but have not yet departed). When event α occurs, the state goes from n to $n+1$, where $n \in \{0, 1, 2, \ldots\}$, while the event σ causes the state to go from n to $n-1$, where $n \in \{1, 2, \ldots\}$. Note that the event σ cannot occur in state 0 (when there are no customers present). Figure 2.3 depicts a possible state trajectory of a queue.

While queues model a wide variety of familiar situations found in daily life, such as waiting in line at the checkout counter of a supermarket, they are also useful in many other application domains, ranging from telecommunication networks to viral epidemics.

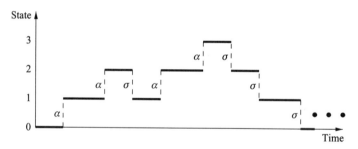

Figure 2.3 State trajectory of a queue.

2.3.2 State Model of a Discrete-Event System

2.3.2.1 State Machines

A discrete-event system with a countable number of states can be represented conveniently by a *state transition diagram*. In such a diagram, a circle represents a state, and an arrow going from one state to another (also called an *arc*) represents a change of state (also called a *state transition*). Each arc has a label that represents the event associated with the state transition. Such a model is also called a *state machine* or *automaton*. The set of all possible states in the system is called the *state space*.

Example (Simple model of computer)

Consider again the previous example of a simple model of a computer. Figure 2.4 shows the state transition diagram of the model. The state space in this example is {on, off, down}.

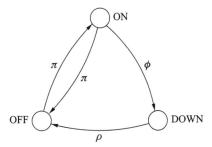

Figure 2.4 State transition diagram for a simple model of a computer.

This example, with a finite number of states, falls within the class of *finite state machines* (or *finite automata*). State machines with an infinite number of states are also possible, as is the case with the single-server queue.

Example (Queue)

Consider the example of a queue. The state, being the number of customers in the system, can be any nonnegative integer. The state space is therefore infinite. Figure 2.5 shows the state transition diagram for the queue.

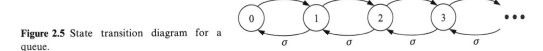

Figure 2.5 State transition diagram for a queue.

To summarize, a state machine is a model consisting of a set of states S (the state space), a set of events E, and a state transition rule δ that specifies the next state when an event occurs at a given state. Specifically, if s is the current state and event e occurs, the state changes from s to s', where $s' = \delta(s, e)$. It is convenient to think of δ as represented by the arcs in the state transition diagram. Usually, a state machine model also includes specification of an *initial state* s_0.

2.3.2.2 Nondeterministic State Machines

The state machine model, though simple, provides a useful basis for modeling many systems found in practice. There are several ways to extend the model to incorporate more flexible features. The most immediate extension is to allow the occurrence of a single event at a given state to cause state transitions to more than just one other state.

Example (Simple model of computer)

Consider again the previous example of a simple model of a computer. Suppose we extend the model by specifying that when the event π (push-switch) occurs at state off, the state can either become on or down. In this case, the state transition diagram of the model is given in Figure 2.6. Note that there is an arc from state off to state on as well as another arc to state down, both labeled with the event π. When event π occurs at state off, the next state is either on or down. We do not specify beforehand which it will be; the next state is therefore not *deterministically* determined by the current state and the event that occurs.

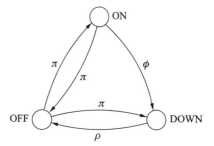

Figure 2.6 Nondeterministic state transition diagram for a simple model of a computer.

State machine models in which a single event at a given state can cause transitions to multiple possible states is called a *nondeterministic* state machine.

2.3.2.3 Markov Chains

Another extension of the state machine model is to incorporate probabilities with events. Specifically, in certain applications, the state transitions occur with certain probabilities. In this case, again we can use a state transition diagram to represent the system, incorporating a probability with each state transition. The resulting model is called a *Markov chain*.

Example (Simple model of computer)

Consider the previous example of the nondeterministic model of a computer. Suppose the probabilities of state transitions are given in Table 2.2. We can represent the Markov chain model for the system using the state transition diagram in Figure 2.7, where, in addition to the event labels, we also incorporate the transition probabilities.

TABLE 2.2 State Transition Probabilities for Markov Chain Model of Computer

Current State	Next State	Event	Probability
OFF	ON	π	0.95
OFF	DOWN	π	0.05
ON	OFF	π	0.9
ON	DOWN	ϕ	0.1
DOWN	OFF	ρ	1.0

Section 2.3 Some Discrete-Event System Models   29

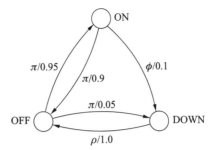

Figure 2.7 State transition diagram for Markov chain model of a computer.

2.3.2.4 State Machines and Discrete-Event Systems

At this stage, it is natural to ask what can be done with models of the kind described earlier. To answer this question, we point out that the study of automata is a mature subject, with many textbooks written on various aspects of their analysis. The same can be said about Markov chains. Although it is beyond the scope of this chapter to discuss the rich theory underlying such models, it should be clear that control engineers can benefit greatly by taking advantage of the many useful techniques from the vast literature on these topics. (We provide some references at the end of the chapter.) In addition to state machines, several alternative discrete-event system models are available that capture different aspects of the modeling process. These include Petri nets, max-plus algebra, and communicating sequential processes. Again, a significant literature is devoted to the study of these models, including their use in applications such as communication protocols, software engineering, transportation systems, and manufacturing. Another line of investigation currently of interest combines discrete-event system models with differential equation models, resulting in what are called *hybrid systems*.

The use of state machines in the study of control of discrete-event systems was initiated by Ramadge and Wonham [18]. Their framework provides a way to explore control theoretic questions in discrete-event systems. More specifically, they incorporate the notion of a control input that can be used to influence the behavior of the system—the control input disables certain events from occurring. Given such a control framework, many control theoretic ideas familiar in the study of control theory can be applied to discrete-event systems. For example, we can explore how to design a "feedback" controller that uses observations of the state trajectory to choose the control input such that the overall behavior of the system satisfies some given design specification. This line of investigation has led to a theory of *supervisory control* for discrete-event systems. We will not discuss this topic any further in this chapter.

2.3.3 State Models with Event Clocks

The previous discussion of discrete-event models does not consider the times at which events occur, but only the order in which they occur and how the state of the system changes as a result. Although many problems of interest can be addressed using only such *untimed* models (also called *logical* models), others require explicitly dealing with the times at which events occur. For example, suppose we are interested in the average waiting time experienced by customers in a system modeled as a single-server queue. The waiting time of each customer is the time duration between the arrival and start of service of a customer—clearly the difference between the occurrence times of two events: a service completion and an arrival. A model capturing waiting times in a queue must include event occurrence times.

2.3.3.1 Event Clocks

We now describe one possible way to extend state machine models to include event occurrence times, leading to a *timed* discrete-event system model. This extension involves introducing the notion of *event clocks*. To proceed, recall that a state machine consists of a set of states S, a set of events E, and a state transition rule δ. At any given state s, consider the set of all events that can occur. We can visualize this set by considering all arcs that leave the state s in the state transition diagram and by listing all event labels for those arcs. We call this set the set of *feasible events* at state s, denoted $E_f(s)$.

Example (Simple model of computer)

Consider again the simple model of a computer shown in Figure 2.4. We have three states, $S = \{\text{on, off, down}\}$, and three events $E = \{\pi, \phi, \rho\}$. At the state off, only event π is feasible; thus, $E_f(\text{off}) = \{\pi\}$. At the state on, there are two feasible events: $E_f(\text{on}) = \{\pi, \phi\}$. At the state down, only one event is feasible: $E_f(\text{down}) = \{\rho\}$.

Note that as the state changes, certain events that were not previously feasible may become feasible. Also, when an event occurs and the state changes, that event may no longer be feasible in the next state.

In our timed discrete-event system model, we associate an event clock with each event in E. When an event that was not previously feasible becomes feasible owing to a state transition, the clock for that event is set to some initial positive value. This clock runs down at unit rate and will eventually reach 0. As soon as it does, the event occurs. Because there may be several events that are feasible at any given state, there are several event clocks running in parallel, each at unit rate. Whichever clock reaches 0 first causes its associated event to occur.

Example (Queue)

Consider the example of a single-server queue, as shown in Figure 2.5. Here, we have state space $S = \{0, 1, 2, \ldots\}$ and event set $E = \{\alpha, \sigma\}$. Note that both events are feasible at all states except in state 0, where only the arrival event α is feasible. In other words, $E_f(0) = \{\alpha\}$, while $E_f(s) = \{\alpha, \sigma\}$ for all $s = 1, 2, \ldots$.

Suppose we start at state 0 (an empty system). The only feasible event is α, so we set the event clock for α at some initial value. As time progresses, the α-clock runs down at unit rate. When this clock reaches 0, the event α occurs (for the first time), and the state changes from 0 to 1. At this stage, both events α and σ are feasible. Therefore, we set both clocks to some initial values, and let them run down at unit rate. Suppose the α-clock reaches 0 first. At that time, α occurs again (a new customer arrives) and the state goes from 1 to 2. Here, both α and σ are again feasible. However, we need only reset the α-clock to some initial value because the σ-clock has not yet reached 0 and still has some positive time remaining. Having set the α-clock, the process continues, with both clocks racing to 0 to determine which event is next to

Section 2.3 Some Discrete-Event System Models

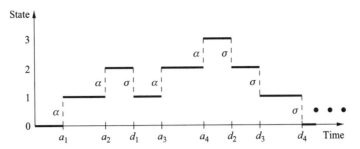

Figure 2.8 Timed state trajectory of a queue.

occur. Figure 2.8 illustrates a possible *timed* state trajectory of the queue, with event times labeled as $a_1, a_2, \ldots$ for the arrival (α) times and $d_1, d_2, \ldots$ for the service completion or departure (σ) times.

Note that the initial clock values in the above timed model have convenient interpretations. Suppose we label the initial setting of the α-clock at the beginning (when the state was 0) as A_0. We can think of this value as the time until the first arrival. When the state first changes from 0 to 1, suppose we denote the initial clock values as A_1 for the α-clock and S_1 for the σ-clock. Then, we can think of A_1 as the *interarrival time* between the first and second customers, and S_1 the *service time* of the first customer (the time it takes to serve the first customer). Note that if $A_1 < S_1$, then the second customer arrives before the first customer has completed service, as is the case in Figure 2.8. In this case, the second customer has to wait in line, and the number of customers in the system increases from 1 to 2. In a similar way, we can define interarrival times $A_2, A_3, \ldots$ and service times $S_2, S_3, \ldots$. In Figure 2.8, the interarrival times are given by $A_n = a_{n+1} - a_n$, $n = 1, 2, \ldots$, these being the times between the occurrences of event α. Note that $S_1, S_2, \ldots$ are *not* the times between the occurrences of event σ because σ is not a feasible event when the state is 0. These periods when the state is 0 are called *idle periods* of the queue. The intervals of time between idle periods are called *busy periods* of the queue.

It is apparent that the (timed) state trajectory of the system is completely determined by the numbers $A_0, A_1, A_2, \ldots$ and $S_1, S_2, \ldots$. These numbers are called the *event lifetimes*. The event lifetimes in a timed model can be viewed as the input to the system, which completely determines the resulting state trajectory.

This description of how event clocks determine the timed state trajectory relies on the assumption that once an event becomes feasible, it remains feasible until it occurs. This property of a system is called *noninterruption*. In systems with this property, once an event clock is set, it continues to run down until it reaches 0. Although many systems in practice, such as queues and networks of queues, satisfy this property, not all systems do; such systems are said to be *interruptive*. For example, the system in Figure 2.4 does not satisfy the noninterruption assumption because when the state changes from on to down, the event π becomes infeasible. Clearly, interruptive systems have to be treated differently. However, many models can be made to satisfy noninterruption by simple modifications. For example, we can modify the system in Figure 2.4 by including an arc from state down to itself with the label π. In other words, event π is feasible in state down but causes a transition back to state down. This model satisfies the noninterruption property and may equally serve our practical purposes.

2.3.3.2 Discrete-Event Simulations

Timed models using event clocks often are used as the basis for computer simulations of discrete-event systems (also called *discrete-event simulations*). Such simulations are useful for computing estimates of performance measures, such as the average waiting time in a queue. To construct the simulation, we first decide how to set the values of the event lifetimes. Typically, the sequence of event lifetimes for each event is assumed to be an independent, identically distributed random sequence with a given distribution. Such a sequence can be generated using a random number generator, a common component of discrete-event simulation software packages. Finally, we implement the previously described mechanism for generating the timed trajectory of the system given the event lifetimes. From the timed trajectory, we can extract whatever information we desire, such as waiting times of customers in a queue.

Discrete-event simulations with random event lifetimes can also be applied to models with nondeterminism. (Recall our previous discussion on nondeterministic state machines.) Here, the occurrence of an event at a given state can lead to several possible next states. The typical approach is to pick one of these next states according to some prespecified probabilities, as is done in Markov chain models. These transitions can also be implemented using a random number generator.

There is a significant literature on discrete-event simulation techniques; see, for example, [3], [9]. It suffices to mention here that simulation tools play an important role in the modeling and performance evaluation of discrete-event systems.

2.3.3.3 Markov and Semi-Markov Processes

The stochastic process resulting from using independent, identically distributed event lifetime sequences and probabilistic state transitions is called a *generalized semi-Markov process*. Researchers have done significant work in characterizing the properties of such processes. These studies provide valuable analytical tools that can be used in the analysis of discrete-event systems modeled by such processes; see references [10] and [11].

In the special case where the distribution functions of the event lifetimes are all exponential, the resulting stochastic process is called a *continuous-time Markov process*. Such processes yield to a rich analytical theory for which a large literature is available. Here, again we encourage the interested reader to take advantage of the many accessible treatments of the theory (see, e.g., [5]).

A special case of significant pedagogical interest in the study of queueing systems is the single-server queue with exponentially distributed interarrival and service times. Such a system is called an $M/M/1$ *queue*. (We shall not discuss the rationale for the notation "M/M/1.") Because such a system yields to Markov process analysis, its properties are easy to derive and are used to provide insight into the behavior of queues. For example, it is easy to derive the formula for the steady-state average waiting time in an M/M/1 queue:

$$W = \frac{\lambda \theta^2}{1 - \lambda \theta}$$

where θ is the mean service time and $1/\lambda$ is the mean interarrival time. The parameter λ is also called the *arrival rate*. Similarly, it is common to express the mean service time as $\theta = 1/\mu$, where μ is called the *service rate*. Note that the steady-state average waiting time exists only if $\lambda < \mu$, in which case we say that the queue is *stable*. The condition $\lambda < \mu$ for stability is intuitively appealing: The queue is stable only if the service rate exceeds the arrival rate, for otherwise the queue will build up indefinitely and customers will have increasingly larger waiting times.

2.4 OPTIMIZATION OF DISCRETE-EVENT SYSTEMS

In the design and operation of discrete-event systems, the designer often has the option of choosing between various alternative systems. Typically, the criterion governing such a choice is the optimality of the system with respect to a certain performance measure. This choice is often exercised through adjusting the values of *control parameters*. For example, in the operation of a communication network, we can often adjust the routing parameters within the network. These routing parameters affect the performance of the network. Naturally, we are interested in choosing their values such that the overall throughput or delay is optimized.

The design and operation of discrete-event systems therefore often center around the problem of performance optimization. In this section, we discuss one possible approach to this problem.

2.4.1 What Is Optimization?

Optimization is the task of making the best choice among a set of given alternatives. To define such a problem, we must first have a way to compare alternative choices. This comparison is usually done via an *objective function*. In the context of our discussion on performance optimization, the objective function is simply the performance measure. In other words, for each choice, the value of the objective function is the performance of the system corresponding to that choice. To be specific, we will assume that our goal is to *minimize* the value of the objective function. In other words, we wish to find the choice with an objective function value that is as small as possible. Maximization problems can be handled simply by multiplying the objective function by -1. Usually, in an optimization problem we also have to specify the set of *feasible* choices, which represents those choices or alternatives over which our minimization is required. This set is also called the *feasible region*.

Example (M/M/1 Queue)

Consider a single-server M/M/1 queue with interarrival rate λ and mean service time θ. Suppose we are interested in the steady-state average *sojourn time*. The sojourn time of a customer in the queue is the duration of time from arrival to service completion (departure) of the customer. In other words, the sojourn time is the sum of the waiting time and the service time. The steady-state average sojourn time is the average sojourn time of all customers, taken over an infinite horizon (i.e., taken over an infinite number of customers, hence the use of the term *steady-state*).

Denote the steady-state average sojourn time by $T(\theta)$, which is a function of the parameter θ. Here, θ is the control parameter that we can adjust. Consider the problem of choosing the parameter θ to minimize the performance measure $J(\theta) = T(\theta) + c/\theta$, where c is a given positive number. The rationale here is that we wish to minimize the sojourn time but with some penalty on choosing small values of θ; if there were no penalty on the choice of θ to minimize the sojourn time, the obvious choice would be $\theta = 0$. The values of θ that are feasible in our problem are those for which the queue is stable. As mentioned before, these are values for which $\theta < 1/\lambda$. In practice, we will need to restrict our set of feasible choices of θ to some subset of the stability region.

Figure 2.9 shows a plot of $J(\theta)$ versus θ for an M/M/1 queue with $\lambda = 1$. In this figure, D denotes the set of feasible parameter values (the feasible region). Because we know the formula for the steady-state average sojourn time in an M/M/1queue, we can analytically compute the solution to this optimization problem: $\theta_* = 0.2$ in the case of the objective function in Figure 2.9. (In fact, the value of $c = 0.0625$ was chosen here to give rise to this convenient solution.)

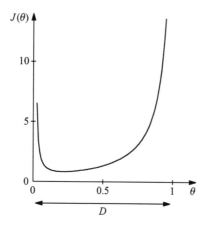

Figure 2.9 Objective function for M/M/1 queue with $\lambda = 1$.

2.4.2 Gradient Algorithms for Optimization

There are many approaches to solving an optimization problem. A common approach is to make use of the *gradient* of the objective function. Given a function $J(\theta)$ where the argument θ is a vector with components $\theta^1, \ldots, \theta^K$, the gradient of J at the point θ, denoted $\nabla J(\theta)$, is a vector with components

$$\frac{\partial J}{\partial \theta^1}(\theta), \ldots, \frac{\partial J}{\partial \theta^K}(\theta).$$

If the argument θ is a scalar parameter, then the gradient is simply the derivative.

The typical method for using gradients for optimization is via a *gradient algorithm*. Such an algorithm has the form

$$\theta_{n+1} = \theta_n - a_n \nabla J(\theta_n),$$

where a_n is a positive scalar called the *step size*. The algorithm is an iterative procedure that produces a sequence of *iterates* $\{\theta_n\} = \{\theta_1, \theta_2, \ldots\}$ with the goal that it converges to the solution of the optimization problem. The rationale behind the form of the gradient algorithm is that the vector $\nabla J(\theta_n)$ at θ_n points in the direction of steepest descent of the function J. The step size simply dictates how large a step to take in that direction to get from θ_n to θ_{n+1}. In applying the algorithm, we have to specify an initial point θ_0, usually a point that represents our best *a priori* guess. For more details on such algorithms, see [8].

The above algorithm may lead to values of iterates θ_n that lie outside of the feasible region D, which is often undesirable. To avoid such a situation, a common approach is to apply a *projection* at each iteration. A projection Π_D is a mapping that takes any point outside of D and gives us a value inside D, but leaves any point inside of D untouched. In other words, if $\theta \notin D$, then $\Pi_D[\theta] \in D$. On the other hand, if $\theta \in D$, then $\Pi_D[\theta] = \theta$. The projected version of the gradient algorithm then has the form

$$\theta_{n+1} = \Pi_D[\theta_n - a_n \nabla J(\theta_n)].$$

Notice that in the projected gradient algorithm, all iterates θ_n lie inside D. A common projection method is to pick the point $\Pi_D[\theta]$ to be the point inside D that is closest to θ.

2.4.3 Gradient Estimation

The standard gradient method does not easily apply to problems involving discrete-event systems because the method relies on being able to compute the value of the gradient at any given point. Discrete-event systems are often too complex to yield analytical expressions for gradients of performance measures. Moreover, such gradients usually depend on certain system parameter values or statistical distributions, which are often unknown. For example, the steady-state average sojourn time in a single-server queue depends on the arrival rate as well as the interarrival and service time distributions. Unless these entities are known, we cannot explicitly compute the gradient.

It turns out that the form of the gradient algorithm can still be used if we have *estimates* of the gradient. In other words, we may consider using the following algorithm:

$$\theta_{n+1} = \Pi_D[\theta_n - a_n h_n],$$

where h_n is an estimate of $\nabla J(\theta_n)$. Of course, for the algorithm to work, the estimate h_n must be a "sufficiently good" estimate of $\nabla J(\theta_n)$. Significant work has been done on such algorithms since the early 1950s. The first paper to study such algorithms rigorously was by Herbert Robbins and Sutton Monro in 1951, who coined the name *stochastic approximation* to describe the method.

Stochastic approximation algorithms are applicable to optimization problems in discrete-event systems only if we have suitable methods to estimate the gradients of performance measures. Since the early 1980s, several methods have been proposed for gradient estimation in discrete-event systems. Foremost among such methods are *perturbation analysis* (see [10], [12]) and the *score function* or *likelihood ratio* method (see [19]).

While it is beyond the scope of this chapter to describe in detail the various gradient estimation techniques for discrete-event systems, here we give a basic description of one such technique: infinitesimal perturbation analysis (IPA). (For a more detailed description of the technique, see [10], [12].) The technique of IPA for discrete-event systems is based on assuming that the event lifetimes are functions of the control parameter and then expressing the performance measure of interest as a function of the event lifetimes.

For example, suppose we are interested in the steady-state mean sojourn time, which is the steady-state average of the sojourn times of the customers in the queue. Note that the sojourn time of a customer is the difference between the service completion time and arrival time of the customer. Because we are interested only in the difference between these two times, we can set the origin of time arbitrarily. So, assume the origin of time is at the beginning of the busy period (recall the definition of a busy period in Section 2.3.3.1). The arrival time of the customer is the sum of interarrival times from the beginning of the busy period to the arrival of the customer; see Figure 2.8. Similarly, the service completion time is the sum of service times from the beginning of the busy period to the departure of the customer. Therefore, the sojourn time of a customer can be expressed as a function of certain interarrival and service times, the event lifetimes of the queueing model.

Suppose the service times are all functions of the control parameter θ, the mean service time. For example, the service time of the nth customer may be $S_n = \theta Y_n$, where Y_n is a positive quantity that does not depend on θ. Note that by writing the sojourn time of a customer as a function of the interarrival and service times, we can express the derivative of the sojourn time with respect to the parameter θ as a function of the derivatives of the service times. (The derivatives of the interarrival times with respect to θ are all zero because the interarrival times are assumed here to be independent of θ.) Because the service time of each customer is an explicit function of θ, the derivative of the service time can also be expressed as an explicit function of θ. For example, if the service time is given by $S_n = \theta Y_n$, then the derivative of the service time with respect to θ is $S'_n = Y_n$.

In IPA, we use the steady-state average of the derivatives of the sojourn times as an estimate of the derivative of the steady-state average sojourn time. Therefore, IPA provides us with a method to estimate the derivative of interest using quantities involving event lifetimes. The method of IPA is useful for derivative estimation in simulations. In addition to estimating quantities such as the sojourn time from simulation, we can also estimate derivatives of such quantities.

Often, the derivative of an event lifetime can further be expressed as a function of the event lifetime itself. For example, if the event lifetime S_n is given by $S_n = \theta Y_n$, then its derivative is given by $S'_n = Y_n = S_n/\theta$. Such an expression allows us to obtain IPA derivative estimates simply by knowing the values of the event lifetimes. This benefit is especially useful when estimating derivatives from empirical observations of a real system. In this case, we can measure the values of event lifetimes and as a consequence also compute their derivatives. These, together with IPA, allow us to estimate derivatives of performance measures from observations of a real system. Thus we can estimate gradients for the purpose of performance optimization during the normal, productive operation of a discrete-event system.

Two limitations of IPA prevent its general applicability. The first is that the method relies on knowing the function relating the event lifetimes with the performance measure. Second, the method works only if this function meets certain technical requirements, which may not hold or may be difficult to check. Nonetheless, these limitations are met in a wide range of applications, such as in many forms of queueing networks (see [10] for examples).

2.4.4 Online Optimization

2.4.4.1 Basic Idea

Our goal is to use derivative estimation techniques together with stochastic approximation (gradient) algorithms to adjust the control parameters so that the system performance is (eventually) optimized. Because techniques such as IPA can be used to estimate derivatives via observations of a real system, the possibility exists to apply such optimization algorithms *on-line* (i.e., while the system is running). Of course, such algorithms can also be used in simulations of a system.

Figure 2.10 illustrates the idea of on-line optimization of a discrete-event system. At each iteration, we take observations of the system (by measuring event lifetimes). Then, we use these observations to form an estimate of the gradient. The gradient estimate is then used to update the control parameter via a stochastic approximation algorithm as described before. The updated control parameter is then fed back to the system, and the process continues in an iterative fashion.

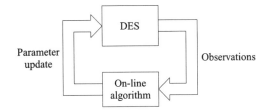

Figure 2.10 On-line optimization of discrete-event system.

Example (Optimization of M/M/1 Queue)

Consider the problem of optimizing the performance measure for an M/M/1 queue described in the previous example: $J(\theta) = T(\theta) + c/\theta$, where $T(\theta)$ is the steady-state average sojourn time and θ is a control parameter associated with the service times. Specifically, the service time of the nth customer is $S_n = \theta Y_n$, where Y_n has mean 1. Because the derivative of J is given by $J'(\theta) = T'(\theta) - c/\theta^2$, the estimation of $J'(\theta)$ involves only estimating $T'(\theta)$. This estimation can be accomplished easily in the single-server queue using IPA (see [6] for details).

We apply the on-line optimization approach described above, driven by IPA estimates of the derivative of J. Figure 2.11 shows plots of the sequence of iterates θ_n versus n (which also counts the number of customers). We used an initial value of $\theta_0 = 0.4$. The dashed line represents a single iterate sequence, while the solid line represents an average over 100 such sequences. Note that the convergence of the algorithm to the optimal value of 0.2 is quite apparent. In fact, we can actually prove that the algorithm converges to the optimal solution in this case (see [6]).

2.4.4.2 Example Application

To further illustrate the on-line optimization approach, we describe an example application. Consider a communication transmitter with total capacity C (bits/s). There are K classes of traffic streams feeding packets to the transmitter, with an infinite buffer to store packets that have arrived but have not yet been transmitted. The length (in bits) of each packet is random, but with unknown distribution. The arrival rate of packets in each class is also unknown and may differ from class to class. The transmitter divides its

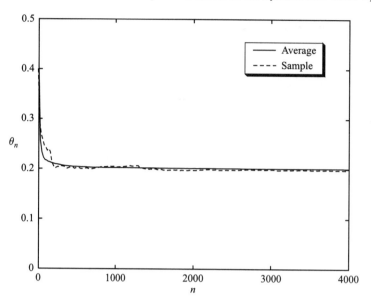

Figure 2.11 Sequence of parameter iterates θ_n for single-server queue.

capacity into K portions, so that the effective capacity experienced by class k is C_k, $k = 1, \ldots, K$. The problem is to choose the values of $C_1, \ldots, C_K$ such that some performance measure is optimized. We assume that the entire capacity is used, so that $C_1 + \cdots + C_K = C$. Such a system is also called a *multiplexer*.

To formulate this problem so that we can apply the techniques described earlier, let $\theta^k = C_k/C$ be the fraction of the total capacity allocated to class k, $k = 1, \ldots, K$. The parameters $\theta^1, \ldots, \theta^K$ are the control parameters that we can adjust. Figure 2.12 illustrates the problem. Let $T_k(\theta^k)$ be the steady-state average packet delay in class k. (The delay experienced by a packet is the time duration from the arrival to the completion of transmission of the packet.) Consider the performance measure

$$J(\theta) = \sum_{k=1}^{K} w_k T_k(\theta^k)$$

where θ is the vector with components $\theta^1, \ldots, \theta^K$, and $w_1, \ldots, w_K$ are positive weights. The performance measure is the weighted sum of the steady-state average packet delays over all classes, where the weights reflect the relative importance of each class. For example, if the traffic classes represent data, video, and speech, we may choose a large

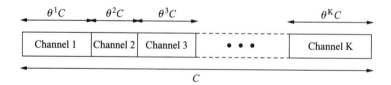

Figure 2.12 Capacity allocation problem.

Section 2.4 Optimization of Discrete-Event Systems

weight for the video class (because video traffic requires small delays), while the weight for the data class can be small (because data traffic is insensitive to delays). The feasible region consists of those vectors θ with positive components summing to 1. The optimization problem can thus be expressed as:

$$\text{minimize} \sum_{k=1}^{K} w_k T_k(\theta^k)$$

$$\text{subject to} \sum_{k=1}^{K} \theta^k = 1$$

$$\theta^k > 0, k = 1, \ldots, K.$$

We can apply the on-line optimization approach described earlier to this problem, using IPA estimators (see [7] for details). Figure 2.13 shows plots of the iterates θ_n^1, θ_n^2, and θ_n^3 for a three-channel system with exponentially distributed interarrival times and service times. (The algorithm does not use any information on the service time distributions or the arrival rates.) As before, each dashed line represents a single iterate sequence, while each solid line represents an average over 100 such sequences. We used $w_1 = 1$, $w_2 = 2$, $w_3 = 3$. Because the exponential interarrival and service time distributions lead to a Markov process, we can derive an expression for the performance measure in this case. Although this expression does not yield to an analytical solution to the optimization problem, we can readily apply a numerical method. Therefore, we can solve the problem independently for the sake of comparison. The solution provided by the numerical method is: $\theta_*^1 = 0.2965$, $\theta_*^2 = 0.3364$, $\theta_*^3 = 0.3671$. From Figure 2.13, it is apparent that the on-line algorithm converges to these values, even though the algorithm does not rely on *a priori* knowledge of the arrival rates or service time distributions. In fact, the convergence of this algorithm can be proved (see [7]).

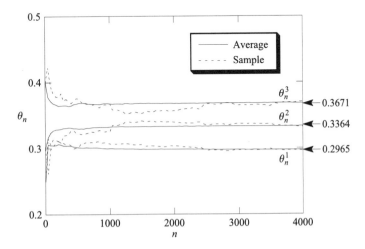

Figure 2.13 Sequences of parameter iterates θ_n^1, θ_n^2, and θ_n^3 for capacity allocation problem.

While the above example illustrates the applicability of the on-line optimization approach, its applicability to other types of problems requires further study. Appropriate models are required for which derivative estimators can be formulated. Much remains to be done along these lines.

2.5 FURTHER READING

For an accessible undergraduate-level textbook on discrete-event systems, see [5]. This reference covers a wide range of topics on discrete-event systems within a single volume. Several books have been written on specific topics related to discrete-event systems, typically at the advanced or research level. These include algebraic models [2], monotone structures [11], and stability analysis [17]. A classic reference on discrete-event simulation is [9]; a more recent book is [3]. A well-used queueing theory text is [13], while [4] is an excellent text on models for data communication networks. The article [18] provides a good overview on the theory of supervisory control of discrete-event systems. An alternative approach is discussed in [14], while [16] describes a similar theory based on Petri nets. More on Petri net models of discrete event systems can be found in [1]. For details on the gradient estimation technique of perturbation analysis, see [10] and [12]. The score function approach is discussed in [19]. For further reading on optimization methods, see [8]. The book [15] provides a treatment of stochastic approximation algorithms and their applications. The use of infinitesimal perturbation analysis and stochastic approximation algorithms for on-line optimization of queues is described in [6].

ACKNOWLEDGMENTS

The author is grateful for the support from the National Science Foundation under grant ECS-9501652 and from DARPA/ITO under grant F19628-98-C-0051.

> **Related Chapters**
>
> - A comprehensive treatment of systems that combine discrete-event and continuous-time dynamics can be found in Chapter 7.
> - The control of communication networks, discussed in Chapter 19, is a particularly important application for the optimization of discrete-event systems.
> - Some modeling methods for continuous-time and sampled data systems are reviewed in Chapter 4.

REFERENCES

[1] R. David and H. Alla, *Petri Nets and Grafcet: Tools for Modeling Discrete Event Systems.* New York: Prentice Hall, 1992.

[2] F. Baccelli, G. Cohen, G. J. Olsder, and J.-P. Quadrat, *Synchronization and Linearity: An Algebra for Discrete Event Systems.* Chichester, England: John Wiley & Sons, 1992.

References

[3] J. Banks, J. S. Carson, and B. N. Nelson, *Discrete-Event System Simulation*, 2nd ed., Upper Saddle River, NJ: Prentice Hall, 1996.

[4] D. Bertsekas and R. Gallager, *Data Networks*, 2nd ed., Englewood Cliffs, NJ: Prentice Hall, 1992.

[5] C. G. Cassandras, *Discrete Event Systems: Modeling and Performance Analysis*. Homewood, IL: Aksen Associates, 1993.

[6] E. K. P. Chong, "On-Line Optimization of Queues using Infinitesimal Perturbation Analysis," in *Discrete Event Systems, Manufacturing Systems, and Communication Networks*, P. R. Kumar and P. P. Varaiya, eds., Vol. 73, IMA Volumes in Mathematics and its Applications. New York: Springer-Verlag, pp. 41–57, 1995.

[7] E. K. P. Chong and P. J. Ramadge, "Convergence of Recursive Optimization Algorithms Using Infinitesimal Perturbation Analysis Estimates." *Discrete Event Dynamic Systems: Theory and Applications*, Vol. 1, no. 4, pp. 339–372, June 1992.

[8] E. K. P. Chong and S. H. Żak, *An Introduction to Optimization*. New York: John Wiley & Sons, 1996.

[9] G. S. Fishman, *Principles of Discrete-Event Simulation*. New York: John Wiley & Sons, 1978.

[10] P. Glasserman, *Gradient Estimation via Perturbation Analysis*. Norwell, MA: Kluwer Academic Publishers, 1991.

[11] P. Glasserman and D. D. Yao, *Monotone Structures in Discrete-Event Systems*. New York: John Wiley & Sons, 1994.

[12] Y.-C. Ho and X.-R. Cao, *Perturbation Analysis of Discrete Event Dynamic Systems*. Norwell, MA: Kluwer Academic Publishers, 1991.

[13] L. Kleinrock, *Queueing Systems, Vol 1: Theory*. New York: John Wiley & Sons, 1975.

[14] R. Kumar and V. K. Garg, *Modeling and Control of Logical Discrete Event Systems*. Norwell, MA: Kluwer Academic Publishers, 1994.

[15] H. J. Kushner and G. G. Yin, *Stochastic Approximation Algorithms and Applications*. New York: Springer-Verlag, 1997.

[16] J. Moody and P. Antsaklis, *Supervisory Control of Discrete Event Systems Using Petri Nets*. Norwell, MA: Kluwer Academic Publishers, 1998.

[17] K. M. Passino and K. L. Burgess, *Stability Analysis of Discrete Event Systems*. New York: John Wiley & Sons, 1998.

[18] P. J. Ramadge and W. M. Wonham, "The control of discrete-event systems." *Proceedings of the IEEE*, Vol. 77, no. 1, pp. 81–97, January 1989.

[19] R. Y. Rubinstein and A. Shapiro, *Discrete Event Systems: Sensitivity Analysis and Stochastic Optimization via the Score Function Method*. New York: John Wiley & Sons, 1992.

Chapter 3 | COMPUTER-AUTOMATED CONTROL SYSTEM DESIGN

Georg Grübel

Editor's Summary

With cost efficiency a technological imperative today, the future impact of control technology does not depend on new algorithms and theories alone. Such developments need to be employed rapidly, reducing the time involved in deploying controllers for new applications. Furthermore, the insertion of advanced control on a large scale cannot require significant numbers of highly skilled (e.g., Ph.D.-degreed) staff.

Thus control technology must be packaged in a form that allows small teams of control engineers to efficiently exploit new research developments and to minimize the cycle time from specification to product. It is a sign of the maturity of control that this packaging is now being accomplished through the development of computer-aided control system design (CACSD) tools. In industries where customized control applications are a frequent demand, CACSD tools that automate several of the steps involved in the practice of control design are now regularly used. At the same time, research continues toward the goal of end-to-end (specification to deployment-ready software) control design automation. No practicing control engineer today can afford to be ignorant of CACSD; it must be considered as much a part of one's professional education as, say, the first course in nonlinear control.

This chapter provides an introduction to CACSD, with specific emphasis on control design automation. System modeling, performance specifications by way of mathematical criteria, and iterative algorithms for realizing an ultimately satisfactory controller are among the topics covered. In other chapters in this volume, the state of the art and current research trends in CACSD are discussed in depth from the perspective of two control application domains—automotive powertrain control (Chapter 12) and flight control (Chapter 13). The latter application is also used to illustrate the general observations in this chapter.

Georg Grübel was with the DLR German Aerospace Center and is a former chair of the IEEE-CSS Technical Committee on CACSD.

3.1 INTRODUCTION

CACSD—computer-aided control system design—is the discipline that allows control engineering methods to be computer executable in a user friendly, reliable, and efficient way. Its activities yield toolboxes and computer-integrated design frameworks that make the broad scope of control methodologies hands-on applicable to the practitioner and that automate the control design and development process as much as possible.

The field started with R. E. Kalman. After Kalman developed the Riccati formalism for linear quadratic gaussian (LQG) control synthesis (1960), he and T. S. Englar developed the first CACSD program suite in 1966. This was the Automatic Synthesis

Program (ASP) [15]. It featured programs for solving the Riccati equation for both continuous and sampled-data systems, computation of time histories by scaled matrix exponentials, stability computations, loss of controllability by sampling, computation of a minimal realization, approximation of an impulse response, multi-rate sampling, continuous time filters, and model-follower control. This program suite made state space optimal control theory applicable to nontrivial control engineering problems, which was not possible just by paper and pencil. Since then, CACSD technology has made tremendous progress in providing fast and reliable numerics [28], the standard functional programming language Matlab, data models for integrated environments based on object-oriented software engineering principles [30], and interactive spreadsheet user interfaces with dynamically coupled information displays for visual exploration, for example, [7].

As illustrated by Figure 3.1, the traditional scope of CACSD covers the triangle of mathematical modeling of input-output connected systems; control system analysis with particular emphasis on feedback stability and robustness; and feedback control synthesis. But the high-powered desktop computation facilities now available enable the control engineer to use a considerably enlarged methodological framework for computer-aided system design: Fast nonlinear simulation allows the engineer to investigate the behavior scenarios of rather complex control systems. Together with control engineering analysis methods, this allows design-embedded assessment of generic feedback properties simultaneously with task-specific system performance. Based on con-

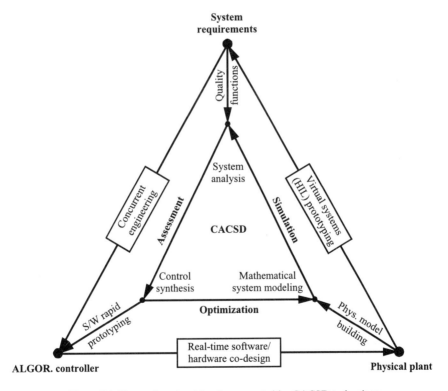

Figure 3.1 The engineering triangle supported by CACSD technology.

current evaluation of all design specifications, multi-objective optimization can be applied to find a best-possible compromise among conflicting design requirements and to automate tuning of the adjustable parameters of a chosen synthesis algorithm. Hence, the control-specific CACSD methods are used in combination with the general-purpose methods of nonlinear simulation, design-embedded assessment, and multi-objective optimization (cf. Figure 3.1). High desktop computation power makes these computation-intensive methods affordable within an interactive design computation process.

The overall perspective of CACSD is to embed the control design process into a simulation-based, "virtual engineering" environment. For that, *modeling* plays a prime role to fit the virtual design objects to reality. First, modeling refers to the physical plant. Computationally affordable symbolic algorithms now allow high-level multidisciplinary system model building, with automatic mathematical model processing, to generate standard linearized CACSD models as well as numerical code for fast nonlinear simulations. Second, modeling refers to requirements capture, that is, the task to develop a proper set of executable quality functions for complete coverage of design requirements. Third, the resulting control law has to be suitably modeled in the form of computer-executable software specifications for seamless transfer of control algorithms to the industrial software engineering process for production-ready real-time software/hardware co-design. This links the "control reality triangle" to CACSD technology (cf. Figure 3.1).

Virtual engineering paves the way for a "first-shot quality" satisfaction design procedure from the early phases of dynamic analysis and control synthesis up to validated specifications of system-embedded control software, which takes into account all performance requirements as well as tolerances and implementation implications. Albeit not yet commonly in use, this will in future be essential for attaining performant, robust control laws, while reducing design engineering cost and development time. This, in particular, holds for the development of competitive "good enough" products whose design is to be optimally tuned to user requirements at the lowest system cost by exploiting the performance potential of a chosen system solution up to its limits.

This contribution deals with CACSD as a technology to support various steps of control design automation. In Section 3.2, design is characterized as an iterative feedback process. This implies that change management in the course of the design process is of prime concern. Hence a model of control design life cycle is assumed, which shows where to start and how to initialize the design process, and when to accept a design or to reiterate on a higher design process level. Section 3.3 deals with object diagrams for physical system modeling, controller parameterization by control synthesis formalisms, and the data computation setup for on-line evaluation in control design tuning. Section 3.4 focuses on executable design specifications in the form of quality functions in time and frequency domains. Quality functions are the basis for detecting the need for change early in the design process. In particular, fuzzy-type interval-quality measures can be used to visualize design satisfaction and conflict detection. Section 3.5 addresses automatic controller tuning by multi-objective multiparameter optimization. This allows interactive design tradeoffs in a noninferior set of design alternatives, meaning that no one quality function can be improved without worsening some other. This compromise is particularly important for designing a balanced multimodel robust controller that not only is good for a nominal design condition but also behaves in a satisfactory manner for a set of operating conditions or system parameter tolerances. Section 3.6 points to the further need to support declarative design of control laws by making best use of automated computations in design-automation machinery. Throughout, the examples to demonstrate conceptual aspects are drawn from [14] and [25] in the application domain of flight control design.

3.2 CONTROL DESIGN LIFE CYCLE TO BE SUPPORTED BY CACSD

In the general domain of systems engineering, control law design is the activity responsible for system dynamics integration to provide precise and stable system functioning, often pushing system operability toward its physical limitations. This task has to be properly handled by the design process. Whereas control theory embodies mathematically formalized expert knowledge of where to start and how to initialize the control law design process, multi-objective optimization-based synthesis tuning provides the quantitative decision clues for when to accept a design iteration versus when to reiterate on a previous design process level. This is depicted as a control design life cycle in Figure 3.2.

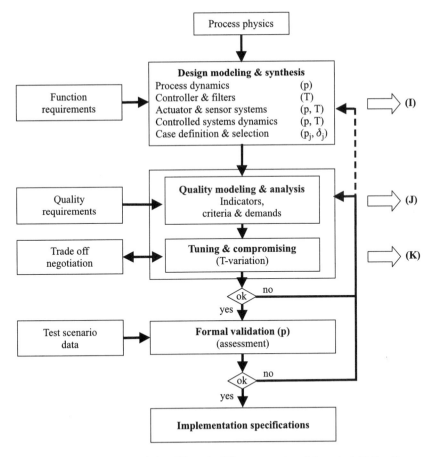

Figure 3.2 Control law design life cycle: Where to start and how to initialize the design process, and when to "ok" or to re-iterate on a higher design process level.

Design Modeling on level (I) makes use of available CACSD tools for system modeling and synthesis of controllers and filters. Object-oriented information modeling relates the accumulated model knowledge in a retraceable way. *Quality Modeling* on level (J) is the task that formally deploys all the requirements on robust stability, dynamic performance, control effort, and implementation constraints in the form of mathematical criteria and satisfaction demands that can be analyzed based on the models of (I). The pertinent information has to be ordered in such a way that different design alternatives can be compared visibly. *Tuning & Compromising* on level (K) parameterizes the feasible control laws to be implemented. If no feasible solution in the set of possible design alternatives can be found to satisfy all quality demands, here the conflicts become visible allowing the designer to reiterate the design process either by relaxing design demands on level (J) or by adding design degrees of freedom via higher-order control law dynamics on level (I). Levels I-J-K also form a hierarchy in terms of the number of iterations the engineer usually has to perform. Design balancing by parametric quality compromising on level K requires the largest number of iterations. This suggests that design changes I,J,K be stored on a hierarchical database, which then allows design iterates to be automatically retraced by pertinent indices i,j,k that mark the corresponding data objects.

Conceptually, control design is a *feedback process* of manipulation and interpretation that reduces the uncertainty of what can be achieved in terms of specifications to attain the desired control function quality attributes of a given dynamic plant object. This is depicted in Figure 3.3. The objects, attributes, and operations one is dealing with are interrelated in Figure 3.4.

The *modification* feedback process between designer and specification in Figure 3.3 corresponds to iterations between level K and level J in Figure 3.2. The *manipulation* feedback process between designer and object corresponds to iterations between level K and level I. The *synthesize-search* feedback process between objects and attributes in Figure 3.4 corresponds to tuning and compromising iterations in level K.

Based on the picture of design as a feedback process, future design environments for control engineering were anticipated in [21], more than 10 years ago. This conceptual framework, combined with pertinent software engineering principles, led to the development of the modular control design software environment ANDECS [9], which is a production implementation of the control design automation concepts dealt with in this chapter. Broadly, three methods of working are distinguished: design by analysis and synthesis (attribute-centered), design by procedure (operation-centered), and design by search (exploration-centered). Attribute-centered design by analysis and synthesis and exploratory design by search may be combined in a

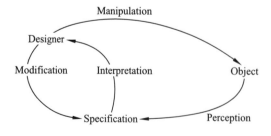

Figure 3.3 Design is a feedback process, (from [21]).

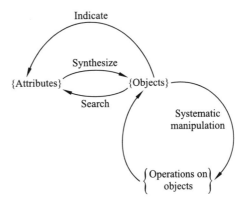

Figure 3.4 Objects, Attributes and Operations, (from [21]).

declarative approach to design. This is possible on the basis of quality functions as executable design specifications and automatic synthesis tuning to attain these specifications.

The design process is a feedback process. That is particularly true for control law design because of its multidisciplinary interrelation with system dynamics. For example, the flight control law of an aircraft determines the basic system architecture to integrate the functional demands of autopilots and piloted flight with flight system dynamics, which encompasses flight mechanics, aerodynamics, structural loads and aeroelasticity, engine dynamics, control sensors and actuators, and control-logic software embedded in the flight control computers. Changes in performance demands and system data originate from the disciplinary domains involved as system development proceeds. This requires a virtual engineering environment, where changes in virtual system prototypes can be executed rapidly and where data integration supports a seamless methodology for retraceable change management. The declarative design approach using interactive multiobjective optimization [23] is believed to be best suited for satisfying this need.

3.3 DESIGN MODELING AND SYNTHESIS ALGORITHMS

Design modeling refers to the control objects to be designed as well as to the design evaluation set up for control law tuning. This encompasses physical system dynamics, control-system sensors and actuators, and controller logic, which incorporates the control law for multivariable control-error dynamics shaping, gain scheduling, mode switching, control redundancy management, and the like. These models, integrated together, represent the *(nonlinear) evaluation model* of the controlled system to assess control performance for various command and disturbance cases. A specific aspect is modeling of the control law itself. This is called control synthesis. For control synthesis, control theory provides a broad spectrum of analytic formalisms, each one requiring a suitable *(linear) synthesis model*. Both types of models, evaluation models and synthesis models, are required as part of the data computation chain for controller tuning.

3.3.1 Physical System Modeling

The multidisciplinary nature of control design becomes most visible in the model-building process of the plant, that is, the physical system that is to be controlled. This requires contributions from different engineering disciplines which are usually represented by different specialized groups in the product development team. To integrate the modeling contributions of different engineering groups, which talk their own domain-specific language, into a computer-executable comprehensive model of system dynamics is not an easy task. It is burdened with high engineering transaction costs since granularity and complexity of system dynamics models evolve as system design proceeds. Declarative system dynamics model building is best suited to handle this problem.

Declarative system dynamics model building using equation-based object libraries is outlined in Figure 3.5. Object diagrams are most appropriate for iterative system model building and allow a computer-processable representation of system equations to be automatically generated. Symbolic formula manipulation then allows generation of an efficient mathematical analysis code and numerical simulation code. Such computer-aided modeling makes best use of vested interests in validated component models and thereby reduces engineering transaction costs in model-based multidisciplinary control design projects. It can be traced back to [2], more than 20 years ago. Nowadays the availability of high desktop computation and visualization power and advanced soft-

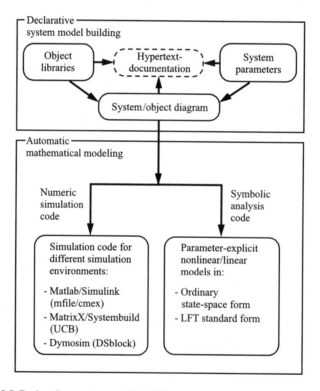

Figure 3.5 Declarative system model building as computer-processable input to automatic mathematical modeling and code generation.

ware [4] makes this approach hands-on applicable to the design engineer. The approach allows the user to automate various steps in the process of system model building and model maintenance as the design life cycle evolves:

- Object-oriented decomposition with respect to basic engineering disciplines allows coding of expert engineering knowledge into validated object class libraries independent of future use in specific functional system interrelations. Object encapsulation of all pertinent modeling information (equations, variables, parameters, units, visualization attributes, etc.) allows structured computer-processed documentation to be presented in interactive hypertext format to the design engineer.
- Objects may contain a well-defined interface that encapsulates model complexity at different levels of granularity. This allows changing the internal model specification without affecting its external interface behavior. In this way, the contents of an object can be specialized to capture more detailed phenomena by making use of the inheritance principle of object-oriented information decomposition. Hence model maintenance can be kept most transparent.
- A "hardware description language" functional composition of a system model from subsystems and constituent objects can be visualized by hierarchical object diagrams of which controller block diagrams are just a special case. Available software allows system composition via object diagrams to be performed interactively by a drag-and-drop graphical user interface. The hierarchical model structure allows the isolation of subsystem design activities within a common system dynamics model. Hence efforts with regard to system dynamics integration and optimization can be kept minimal. Since domain-specific description icons for the various system components can be used, a hierarchically structured object diagram is equally expressive for, among other things, mechanical, electrical, and hydraulic components, and for analog/digital control elements.
- Integrated symbolic equation manipulation yields efficient mathematical system models (e.g., all equations that are not necessary for a specific task are automatically removed) from which compilers are able to generate numerically efficient simulation code for different simulation run-time environments. Hence reuse of the same system model in different simulators is automated.

Aircraft dynamics modeling for flight control design demonstrates the feasibility of this approach (cf. Figure 3.6). By means of a graphical object editor, one can zoom in on objects and display their internal structure. Zooming in on the aerodynamics model results in the object diagram displayed in the top left of Figure 3.6. Zooming in on the aerodynamics object results in the parameters and equations window of this object, part of which is displayed in the top right of Figure 3.6. By interactively augmenting the object diagram, the flight mechanics aircraft model can be visibly changed to an aircraft flight system dynamics model including the structural flexibility effects of the aircraft body ("flexBody"), motivators (e.g., longitudinal motion elevators with electro-hydraulic actuators), and sensors and controls to close the control feedback loop, as depicted in Figure 3.7.

50 Chapter 3 Computer-Automated Control System Design

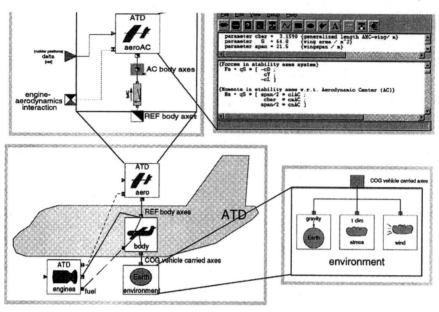

Figure 3.6 Declarative system model building by example of a flight mechanics object diagram.

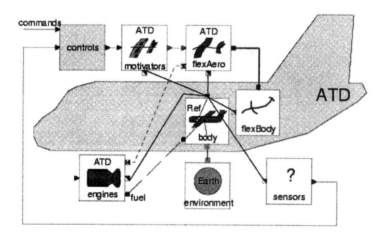

Figure 3.7 Aircraft flight-system dynamics object diagram including the feedback control loop via sensors, controls, and motivators.

3.3.2 Synthesis Algorithms and Controller Modeling

The core of control design is synthesis of performant, stability-robust control laws. Control law synthesis consists of three activities: development or adoption of a proper control law structure, which defines the feedback control system architecture with feedback and actuation variables, dynamic compensators as well as signal estimation and filtering; parameterization of the control law with respect to adjustable tuning parameters; and tuning of the control parameters to satisfy all control requirements for the given physical plant. Tuning is the means to properly adapt a generic synthesis approach to the given plant at hand. Actually, the same control structure generates quite different stable or unstable control system behavior depending on how the control law parameters are tuned. To best cope with the given requirements, the control law structure chosen has to be sufficiently rich in independent tuning parameters. This may necessitate design iterations to change the controller structure by increasing the order of the controller dynamics or by augmenting the feedback information structure through additional control feedback loops.

Control law structures can be developed through different basic approaches, for example, the proportional-integral-derivative (PID) compensator approach, the model-based analytic approach, or the rule-based fuzzy control approach. In any case, the result is a controller model that can be represented as an algorithm diagram interrelating various algorithmic blocks. As an example, part of a control law for piloted flight is shown in Figure 3.8. Note that this belongs to the controls part of the object diagram of aircraft flight system dynamics of Figure 3.7. In addition, control logic with discrete-event models can be handled in the declarative framework of an object diagram. For instance, Petri nets are used to model a simple form of redundancy management for two elevator actuators in Figure 3.9. Such computer-processable control logic and control

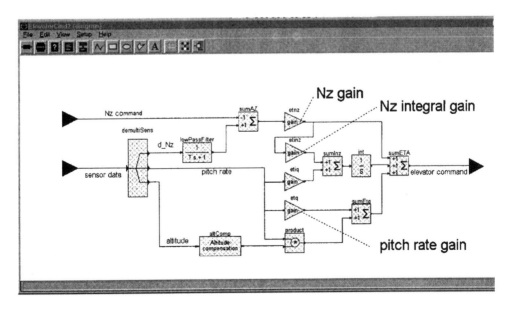

Figure 3.8 Part of a flight control law as executable algorithm diagram.

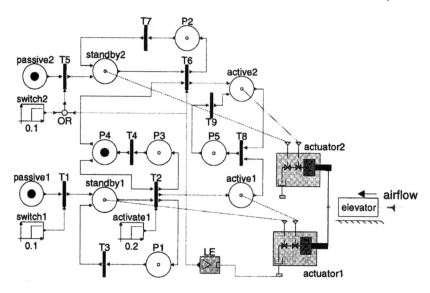

Figure 3.9 Object diagram of redundancy control logic for two elevator actuators.

algorithm diagrams also serve as computer-executable specifications to develop pertinent real-time embedded control software by "autocode" software generators.

Control synthesis algorithms parameterize a controller model, or part of it, with respect to adjustable tuning parameters. Hence, if the parameters of the controller model are denoted by K and the adjustable tuning parameters are denoted by T, then one may distinguish between *direct* control law parameterization $K = T$ and *indirect* parameterization via a synthesis algorithm $K = f(T)$:

PID compensator control	$K = T$
model-based control	$K = f\,(T, \text{synthesis model})$
rule-based control	$K = f\,(T, \text{fuzzy control rules})$.

PID control design deals with proportional, integral, and derivative action on the feedback control error as well as the shaping of dynamic compensation filters to cope with feedback stability. Tuning parameters are the P, I, D gains and compensator time constants. This approach is commonly called the classical approach.

The model-based analytical approach of mathematical control theory provides a broad spectrum of different synthesis methods. They may be classified mainly as eigenstructure methods and Lyapunov-type methods in state space domain, as optimal control methods in time domain, and as H_∞ loop-shaping methods in frequency domain. The solvability requirements of the underlying mathematical synthesis problem induce a specific control law structure (e.g., state or observer-type output feedback). The synthesis formalism parameterizes this control structure as a function of free synthesis-tuning parameters such as elements of the desired eigenstructure, parameters of positive definite weighting matrices in a quadratic integral criterion, or the corner frequencies of weighting filters for loop shaping. Hence, by this approach the control law is parameterized not explicitly but implicitly as a function of "synthesis tuning" parameters.

Rule-based fuzzy control yields a nonlinear-gain feedback control structure specified linguistically by if-then rules on fuzzified error and actuation variables. Tuning parameters, for example, are the scaling coefficients of membership functions and the weights among the defining rules, see [13].

Direct control law parameterization is visible with respect to control law structure and a dynamic augmentation thereof. Indirect control law parameterization via an analytic synthesis algorithm is aimed at visibility with respect to some type of generic feedback property (e.g., stability). Nonlinear control law parameterization by fuzzy control is visible with respect to the functional control-error behavior, for instance. Note that an admissible control law structure can be composed of individual parts that are parameterized directly or indirectly, due to a combination of different synthesis approaches.

3.3.3 Performance Evaluation Setup

Control performance has to be tuned and evaluated with respect to various command and disturbance cases of assumed plant operation. This requires that the plant model be integrated with the controller model and that the resulting control system model be analyzed by suitable time and frequency domain methods for pertinent command and disturbance inputs. *Robust control design* asks for a controller with good command and disturbance performance over a range of operating conditions and within a tolerance band of system parameters. This is handled by selecting a number of *evaluation cases* that are characterized by a parameter set $\{p_j\}$, which "discretizes" the domain of operational and parametric uncertain system dynamics behavior over the continuum interval set $[p]$ of plant operation and system parameters.

Figure 3.10 shows an example of discretizing the altitude/Mach flight envelope by several evaluation cases, together with the corresponding pitch rate responses of the uncontrolled aircraft due to an elevator step disturbance. For robust control *tuning* an average system operation case is dealt with concurrently with operation cases that characterize extreme variations of the dynamic system response, for example, the cases 6, 1, 5. This is called robust "multi-case tuning." For robustness *assessment*, one takes all the vertices of the flight envelope to initialize the search for worst-case nonlinear behavior within the permissible system operation domain.

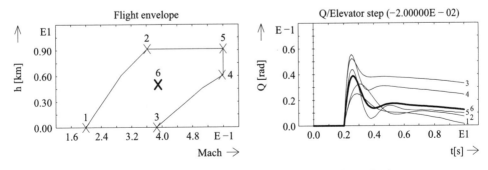

Figure 3.10 Discretization visualization of the flight envelope by evaluation cases.

This kind of CACSD control performance evaluation requires a data computation chain as depicted in Figure 3.11, which simultaneously involves multiple plant model instantiations $\{p_j\}$ for the various evaluation cases, the controller model, and a synthesis model in case an analytic synthesis formalism is used. The conceptual separation of control synthesis into three design tasks (i.e., representing controller structure by a model, controller model parameterization by a synthesis algorithm, and controller tuning based on a data computation chain for performance evaluation) lends itself to a *generic CACSD computation setup* to handle all kinds of control design problems. This setup encompasses design modeling as dealt with in this section as well as quality modeling, which is discussed in the next section.

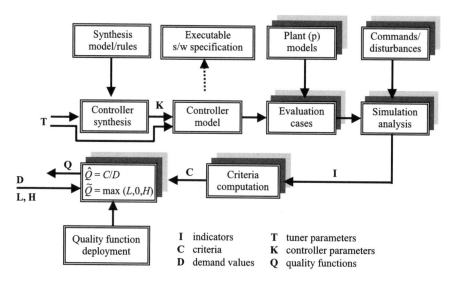

Figure 3.11 Generic data computation set up for (multi-case) synthesis tuning.

3.4 QUALITY MODELING FOR DESIGN ANALYSIS AND DECISION MAKING

Assessing design satisfaction, comparing design alternatives, and detecting and negotiating design conflicts require a suitable formulation that will capture noncommensurable design requirements formally. This is done by *quality functions*. A quality function is a tuple of a requirement evaluation criterion $c(i)$, which is defined as a real-valued mathematical function of system performance indicators i, together with a demand d to assess performance satisfaction. In the specific case of hardware-in-the-loop control tuning, often called control calibration, indicator data are measured as real operation data. An *executable* quality function is a quality function with a system model associated with it that allows generation of high-fidelity performance indicator data. Such "virtual" data obtained from high-fidelity evaluation models have to be used in earlier design stages where hardware-in-the-loop control tuning is not yet feasible. This links quality modeling to the previously treated design modeling (cf. Figure 3.11). Model-based control performance indicators typically are closed-loop system eigenvalues, step

Section 3.4 Quality Modeling for Design Analysis and Decision Making

time-responses of linearized system models, frequency responses of various types of transfer functions, and simulated nonlinear system time responses.

Evaluation criteria can be deterministic or stochastic. Besides *generic control* performance criteria, for example [5], *synthesis-specific* criteria can be considered in addition to monitoring characteristic properties of the control synthesis process [29]. Furthermore, Table 3.1 shows various examples of how *task-specific* control requirements may be expressed by quality functions. The table also exemplifies another aspect:

TABLE 3.1 Control Design Requirements and Quality Functions (Criteria and Demands) for the Aircraft Landing Approach as Specified in the Control Design Challenge [22]

	Requirements	Mathematical Criteria	Demands
1	Altitude unit step: settling time < 45 s	$c = \int_{t_1}^{t_2}(h(t) - 1)^2 dt$ $t_1 = 10s, \ t_2 = 30s$	min
2	Altitude unit step: rise-time < 12 s	$c = t_1 - t_2$ $h(t_1) = 0.1, \ h(t_2) = 0.9$	< 12
3	Cross-coupling altitude airspeed: for a step in commanded altitude of 30 m, the peak value of the transient of the absolute error between V_A and commanded airspeed should be smaller than 0.5 m/s	$c = \max_t |V_a(t) - V_{cmd}|$	< 0.5/30
4	Airspeed unit step: settling time < 45 s	$c = \int_{t_1}^{t_2}(V_A(t) - 1)^2 dt$ $t_1 = 10s, \ t_2 = 30s$	min
5	Airspeed unit step: rise time < 12 s	$c = t_2 - t_1$ $V_A(t_1) = 10s, \ V_A(t_2) = 30s$	< 12
6	Cross-coupling airspeed altitude: for a step in commanded airspeed of 13 m/s, the peak value of the transient of the absolute error between h and commanded h_c should be smaller than 10 m.	$c = \max_t |h(t)|$	< 10/13
7	Altitude unit step: overshoot < 5%	$c = \max_t h(t)$	< 1.05
8	Airspeed unit step: overshoot < 5%	$c = \max_t V_A(t)$	< 1.05
9	Airspeed wind disturbance: for a wind step with amplitude of 13 m/s, there should be no deviation in the airspeed larger than 2.6 m/s for more than 15 s.	$c = \max_{t > 15} |V_A(t)|$	< 2.6
10	Altitude wind disturbance: no explicit specification given	$c = \int_0^{t_2} h^2(t) dt$ $t_2 = 30s$	min
	Control activity criteria, effort minimization for:		
11	tailplane, altitude command		min
12	throttle, altitude command	$c = \int_0^{t_2} u^2(t) dt$	min
13	tailplane, airspeed command		min
14	throttle, airspeed command		min
15	throttle, wind step	$c = \int_0^{t_2} \dot{u}^2(t) dt$	min
16	throttle rate, wind step		min
17	Relative stability of eigenvalues λ_i: no explicit specification	$c = 1 - \min_i \left(\frac{-Re(\lambda_i)}{|\lambda_i|}\right)$	< 0.6
18	Absolute stability of eigenvalues λ_i: no explicit specification	$c = \exp\left(\max_i (Re(\lambda_i))\right)$	< 0.95

To cover control-task performance in due detail may require quite a number of quality functions associated with different command and disturbance cases. In robust multi-case tuning, this number even increases because a set of different plant model instantiations is to be considered within a tolerance band of parameter uncertainties or operating conditions. For instance, taking three operating conditions simultaneously, to achieve robust control tuning, amounts to $3 \times 18 = 54$ quality functions based on Table 3.1.

Computationally, quality functions can be evaluated in parallel to attain reasonable computer processing times, but a great number of noncommensurable quality functions pose a complexity burden on decision evaluation for design tradeoffs. This problem may be alleviated by interactive visual decision support, as discussed later.

3.4.1 Quality Functions

Without loss of generality, a quality function criterion can be mathematically formulated as a real-valued function which assumes that the smaller a value, the better the requirement is satisfied. Table 3.1 shows various examples. Then, design satisfaction can be assessed either by the demand that criteria values are lower than given upper bounds or that they are as low as possible. Table 3.1 also gives examples for such demands denoted either by < or min. The min demand can be interpreted as an inequality demand with a yet undefined upper bound α as low as possible.

This allows definition of commensurable, normalized quality measures to assess requirement satisfaction:

$$\hat{q}_j := c_j / d_j,$$

$\hat{q}_j \leq 1$:   requirement j is satisfied,

$\hat{q}_j > 1$:   requirement j is not satisfied, and (3.1)

$\hat{q}_j \leq \alpha$: $\alpha =$ "min" denotes best possible requirement satisfaction.

In practice, requirements most often are formulated by indicator *intervals* to judge system behavior by *quality levels*. For example, for piloted flight, handling quality levels are to be satisfied according to various interval-quality criteria such as the C^*, Phase Rate, Open-Loop-Outset Point (OLOP), and Neal Smith criteria (cf. Figure 3.15). Thereby level 1 may be "good", level 2 may be "acceptable" and level 3 may be "not acceptable" (bad). This kind of specification of design requirements can be treated by a suitable fuzzy definition of interval quality, for example [18]: Requirement satisfaction as a function of a scalar indicator i is measured by an interval quality function $\bar{q}(i)$ that is characterized by means of at most four good/bad values $b_l < g_l < g_h < b_h$

$$\bar{q}(i) := \max\{L(i), 0, H(i)\},$$

$$L(i) = (i - g_l)/(b_l - g_l), \qquad b_l < g_l \qquad (3.2)$$

$$H(i) = (i - g_h)/(b_h - g_h), \qquad g_l < g_h < b_h.$$

The graph of this mapping is depicted in Figure 3.12. Such interval quality functions are also appropriate to quantify robustness requirements: If system behavior is known to change within an interval set of operation conditions and system parameter

Section 3.4 Quality Modeling for Design Analysis and Decision Making   57

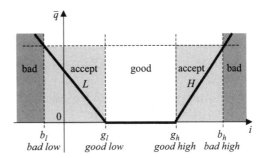

Figure 3.12 Mapping of indicator values to an interval quality function.

tolerances, then it makes no sense to specify performance requirements to the point. Rather, only intervals of "good" and "acceptable" performance satisfaction are meaningful requirements, and only by such performance intervals can the achieved degree of design robustness be formally assessed.

The interval mapping Eq (3.2) is normalized so that $\bar{q} = 1$ defines a separation between feasible (good, acceptable) and not acceptable. This correlates to $\hat{q} = c/d = 1$ in Eq (3.1). An additional feature is that all "good" indicator values are mapped to zero, that is, to the lowest possible value. This is of advantage later, when multi-objective optimization techniques are used for tuning.

Definition of a quality measure $\bar{q}$ by using the max-operator (3.2) yields a non-smooth function (cf. Figure 3.12). For numerical treatment in an optimization algorithm, such a function can be smoothed mathematically by the following approximation, cf. [20], which yields a numerically well-behaved soft interval quality function, $\tilde{q}(i)$:

Take the general case of a time- (or frequency-) discretized indicator function $i_k := i(t_k)$; then $\tilde{q}(i_k)$ is a smooth approximation of the max-interval function $\bar{q}(i_k)$

$$\tilde{q}(i_k) \approx \bar{q}(i_k) + (1/\rho)\log \sum_k \{\exp[\rho(L_k(i_k) - \bar{q}(i_k))] + \exp[\rho(H_k(i_k) - \bar{q}(i_k))]\} \quad (3.3)$$

To get a good approximation around $\bar{q} = 1$, a value of ρ which is about 20 yields an error of less than 1%. The exponents in (3.3) are always less than or equal to zero, and hence unfavorably large values are avoided in evaluating the exponential functions. Summation is also numerically stable since all addends are positive and less than or equal to one. Obviously, this type of smoothing approximation can also be applied to the max-functions in Table 3.1.

Compound quality functions can be formulated by using the maximum function or a smooth approximation thereof

$$q(i_1, \ldots i_j) = \max\{q_1(i_1), \ldots, q_k(i_1), q_{k+1}(i_2), \ldots, q_n(i_j)\} \quad (3.4)$$

If the individual members of the maximum function are interval quality functions of type (3.2), a linguistic interpretation in the vein of fuzzy logic is [18]:

(q satisfies property s) if

(i_1 has property 1) AND (i_2 has property 2) $AND\ldots AND$ (i_j has property n),
(3.5)

where "i_j has property$\ldots$" means that the indicator value i_j is good or acceptable with respect to its membership function.

As an example, consider the stability indicator "eigenvalue damping ζ" defined as

$$\zeta_j = -Re\lambda_j / \sqrt{Re^2\lambda_j + Im^2\lambda_j}, \qquad (3.6)$$

where values greater than 0.7 are considered as good and values less than 0.3 are considered as bad. To map ζ to a compliant interval quality function $\bar{q}(\zeta)$, the following good/bad values are appropriate:

$$b_l = 0.3, \quad g_l = 0.7, \quad \text{and} \quad H(\zeta) := 0.$$

That is, with demand upper bound $d = 1$ damping values greater than $b_l = 0.3$ are mapped to a "satisfactory" interval, and those greater than $g_l = 0.7$ are mapped to the "good" interval, which in this case is open to the right ($H = 0$). This interval quality function is depicted in Figure 3.13.

Relation (3.5) can be used to contend with all n system eigenvalues simultaneously; that is, ($\bar{q}(\zeta)$ is at least satisfactory) if ($\bar{q}(\zeta_1)$ is at least satisfactory) $AND \ldots AND$ ($\bar{q}(\zeta_n)$ is at least satisfactory). Hence "$\bar{q}(\zeta)$ is satisfactory" means "the system is well damped."

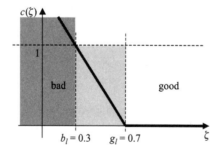

Figure 3.13 Example: Interval quality function for damping values ζ.

3.4.2 Feasible Design and Competing Requirements

Quality measures with positive *the smaller the better* criteria and quality limiting upper bounds yield a visible way to assess requirement satisfaction by design alternatives. Define for the set of all quality functions

$$\alpha := \max_j \{q_j\}, \quad j = 1, \ldots, J. \qquad (3.7)$$

Then a design alternative with $\alpha \leq 1$ is a feasible design that satisfies all requirements within the demanded bounds. In particular, a design alternative $a^{(II)}$ is said to be better than a design alternative $a^{(I)}$ if $\alpha^{(II)} < \alpha^{(I)} \leq 1$, and a "best-feasible" design over all alternatives is characterized by

$$\alpha^* = \min\{\alpha\} \qquad (3.8)$$

In well-posed engineering design problems, one always encounters competing requirements of *performance versus cost*, and one has to search for a suitable tradeoff. This search is to be confined to the set of "best achievable" compromise solutions, known as *Pareto-optimal* solutions, where improvement in any one quality measure can be achieved only by deterioration in at least one other quality measure. Generally, a design alternative $a^{(II)}$ is said to be *Pareto preferred*, or noninferior, to an alternative $a^{(I)}$ if all quality measures of $a^{(II)}$ are better (smaller) than or equal to those of $a^{(I)}$, with at least one being strictly better. Hence as a best choice one may select the best-feasible candidate (3.8) out of a Pareto preferred set.

3.4.3 Visualization for Comparative Design Exploration

Given a set of feasible design alternatives, one has to compare them and select a suitable tradeoff candidate out of this set, or one has to decide *how* to improve design further, that is, to generate a further design alternative with a "better" tradeoff. For this *decision process* advanced CACSD environments adopt the paradigm of "vision to think" to explore design patterns by interactive information steering. This is most intelligible if organized as a multilayered information spreadsheet graphical user interface (GUI), for example [7].

Comparative design evaluation in view of tradeoff decisions requires that many quality functions be simultaneously considered. This needs a high-dimensional kind of display to visibly compare different design alternatives. The means to do this is a display in "parallel coordinates": A high-dimensional space is spanned by parallel coordinate axes and a polygonal line represents a point in this space. For example, Figure 3.14 depicts such a parallel coordinates display of quality functions showing five different design alternatives of a flight control law. Feasibility assessment is visible: all polynomial lines below a border line of value 1 indicate requirements satisfaction, and values above this line indicate design deficiencies with respect to the adopted specifications. This also allows detection of competing requirements: For any two design alternatives, which belong to a Pareto-optimal set, competing requirements are visualized by polygonal lines that are crossing. As an example, in Figure 3.14 a strong conflict can be immediately detected between maximum elevator rate (ELEVRATE) and satisfaction of the C* handling quality criterion (CSTAR).

The parallel coordinates of Figure 3.14 span a seven-dimensional design-response surface. The sequential ordering of the coordinate axes is not unique. Coordinates may be ordered to focus on hot spots of high design sensitivity; they may be ordered with respect to different classes of requirements (e.g., the classes' automatic control requirements, handling quality requirements, control effort); or they may be clustered with respect to different operation and parameter tolerance conditions handled as a multi-

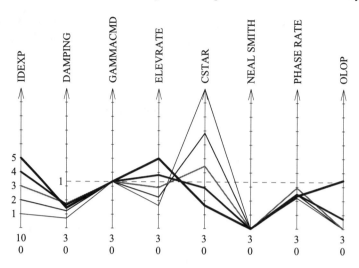

Figure 3.14 Design-satisfaction visualization by a parallel coordinates display: five control law design alternatives are visualized by seven quality functions each. (The first coordinate orders the design alternatives 1 ... 5.)

model design problem. For the latter, one may line up the different operation cases along a third, orthogonal axis in a three-dimensional parallel coordinates display.

An interactive parallel coordinates editor as described in [6] is most helpful to explore the design response surface. It also serves as a graphical steering aid to activate different views on different information levels. For a particular quality function in this display one may zoom in to visualize of the pertinent indicators' behavior within their intrinsic quality bounds as shown in Figure 3.15. Or one may interactively select any two competing quality functions and display them in a two-dimensional Cartesian view to reveal the "compromise gradient" between available design alternatives (cf. Figure 3.17).

The paradigm "use vision to think" is brought to bear most apparently by engineer-in-the-loop virtual reality system simulations whereby complete operation scenarios can be explored interactively. For flight control, for example, the Aviator Visual Design Simulator (AVDS), software [26] allows full-envelope virtual flight tests at the control engineer's desktop computer. This integrated operational view deepens insight into the system dynamics problem at hand and allows design validation. It eases detection of design deficiencies that might be hidden by a mere view on discrete design points, Figure 3.10, which, for design purposes, abstract system operation just by a number of different evaluation cases.

3.5 AUTOMATIC TUNING AND DECLARATIVE COMPROMISING

Analytic synthesis methods rely on the assumption that all design requirements can be translated into commensurable measures and analytically treatable tuning rules. Thus analytic synthesis tuning struggles with conservatism in exploiting the full performance potential of the underlying controller structure when applied to multi-objective, non-

Section 3.5 Automatic Tuning and Declarative Compromising

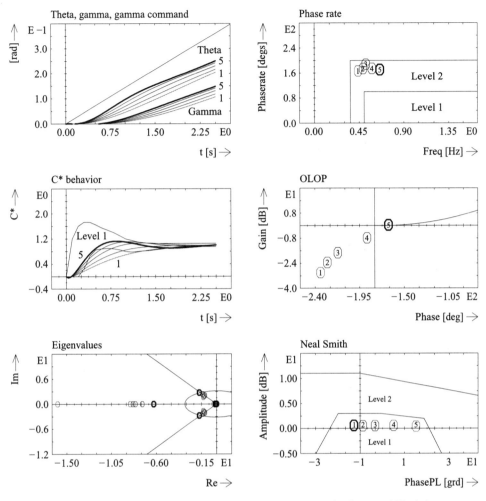

Figure 3.15 Quality-behavior visualization by various types of indicators within their intrinsic quality level bounds, where the marked 5 design alternatives correspond to the quality function display in Figure 3.14.

commensurable design demands as they naturally occur in practice. Nevertheless, a multitude of CACSD toolboxes is available to support multivariable control system design problems by various kinds of analytic synthesis methods. To make the best use of their analytic synthesis strength, such toolboxes ought to be combined with the versatility of data-driven search algorithms in order to make them applicable to general, quantitative design requirements. Various case studies with the ANDECS CACSD environment can be found in [11].

The generic picture of how to embed an automatic search algorithm in the process of control law synthesis has already been developed in Figure 3.11. It shows the data computation chain for quality functions, Q, that measure noncommensurable requirements, to feed a search algorithm for attaining proper tuning parameter values (T, K). This computation chain is generic to allow indirect tuning parameterization (T), as it is used in a particular synthesis toolbox, as well as direct tuning (K) of explicit controller parameters.

3.5.1 Automated Tuning by Multi-Objective Parameter Optimization

Generally, multiple parameters need to be tuned simultaneously. In addition, these parameters may be of different types since both synthesis parameters and additional control law parameters may have to be tuned concurrently if in an incremental design process an analytically generated control law structure is augmented by additional dynamic compensators, filters, and signal limiters. Manual sequential tuning of one parameter after another is not very efficient either in the engineering time required or in the result that can be achieved. Hence an *algorithmic tuning* procedure is sought, which can be used for *automated* tuning of *multiple parameters* of different types. Moreover, in view of Section 3.4.2, automatic tuning should find Pareto-optimal solutions. To find Pareto-optimal tuning parameters $T = f(Q)$, multi-objective evolutionary algorithms as well as nonlinear mathematical programming algorithms can be applied.

Multi-objective evolutionary algorithms [8] directly use Pareto-preference ranking. The fitness of a population's individual is measured by how many other individuals it is inferior to. According to this criterion, populations are ranked, where the best solutions will be the noninferior ones. Thus, noninferior solutions will always be most likely to be selected, leading to convergence to a Pareto set. Methods like multiple subpopulations and Pareto-fitness sharing are applied to force individuals of the same Pareto rank to spread out evenly over the entire Pareto set. Evolutionary strategies cope well with large numbers of parameters as well as with a large search space, which makes them likely to find the global instead of a local solution in multimodal problems. They require a large number of function evaluations, but on return they yield multiple solutions that are well dispersed in or near to the entire Pareto-optimal set.

Nonlinear programming algorithms use an analytical optimality condition for Pareto optimality and for attaining a numerical convergence condition. If suitably parameterized, nonlinear programming can be used to systematically find a sequence of Pareto-optimal solutions one by one. Thus this approach fits well with interactively exploring the compromise nature of a Pareto-optimal set in the *engineer's search* for a "best" tradeoff. Efficient nonlinear programming algorithms require smooth functions and are bound to a local solution in the neighborhood of the starting condition that has to be provided to initialize the algorithm. Since the run-time of such algorithms increases more than linearly with the number of parameters to be optimized, for interactive application of nonlinear programming algorithms, the number of parameters should be kept low.

Hence a two-phase tuning procedure is appropriate. In the first phase, a multi-objective evolutionary algorithm is used to globally optimize all available tuning parameters. In the second phase, a nonlinear-programming interactive tuning system is applied for engineering tradeoff search, which starts with a global (near-) Pareto-optimal design alternative found by an evolutionary algorithm in the first phase. For computing efficiency, one may confine optimization in the second phase to a reduced (segmented) tuning parameter set.

To check whether a feasible solution can be attained by proper tuning, the following constrained minimization problem with an auxiliary variable $\alpha \geq 0$ is considered:

Section 3.5 Automatic Tuning and Declarative Compromising 63

$$\min_{T,\alpha} \alpha(T) \; s.t. \; q_j(T) \leq \alpha, \quad (3.9)$$

for all quality functions, $j = 1, \ldots, J$.

The constrained parameter optimization problem (3.9) can be solved by standard nonlinear programming algorithms [16].

If $T = T^*$ minimizes (3.9), then T^* is a noninferior, Pareto-optimal, solution [23], and with $\alpha^* \leq 1$ this is also a best feasible solution. If $\alpha^*(T^*) > 1$, no feasible solution is possible for the stated quality limits and the chosen controller structure. Hence one has to reiterate design on a higher process level (see the design life cycle depicted in Figure 3.2).

With (3.7), optimization problem (3.9) is equivalent to the min-max optimization problem

$$\min_{T} \max_{j} \{q_j(T)\} = \alpha^*(T^*). \quad (3.10)$$

A solution (3.10) characterizes the particular Pareto-optimal tuning alternative where two (or more) competing quality functions become equal with minimized maximum value

$$q_i(T^*) = q_k(T^*) = \alpha^*(T^*) > 0, \quad (3.11)$$

thus revealing the main conflicting requirements. If $\alpha^*(T^*) < 1$, obviously there is room for a feasible tradeoff in compromising conflicting requirements.

If algorithmic tuning is used to find feasible tuning parameters, then according to Figure 3.16, with $T = f(Q)$, this closes the CACSD computation chain of Figure 3.11 to yield an automated *tuning loop*. A CACSD environment that implements this kind of automated tuning should support and integrate both kinds of Pareto optimization approaches, that is, evolutionary algorithms, for example, [29], and nonlinear programming algorithms, for example, [10]. Seamless, versatile, and user friendly integration, however, is not a simple task and so far no such integrated CACSD environment is readily available.

3.5.2 Declarative Compromising

In the previous section automatic tuning is used to find Pareto-optimal tuning parameters; now declarative compromise search within the space of feasible quality

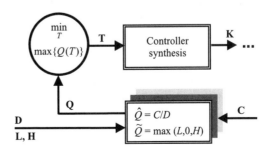

Figure 3.16 Min-max optimization $T = F(Q)$ closes the multi-objective tuning loop, which incorporates the computation chain of Figure 11.

function values is dealt with to attain a "best" engineering tradeoff. For this purpose, compromising demands d_c are interactively declared on conflicting requirements, whereby d_c is chosen to restrain the amount of degradation one is willing to pay on a requirement's quality function q_c in order to improve satisfaction of all conflicting requirements in the best possible way. Starting an interactive iteration process with an already known Pareto-optimal solution $Q^{(v-1)} = \{q_j^{(v-1)}, q_c^{(v-1)}\}$, one chooses an upper bound $d_c^{(v)}$ for the next design step v such that

$$q_c^{(v-1)} < d_c \leq 1 \tag{3.12}$$

and looks for a solution of the constrained minimization problem

$$\min_{T,\alpha} \alpha(T)$$

$$\text{s.t.} \quad q_j(T) \leq \alpha, \tag{3.13}$$

$$q_c(T) \leq d_c^{(v)},$$

for all quality functions, $j = 1, \ldots, J; j \neq c$.

This corresponds to the a posteriori min-max tuning approach developed by Kreisselmeier and Steinhauser [10, 20]. Compromising is restricted to the set of Pareto-optimal alternatives. Min-max optimization, *constrained* by d_c, then attains the "best-possible" solution in the sense that all criteria of interest are minimized up to the constraint of a prequantified limit of degradation one declares to be acceptable for conflicting quality functions.

Consider an example from flight control: By manual tuning according to current industrial practice, control parameters $K = T^0$ for the control law of Figure 3.8 have been determined. Some analysis shows that satisfaction of the C* flight-handling criterion is in strong conflict with maximum control rate. Now, constrained min-max optimization is used to quantitatively explore possible compromises. To achieve this, start with values T^0 to compute a first Pareto-optimal design alternative. This improves both C* and maximum control rate without degradation of the other criteria beyond their already achieved level of satisfaction. Figure 3.17 shows criteria values marked by "0" for the start value corresponding to T^0 and marked by "1" for the first achieved Pareto-optimal solution, T^1. For this first step, a multi-objective evolutionary algorithm might have been used to attain a global Pareto-optimal solution.

An "improved" Pareto-optimal solution, which does not exeed the maximum control rate of the start design T^0, can now be obtained as follows: The demand for C* is set to α, which is to be minimized, and the demand value d_c for control rate is relaxed to value 1.1, which was the value attained by industrial practice to start with. The previous (Pareto-optimal) tuning values T^1 are chosen as attainable start values for the optimization algorithm. After three to four optimization iterations, which take only a couple of seconds of computation time, the optimizer reaches a new Pareto-optimal design alternative "2," which is characterized by having attained the specified upper bound for control rate, while decreasing C* as much as possible. Repeating this procedure accordingly four times results in the compromise set of Figure 3.17, which

Section 3.5 Automatic Tuning and Declarative Compromising

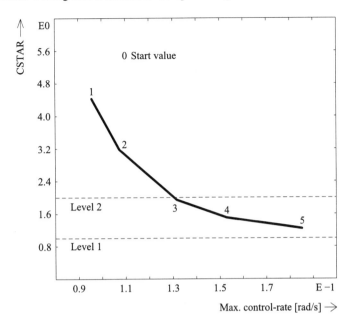

Figure 3.17 Compromise-gradient visualization in the C^*/control-rate plane.

actually is just a different visualization of the same five design alternatives displayed in Figure 3.14 and Figure 3.15. The shape of the Pareto-optimal compromise set can be used to negotiate C^*-quality versus control-effort. For example, an allowed increase of maximum elevator rate by 0.02 [rad/s] from "case 2," which itself corresponds to the starting result, will improve the C^* quality measure by about 30% to reach level 2 of handling quality. But a much higher maximum control rate has to be allowed to reach level 1. This is now a question of a tradeoff to be negotiated by the control design engineer with the cost-responsible systems engineer.

Table 3.2 lists the iteratively compromised controller gains. Automated tuning of the three parameters by constrained parameter optimization takes a couple of seconds for execution of each declarative design command, whereas manual tuning of all three parameters concurrently would have been a much more cumbersome and time-consuming trial-and-error task since it requires a nonlinear change to attain the appropriate gains.

As demonstrated by this example, compromising is an active "what-if" exploration of a Pareto-optimal solution set in quality-function space. Incrementally relaxing upper bounds on conflicting criteria by means of graphical-interactive input on a visualization display like Figure 3.17 allows exploration of the compromise gradient for design tradeoff negotiations.

TABLE 3.2 Controller Gains: Start Values and Compromising Alternatives of Figure 3.17.

	start (0)	1	2	3	4	5
Nz gain	3.76	3.29	3.69	4.52	5.25	6.37
Nz integral gain	2.00	0.96	0.84	0.67	0.60	0.57
Pitch rate gain	0.20	0.00	0.00	0.00	0.03	0.16

3.5.3 Robust Control Laws by Multimodel Compromising

Quality-function control law tuning by parameter optimization is not restricted to a single evaluation model. Any type of control law parameterization can be used together with a set of linear/nonlinear evaluation models. Several operating conditions, for instance, the ones marked in Figure 3.10, and several "quality vertex" dynamics models for the most stable (slowest) and the least stable (fastest) behavior within the system parameter tolerance range can be handled in this way. The idea is to deploy individual quality functions for the different evaluation models and to concatenate all the quality functions to one quality-function set, which then is to be compromised within the set of Pareto-optimal solutions. The goal of robustness-aimed compromising is to achieve a *balanced* control behavior so that off-nominal cases of system behavior are also controlled well within given quality intervals.

This approach turns out to be design efficient. For the GARTEUR[1] design challenge on robust flight control design [22], large parameter tolerances in mass, center of gravity, airspeed, and computation time delay of the digital flight control computer have been taken into account by two suitably chosen, worst-case models, in addition to a model with an average parameter instantiation. This kind of multi-model compromising has resulted in the most robust control law among 12 design competition entries developed along different design methodologies. For this multi-model design case, the *overall* number of quality functions amounted to $3 \times 18 = 54$, based on Table 3.1. But since only those quality functions have to be actively compromised which are in strong conflict, not all quality functions of each evaluation model have to be treated actively by the optimization algorithm. Rather, weak conflicts, where quality functions remain within their "good" level (cf. Section 3.4.1) have only to be monitored after an optimization cycle and hence need no evaluation during an optimization run. For this interactive search, the data structure of the optimization environment should allow activation/deactivation of quality functions at run-time [7].

3.6 FURTHER CACSD TECHNOLOGY

The common control engineer's daily life is much more occupied with tuning and incremental improvement of already available controllers to adapt them to changing product needs than with designing a new controller from scratch. Thus CACSD technology to support computer *automated* control system design within the control system life cycle has been emphasized in this chapter. The processing power and visualization capabilities of desktop computing platforms today, together with advanced modeling, simulation, and multi-objective optimization methodologies, opens the perspective towards a new control engineering lifestyle of virtual product engineering in a system dynamics context, where a validated "first-shot quality" design of executable specifications for "autocode" control software development and maintenance becomes feasible.

---

[1] GARTEUR = Group for Aeronautical Research and Technology in EURope.

Three levels in the control design life cycle have been dealt with in view of design automation: design modeling; quality modeling; and tuning and compromising. The follow-on design activity of formal validation (assessment) (cf. Figure 3.2) seamlessly fits into this conceptual framework. *Assessment* by a systematic, formal procedure is to be performed with respect to a relevant subset of the quality functions that capture the given control requirements. In particular, this refers to the requirement of stability robustness. The design optimization environment (cf. Figures 3.11 and 3.16) can also be used for model-based assessment: Now the optimizer has to search for the worst values of quality functions, which can be attained within an assumed system model tolerance range. That is, instead of minimizing over the tuning parameters T, the optimizer is used to maximize over the tolerance range of system parameters p. To detect all hidden design deficiencies a global search procedure, for example, employing evolutionary algorithms [17], is required. A different type of approach is model-free validation based on machine learning, for example, the Unfalsified Control Concept [27]. As an advantage of this data-driven approach, experimental data can be used, as well as simulated data, which allows unmodeled hardware in the validation loop. Another aspect is validation of control logic in discrete-event/continuous-time control systems. This relies on formal methods of computer science. How to apply qualitative modeling techniques to the continuous-time part to allow formal validation ("model checking") by computer science methods is presently a topic of active research.

Declarative control design with explicitly treated quality functions is apparently better suited to handle the design complexity problem than procedural design in that it allows interactive exploration of visualized design conflicts and the achievable design potential. The methodology fosters a design process, in which a full set of quality functions is taken care of from the very beginning and parametric and operational tolerances are handled on-line in the design loop by suitably chosen worst-case evaluation models. This provides the quantified comparison baseline for design decisions and change management. In future control design environments, this methodology has to be supported by a *uniform information model* for CACSD versioning. Developing the information model within the ISO Standard To Exchange Product Data (STEP) would formally link the computer-aided control design process with overall product design (cf. [30]). This should be paralleled by an automatic translation of system model components into the object-oriented STEP/EXPRESS language.

The optimization-based approach is computationally demanding and asks for further development of *superfast algorithms* for synthesis, analysis, and simulation. In particular, automated separation of linear from nonlinear model equations for so-called inline integration [3] in system simulation ought to be advanced by symbolic equation handling and pertinent numerical solvers. The development of efficient model reduction algorithms to generate high-fidelity system dynamics models from high-granularity disciplinary models, for example, finite-element models of structural dynamics, is also of concern.

The various evaluation cases in multi-objective tuning and compromising can be executed in parallel by meta-computing, that is, by sharing distributed computation power in a computer network. This means that *information systems interoperability* [19] using object-based middleware techniques for a distributed engineering-software operating system on top of different computer operating systems, or for a federation of Java virtual machines, ought to be adopted for CACSD environments. This implies that parameterized quality functions as well as the synthesis, analysis, and simulation modules in Figure 3.11 are provided as data-typed computation components. To support *interworking* via a network, the *semantics* of the quality functions and evaluation cases have to be made transparent to information brokers such that multidisciplinary users of an interoperable CACSD system get ready access to this design information. This holds in particular for the multidisciplinary development and maintenance of system models, where a standard modeling language like Modelica [24] for both physical and algorithmic control components should be used in future. Computer-supported information sharing is important to keep track of design consistency by all stakeholders who experience long engineering transactions in their project work.

For initialization of automated design computations, CACSD environments have to support a manual, procedural process mode as well. This is usually called a Computer-Aided Control Engineering (CACE) environment. It should rely on a repository that provides the following *layered* services: engineering-database services, model/data definition services, algorithmic services, tool-control services, task-control services, user-interaction services, and process communication services [12], in which every layer connects only to interfaces that are no more than one level above or below it. Layering is an architectural provision to ease future upgrades. Such an "open," layered framework is not yet commercially available. Its development and implementation with production-quality software on all layers remain a challenge.

Nonlinear parameter optimization by algorithmic search techniques is the key asset in automated control design tuning. In a CACSD design-automation environment, a generic optimization setup (cf. Figure 3.16) should be available, in which the generic optimization task is defined once and execution can be performed by any suitable solver from both the field of nonlinear mathematical programming and the field of evolutionary computation. This requires development of a proper data structure to interactively switch among different kinds of solvers and to activate/deactivate any quality function at run-time. In the set of solvers, "hybrid" solvers combining evolutionary algorithms with nonlinear programming algorithms should be available for global/local search. Conceptually, data-driven search is used to invert the design-response surface spanned by the attainable values of quality functions (cf. Figure 3.14), with respect to the tuning parameters. One may investigate the use of neural networks as a general learning approach for approximating this map as the set of feasible design alternatives evolves by the various optimization search iterations. A neural network approximation is analytically differentiable and allows gradient computation by applying the chain rule of differentiation. Such an approximation of the design-response surface can save evaluation time in initializing retuning of a controller later on.

At a glance, CACSD technology developments and commercial-off-the-shelf CACSD software products are scattered over a broad range of activities. Declarative design in view of *control design automation*, as outlined in this contribution, to a great extent makes these developments coherent with a suitable computation machinery for "virtual engineering." This is a step towards increasing control engineering competitiveness by a computer-aligned design process for better balanced, performance-reliable controllers achieved in shorter engineering time. An information port to CACSD developments is provided by the home page of the IEEE Technical Committee on CACSD [1].

ACKNOWLEDGMENT

The contributions of the author's former colleagues in the DLR—Control Design Engineering Group are gratefully acknowledged—in particular, H.-D. Joos on multi-objective flight control design issues; D. Moormann, G. Looye, and P. Mosterman on object-oriented flight-system dynamics modeling; K.H. Kienitz on fuzzy-type specifications in goal attainment; R. Finsterwalder on interactive exploration of design-information patterns; and A. Varga on performant control numerics software.

Related Chapters

- An in-depth discussion of how CACSD tools are being used for automotive powertrain controller development can be found in Chapter 15.
- Applications of CACSD to flight control are also described in Chapter 11.
- A variety of modeling and simulation methods for control systems is outlined in Chapter 4.

REFERENCES

[1] IEEE TC on CACSD: : http://www-er.df.op.dlr.de/cacsd/.

[2] H. Elmqvist, "A structured model language for large continuous systems." Ph.D. Thesis, Department of Automatic Control, Lund Institute of Technology, Sweden, 1978.

[3] H. Elmqvist, M. Otter, and F. E. Cellier, "Inline integrations: A mixed symbolic/numeric approach for solving differential-algebraic equation systems." *Proc. European Simulation Multiconference*, Prague, June 5–8, pp. xxiii–xxxiv, 1995.

[4] H. Elmqvist, S. E. Mattson, and M. Otter, "Modelica—a language for physical system modeling, visualization and interaction." *Proc. 10th IEEE Int. Symposium on Computer Aided Control System Design*, Hawaii, August 22–27, pp. 630–639, 1999.

[5] W. Feng and Y. Li, "Performance indices in evolutionary CACSD automation with application to batch PID generation." *Proc. 10th IEEE Int. Symposium on Computer Aided Control System Design*, Hawaii, August 22–27, pp. 486–491, 1999.

[6] R. Finsterwalder, "A 'parallel coordinate' editor as visual decision aid in a multi-objective concurrent control engineering environment." *Proc. IFAC Symposium on Computer Aided Design in Control Systems*, Swansea, UK, July 15–17, pp. 118–122, 1991.

[7] R. Finsterwalder, H.-D. Joos, and A. Varga, "A graphical user interface for flight control development." *Proc. 10th IEEE Int. Symposium on Computer Aided Control System Design*, Hawaii, August 22–27, pp. 439–444, 1999.

[8] C. M. Fonseca and P. J. Fleming, "An overview of evolutionary algorithms in multiobjective optimization." *Evolutionary Computing*, Vol. 3, no. 1, pp. 1–16, 1995.

[9] G. Grübel, H.-D. Joos, M. Otter, and R. Finsterwalder, "The ANDECS design environment for control engineering". *Proc. 12th IFAC World Congress*, Sydney, Australia, Vol. 6, pp. 447–454, 1993.

[10] G. Grübel, R. Finsterwalder, G. Gramlich, H.-D. Joos, and S. Lewald, "ANDECS: A computation environment for control applications of optimization." In R. Bulirsch and D. Kraft (eds.), *Control Applications of Optimization, Int. Series of Numerical Mathematics*, Vol. 115, Birkhäuser Verlag Basel, pp. 237–254, 1994.

[11] G. Grübel (ed.), "Case study: Applied multidisciplinary dynamics design experimenting." *Proc. IFAC Conf. on Integrated Systems Engineering*, Baden-Baden, Germany, September 27–29, pp. 89–117, 1994.

[12] G. Grübel, "The ANDECS CACE framework." *IEEE Control Systems Magazine*, pp. 8–13, April 1995.

[13] H.-D. Joos, M. Schlothane, and G. Grübel, "Multi-objective design of controllers with fuzzy logic." *Proc. IEEE/IFAC Joint Symposium on Computer-Aided Control System Design*, Tucson, AZ, March 7–9, pp. 75–82, 1994.

[14] H.-D. Joos, "A methodology for multi-objective design assessment and flight control synthesis tuning." *J. Aerospace Science and Technology*, Vol. 3, no. 3, pp. 161–176, 1999.

[15] R. E. Kalman and T. S. Englar, *A User's Manual for the Automatic Synthesis Program (Program C)*. NASA CR-475, 1966.

[16] C. T. Kelley, *Iterative Methods for Optimization*. Frontiers in Applied Mathematics, No. 18, Society for Industrial and Applied Mathematics, 1999.

[17] J.-H. Kim and H. Myung, "Evolutionary programming techniques for constrained optimization problems." *IEEE Trans. on Evolutionary Computation*, Vol. 1, no. 2, pp. 129–140, July 1997.

[18] K. H. Kienitz, "Controller design using fuzzy logic—A case study." *Automatica*, Vol. 29, no. 2, pp. 549–554, 1993.

[19] B. Krämer, M. Papazoglou, and H.-W. Schmidt, *Information Systems Interoperability. Advanced Software Development Series*, Research Studies Press. New York: John Wiley & Sons, 1998.

[20] G. Kreisselmeier and R. Steinhauser, "Application of vector performance optimization to robust control loop design for a fighter aircraft." *Int. Journal Control*, Vol. 37, no. 2, pp. 251–284, 1983.

[21] A. G. J. MacFarlane, G. Grübel, and J. Ackermann, "Future design environments for control engineering." *Automatica*, Vol. 25, no. 2, pp. 165–176, 1989.

[22] J. F. Magni, S. Bennani, and J. C. Terlouw (eds.), "Robust flight control—A design challenge." *Lecture Notes in Control and Information Sciences*, Vol. 224, New York: Springer, 1997.

[23] K. M. Miettinen, *Nonlinear Multiobjective Optimization*. Norwell, MA: Kluwer Academic Publishers, 1999.

[24] Modelica: http://www.Modelica.org

[25] D. Moorman, P. J. Mosterman, and G. Looye, "Object-oriented model building of aircraft flight dynamics and systems." *J. Aerospace Science and Technology*, Vol. 3, no. 3, pp. 115–126, 1999.

[26] S. J. Rasmussen and S. G. Breslin, "AVDS: A flight system design tool for visualization and engineer-in-the-loop simulation." *Proc. AIAA Guidance and Control Conference*, AIAA- 97-3467, pp. 135–143, 1997.

[27] M. G. Safonov and T.-C. Tsao, "The unfalsified control concept and learning." *IEEE Transactions on Automatic Control*, Vol. 42, no. 6, pp. 841–843, June 1997.

[28] V. Sima and S. Van Huffel, "High-performance algorithms and software for systems and control computations." *Proc. 10th IEEE Int. Symposium on Computer Aided Control System Design*, Hawaii, August 22–27, pp. 85–90, 1999.

[29] K. C. Tan, T. H. Lee, and E. F. Khor, "Control system design automation with robust tracking thumbprint performance using a multi-objective evolutionary algorithm." *Proc. 10th IEEE Int. Symposium on Computer Aided Control System Design*, Hawaii, August 22–27, pp. 498–503, 1999.

[30] T. Varsamidis, "Object-oriented information modelling for computer-aided control engineering." Ph.D. Thesis, School of Electronic Engineering and Computer Systems, University of Wales, Bangor, UK, 1998.

Chapter 4

SYSTEM MODELING

Pradeep Misra

Editor's Summary

Models—in the sense of mathematical representations of systems—are critical to all advanced control. Not only is it a truism that we can only control or optimize a system to the extent that we understand it, but also virtually all advanced control techniques rely on an explicit representation of the system. In some cases, a model is used in the design process alone; in others, the on-line controller contains a model; in yet others models of different fidelity and complexity are used for design, analysis, and operation.

While control engineers use models on a regular basis, the full variety of models as relevant to control is not always appreciated. Models in control engineering can be distinguished along several dimensions, such as static and dynamic, linear and nonlinear, first principles and empirical, lumped and distributed. This chapter provides a broad overview of many of the popular modeling methodologies. It does not attempt to discuss any one approach in complete detail, but rather to explain key concepts and to present a number of techniques that are seldom treated as parts of one unified topic. Subjects of specific importance for control that are also discussed include model reduction and linearization of nonlinear models. More in-depth discussions of specific methods are included in other chapters in this volume—see Chapter 2 for discrete-event system models, Chapter 7 for hybrid models that combine continuous-time and discrete-event behaviors, and Chapter 6 for nonlinear "approximators." Examples of models for different applications can be found in several chapters in the second part of this book.

As the systems that we attempt to control become increasingly more large-scale and complex, no one type of modeling approach will be sufficient. A future trend in system modeling is the developing of macromodels or multimodels, integrations of disparate models in one framework. As today's systems become tomorrow's subsystems, the control engineer's comprehensive understanding of system modeling will only continue to become more important.

Pradeep Misra is an associate professor in the Department of Electrical Engineering at Wright State University. He also serves as the secretary/administrator of the IEEE Control Systems Society.

4.1 INTRODUCTION

A system may be defined as a mechanism comprised of a collection of objects (physical or abstract) related through physical relationships, along with mathematical rules that govern the behavior of the mechanism. Only for the most simplistic systems is it possible to determine the exact relationships and rules that characterize their behavior.

Present technology has enabled engineers to build increasingly complex systems, which in turn have provided us with the means to perform increasingly difficult tasks. This increased ability is achieved at the price of distancing ourselves from compact analytical models, typical of systems a few decades ago. Therefore, modeling and simulation in a broader sense have taken on extremely important roles in modern-day system analysis and design.

Use of models has been prevalent from ancient times. Although modeling techniques have changed with the advent of technology, the paradigm remains the same. The process of developing models of complex systems is, by nature, iterative. Typical iterative phases during the modeling process are illustrated in Figure 4.1 [31].

For a reasonably complex system, a model is a simplified rendering of underlying mechanisms and rules that capture the essence of the system. Typically, regardless of its complexity, the model will not be able to replicate the actual system exactly when first derived. Therefore the model must be verified through simulation by comparison of the system's response with that of the model and refinement of the model iteratively, until the model accurately mimics the physical system's behavior. Occasionally, this leads to taking a second look at the underlying principles and redefinition of the model. The degree of accuracy required is, of course, application-dependent.

4.1.1 Historical Perspective

From the earliest times, the driving force for developing models has been to explain, in a comprehensive manner, the physical world around us. For example, several models of our solar system were developed over the centuries. It was assumed that Earth was at the center of the solar system and these models were only partly successful in describing and predicting the behavior of the system. Only in the early seventeenth century, when Johannes Kepler proposed the three laws (modeling assumptions, really) of planetary motion—the law of orbits, the law of areas, and the law of periods—did the model of the solar system begin to be more reliable. These laws were based on data gathered by physical observation of the motion of planets; hence the model may be considered an empirical model. Later in the same century, Isaac Newton provided a mathematical and physical basis for Kepler's laws by deriving the three laws from the concepts of conservation of energy and momentum.

The solar system example clearly shows the evolutionary process in developing a sophisticated/reliable model. Earlier models assumed Earth, and after Copernicus, the Sun as the center, with other bodies in the solar system going around in circular orbits. The models were based on data gathered with low-fidelity sensors (human eyes), and there was no evidence that the models were incorrect until, by use of higher fidelity

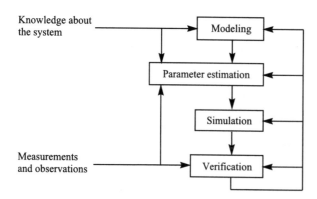

Figure 4.1 Iterative modeling process.

instruments (telescopes), it was determined that these models did not satisfactorily explain the system. The underlying principles were reexamined, leading to a more sophisticated model with elliptical orbits. The model was then refined and validated by alternative means.

Mathematical maturity has enabled the present-day engineer to develop sophisticated and reliable models. This advance, coupled with significant improvements in technology, has enabled us to construct and employ highly complex systems. Lest we become complacent, however, nature continues to humble us by categorically illustrating the limitations of even the best of the current technology. Conspicuous illustrations of these limits are the sinking of the *Titanic* in 1912, the collapse of the Tacoma Narrows suspension bridge in 1940, and the destruction of the Ariane 5 launcher in 1996, to name a few. Some of these disasters could have been avoided with better modeling and testing. But before we become too disheartened by the failures, let us hasten to point out the successes in which we can take pride. From earlier times, we can cite the Egyptian pyramids, the European cathedrals, and the Taj Mahal, and from modern times, the Golden Gate bridge, the Apollo project, the Boeing 777 airplane, the space shuttle, and the space station. The earlier accomplishments utilized empirical models, whereas the later ones used sophisticated mathematical models of various subsystems, decomposed mainly hierarchically but not exclusively, and then integrated for the final design—a multimodel design approach [27].

4.1.2 Modeling and Control

Modeling of a system is seldom, if ever, a goal in itself; rather, it is the means to attain some goal. From the viewpoint of a control systems engineer, the goal is to control the response of the underlying system. The successful operation of a system under changing (and often not fully predictable) conditions often requires a feedback or closed-loop control system. The response of the system is compared with the desired operating conditions, and the difference between the two is used to adjust the response through a controller. The purpose of the controller is to minimize and, ideally, to eliminate the effect of external disturbances, steady-state errors, transient errors and variations in plant parameters on the output of the system.

A typical feedback control problem involves the selection of sensors to measure and monitor the system outputs; selection of actuators to drive the system; mathematical formulation of the desired output response; development of mathematical models of sensors, actuators, and the plant; design of the controller based on the plant model and the control criteria; and evaluation of the feedback control system by extensive simulation or, where possible, applying it to the actual system. These steps are iterated until the feedback control system exhibits the desired response. It is, therefore, evident that modeling plays a key role in the effective control of a physical system or a process.

4.1.3 Classification

The vastness of the field of modeling and simulation, which spans all aspects of science and engineering, makes it virtually impossible to provide a synoptic coverage of the subject. It is difficult to pin down a specific classification scheme because distinctions arise from a variety of factors. Nonetheless, it can be safely said that a useful classification must depend on mathematical concepts that are *required* to accurately represent the physical phenomenon. Some obvious distinctions would be static versus dynamical models, time-invariant versus time-varying models, linear versus nonlinear models, deterministic versus stochastic models, "crisp" versus "fuzzy" models [1], and so on. Within each such classification, there could be continuous, discrete, or hybrid (a mixture of continuous and discrete variables) models. System models may also be classified based on their representation, for example, state variable models versus input-output models. Various model types can be loosely represented in the treelike structure shown in Figure 4.2 [10]. In the figure, each branch may be continued to have further subclassifications, as shown for the linear constant coefficient models.

The layout of the rest of this chapter is as follows. In Section 4.2, we discuss linear and nonlinear static models and estimation of their parameters. In Section 4.3, we focus on linear lumped parameter dynamical models obtained from mathematical descriptions of electrical, mechanical (translational and rotational), and thermal systems. We also discuss linear dynamical models obtained by parameter estimation from input-output data. Finally, we discuss model order reduction by modal truncation, singular perturbation, and balanced realizations. In Section 4.4, we cover nonlinear system models, as well as some of the commonly encountered nonlinearities and several techniques to obtain linearized approximate models of nonlinear systems. In Section 4.5, we consider classification and finite difference models of distributed parameter systems. Finally, Section 4.6 contains a general discussion of the scope and future of mathematical modeling.

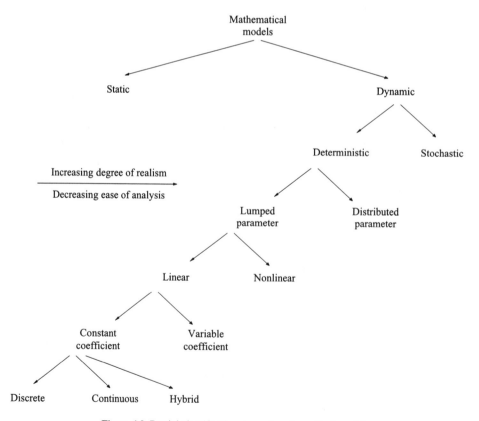

Figure 4.2 Partial classification tree of mathematical models.

4.2 STATIC MODELS

Models described by mathematical equations may be stated in a fairly general form as

$$F(\mathcal{D}(x(t)), x(t), u(t), t) = 0 \qquad (4.1)$$

where t indicates that the underlying system is a continuous-time system, F represents a system of interrelated equations, $x(t)$ represents states, $\mathcal{D}(x(t))$ their first derivatives, and $u(t)$ external inputs. In a similar manner, discrete-time systems may be represented as

$$F(\mathcal{D}(x[k]), x[k], u[k], k) = 0 \qquad (4.2)$$

where k represents the discrete-time axis, $x[k]$ represents states, $\mathcal{D}(x[k])$ their first difference or time shift, and $u[k]$ external inputs. In Eq. (4.1) or (4.2), if the term $\mathcal{D}(x(t))$ (respectively, $\mathcal{D}(x[k])$) is uniformly zero, the model is static. In that case, the relationship between state variables is purely algebraic. Whether or not the resulting model is linear will depend on how the states $x(t)$ (respectively, $x[k]$) are related. In the remainder of this section, we will assume that the models are deterministic. Furthermore, we will deal with continuous models unless otherwise warranted.

A scientist or an engineer tasked with developing a model of a system is generally privy to some knowledge about the system. This prior understanding can often be captured in the form of a model structure. The problem that then remains is the identification of the parameters associated with the model form. Parameter estimation techniques have been developed that can estimate the *best-fit* values of these parameters based on data collected during the operation of the system. Clearly, the greater the quantity and the better the quality of the collected data, the more accurately the parameter values can be determined.

Notions of model structure and parameter estimation arise in static as well as dynamic models. In the next few paragraphs, we discuss one popular class of approaches to parameter estimation in static models, both linear and nonlinear. The field of system identification specifically focuses on parameter estimation for dynamical models.

4.2.1 Linear Models

Assume that we are given the data points $\{(x_i, y_i), i = 1, \ldots n\}$, where x and y are both scalars and x_i is monotonically increasing. Often the data lends itself to models of the form

$$y = f(x) = \sum_{j=1}^{m} C_j f_j(x), \qquad (4.3)$$

where $f_j(x)$ are known functions. Then, the best values of the unknowns C_j that will fit the model are given by minimization of the least-squares error $E(C_j, j = 1, \ldots, m)$, defined as

$$E(C_j, j = 1, \ldots, m) = \sum_{k=1}^{n}\left[\left[\sum_{j=1}^{m} C_j f_j(x_k)\right] - y_k\right]^2. \tag{4.4}$$

For models of the type described by (4.3), the computation of the unknown parameters (C_j) becomes rather straightforward. To minimize the sum of the square of the error $E(C_j, j = 1, \ldots, m)$, we take partial derivatives of the error with respect to $C_j, j = 1, \ldots, m$. Setting $\partial E/\partial C_j = 0, j = 1, \ldots, m$, will yield normal equations

$$\sum_{j=1}^{m}\left[\sum_{k=1}^{n} f_i(x_k) f_j(x_k)\right] C_j = \sum_{k=1}^{n} f_i(x_k) y_k, \quad i = 1, \ldots, m. \tag{4.5}$$

The system of equations in (4.5) can be easily formulated in matrix–vector form as

$$F^T F C = F^T Y,$$

where superscript $(\cdot)^T$ denotes matrix transposition, $C = [C_1\ C_2\ \cdots\ C_m]$ is the vector of unknowns to be determined, $Y = [y_1\ y_2\ \cdots\ y_n]$ is defined from the given data, and the matrix F is defined as:

$$F = \begin{bmatrix} f_1(x_1) & f_1(x_2) & \cdots & f_1(x_n) \\ f_2(x_1) & f_2(x_2) & \cdots & f_2(x_n) \\ \vdots & \vdots & \ddots & \vdots \\ f_m(x_1) & f_m(x_2) & \cdots & f_m(x_n) \end{bmatrix}.$$

The system of equations in (4.5) may be solved quite efficiently using Cholesky decomposition, QR decomposition or singular value decomposition (SVD) [9], [26].

The linear least-squares formulation in (4.5) can be readily specialized for a few frequently encountered situations. For example, if the functions $f_j(x) = x^{j-1}$, one gets a polynomial model in x. For $j = 1$ the result would be a straight line, for $j = 2$, a parabola, and so forth. Furthermore, the least-squares line can account for several models through linearization to $Y = CX + D$. Table 4.1 shows a few models in which a linearized representation enables us to obtain the model in a straightforward

TABLE 4.1 Linearization for Linear Least-Squares Approximation

Model	Linearization	Change of Variable	Constants
$y = \dfrac{A}{x} + B$	$Y = CX + D$	$X = \dfrac{1}{x},\ Y = y$	$A = C,\ B = D$
$y = \dfrac{1}{Ax + B}$	$Y = CX + D$	$X = x,\ Y = \dfrac{1}{y}$	$A = C,\ B = D$
$y = B e^{Ax}$	$Y = CX + D$	$X = x,\ Y = \log_e(y)$	$A = C,\ B = e^D$
$y = B x^A$	$Y = CX + D$	$X = \log_e(x),\ Y = \log_e(y)$	$A = C,\ B = e^D$

manner. One can, of course, find several other models that may be linearized in similar fashion.

The simplicity of linear least-squares approximation may tempt one to use higher degree polynomial approximations to fit nonlinear data. Naturally, in theory, there is no limit to the degree of polynomial used in representing the data, as long as the underlying model lends itself to it. However, if the data do not correspond to a high-degree polynomial model, the fitted polynomial may exhibit highly oscillatory behavior. The oscillatory behavior may be reduced to some extent by using the least-squares fit with orthogonal polynomials such as Chebyshev polynomials, but it cannot be eliminated altogether.

In Eq. (4.3), the system model $y = f(x)$ was dependent on a single variable x. From a computational viewpoint, the modeling problem becomes considerably more challenging if the number of independent variables is more than one. Fortunately, from a mathematical standpoint, the result is often a simple extension of the single variable case. The expressions for the least-squares error (4.4) and normal equations (4.5) takes on vector forms; that is, x is replaced by $\mathbf{x}$, where $\mathbf{x}$ represents a vector of variables. The modified system of equations takes the following form:

$$E(C_j, j = 1, \ldots, m) = \sum_{k=1}^{n} \left[\left[\sum_{j=1}^{m} C_j f_j(\mathbf{x}_k) \right] - y_k \right]^2.$$

Setting $\partial E / \partial C_j = 0, j = 1, \ldots, m$, will yield normal equations

$$\sum_{j=1}^{m} \left[\sum_{k=1}^{n} f_i(\mathbf{x}_k) f_j(\mathbf{x}_k) \right] C_j = \sum_{k=1}^{n} f_i(\mathbf{x}_k) y_k, \qquad i = 1, \ldots, m,$$

leading to a convenient matrix–vector form for determining the unknowns C_j.

4.2.2 Nonlinear Models

For the cases discussed in Section 4.2.1, the choice of the model enabled us to reformulate the least-squares parameter estimation problem such that the model was linearly dependent on unknown parameters $C_j, j = 1, \ldots, m$. Next, we generalize the parameter estimation problem to the cases where such a simplification is not possible. The basic outline of the parameter estimation technique remains unchanged; that is, we define a least-squares error and compute the parameter values (values of C_j) so that the least-squares error is minimized. Unfortunately, because of nonlinear relationships among parameters, the minimization procedure for determining C_js becomes iterative.

We may define the model to be fitted as

$$y = f(x, C)$$

where nonlinearities are embedded in $f(x, C)$ and C is a vector of model parameters $\{C_j, j = 1, \ldots, m\}$ to be determined. As before, the least-squares error may be defined as the sum of the squares of errors between the model and the measured data

$$E(C_j, j = 1, \ldots, m) = \sum_{k=1}^{n} [y_k - f(x_k, C)]^2.$$

Expanding the error to a quadratic form (through the use of Taylor series expansion) at the rth iteration, we get

$$E(C) \approx E(C^r) + [C - C^r]^T S + \frac{1}{2}[C - C^r]^T T [C - C^r] \tag{4.6}$$

where C^r is the estimate of the parameter vector C at the rth iteration, S is a column vector of length m containing the first derivatives of E with respect to C, and T is an $m \times m$ matrix containing second-order derivatives of E with respect to elements of the parameter vector C, evaluated at C^r. S and T are, respectively, known as the gradient vector and the Hessian matrix of E. Mathematically, the former is given by

$$S_j = \frac{\partial E(C)}{\partial C_j}$$
$$= -2 \sum_{k=1}^{n} [y_k - f(x_k, C)] \frac{\partial f(x_k, C)}{\partial C_j}, \quad j = 1, \ldots, m, \; C = C^r$$

and the latter by

$$T_{ij} = \frac{\partial^2 E(C)}{\partial C_i \partial C_j}$$
$$= 2 \sum_{k=1}^{n} \left[\frac{\partial f(x_k, C)}{\partial C_i} \frac{\partial f(x_k, C)}{\partial C_j} - [y_k - f(x_k, C)] \frac{\partial^2 f(x_k, C)}{\partial C_i \partial C_j} \right] \quad i, j = 1, \ldots, m, \; C = C^r$$

where the second derivative is often ignored.

It is easily seen by setting $E(C) - E(C^r) = 0$ in (4.6), that knowing S and T, we can iteratively refine the model parameter C, using the *inverse Hessian iteration*

$$C_{r+1} = C_r + T^{-1} S. \tag{4.7}$$

If the second-order approximation in (4.6) is a poor local approximation or the computed parameter vector C is far from the optimal value, the inverse Hessian iteration fails to give a good fit of the model to the data. In that situation, an approach similar to *steepest descent* [4, 26], given by

$$C_{r+1} = C_r + \alpha S$$

where α is a small constant, is used to refine the approximation. Furthermore, when successive iterations using the steepest descent produce relatively small improvement, one can switch back to inverse Hessian iteration in (4.7). The resulting elements of vector C are the required parameters of the nonlinear model. Of course, the above optimization scheme is one of several that may be used for obtaining optimal C. Several others may be found in [4].

The preceding discussion addresses only models obtained through unconstrained optimization. In practice, model parameters may be constrained to lie within certain predefined regions, making the problem considerably more difficult. A commonly employed approach to solve such problems employs Lagrange multipliers [2].

4.3 DYNAMICAL MODELS

A very large variety of physical systems exhibit behavior that evolves over time. For such systems, which are of considerable interest for control, the models and techniques discussed in the previous section are no longer adequate. Instead, dynamical models are used to express their changing behavior.

Recall the general functional representation for a system model:

$$(Continuous) \quad F(\mathcal{D}(x(t)), x(t), u(t), t) = 0$$
$$(Discrete) \quad F(\mathcal{D}(x[k]), x[k], u[k], k) = 0. \quad (4.8)$$

For static models, we assumed that the term $(\mathcal{D}(x(\cdot))$ was zero. For dynamical models, this term is no longer uniformly zero. If in (4.8), we can rewrite the differential or difference relationship as

$$\mathcal{D}(x(t)) = \hat{F}(x, u, t), \quad (4.9)$$

then we have state space models. On the other hand, when it is not possible to express (4.8) as an explicit system of differential equations (4.9), the resulting system is known as a *singular*, *implicit*, or *differential-algebraic* system. Such models represent dynamical systems with algebraic constraints. Note that, while (4.9) refers to the continuous-time case only, as shown in Figure 4.3, it is always possible to switch between continuous-time and discrete-time models.

For continuous-time systems, if the function representation exhibits dynamical dependence on *time t* (temporal variable) as well as *space x* (spatial variable), then the system is defined as a *distributed parameter system*. Such systems, for example, structures, heat flow, and fluid flow, are modeled using partial differential equations. On the other hand, if the dependency on a space variable is absent or negligibly small, then the system model may be expressed using finitely many differential equations, and the system is then said to be a *lumped parameter system*. The latter is perhaps the most studied class of dynamical systems, especially for linear cases such as electrical circuits, translational or rotational mechanical systems, and so on. A somewhat more exotic variety would include systems that are continuous, but may trigger a controlled switch in a continuous process or set off a timing mechanism in a discrete system. Such systems are known as *state event systems*. On the discrete side, dynamical systems include

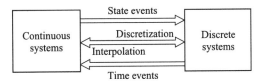

Figure 4.3 Continuous and discrete interconversion.

sampled data systems, represented by difference equations. Depending on the sampling scheme, they may be single-rate or *multi-rate* sampled data systems. If the sampling period is not fixed, the systems are called *discrete-event systems*. In these systems, the system dynamics is affected only at the occurrence of events; in between events, the system parameters retain their values. Examples include systems such as traffic networks and communication networks. If we interface discrete-event systems with continuous dynamical systems, we get *hybrid* systems [31].

4.3.1 Lumped Parameter Models

The major components of lumped parameter models are energy sources, passive energy storage elements, and passive dissipative elements. Along with the above basic components used in electrical, mechanical, hydraulic, or thermal systems, there are also transduction elements that transfer energy between various physical systems. Some commonly encountered transduction elements include electric motors (electrical energy to mechanical energy), generators (mechanical or fluid energy to electrical energy), heating coils (electrical energy to thermal energy), and so on.

A lumped parameter model is described by a system of differential equations. These equations are obtained by formulating a set of mathematical equations by summation of *through*-variables at any junction, summation of *across*-variables within any closed loop, and a mathematical representation of each element of the system. The dynamical order of the system is governed by the number of independent energy storing elements.

Nonlinearities will be addressed in the next section; for the time being, we will review idealized linear elements. Variables required to formulate various lumped parameter models and symbols commonly used to denote them are listed in Table 4.2. The differential and algebraic equations describing individual elements are listed in Table 4.3.

TABLE 4.2 Common Variables Used in System Modeling

System	Through-Variable	Across-Variable
Electrical	Current (I)	Voltage (V)
Hydraulic	Fluid flow rate (q)	Pressure (P)
Rotational	Torque (T)	Angular Velocity (ω)
Translational	Force (F)	Velocity (v)
Thermal	Heat flow rate (Q)	Temperature (T)
Thermodynamical	Entropy flow ($\frac{dS}{dt}$)	Temperature (T)

As mentioned earlier, transduction is used to transfer energy from one type of system to another; this does not preclude a transfer of energy from (say) electrical to electrical form. Transducers may be divided into two major categories:

- *Transformers*: Transformers relate through-variables to through-variables and across-variables to across-variables, for example, electric transformers relating voltage to voltage and current to current with an appropriate proportionality constant. Another common example would be an electric motor that relates armature current to torque and voltage across the armature to angular velocity.

Section 4.3 Dynamical Models

- *Gyrators*: Gyrators relate through-variables to across-variables and vice versa, for example, a hydraulic ram that relates fluid pressure (across variable) to linear force (through variable) and fluid flow rate (through variable) to linear velocity of the piston (across variable).

One may now model the system through differential relationships in the time domain directly to obtain *state space models* or use Laplace transforms of various differential relations to obtain *transfer function models* [5, 15, 24]. As discussed later in Section 4.4, one may also obtain linear lumped parameter models by local linearization of nonlinear systems.

TABLE 4.3 Relationships between Through and Cross Variables

Element	Symbol	Element Equation
Electrical inductance	L	$V = L \dfrac{dI}{dt}$
Electrical capacitance	C	$I = C \dfrac{dV}{dt}$
Electrical resistance	R	$R = \dfrac{V}{I}$
Translational spring	K	$v = \dfrac{1}{K} \dfrac{dF}{dt}$
Translational mass	M	$F = M \dfrac{dv}{dt}$
Translational damper	B	$B = \dfrac{F}{v}$
Rotational spring	K	$\omega = \dfrac{1}{K} \dfrac{dT}{dt}$
Rotational mass (inertia)	I	$T = I \dfrac{d\omega}{dt}$
Rotational damper	B	$B = \dfrac{T}{\omega}$
Fluid inertia	I	$P = I \dfrac{dq}{dt}$
Fluid capacitance	C	$q = C \dfrac{dP}{dt}$
Fluid resistance	R	$R = \dfrac{P}{q}$
Thermal capacitance	C	$Q = C \dfrac{dT}{dt}$
Thermal resistance	R	$R = \dfrac{T}{Q}$

4.3.2 System Identification

In many practical situations, models of components of the system and their mathematical interrelationships are not known precisely. In these cases, the strategy described in the previous section becomes unsuitable. Instead, one must construct a mathematical model of the system from measurement of the system's response to known signals. The process of constructing mathematical models of dynamical systems from measured data is known as identification. We will restrict the discussion to linear, lumped, time-invariant, deterministic systems. Although methods for identifying models for more general classes of systems exist, they are beyond the scope of the present discussion. We will consider discrete-time identification because regardless of whether or not the system under consideration is continuous, the input and the output measurements are obtained at discrete-time instances.

For deterministic systems, models obtained through identification methods may be broadly classified into nonparametric and parametric models. If the structure of the model is not defined *a priori*, then the system model is referred to as a *nonparametric* model. Some commonly used nonparametric models include impulse response and frequency response models. On the other hand, if the structure of the model is predetermined, the model is said to be a *parametric* model. For example, the model may be restricted to be a sum of a fixed number of decaying exponentials. The identification process then determines the coefficient and the decay rate for each term in the summation. Some commonly used parametric models include differential or difference equations, transfer functions, and state space descriptions. Paradoxically, so-called nonparametric models have many more parameters generally than parametric models. It is interesting to note that parametric models can be deduced from nonparametric models. Because parametric models are more compact (fewer parameters to identify), it is not surprising that their identification has garnered more attention. Among parametric models, the majority of the research has been devoted to identification of models expressed by difference equations.

Assume the following general model:

$$E(q)y(k) = \frac{A(q)}{B(q)}u(k) + \frac{C(q)}{D(q)}e(k)$$

where $A(q)$, $B(q)$, $C(q)$, $D(q)$, and $E(q)$ are polynomials in the delay operator q, $y(k)$ and $u(k)$ represent the output and the input, respectively, and $e(k)$ represents external disturbances. Then, depending on the elements of various polynomials, the following system models are commonly studied.

$$
\begin{aligned}
y(k) &= A(q)u(k) + e(k) &\text{(FIR)} \\
E(q)y(k) &= A(q)u(k) + e(k) &\text{(ARX)} \\
E(q)y(k) &= A(q)u(k) + C(q)e(k) &\text{(ARMAX)} \\
y(k) &= \frac{A(q)}{B(q)}u(k) + e(k) &\text{(OE)} \\
y(k) &= \frac{A(q)}{B(q)}u(k) + \frac{C(q)}{D(q)}e(k) &\text{(BJ)}
\end{aligned}
\quad (4.10)
$$

Section 4.3 Dynamical Models

The model acronyms in (4.10) stand for *finite impulse response* (FIR), *auto-regressive with exogenous input* (ARX), *auto-regressive moving average with exogenous input* (ARMAX), *output error* (OE), and *Box-Jenkins* (BJ) [3].

4.3.2.1 Transfer Function Models

We first consider the output error model. Taking the z transform of the output error model under ideal conditions ($e(k) = 0$), we get a proper *stable* z-domain transfer function:

$$H(z) = \frac{a_0 + a_1 z^{-1} + \cdots + a_{n-1} z^{-(n-1)}}{1 + b_1 z^{-1} + \cdots + b_{n-1} z^{-(n-1)} + b_n z^{-n}},$$

where the coefficient of the z^0 term in the denominator is assumed to be unity without loss of generality and n is the order of the system model and the degree of the numerator polynomial has been assumed to be $(n-1)$. The above transfer function can also be written as the infinite series

$$H(z) = h_0 + h_1 z^{-1} + \cdots + h_k z^{-k} + h_{k+1} z^{-(k+1)} + \cdots$$

Let f, h represent column vectors of length $N(\gg n)$, where

$$f = \begin{bmatrix} f_0 & f_1 & \cdots & f_{N-1} \end{bmatrix}^T \text{ and}$$
$$h = \begin{bmatrix} h_0 & h_1 & \cdots & h_{N-1} \end{bmatrix}^T$$

denote, respectively, the measured and the actual values of the impulse response. The latter are also known as *Markov parameters* of the system. Then, the least-squares identification problem can be stated as

$$\min_{a,b} \|e\| = \min_{a,b} \left[\sum_{i=0}^{N-1} e_i^2 \right]^{1/2}$$

where $e = f - h$ and

$$a = \begin{bmatrix} a_0 & a_1 & \cdots & a_{n-1} \end{bmatrix}^T$$
$$b = \begin{bmatrix} 1 & b_1 & \cdots & b_n \end{bmatrix}^T \quad (4.11)$$

The transfer function coefficients are related to $H(z)$ as

$$\begin{bmatrix} a \\ \cdots \\ 0 \end{bmatrix} = \begin{bmatrix} H_1 \\ \cdots \\ H_2 \end{bmatrix} b \quad (4.12)$$

where a and b were defined in (4.11) and the elements of H_1 and H_2 are obtained by equating like powers of z^{-1} on the two sides in the following relationship

$$a_0 + a_1 z^{-1} + \cdots + a_{n-1} z^{-(n-1)} = (1 + b_1 z^{-1} + \cdots + b_{n-1} z^{-(n-1)} + b_n z^{-n})$$
$$\times (h_0 + h_1 z^{-1} + \cdots + h_r z^{-r} + h_{r+1} z^{-(r+1)} + \cdots)$$

Clearly, if b and H_1 are known, then a can be found from $a = H_1 b$. However, because of measurement errors, it is virtually impossible to determine the exact h. Hence, we replace H_1 and H_2 by matrices F_1 and F_2 formed from the measured impulse response data f. Since f is an estimate of h,

$$F_2 b = d(b)$$

where $d(b)$ is the equation error.

The elements b_1 to b_n of b can be found by minimizing $\|d(b)\|$. Subsequently, substituting b in (4.12), we can compute a. Further discussion and extensions of the above least-squares parameter identification scheme can be found in [6, 17, 23] and the references therein.

4.3.2.2 State Space Models

Although the identification of a transfer function model is mathematically rather straightforward, the resulting model can be sensitive to parameter variation. A small change in the coefficients of the model can affect the system response considerably. In general, state space models identified from the measured input-output data are less sensitive to small perturbations in their parameters. A state space model for a linear time-invariant discrete system with n states, m inputs, and p outputs can be described by

$$\begin{aligned} x(k+1) &= Ax(k) + Bu(k) \\ y(k) &= Cx(k) + Du(k), \end{aligned} \quad (4.13)$$

where A is an $(n \times n)$ state matrix, B is an $(n \times m)$ input matrix, C is a $(p \times n)$ output matrix, and D is a $(p \times m)$ input-output matrix. The vectors x, u, and y are known as the state, input, and output vectors, respectively. We now outline one popular approach to state space identification.

Assuming that $x(0)$ is zero, we can show the output of the system (4.13) to a known input sequence $u(0), u(1), \ldots,$ to be

$$\begin{aligned} y(0) &= Du(0) \\ y(1) &= CBu(0) + Du(1) \\ y(2) &= CABu(0) + CBu(1) + Du(2) \\ &\vdots \\ y(r-1) &= CA^{r-2}Bu(0) + CA^{r-3}Bu(1) + \cdots + CBu(r-2) + Du(r-1) \end{aligned}$$

These relations can be expressed in a matrix-vector form as

Section 4.3 Dynamical Models

$$[y(0) \quad y(1) \quad \cdots \quad y(r-1)] =$$

$$[H(0) \quad H(1) \quad \cdots \quad H(r-1)] \begin{bmatrix} u(0) & u(1) & \cdots & u(r-1) \\ 0 & u(0) & \cdots & u(r-2) \\ \vdots & \vdots & \ddots & \vdots \\ 0 & 0 & \cdots & u(0) \end{bmatrix}.$$

For the purpose of explanation, assume that the underlying system has a single input and a single output and that $u(0)$ is nonzero. We define the three matrices above as Y, H_r, and U, respectively, and we get $H_r = YU^{-1}$. Note that $D = H(0)$ and $CA^{i-1}B = H(i)$, $i = 1, 2, \ldots$, are the Markov parameters of the system [13]. The computed Markov parameters are then arranged in a *Hankel* matrix defined as

$$H_r = \begin{bmatrix} CB & CAB & \cdots & CA^{r-1}B \\ CAB & CA^2B & \cdots & CA^rB \\ \vdots & \vdots & \ddots & \vdots \\ CA^{r-1}B & CA^rB & \cdots & CA^{2(r-1)}B \end{bmatrix} = \begin{bmatrix} H(1) & H(2) & \cdots & H(r) \\ H(2) & H(3) & \cdots & H(r+1) \\ \vdots & \vdots & \ddots & \vdots \\ H(r) & H(r+1) & \cdots & H(2r-1) \end{bmatrix}.$$

The size of the Hankel matrix is increased with additional measured input-output data until its rank stops growing, that is, $n = \text{rank}(H_r) = \text{rank}(H_{r+1})$.

Once the Hankel matrix of rank n is found, the state space model of order n is constructed as follows:

- Perform the singular value decomposition on H_k as

$$H_k = \begin{bmatrix} U_{11} & U_{12} \\ U_{21} & U_{22} \end{bmatrix} \begin{bmatrix} S & 0 \\ 0 & 0 \end{bmatrix} \begin{bmatrix} V_{11} & V_{12} \\ V_{21} & V_{22} \end{bmatrix}^T.$$

- Define a modified Hankel matrix $\hat{H}_r$ as

$$\hat{H}_r = \begin{bmatrix} CAB & CA^2B & \cdots & CA^rB \\ CA^2B & CA^3B & \cdots & CA^{r+1}B \\ \vdots & \vdots & \ddots & \vdots \\ CA^rB & CA^{r+1}B & \cdots & CA^{2r}B \end{bmatrix}.$$

- Define matrices associated with the state space model as

$$A = S^{-1/2} \begin{bmatrix} U_{11} \\ U_{21} \end{bmatrix}^T \hat{H}_r \begin{bmatrix} V_{11} \\ V_{21} \end{bmatrix} S^{-1/2}$$

$$B = S^{1/2} V_{11}^T$$

$$C = U_{11} S^{1/2}$$

$$D = H(0).$$

Note that the techniques for model identification described here illustrate the basic principles. For more realistic situations, one must take into account several practical constraints [16, 20].

4.3.3 Model Reduction

Often, in their zeal to capture all possible details, engineers *overmodel* the physical system. This leads to fairly large-order system models. Naturally, the higher the dynamical order of the system model, the more complex and usually inefficient is the controller derived from it. It is, therefore, natural to seek lower order approximate models to closely describe the actual plant. This is known as the *model reduction problem*.

The earlier approaches to model reduction were based on truncation of less important states from state space models. Essentially, the relatively fast-decaying modes of the system can be ignored as their influence on the performance of the system is less pronounced than that of slowly decaying modes.

4.3.3.1 Modal Truncation

Given the state space description of a linear time invariant system,

$$\frac{dx}{dt} = Ax + Bu$$
$$y = Cx + Du,$$

we assume that the state matrix is in its *Jordan form*, that is, a matrix with elements along the main diagonal and possibly a few 1s along the first super diagonal when the state matrix A has dependent eigenvectors for repeated eigenvalues. Assume that A can be partitioned as

$$A = \begin{bmatrix} \Lambda_1 & 0 \\ 0 & \Lambda_2 \end{bmatrix},$$

where the elements along the diagonal are arranged in increasing magnitude of the negative real parts. Furthermore, the magnitude of the negative real part of the last element of Λ_1 is much smaller than the magnitude of the negative real part of the first element of Λ_2. Then, the modes in the block Λ_2 will decay much faster than those in Λ_1. The faster modes may be truncated to get a reduced-order model. Clearly, modes of the lower order approximation are a subset of modes of the original model. In addition, all reduced-order models obtained by modal truncation match perfectly with the original model at infinite frequency.

4.3.3.2 Singular Perturbation

Singular perturbation is a well-studied alternative to the modal method discussed above. Here again, the underlying principle is to partition the original system as a fast and a slow subsystem. The states of the fast subsystem are set to their steady-state values to obtain a reduced-order model. Briefly, the singular perturbation approach may be described as follows. Given the state space model:

Section 4.3 Dynamical Models

$$\frac{d}{dt}\begin{bmatrix} x_1(t) \\ \epsilon x_2(t) \end{bmatrix} = \begin{bmatrix} A_{11} & A_{12} \\ A_{21} & A_{22} \end{bmatrix}\begin{bmatrix} x_1(t) \\ x_2(t) \end{bmatrix} + \begin{bmatrix} B_1(t) \\ B_2(t) \end{bmatrix} u(t)$$

$$y(t) = \begin{bmatrix} C_1 & C_2 \end{bmatrix}\begin{bmatrix} x_1(t) \\ x_2(t) \end{bmatrix} + [D]u(t)$$

(4.14)

where ϵ is a small number [28]. Then setting $\epsilon = 0$, and assuming that A_{22} is invertible, yields

$$0 = A_{21}x_1(t) + A_{22}x_2(t) + B_2u(t) \quad \text{or,}$$

$$x_2(t) = -A_{22}^{-1}A_{21}x_1(t) - A_{22}^{-1}B_2u(t).$$

Substituting $x_2(t)$ in (4.14), one obtains a singular perturbation approximation model

$$\frac{dx_1(t)}{dt} = [A_{11} - A_{12}A_{22}^{-1}A_{21}]x_1(t) + [B_1 - A_{12}A_{22}^{-1}B_2]u(t)$$

$$y(t) = [C_1 - C_2A_{22}^{-1}A_{21}]x_1(t) + [D - C_2A_{22}^{-1}B_2]u(t).$$

Singular-perturbation-based order reduction yields models with a good match with the original system at low frequencies.

4.3.3.3 Balanced Reduction

Often the *overmodeled* system may have states that are either uncontrollable or unobservable. Because these states do not contribute to transfer of signals from the input to the output of the system, they may be eliminated from the system dynamics. In addition, there may be states that are *weakly* controllable and/or observable; that is, these states make a relatively small contribution in the transfer of signals from the input to the output. Internally balanced state space realizations have a basis such that each basis vector is equally controllable and observable. Moreover, it is possible to quantify the degree of controllability and observability of each basis vector for such realizations. Given a stable state space realization (A, B, C, D), one can obtain its controllability grammian (P) and observability grammian (Q), respectively, through the solution of Lyapunov equations

$$AP + PA^T + BB^T = 0$$

$$A^TQ + QA^T + C^TC = 0.$$

Then, the realization (A, B, C, D) is said to be internally balanced if $P = Q = \Sigma$ and

$$\Sigma = \begin{bmatrix} \sigma_1 I_{\ell_1} & 0 & \cdots & 0 \\ 0 & \sigma_2 I_{\ell_2} & \cdots & 0 \\ \vdots & \vdots & \vdots & \vdots \\ 0 & 0 & \cdots & \sigma_r I_{\ell_r} \end{bmatrix} = \begin{bmatrix} \Sigma_1 & 0 \\ 0 & \Sigma_2 \end{bmatrix}$$

where all diagonal elements can be ordered so that $\sigma_1 > \sigma_2 > \cdots > \sigma_r > 0$. The elements σ_i represent the degree of controllability and observability of each basis vector of the state space.

Assume that Σ can be split into two submatrices Σ_1 and Σ_2, with a clear difference between the relative values of the magnitudes of their diagonal elements. Then a reduced-order model may be obtained by

1. *Direct truncation*: This is easily achieved by splitting the state vector of balanced realization conformable to the dimensions of the two blocks, Σ_1 and Σ_2, and truncating the states that correspond to Σ_2 [8, 21].
2. *Singular perturbation approximation*: For some applications, direct truncation is not desirable because the corresponding reduced-order models incur the greatest approximation errors in the low-frequency range. Balanced singular perturbation ensures that the error in the low-frequency range is eliminated [7].

4.4 NONLINEAR DYNAMICAL SYSTEMS

As noted in the previous section, the following state description may be used to represent nonlinear dynamical systems

$$\frac{dx}{dt} = \hat{F}(x, u, t). \tag{4.15}$$

where the states in (4.15) are nonlinearly related. The system is referred to as *autonomous* if (4.15) does not have explicit dependence on time; otherwise it is said to be *nonautonomous*. Furthermore, if $u(t) = 0$, the system (4.15) is called *unforced*; otherwise it is a referred to as a *forced* system. Note that in the literature autonomous is often replaced by time-invariant and nonautonomous by time-varying.

Several inherent properties of linear systems, which make them easier to model and analyze, become invalid for nonlinear systems. Properties such as superposition and commutativity (the linearity of a cascade of two linear subsystems) do not carry over to nonlinear systems. The response of nonlinear systems is not as predictable as for linear systems; for example, sinusoidal excitation of linear systems, but not nonlinear ones, produces sinusoidal output with the same frequency as the excitation signal. Another striking difference is the number of equilibrium points. An equilibrium point of the system (4.15) is defined as a state in which the velocity of the state is zero; that is, for an autonomous system $F(x, t) = 0$. Linear systems have no more than a single *isolated* equilibrium point. In contrast, a nonlinear system may have multiple isolated equilibrium points. As a straightforward example, $\frac{dx}{dt} = 2\pi x$ has a single equilibrium point at the origin, whereas $\frac{dx}{dt} = x - x^2$ has equilibrium points at 0 and 1.

A typical feedback control system will have a plant and a controller, together with sensors and actuators. If all elements of these four components are linear, the resulting system is linear. If any one of the four components exhibits nonlinearity, the overall system becomes nonlinear. A viable approach to work with such systems is to identify, and where possible, model the nonlinearity.

Section 4.4 Nonlinear Dynamical Systems

4.4.1 Common Effects of Nonlinearities

Next, a few of the more commonly encountered effects of nonlinearities in control systems are reviewed.

Limit Cycles. Consider a simple harmonic oscillator represented by the following linear unforced system:

$$\frac{d^2x}{dt^2} + x = 0, \quad x(0) = x_0, \quad \left.\frac{dx}{dt}\right|_{t=0} = x_1.$$

Its solution to the initial conditions x_0 and x_1 is given by

$$x(t) = r\cos(t + \phi_0), \quad \frac{dx}{dt} = -r\sin(t - \phi_0)$$

where $r = (x_0^2 + x_1^2)^{1/2}$ and $\phi_0 = \tan^{-1}(x_0/x_1)$, which is periodic for any choice of nonzero initial conditions. Clearly, the solution represents a circle of radius r in the $(x, dx/dt)$-plane, also known as the *phase-plane*. Furthermore, the entire phase-plane is covered with periodic solutions.

On the other hand, nonlinear systems can exhibit periodic solutions that are *isolated*; that is, there exists a finite neighborhood around the periodic solution that does not contain any other periodic solutions. These isolated periodic solutions are called *limit cycles*. These are closed trajectories in the phase-plane, and a nearby trajectory would either converge to or diverge from the limit cycle. A limit cycle is defined as a *stable limit cycle* if the state trajectory converges to the limit cycle and an *unstable limit cycle* if the state trajectory diverges from the limit cycle. A nonlinear system may possess both stable and unstable multiple limit cycles as shown in Figure 4.4.

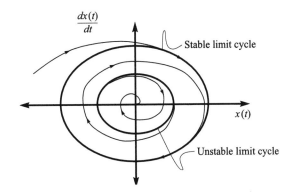

Figure 4.4 Multiple limit cycles: stable and unstable.

Bifurcation. As mentioned earlier, linear systems possess at most one isolated equilibrium point, whereas nonlinear systems may possess several isolated equilibrium points. Moreover, in nonlinear systems, if there is a parameter that is varying, then it is conceivable that the number of isolated equilibrium points may change as the parameter value changes. This phenomenon of qualitative change in system dynamics is known as *bifurcation*. The bifurcation point is the parameter value at which the change occurs. A famous example of a nonlinear system that exhibits bifurcation is the Duffing's equation, given by

$$\frac{d^2x}{dt^2} + \gamma \frac{dx}{dt} + \alpha x = \beta x^3.$$

The equilibrium point is at $x = 0$, if $\alpha/\beta < 0$. However, when $\alpha/\beta \geq 0$, the equilibrium point splits into three equilibrium points at 0 and $\pm\sqrt{\alpha/\beta}$. This type of bifurcation is known as a *pitchfork* bifurcation.

It is also possible that as a control parameter is varied, a stable fixed point becomes unstable to form a limit cycle or a stable limit cycle becomes unstable to a fixed point. This phenomenon is called *Hopf* bifurcation. A typical representation of pitchfork and Hopf bifurcation is shown in Figure 4.5.

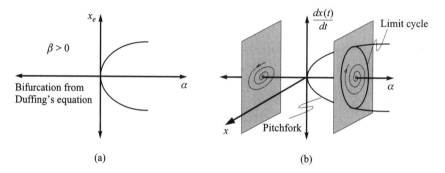

Figure 4.5 (a) Pitchfork (x_e: steady-state equilibrium point) and (b) Hopf bifurcation.

Chaos. Nonlinear dynamical systems can exhibit equilibrium behavior other than fixed points (points mapping onto themselves) and periodic (limit cycles). Specifically, some nonlinear systems exhibit aperiodic dynamics. The output can appear to be completely random, even though the system is deterministic. Such behavior is usually referred to as chaos and represents unpredictability of the system output. The same dynamical system can exhibit fixed point, periodic, and chaotic characteristics for different values of system parameters. Chaos often arises through changes in the system's qualitative behavior from fixed points to limit cycles, which then undergo a series of period doublings, to a chaotic regime, as a parameter of the system is steadily changed. Chaotic systems are also extremely sensitive to changes in initial conditions—two infinitesimally close initial conditions can result in arbitrarily different temporal evolution.

This is exhibited by Duffing's system with a sinusoidal input

$$\frac{d^2x}{dt^2} + \gamma \frac{dx}{dt} + \alpha x = \beta x^3 + \theta \cos \omega t.$$

If the system is simulated with two close but distinct initial conditions $(x(0), \frac{dx}{dt}|_{x=0}) = (x_1(0), x_2(0)) = (3, 4)$ and $(4.01, 4.01)$, it gives completely different responses [30]. The results of the two simulations are shown in Figure 4.6 in the (x_1, x_2) plane.

Section 4.4 Nonlinear Dynamical Systems

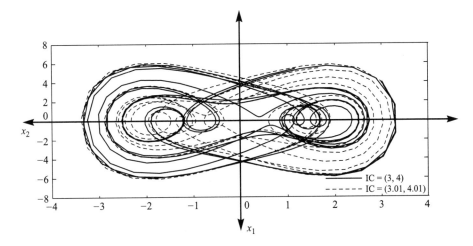

Figure 4.6 Chaotic behavior.

The chaos phenomenon is exhibited by a rich variety of physical systems such as turbulent flow, mechanical systems with backlash, and meteorological phenomena. An interesting aspect of this seemingly random variation in the output response is the fact that chaos is strictly deterministic in nature. Chaotic systems are inherently unpredictable despite being deterministic.

Hysteresis. Hysteresis is a nonlinear phenomenon commonly observed in the backlash of gear trains, mechanical bearings, and a chain of cars connected through links in a train. Mechanical backlash is the motion lost when the direction of motion is reversed. Hysteresis is also common in electromagnetic applications where the flux density follows different paths depending on whether the magnetization force is increasing or decreasing. Typical examples are shown in Figure 4.7.

In addition, there are other behaviors such as *saturation, dead zone, subharmonic generation, discretization, asynchronous quenching,* and *jump hysteresis* that add to the richness of nonlinear models [29, 30].

A commonly employed technique to characterize models and visualize their behavior is through *phase-plane* representation. A phase-plane represents a two-dimensional state space for a given second-order dynamical system. Each second-order system may

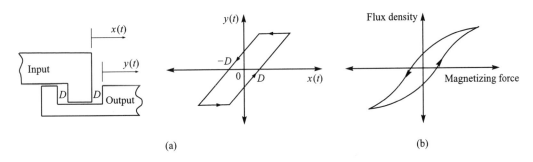

Figure 4.7 (a) Backlash and (b) hysteresis.

be modeled as a system of two coupled first-order systems with two states. Specifically, writing a second-order system as

$$\frac{dx_1}{dt} = f_1(x_1, x_2)$$
$$\frac{dx_2}{dt} = f_2(x_1, x_2),$$

we get

$$\frac{dx_2}{dx_1} = \frac{f_2(x_1, x_2)}{f_1(x_1, x_2)}.$$

The corresponding trajectories when plotted in the (x_1, x_2) plane yield a phase plane. The trajectories may be drawn using a slope at each point of the phase trajectory or via the method of isoclines [29]. The phase-plane trajectories can now be used to study properties such as limit cycles, chaos, and bifurcation. A major drawback of this approach is the difficulty in visualizing the solution in dimensions greater than two.

4.4.2 Linearization

In general, the qualitative behavior of a nonlinear system at a point near one of its equilibrium points is adequately captured by its linearized approximation at that point. Hence, it is natural to consider ways to obtain a linearized representation of nonlinear systems. A linear model may be obtained as a local linearized approximation model obtained through Taylor series expansion or (more recently) as a global linearized model obtained through feedback linearization, where the former is an approximate representation and the latter is exact. In addition, we may obtain reasonably accurate linear models of some nonlinearities through a Fourier series based approximation technique. In the rest of this section, these models are briefly discussed.

4.4.2.1 Local Linearized Approximation

We assume that the system is expressed by the following nonlinear vector differential equations

$$\frac{dx}{dt} = F(x, u) \qquad (4.16)$$

where x is the state vector of length n and u is the input vector of length m. Furthermore, F is a differentiable vector function of x and u. Clearly, its isolated equilibrium points can be determined from $F(x, 0) = 0$. By a shift of axis, an equilibrium point can be made to lie on the origin in the shifted system. Specifically, if an isolated equilibrium point lies at $x = x_0$, then the system may be modified to $\frac{dz}{dt} = F_z(z, u)$ with an isolated equilibrium point at the origin of the new state space. Hence we can assume, without any loss of generality, that $x = 0$ is an equilibrium point of the system.

Next, using Taylor series expansion of $F(x, u)$ about $x = 0$, $u = 0$ gives

Section 4.4 Nonlinear Dynamical Systems

$$F(x, u) = F(0, 0) + \left\{\frac{\partial F}{\partial x}\right\}_{x=0, u=0} x + \left\{\frac{\partial F}{\partial u}\right\}_{x=0, u=0} u + \frac{1}{2!}\left\{\frac{\partial^2 F}{\partial u^2}\right\}_{x=0, u=0} x^2$$

$$+ \frac{1}{2!}\left\{\frac{\partial^2 F}{\partial u^2}\right\}_{x=0, u=0} u^2 + \frac{2}{2!}\left\{\frac{\partial^2 F}{\partial x \partial u}\right\}_{x=0, u=0} xu + \cdots.$$

Neglecting terms beyond first derivatives and noting that by definition, $F(0, 0) = 0$, the linearized *approximation* of the system (4.16) may be expressed as the linear state variable model

$$\frac{dx}{dt} = Ax + Bu,$$

where A and B are, respectively, $n \times n$ and $n \times m$ Jacobian matrices defined as

$$A = \left[\frac{\partial F}{\partial x}\right]_{x=0, u=0} \quad \text{and} \quad B = \left[\frac{\partial F}{\partial u}\right]_{x=0, u=0}.$$

This seemingly straightforward method has one major drawback. The dynamics represented by it are valid only in a small neighborhood of the equilibrium point. Thus, the method becomes less appealing for highly nonlinear systems.

4.4.2.2 Describing Function Approximation

Use of describing functions attempts to extend transfer function-based modeling, analysis, and control techniques from linear to nonlinear systems. Essentially, a describing function gives a linear approximation of a nonlinear system. The premise is that if a sinusoidal input is applied to a system with a nonlinearity (symmetric about the origin) and it is assumed that the output has the same fundamental frequency as the input, then provided that the term containing the fundamental frequency is most significant, all higher order harmonics, subharmonics, and the dc component may be ignored [29]. The resulting model will be a reasonable linear approximation of the nonlinearity.

In performing describing function analysis, it is assumed that there is only one nonlinearity in the system. If the system contains more than one nonlinearity, they are all lumped into a single one. Clearly, the approach has some limitations, yet the simplicity afforded by the method makes it quite popular for modeling nonlinearities. A common procedure to obtain the describing function model of a nonlinearity is to find the Fourier series representation of the output and use the fundamental harmonic to construct the describing function.

Let the input to the nonlinear element $N(M, \omega)$ be

$$x(t) = M \sin \omega t$$

and let the steady-state output of the nonlinear device be given by

$$y(t) = \sum_{\ell=1}^{\infty} N_\ell \sin(\omega t + \phi_\ell). \tag{4.17}$$

with the $\ell = 0$ term being zero and $\ell = 1$ representing the term with the most significant contribution. Then, by definition, the describing function of the nonlinearity is

$$N(M, \omega) = \frac{N_1}{M} e^{j\phi_1}. \tag{4.18}$$

Describing function descriptions are obtained by Fourier series expansion of the output waveform emerging from the nonlinear element when excited by a sinusoidal input with a fixed frequency. Clearly, (4.17) may be expressed as a trigonometric Fourier series expansion

$$y(t) = a_0 + \sum_{k=1}^{\infty} a_k \cos(k\omega t) + b_k \sin(k\omega t)$$

where a_k and b_k are the Fourier series coefficients computed as

$$a_0 = \frac{1}{T} \int_{-T/2}^{T/2} y(t) d(\omega t)$$

$$a_k = \frac{2}{T} \int_{-T/2}^{T/2} y(t) \cos(k\omega t) d(\omega t)$$

$$b_k = \frac{2}{T} \int_{-T/2}^{T/2} y(t) \sin(k\omega t) d(\omega t)$$

where T is the period of the input signal. Then the describing function defined in (4.18) can be reduced to the complex expression

$$N(M, \omega) = \frac{b_1}{M} + j\frac{a_1}{M} = \sqrt{\left(\frac{a_1}{M}\right)^2 + \left(\frac{b_1}{M}\right)^2} \angle \tan^{-1}(a_1/b_1).$$

A reasonably large variety of nonlinearities satisfy the following conditions: (a) the nonlinearity exhibits odd function behavior, that is, $f(-t) = -f(-t)$, (b) there is only one nonlinear component in the system, (c) the nonlinearity is time invariant, and (d) all higher-order harmonics are filtered owing to the low-pass property of the controller in the feedback configuration. Therefore in many, *but not all*, cases the fundamental term is the only significant component of the output $y(t)$, justifying the corresponding describing function as a reasonable approximation of the underlying nonlinearity. Describing functions for several of the nonlinearities discussed above may be found in advanced control texts, for example, [29] and [30].

4.4.2.3 Feedback Linearization

Both local linearizations and describing functions lead to approximate linear models. Although quite adequate for some applications, they fall short in others because of their inherent limitations. An alternative to approximate linearization is feedback linearization. This approach to obtain linearized models has garnered considerable attention in the last few years. Naturally, the feedback linearization approach has its own limitations and shortcomings, and overcoming them is very much a topic of current research. Feedback linearization is based on differential geometry and requires some mathematical sophistication; hence readers are referred to [11, 14, 19, 30] and other texts.

4.5 MODELS OF DISTRIBUTED PARAMETER SYSTEMS

In the models discussed thus far, the dynamical changes were limited to the time variable. A very large class of physical systems lends itself to a change of dynamical behavior in both time and one or several space variables. Such systems are modeled using partial differential equations (PDEs). As a quick illustration, we model the vibration of a clamped string. To simplify matters, it is assumed that the string is homogeneous and perfectly elastic, that gravitational forces compared to lateral tension on the string are negligibly small, and that every infinitesimally small section of the string traverses in a vertical plane. Using the notation in Figure 4.8, it can be easily seen that

$$T_A \cos \alpha = T_B \cos \beta = T$$

where T_A and T_B are tensions at the endpoints of the segment under consideration and T is a constant. Note that the net lateral tension is zero since there is no lateral deflection by assumption. By Newton's law, the net vertical tension is

$$T_B \sin \beta - T_A \sin \alpha = \rho \Delta x \frac{\partial^2 f}{\partial t^2}$$

where $\rho \Delta x$ is the mass of the segment, $f(x, t)$ is the vertical displacement, and $\frac{\partial^2 f}{\partial t^2}$ is the acceleration. Dividing the above equation by T, we get the force equation

$$\frac{\rho \Delta x}{T} \frac{\partial^2 f}{\partial t^2} = \tan \beta - \tan \alpha \qquad (4.19)$$

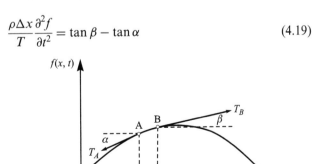

Figure 4.8 Deflection in a stretched string.

where $\tan\alpha$ and $\tan\beta$ are the slopes at x and $x + \Delta x$, respectively. Since $f(x, t)$, the vertical displacement, is a function of x as well as t, we have

$$\tan\alpha = \left.\frac{\partial f}{\partial x}\right|_x, \quad \tan\beta = \left.\frac{\partial f}{\partial x}\right|_{x+\Delta x}.$$

Rewriting the force equation (4.19), we get

$$\frac{\rho}{T}\frac{\partial^2 f}{\partial t^2} = \frac{1}{\Delta x}\left[\left.\frac{\partial f}{\partial x}\right|_{x+\Delta x} - \left.\frac{\partial f}{\partial x}\right|_x\right].$$

In the limit, as $\Delta x \to 0$, the force equation becomes the partial differential equation

$$\frac{\partial^2 f}{\partial t^2} = \frac{T}{\rho}\frac{\partial^2 f}{\partial x^2}.$$

4.5.1 Classification of PDEs

Depending on the underlying process, partial differential equations may be linear or nonlinear. To keep the discussion concise, we limit the scope to only linear PDEs. The following classification, though not exhaustive, captures a very large class of practical models. The classification is based on special cases of the general three-dimensional convection-diffusion equation or the advection-diffusion equation, given by

$$\frac{\partial f}{\partial t} + U_x\frac{\partial f}{\partial x} + U_y\frac{\partial f}{\partial y} + U_z\frac{\partial f}{\partial z} = \kappa\left[\frac{\partial^2 f}{\partial x^2} + \frac{\partial^2 f}{\partial y^2} + \frac{\partial^2 f}{\partial z^2}\right], \quad \kappa \geq 0 \quad \Leftrightarrow$$

$$\frac{\partial f}{\partial t} + U \cdot \nabla f = \kappa \nabla^2 f, \quad \text{where,}$$

$$U \cdot \nabla f = U_x\frac{\partial f}{\partial x} + U_y\frac{\partial f}{\partial y} + U_z\frac{\partial f}{\partial z}$$

$$\nabla^2 f = \frac{\partial^2 f}{\partial x^2} + \frac{\partial^2 f}{\partial y^2} + \frac{\partial^2 f}{\partial z^2}.$$

(4.20)

Furthermore, $f = f(x, y, z, t)$ is a function of the three space variables as well as time, and U is the velocity vector (U_x, U_y, U_z). The three scalar elements in U may be functions of one or more of the function f, the spatial variables x, y, z, and the temporal variable t. If U is a constant or an explicit function of (x, y, z) and t, we obtain a *linear* convection-diffusion equation. On the other hand, if U also depends on f, then the corresponding convection-diffusion equation is *nonlinear*.

Depending on the underlying physical problem, some of the components in (4.20) may be zero, leading to special forms of a convection-diffusion equation [25]:

Parabolic PDEs. When all three components of the velocity vector U are uniformly zero, Eq. (4.20) simplifies to

$$\frac{\partial f}{\partial t} = \kappa \nabla^2 f. \tag{4.21}$$

Equation (4.21) represents a prototypical parabolic partial differential equation in spatial and temporal variables. It is also known as the *unsteady diffusion equation*.

Elliptic PDEs. When all three components of the velocity vector U are uniformly zero, and in addition f is independent of t, Eq. (4.21) simplifies to

$$\nabla^2 f = 0. \tag{4.22}$$

Equation (4.22) represents a prototypical elliptic partial differential equation in spatial variables. It is also known as *Laplace's equation*. Following are more general formulations of elliptic PDEs:

$$\nabla^2 f = g(x, y, z) \quad \text{Poisson's equation}$$

$$\nabla^2 f + h(x, y, z)f = g(x, y, z) \quad \text{Helmholtz's equation.}$$

Hyperbolic PDEs. When the diffusivity $\kappa = 0$, Eq. (4.20) simplifies to

$$\frac{\partial f}{\partial t} + U \cdot \nabla f = 0. \tag{4.23}$$

Equation (4.23) represents a prototypical hyperbolic partial differential equation in spatial and temporal variables. It is also known as the *convection equation*.

It is easily seen that the above classification is based on three-dimensional flow models. If the flow is limited to two of the three spatial variables, then Eq. (4.20) represents two-dimensional flow models. Similarly, if the flow is restricted to a single spatial variable, then (4.20) represents one-dimensional flow models.

4.5.2 Finite Difference Models of PDEs

Because of their inherently complex nature, the best technique for solution of most PDEs is numerical. To obtain a numerical solution, the solution space of the PDE is discretized into rectangles of appropriate dimensions for time and one-space dimensional problems (or two-space dimensional problems), rectangular parallelepipeds for time and two-space dimensional problems (or three-space dimensional problems), and rectangular *hyper* parallelepipeds in higher dimensions. Development of finite difference models is most easily illustrated by means of time and one-space dimension, for example, the one-dimensional diffusion equation

$$\frac{\partial f(x, t)}{\partial t} = \kappa \frac{\partial^2 f(x, t)}{\partial x^2}. \tag{4.24}$$

Discretization of its solution space for $x_0 \leq x \leq x_f$ and $t^0 \leq t \leq t^f$ involves dividing the time and space axes into *small* steps. Step sizes are governed by the nature of the problem and desired accuracy. If the solution space exhibits large variations, it is advisable to use small step sizes and vice versa. The discretized solution space for the above problem is shown in Figure 4.9.

Based on the approximation model used for derivatives in the PDE, we get several distinct finite difference models of the system described by the PDEs.

4.5.2.1 Explicit Models (Forward Differences)

We discretize the spatial axis $x \in (x_0, x_f)$ in steps of Δx and denote the point $x_0 + (i-1)\Delta x$ by x_i. Similarly, we discretize the temporal axis with increments Δt such that t^j denotes $t^0 + j\Delta t$. Then, the point f_i^j represents the solution at the point (x_i, t^j) of the solution space (as shown in Figure 4.9). Using finite difference approximations of derivatives in (4.24),

$$\frac{\partial f(x,t)}{\partial t} = \frac{f_i^{j+1} - f_i^j}{\Delta t} + \mathcal{O}(\Delta t)$$
$$\frac{\partial^2 f(x,t)}{\partial x^2} = \frac{f_{i+1}^j - 2f_i^j + f_{i-1}^j}{(\Delta x)^2} + \mathcal{O}((\Delta x)^2), \qquad (4.25)$$

we get a finite difference model that is first-order accurate in time and second-order accurate in space. In (4.25), the temporal variable derivative is approximated by *forward differences* and the space derivative by *central differences*. The corresponding stencil is shown in Figure 4.9. The model can be solved for f_i^{j+1} as

$$f_i^{j+1} = \alpha f_{i-1}^j + (1 - 2\alpha)f_i^j + \alpha f_{i+1}^j, \quad \alpha = \frac{\kappa \Delta t}{(\Delta x)^2}.$$

Since this is a discrete system, for stability the eigenvalues of the state matrix must be less than unity in magnitude. Note that it is also possible to use a forward difference approximation of the time derivative with second-order accuracy in time.

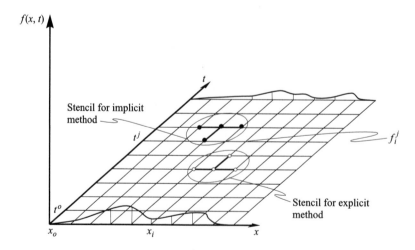

Figure 4.9 Discretization of space-time solution space.

4.5.2.2 Implicit Models (Backward Differences)

Explicit models based on forward differences are very efficient to solve because they require a simple sparse matrix multiplication. However, they are not unconditionally stable; stability depends on the step size chosen to discretize the grid. An alternative approach is to use backward differences in the temporal variable and central differences in the spatial variable. As will be seen shortly, to propagate the solution, this approach requires solution of a sparse system of equations.

Assume that the point under consideration is (x_i, t^{j+1}). Approximating the PDE at this point using backward differences in temporal and central differences in spatial variables, we get

$$\frac{f_i^{j+1} - f_i^j}{\Delta t} + \mathcal{O}(\Delta t) = \kappa \frac{f_{i+1}^{j+1} - 2f_i^{j+1} + f_{i-1}^{j+1}}{(\Delta x)^2} + \mathcal{O}((\Delta t)^2). \tag{4.26}$$

On rearranging Eq. (4.26), we get

$$-\alpha f_{i-1}^{j+1} + (1 + 2\alpha)f_i^{j+1} - \alpha f_{i+1}^{j+1} = f_i^j. \tag{4.27}$$

The stencil that corresponds to (4.27) is shown in Figure 4.9. It has been shown that the implicit model in (4.27) is *unconditionally stable*. Furthermore, to improve the accuracy in the temporal variable to $\mathcal{O}((\Delta t)^2)$, one could use the Crank-Nicholson method [25].

Following a similar approach, it is easy to see that both implicit and explicit finite difference models for other PDEs can be obtained. However, one must verify that the solution will converge by ensuring that the spectral radius of the state matrix of the underlying dynamical system is less than unity.

4.6 MACROMODELS: SCOPE AND FUTURE

Although the discussion thus far has been limited to techniques for a specific type of system models, it would only be appropriate to finally turn our attention to more general questions. It is true that, to a large extent, present technology permits the engineer to arrive at an acceptably accurate model of a specific component of the system. We refer to these models as *micromodels*. A reasonably complex system may consist of a generous mix of components that may be linear, nonlinear, time driven, event driven, and so forth. From a modeling viewpoint, a complex system has a highly amorphous structure. Therefore, no single modeling technique is either suited or sufficient for representation of a system. A challenging aspect of research in modeling always has been and continues to be tactical integration of information (component models) to achieve a specific objective. We will refer to these heterogeneous systems as *macromodels*.

Perhaps it would be fair to say that because of the nature of the subject, the field of communication networks has had to deal with coordination and control of distributed information the most. It is also interesting to note that there is much commonality between some of the macromodeling techniques proposed in the control system literature and those in use in communication networks. Some commonly used architectures for local area networks in communication network systems are shown in Figure 4.10.

Over the years, several macromodels have been proposed in the systems literature. Most continue to be used in their largely original form with suitable modifications or enhancements as warranted by the system or situation under consideration. These include hierarchical models, decentralized models, multimodels, and so on.

Hierarchical Models. Hierarchical models represent a top-down control/coordination protocol. There may be several levels of hierarchy within a system. The modules at each level of hierarchy control the modules at the next lower level and are in turn controlled by the modules at the next higher level. The decision-making structure is pyramidal; the various levels of hierarchy in the system share information vertically. It is also evident that the decision process has a larger significance when made at higher levels of hierarchy. Hierarchical models correspond to the *star* topology in communication networks. A mathematically sophisticated treatment of hierarchical control may be found in [18].

Decentralized Models. A decentralized model represents a more distributed information and control structure. Although the system remains structurally intact, its output information is shared among several controllers. These controllers collectively contribute to the control of the system. Unlike hierarchical models, where the structure is top-down, the structure of decentralized models is more lateral. The control decisions are made locally, and the effect of local control decisions *may* be coordinated through a centralized coordination module. The model provides sufficient flexibility for reorganization and lends itself naturally to building redundancy into the control system. In decentralized models, although the information is easily accessible throughout the system, the control architecture is decentralized. In fact, the local controllers are given considerable autonomy. The *tree* protocol in communication networks has considerable commonality with decentralized systems, where the system buses carry the information and subsystems are controlled by local controllers. The reader is referred to [12] for more details on decentralized models.

Supervisory Control Models. Hierarchical and decentralized models have a unique characteristic: They provide a uniform structure to the entire system model. This prop-

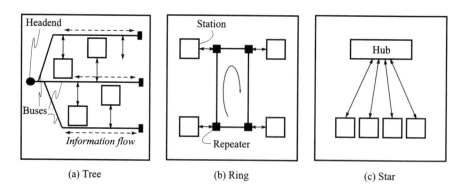

Figure 4.10 Common network topologies.

erty can be a strong point for situations where there is little or no ambiguity about micromodels and leads to mathematically well-defined models. On the other hand, if the system is sufficiently complex, it may not be the best strategy to use a uniform modeling technique throughout the system. Supervisory control models provide a framework to address the situation where a multitude of nonuniform micromodels represent low-level systems. The task of the supervisory control module is then to assign an appropriate weight to each micromodel and to develop a composite model output to achieve the control objective.

The control flow outlined in Figure 4.11 represents the general framework within which supervisory control models may operate. For more information on supervisory control (also referred to as a multiple model approach), the readers are referred to [22].

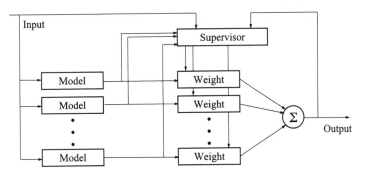

Figure 4.11 Supervisory control.

4.7 REMARKS

The preceding discussion was deliberately limited to deterministic models. In addition to models and techniques discussed in this chapter, there are mathematically mature techniques for several other modeling-related issues. For example, an extensive literature addresses the problem of *stochastic* modeling, briefly alluded to in Section 4.3.2, where an explicit representation of random uncertainty is taken into account.

As was mentioned at the very beginning of this chapter, the field of modeling is vast and ever growing. It would be a daunting task to encapsulate the entire subject into a few short pages. The primary purpose of this chapter was to expose the reader to essentials of modeling rather than to provide an exhaustive treatment of models and modeling techniques. Some of the topics discussed in this chapter are covered in considerably greater detail in related chapters listed in the box below. For the rest, the reader is urged to refer to the references listed at the end of the chapter and to research journals, where this ever-evolving subject gets continued attention.

ACKNOWLEDGMENT

The author gratefully acknowledges numerous constructive suggestions by Tariq Samad in writing this chapter.

Related Chapters

- A discussion on models for discrete-event systems is included in Chapter 2.
- Models for systems that combine continuous-time and discrete-event behavior are described in Chapter 7.
- A detailed treatment of nonlinear approximators as system models can be found in Chapter 6.
- Compositional approaches to system modeling that rely on computational tools are presented in Chapter 3.

REFERENCES

[1] D. Y. Abramovitch and L. G. Bushnell, "Report on the fuzzy versus conventional control debate." *IEEE Control Systems Magazine*, Vol. 19, no. 3, pp. 88–91, 1999.

[2] A. E. Bryson, *Dynamic Optimization*. Reading, MA: Addison-Wesley, 1999.

[3] P. E. Caines, *Linear Stochastic Systems*. New York: John Wiley & Sons, 1988.

[4] J. E. Dennis and R. B. Schnabel, *Numerical Methods for Unconstrained Optimization and Nonlinear Equations*. Englewood Cliffs, NJ: Prentice-Hall, 1983.

[5] R. C. Dorf and R. H. Bishop, *Modern Control Systems*. Reading, MA: Addison-Wesley, 1998.

[6] A. G. Evans and R. Fischl, "Optimal least squares time-domain synthesis of digital filters." *IEEE Trans. Audio and Electroacoustics*, Vol. AEA-21, pp. 61–65, 1973.

[7] K. V. Fernando and H. Nicolson, "Singular perturbation model reduction of balanced systems." *IEEE Trans Automat. Control*, Vol. AC-27, pp. 466–468, 1982.

[8] K. Glover, "All optimal Hankel norm approximations of linear multivariable systems, and their L_∞ error bounds." *Int. J. Contr.*, Vol. 39, pp. 1115–1193, 1984.

[9] G. H. Golub and C. Van Loan, *Matrix Computations*, 2nd ed. Baltimore, MD: Johns Hopkins University Press, 1989.

[10] W. J. Palm III, *Modeling, Analysis and Control of Dynamic Systems*. New York: John Wiley & Sons, 1983.

[11] A. Isidori, *Nonlinear Control Systems*. New York: Springer-Verlag, 1989.

[12] M. Jamshidi, *Large Scale Systems—Modeling, Control and Fuzzy Logic*. Englewood Cliffs, NJ: Prentice-Hall, 1996.

[13] T. Kailath, *Linear Systems*. Englewood Cliffs, NJ: Prentice-Hall, 1980.

[14] H. K. Khalil, *Nonlinear Systems*. New York: Macmillan, 1992.

[15] B. Kuo, *Automatic Control Systems*. Englewood Cliffs, NJ: Prentice Hall, 1995.

[16] L. Ljung, *System Identification: Theory for the User*. Englewood Cliffs, NJ: Prentice-Hall, 1987.

[17] L. E. McBride, H. W. Schafgen, and K. Steiglitz, "Time-domain approximation by iterative methods." *IEEE Trans. Circuit Theory*, Vol. CT-13, pp. 318–387, 1966.

[18] M. C. Mesarovic, D. Macko, and Y. Takahara, *Theory of Hierarchical Multilevel Systems*. New York: Academic Press, 1970.

References

[19] R. R. Mohler, *Nonlinear Systems, Vol. 1, Dynamics and Control*. Englewood Cliffs, NJ: Prentice-Hall, 1991.

[20] M. Moonen and J. Vandewalle, "A QSVD approach to on- and off-line state-space identification." *Int. J. Contr.*, Vol. 51, pp. 1133–1146, 1990.

[21] B. C. Moore, "Principal component analysis in linear systems: Controllability, observability, and model reduction." *IEEE Trans. Automat. Control*, Vol. AC-26, pp. 17–31, 1981.

[22] R. Murray-Smith and T. A. Johansen, *Multiple Model Approaches to Modelling and Control*. London: Taylor & Francis Ltd., 1997.

[23] J. P. Norton, *An Introduction to Identification*. New York: Academic Press, 1986.

[24] K. Ogata, *Modern Control Engineering*. Englewood Cliffs, NJ: Prentice Hall, 1997.

[25] C. Pozrikidis, *Numerical Computation in Science and Engineering*. London: Oxford University Press, 1998.

[26] W. H. Press, S. A. Teukolsky, W. T. Vetterling, and B. P. Flannery, *Numerical Recipes in C, The Art of Computer Programming*. Cambridge, MA: Cambridge University Press, 1997.

[27] T. Samad, "Complexity management: Multidisciplinary perspectives on automation and control." Technical Report CON-R98-001, Honeywell Technology Center, 1998.

[28] V. Saxena, J. O'Reilly, and P. V. Kokotovic, "Singular perturbation and time scale methods in control theory: Survey 1976–1983." *Automatica*, Vol. 20, pp. 272–293, 1984.

[29] S. M. Shinners, *Modern Control System Theory and Design*. New York: John Wiley & Sons, 1992.

[30] J. J. E. Slotine and W. Li, *Applied Nonlinear Control*. Englewood Cliffs, NJ: Prentice Hall, 1991.

[31] P. P. J. van den Bosch and A. C. van der Klauw, *Modeling, Identification and Simulation of Dynamical Systems*. Boca Raton, FL: CRC Press, 1994.

Chapter 5
INTELLIGENT CONTROL: AN OVERVIEW OF TECHNIQUES

Kevin M. Passino

Editor's Summary

In many established fields, the label "intelligent" heralds new developments that take issue with some traditional assumptions in research. In the case of intelligent control, an explicit attempt is made to draw inspiration from nature, biology, and artificial intelligence, and a methodology is promoted that is more accepting of heuristics and approximations—and is less insistent on theoretical rigor and completeness—than is the case with most research in control science.

Beyond such general and abstract features, succinct characterizations of intelligent control are difficult. Extensional treatments are an easier matter. Fuzzy logic, neural networks, genetic algorithms, and expert systems constitute the main areas of the field, with applications to nonlinear identification, nonlinear control design, controller tuning, system optimization, and encapsulation of human operator expertise. Intelligent control is thus no narrow specialization; it furnishes a diverse body of techniques that potentially addresses most of the technical challenges in control systems. It is also important to emphasize that intelligent control is by no means methodologically opposed to theory and analysis. Chapter 6 of this book, for example, discusses some theoretical results for neural networks and fuzzy models as nonlinear approximators.

Introductory tutorials to the key topics in intelligent control are provided in this chapter. No prior background in these topics is assumed. Examples from ship maneuvering, robotics, and automotive diagnostics help motivate the discussion. (Other chapters in this volume, notably Chapter 16, also outline applications of intelligent control.) General observations on autonomy and adaptation—two characteristics that are often considered essential to any definition of intelligence—are also included.

Kevin Passino is an associate professor in the Department of Electrical Engineering at The Ohio State University, past chair of the IEEE-CSS Technical Committee on Intelligent Control, and current vice president of Technical Activities for CSS.

5.1 INTRODUCTION

Intelligent control achieves automation via the emulation of biological intelligence. It either seeks to replace a human who performs a control task (e.g., a chemical process operator), or it borrows ideas from how biological systems solve problems and applies them to the solution of control problems (e.g., the use of neural networks for control). In this chapter we provide an overview of several techniques used for intelligent control and discuss challenging industrial application domains where these methods may provide particularly useful solutions.

This chapter should be viewed as a resource for those who are in the early stages of *considering* the development and implementation of intelligent controllers for industrial applications. It is impossible to provide the full details of a field as large and diverse as intelligent control in a single chapter. Hence, the focus here is on presenting the main ideas that have been found most useful in industry. Examples of how these methods have been used are given, and references for further study are provided.

The chapter begins with a brief overview of the main (popular) areas in intelligent control, notably, fuzzy control, neural networks, expert and planning systems, and genetic algorithms. In addition, complex intelligent control systems, in which the goal is to achieve autonomous behavior, are summarized. In each case, applications are used to motivate the need for the technique. Moreover, we explain in broad terms how to apply the methods to challenging problems. We summarize the advantages and disadvantages of the approaches and provide some comparative analyses with conventional control methods.

Overall, this chapter should be viewed as a practitioner's *first* introduction to intelligent control. The focus is on challenging problems and their solutions. The reader should be able to gain novel ideas about how to solve challenging problems and will find resources to carry these ideas to fruition.

5.2 INTELLIGENT CONTROL TECHNIQUES

In this section we provide brief overviews of the main areas of intelligent control. The objective here is not to provide a comprehensive treatment; rather, we seek only to present the basic ideas to give a flavor of the approaches.

5.2.1 Fuzzy Control

Fuzzy control is a methodology that represents and implements a (smart) human's knowledge about how to control a system. A fuzzy controller is shown in Figure 5.1; it has several components:

- The rule base is a set of rules about how to control.
- Fuzzification is the process of transforming the numeric inputs into a form that can be used by the inference mechanism.

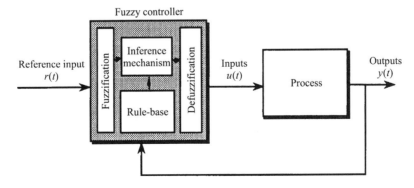

Figure 5.1 Fuzzy control system.

- The inference mechanism uses information about the current inputs (formed by fuzzification), decides which rules apply in the current situation, and forms conclusions about what the plant input should be.
- Defuzzification converts the conclusions reached by the inference mechanism into a numeric input for the plant.

5.2.1.1 Fuzzy Control Design

As an example, consider the tanker ship-steering application in Figure 5.2 in which the ship is traveling in the x direction at a heading ψ and is steered by the rudder input δ. Here, we seek to develop the control system in Figure 5.3 by specifying a fuzzy controller that would emulate how a ship captain would steer the ship. Here, if ψ_r is the desired heading, $e = \psi_r - \psi$ and $c = \dot{e}$.

The design of the fuzzy controller essentially amounts to choosing a set of rules (rule base) in which each rule represents the captain's knowledge about how to steer. Consider the following set of rules:

1. If e is **neg and** c is **neg then** δ is poslarge.
2. If e is **neg and** c is **zero then** δ is possmall.
3. If e is **neg and** c is **pos then** δ is zero.
4. If e is **zero and** c is **neg then** δ is possmall.
5. If e is **zero and** c is **zero then** δ is zero.
6. If e is **zero and** c is **pos then** δ is negsmall.
7. If e is **pos and** c is **neg then** δ is zero.
8. If e is **pos and** c is **zero then** δ is negsmall.
9. If e is **pos and** c is **pos then** δ is neglarge.

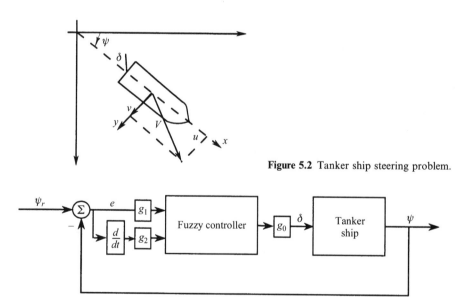

Figure 5.2 Tanker ship steering problem.

Figure 5.3 Control system for tanker.

Here, "neg" means negative, "poslarge" means positive and large, and the others have analogous meanings. What do these rules mean? Rule 5 says that the heading is good, so let the rudder input be zero. For Rule 1:

- "e is neg" means that ψ is greater than ψ_r.
- "c is neg" means that ψ is moving away from ψ_r (if ψ_r is fixed).
- In this case we need a large positive rudder angle to get the ship heading in the direction of ψ_r.

The other rules can be explained in a similar fashion.

What, precisely, do we (or the captain) mean by, for example, "e is pos," or "c is zero," or "δ is poslarge"? We quantify the meanings with "fuzzy sets" ("membership functions"), as shown in Figure 5.4. Here, the membership functions on the e axis (called the e "universe of discourse") quantify the meanings of the various terms (e.g., "e is pos"). We think of the membership function having a value of 1 as meaning "true," while a value of 0 means "false." Values of the membership function in between 0 and 1 indicate "degrees of certainty." For instance, for the e universe of discourse the triangular membership function that peaks at $e = 0$ represents the (fuzzy) set of values of e that can be referred to as "zero." This membership function has a value of 1 for $e = 0$ (i.e., $\mu_{zero}(0) = 1$) which indicates that we are absolutely certain that for this value of e we can describe it as being "zero." As e increases or decreases from 0, we become less certain that e can be described as "zero," and when its magnitude is greater than π we are absolutely certain that it is *not* zero, so the value of the membership function is zero. The meaning of the other two membership functions on the e universe of discourse (and the membership functions on the change-in-error universe of discourse) can be

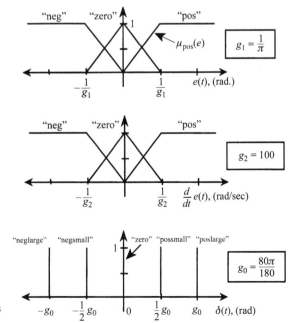

Figure 5.4 Membership functions for inputs and output.

described in a similar way. The membership functions on the δ universe of discourse are called "singletons." They represent the case where we are only certain that a value of δ is, for example, "possmall" if it takes on only one value, in this case $\frac{40\pi}{180}$, and for any other value of δ we are certain that it is not "possmall." Finally, notice that Figure 5.4 shows the relationship between the scaling gains in Figure 5.3 and the scaling of the universes of discourse. (Notice that for the inputs there is an inverse relationship since an increase an input scaling gain corresponds to making, for instance, the meaning of "zero" correspond to smaller values.)

It is important to emphasize that other membership function types (shapes) are possible; it is up to the designer to pick ones that accurately represent the best ideas about how to control the plant. Fuzzification (in Figure 5.1) is simply the act of finding, for example, $\mu_{pos}(e)$ for a specific value of e.

Next, we discuss the components of the inference mechanism in Figure 5.1. First, we use fuzzy logic to quantify the conjunctions in the premises of the rules. For instance, the premise of Rule 2 is

"e is neg **and** c is zero."

Let $\mu_{neg}(e)$ and $\mu_{zero}(c)$ denote the respective membership functions of each of the two terms in the premise of Rule 2. Then, the premise certainty for Rule 2 can be defined by

$$\mu_{premise(2)} = \min\{\mu_{neg}(e), \mu_{zero}(c)\}.$$

Why? Think about the conjunction of two uncertain statements. The certainty of the assertion of two things is the certainty of the least certain statement.

In general, more than one $\mu_{premise(i)}$ will be nonzero at a time, so more than one rule is "on" (applicable) at every time. Each rule that is "on" can contribute to making a recommendation about how to control the plant and generally ones that are more on (i.e., have $\mu_{premise(i)}$ closer to one) should contribute more to the conclusion. This completes the description of the inference mechanism.

Defuzzification involves combining the conclusions of all the rules. "Center-average" defuzzification uses

$$\delta = \frac{\sum_{i=1}^{9} b_i \mu_{premise(i)}}{\sum_{i=1}^{9} \mu_{premise(i)}}$$

where b_i is the position of the center of the output membership function for the ith rule (i.e., the position of the singleton). This is simply a weighted average of the conclusions. It completes the description of a simple fuzzy controller (and notice that we did not use a mathematical model in its construction).

There are many extensions to the fuzzy controller that we describe above. There are other ways to quantify the "and" with fuzzy logic, other inference approaches, other defuzzification methods, "Takagi-Sugeno" fuzzy systems, and multi-input multi-output fuzzy systems. See [7, 25, 26, 31] for more details.

5.2.1.2 Ship Example

Using a nonlinear model for a tanker ship [3], we get the response in Figure 5.5 (tuned using ideas from how you tune a proportional-derivative controller; notice that the values of $g_1 = 2/\pi$, $g_2 = 250$, and $g_0 = 8\pi/18$ are different from the first guess values shown in Figure 5.4) and the controller surface in Figure 5.6. The control surface shows that there is nothing mystical about the fuzzy controller! It is simply a static (i.e., memoryless) nonlinear map. For real-world applications, most often the surface will have been shaped by the rules to have interesting nonlinearities.

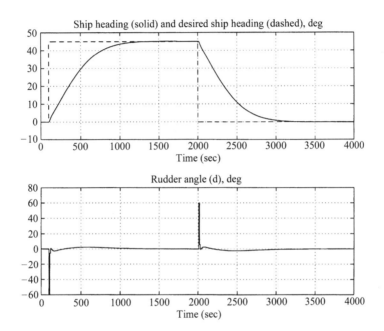

Figure 5.5 Response of fuzzy control system for tanker heading regulation ($g1 = 2/\pi$; $g2 = 250$; $g0 = 8\pi/18$).

5.2.1.3 Design Concerns

One encounters several design concerns when constructing a fuzzy controller. First, it is generally important to have a very good understanding of the control problem, including the plant dynamics and closed-loop specifications. Second, it is important to construct the rule base very carefully. If you do not tell the controller how to properly control the plant, it cannot succeed! Third, for practical applications, you can run into problems with controller complexity since the number of rules used grows exponentially with the number of inputs to the controller, if you use all possible combinations of rules. (However, note that the number of rules on at any one time grows much slower for the ship example.) As with conventional controllers there are always concerns about the effects of disturbances and noise on, for example, tracking error. (Just because it is a fuzzy controller does not mean that it is automatically a "robust" controller.) Indeed, analysis of robustness properties, along with

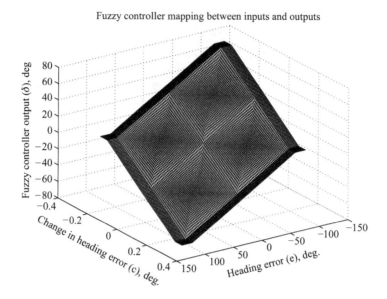

Figure 5.6 Fuzzy controller surface.

stability, steady-state tracking error, and limit cycles, can be quite important for some applications. As already mentioned, since the fuzzy controller is a nonlinear controller, the current methods in nonlinear analysis apply to fuzzy control systems also. (See [7, 24, 25, 31] to find out how to perform stability analysis of fuzzy control systems.)

In summary, the main advantage of fuzzy control is that it provides a heuristic (not necessarily model-based) approach to nonlinear controller construction. In the next section, we will discuss why this advantage can be useful in the solution to challenging industrial applications.

5.2.2 Neural Networks

Artificial neural networks are circuits, computer algorithms, or mathematical representations loosely inspired by the massively connected set of neurons that form biological neural networks. Artificial neural networks are an alternative computing technology that have proven useful in a variety of pattern recognition, signal processing, estimation, and control problems. In this chapter we focus on their use in estimation and control.

5.2.2.1 Multilayer Perceptrons

The feedforward multilayer perceptron is the most popular neural network in control system applications, and so we limit our discussion to it. The second most popular network is probably the radial basis function neural network (of which one form is identical to one type of fuzzy system).

The multilayer perceptron is composed of an interconnected set of neurons, each of which has the form shown in Figure 5.7. Here,

Section 5.2 Intelligent Control Techniques

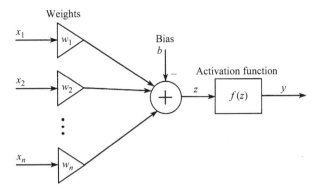

Figure 5.7 Single neuron model.

$$z = \sum_{i=1}^{n} w_i x_i - b$$

and the w_i are the interconnection "weights" and b is the "bias" for the neuron. (These parameters model the interconnections between the cell bodies in the neurons of a biological neural network.) The signal z represents a signal in the biological neuron, and the processing that the neuron performs on this signal is represented with an "activation function" f where

$$y = f(z) = f\left(\sum_{i=1}^{n} w_i x_i - b\right). \tag{5.1}$$

The neuron model represents the biological neuron that "fires" (turns on) when its inputs are significantly excited (i.e., z is big enough). *Firing* is defined by an activation function f where two (of many) possibilities for its definition are:

- Threshold function:

$$f(z) = \begin{cases} 1 & \text{if } z \geq 0 \\ 0 & \text{if } z < 0 \end{cases}$$

- Sigmoid (logistic) function:

$$f(z) = \frac{1}{1 + \exp(-z)}. \tag{5.2}$$

There are many other possible choices for neurons, including a linear neuron that is simply given by $f(z) = z$.

Equation (5.1), with one of the above activation functions, represents the computations made by one neuron. Next, we interconnect them. Let circles represent the neurons (weights, bias, and activation function), and lines represent the connections between the inputs and neurons and the neurons in one layer and the next layer. Figure 5.8 is a three-"layer" perceptron since there are three stages of neural processing between the inputs and outputs.

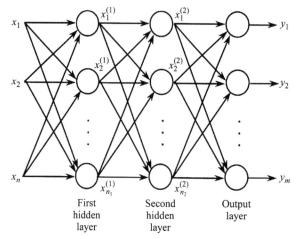

Figure 5.8 Multilayer perceptron model.

Here, we have

- Inputs: x_i, $i = 1, 2, \ldots, n$
- Outputs: y_j, $j = 1, 2, \ldots, m$
- Number of neurons in the first hidden layer, n_1, in the second hidden layer n_2, and in the output layer, m
- In an N layer perceptron there are n_i neurons in the ith hidden layer, $i = 1, 2, \ldots, N - 1$.

We have

$$x_j^1 = f_j^1 \left(\sum_{i=1}^{n} w_{ij}^1 x_i - b_j^1 \right)$$

with $j = 1, 2, \ldots, n_1$. We have

$$x_j^2 = f_j^2 \left(\sum_{i=1}^{n_1} w_{ij}^2 x_i^1 - b_j^2 \right)$$

with $j = 1, 2, \ldots, n_2$. We have

$$y_j = f_j \left(\sum_{i=1}^{n_2} w_{ij} x_i^2 - b_j \right)$$

with $j = 1, 2, \ldots, m$. Here, we have the following:

- w_{ij}^1 (w_{ij}^2) are the weights of the first (second) hidden layer.
- w_{ij} are the weights of the output layer.
- b_j^1 are the biases of the first hidden layer.

- b_j^2 are the biases of the second hidden layer.
- b_j are the biases of the output layer.
- f_j (for the output layer), f_j^2 (for the second hidden layer), and f_j^1 (for the first hidden layer) are the activation functions (all can be different).

5.2.2.2 Training Neural Networks

How do we construct a neural network? We train it with examples. Regardless of the type of network, we will refer to it as

$$y = F(x, \theta)$$

where θ is the vector of parameters that we tune to shape the nonlinearity it implements. (F could be a fuzzy system too in the discussion below.) For a neural network, θ would be a vector of the weights and biases. Sometimes we call F an approximator structure. Suppose that we gather input-output training data from a function $y = g(x)$ that we do not have an analytical expression for (e.g., it could be a physical process).

Suppose that y is a scalar but that $x = [x_1, \ldots, x_n]^\top$. Suppose that $x^i = [x_1^i, \ldots, x_n^i]^\top$ is the ith input vector to g and that $y^i = g(x^i)$. Let the training data set be

$$G = \{(x^i, y^i) : i = 1, \ldots, M\}.$$

The function approximation problem is how to tune θ using G so that F matches $g(x)$ at a test set Γ. (Γ is generally a much bigger set than G.) For system identification the x^i are composed of past system inputs and outputs (a regressor vector) and the y^i are the resulting outputs. In this case, we tune θ so that F implements the system mapping (between regressor vectors and the output). For parameter estimation, the x^i can be regressor vectors, but the y^i are parameters that you want to estimate. In this way we see that by solving the above function approximation problem we are able to solve several types of problems in estimation (and control, since estimators are used in, for example, adaptive controllers).

Consider the simpler situation in which it is desired to cause a neural network $F(x, \theta)$ to match the function $g(x)$ at only a single point $\bar{x}$ where $\bar{y} = g(\bar{x})$. Given an input $\bar{x}$, we would like to adjust θ so that the difference between the desired output and neural network output

$$e = \bar{y} - F(\bar{x}, \theta) \tag{5.3}$$

is reduced (where $\bar{y}$ may be either vector or scalar valued). In terms of an optimization problem, we want to minimize the cost function

$$J(\theta) = e^\top e. \tag{5.4}$$

Taking infinitesimal steps along the gradient of $J(\theta)$ with respect to θ will ensure that $J(\theta)$ is nonincreasing. That is, choose

$$\dot{\theta} = -\bar{\eta} \nabla J(\theta), \tag{5.5}$$

where $\bar{\eta} > 0$ is a constant and if $\theta = [\theta_1, \ldots, \theta_p]^\top$,

$$\nabla J(\theta) = \frac{\partial J(\theta)}{\partial \theta} = \begin{bmatrix} \frac{\partial J(\theta)}{\partial \theta_1} \\ \vdots \\ \frac{\partial J(\theta)}{\partial \theta_p} \end{bmatrix}. \tag{5.6}$$

Using the definition for $J(\theta)$, we get

$$\dot{\theta} = -\bar{\eta} \frac{\partial (e^\top e)}{\partial \theta}$$

or

$$\dot{\theta} = -\bar{\eta} \frac{\partial}{\partial \theta} (\bar{y} - F(\bar{x}, \theta))^\top (\bar{y} - F(\bar{x}, \theta))$$

so that

$$\dot{\theta} = -\bar{\eta} \frac{\partial}{\partial \theta} (\bar{y}^\top \bar{y} - 2F(\bar{x}, \theta)^\top \bar{y} + F(\bar{x}, \theta)^\top F(\bar{x}, \theta)).$$

Now, taking the partial we get

$$\dot{\theta} = -\bar{\eta} \left(-2 \frac{\partial F(\bar{x}, \theta)^\top}{\partial \theta} \bar{y} + 2 \frac{\partial F(\bar{x}, \theta)^\top}{\partial \theta} F(\bar{x}, \theta) \right).$$

If we let $\eta = 2\bar{\eta}$, we get

$$\dot{\theta} = \eta \frac{\partial F(\bar{x}, z)}{\partial z}^\top \bigg|_{z=\theta} (\bar{y} - F(\bar{x}, \theta))$$

so

$$\dot{\theta} = \eta \zeta(\bar{x}, \theta) e \tag{5.7}$$

where $\eta > 0$, and

$$\zeta(\bar{x}, \theta) = \frac{\partial F(\bar{x}, z)}{\partial z}^\top \bigg|_{z=\theta}, \tag{5.8}$$

Using this update method, we seek to adjust θ to try to reduce $J(\theta)$ so that we achieve good function approximation.

In discretized form and with nonsingleton training sets, updating is accomplished by selecting the pair (x^i, y^i), where $i \in \{1, \ldots, M\}$ is a random integer chosen at each iteration, and then using Euler's first-order approximation the parameter update is defined by

$$\theta(k+1) = \theta(k) + \eta \zeta^i(k) e(k), \tag{5.9}$$

where k is the iteration step, $e(k) = y^i - F(x^i, \theta(k))$ and

$$\zeta^i(k) = \left.\frac{\partial F(x^i, z)}{\partial z}\right|_{z=\theta(k)}^\top. \tag{5.10}$$

When M input-output pairs, or patterns, (x^i, y^i) where $y^i = g(x^i)$ are to be matched, "batch updates" can also be done. In this case, let

$$e^i = y^i - F(x^i, \theta), \tag{5.11}$$

and let the cost function be

$$J(\theta) = \sum_{i=1}^{M} e^{i\top} e^i, \tag{5.12}$$

and the update formulas can be derived similarly. This is actually the backpropagation method (except we have not noted that because of the structure of the layered neural networks certain computational savings are possible). In practical applications the backpropagation method, which relies on the steepest descent approach, can be very slow since the cost $J(\theta)$ can have long low-slope regions. It is for this reason that in practice numerical methods are used to update neural network parameters. Two of the methods that have proven particularly useful are the Levenberg-Marquardt and conjugate-gradient methods. For more details, see [5, 8, 12, 13, 14, 16, 17, 21, 32].

5.2.2.3 Design Concerns

You encounter several design concerns in solving the function approximation problem using gradient methods (or others) to tune the approximator structure. First, it is difficult to pick a training set G that you know will ensure good approximation. (Indeed, most often it is impossible to choose the training set; often some other system chooses it.) Second, the choice of the approximator structure is difficult. Although most neural networks (and fuzzy systems) satisfy the universal approximation property, so that they can be tuned to approximate any continuous function on a closed and bounded set to an arbitrary degree of accuracy, this generally requires that you be willing to add an arbitrary amount of structure to the approximator (e.g., nodes to a hidden layer of a multilayer perceptron). Because of finite computing resources, we must then accept an approximation error. How do we pick the structure to keep this error as low as possible? This is an open research problem, and algorithms that grow or shrink the structure automatically have been developed. Third, it is generally impossible to guarantee convergence of the training methods to a global minimum owing to the presence of many local minima. Hence it is often difficult to know when to terminate the algorithm. (Often tests on the size of the gradient update or measures of the approximation error are used to terminate.) Finally, there is the important issue of generalization, in which the neural network is hopefully trained to nicely interpolate between similar inputs. It is very difficult to guarantee that good interpolation is achieved. Normally, all we can do is use a rich data set (large, with some type of uniform and dense spacing of data points) to test that we have achieved good interpolation. If we have not, then we may not have used enough complexity in our model structure, or we may have too much complexity that resulted in "overtraining" where we match very well at the training data but there are large excursions elsewhere.

In summary, the main advantage of neural networks is that they can achieve good approximation accuracy with a reasonable number of parameters by training with data. (Hence there is a lack of dependence on models.) We will show how this advantage can be exploited in the next section for challenging industrial control problems.

5.2.3 Genetic Algorithms

A genetic algorithm (GA) is a computer program that simulates the characteristics of evolution, natural selection (Darwin), and genetics (Mendel). It is an optimization technique that performs a parallel (i.e., candidate solutions are distributed over the search space) and stochastic but directed search to evolve the most fit population. Sometimes when it "gets stuck" at a local optimum, it is able to use the multiple candidate solutions to try to simultaneously find other parts of the search space that will allow it to "jump out" of the local optimum and find a global one (or at least a better local one). GAs do not need analytical gradient information, but with modifications they can exploit such information if it is available.

5.2.3.1 The Population of Individuals

The fitness function of a GA measures the quality of the solution to the optimization problem (in biological terms, the ability of an individual to survive). The GA seeks to maximize the fitness function $J(\theta)$ by selecting the individuals that we represent with the parameters in θ. To represent the GA in a computer, we make θ a string (called a chromosome) as shown in Figure 5.9.

In a base-2 representation, alleles (values in the positions, genes on the chromosome) are 0 and 1. In base-10, the alleles take on integer values between 0 and 9. A sample binary chromosome is given by: 1011110001010, while a sample base-10 chromosome is: 8219345127066. These chromosomes should not necessarily be interpreted as the corresponding positive integers. We can add a gene for the sign of the number and fix a position for the decimal point to represent signed reals. In fact, representation via chromosomes is generally quite abstract. Genes can code for symbolic or structural characteristics, not just for numeric parameter values, and data structures for chromosomes can be trees and lattices, not just vectors.

Chromosomes encode the parameters of a fuzzy system or neural network, or an estimator or controller's parameters. For example, to tune the fuzzy controller dis-

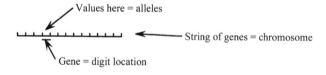

Figure 5.9 String for representing an individual.

cussed earlier for the tanker ship, you could use the chromosome:

$$b_1 b_2 \cdots b_9$$

(These are the output membership function centers.) To tune a neural network, you can use a chromosome that is a concatenation of the weights and biases of the network. Aspects of the structure of the neural network, such as the number of neurons in a layer, the number of hidden layers, or the connectivity patterns can also be incorporated into the chromosome. To tune a proportional-integral-derivative (PID) controller, the chromosome would be a concatenation of its three gains.

How do we represent a set of individuals (i.e., a population)? Let $\theta_i^j(k)$ be a single parameter at time k (a fixed-length string with sign digit), and suppose that chromosome j is composed of N of these parameters that are sometimes called traits. Let

$$\theta^j(k) = \left[\theta_1^j(k), \theta_2^j(k), \ldots, \theta_N^j(k)\right]^\top$$

be the jth chromosome.

The population at time ("generation") k is

$$P(k) = \{\theta^j(k) : j = 1, 2, \ldots, S\} \qquad (5.13)$$

Normally, you try to pick the population size S to be big enough so that broad exploration of the search space is achieved, but not too big or you will need too many computational resources to implement the genetic algorithm.

Evolution occurs as we go from a generation at time k to the next generation at time $k+1$ via fitness evaluation, selection, and the use of genetic operators such as crossover and mutation.

5.2.3.2 Genetic Operators

Selection follows Darwin's theory that the most qualified individuals survive to mate. We quantify "most qualified" via an individual's fitness $J(\theta^j(k))$. We create a "mating pool" at time k:

$$M(k) = \{m^j(k) : j = 1, 2, \ldots, S\}. \qquad (5.14)$$

Then, we select an individual for mating by letting each $m^j(k)$ be equal to $\theta^i(k) \in P(k)$ with probability

$$p_i = \frac{J(\theta^i(k))}{\sum_{j=1}^{S} J(\theta^j(k))}. \qquad (5.15)$$

With this approach, more fit individuals will tend to end up mating more often, thereby providing more offspring. Less fit individuals, on the other hand, will have contributed less of the genetic material for the next generation.

Next, in the reproduction phase that operates on the mating pool, there are two operations: crossover and mutation. Crossover is mating in biological terms (the process of combining chromosomes), for individuals in $M(k)$. For crossover, you first specify the crossover probability p_c (usually chosen to be near unity). The procedure for crossover is: Randomly pair off the individuals in the mating pool $M(k)$. Consider chromosome pair θ^j, θ^i. Generate a random number $r \in [0, 1]$. If $r \geq p_c$, then do not crossover (just pass the individuals into the next generation). If $r < p_c$, then crossover θ^j and θ^i. To crossover these chromosomes, select at random a cross site and exchange all the digits to the right of the cross site of one string with the other (see Figure 5.10). Note that multipoint (multiple cross sites) crossover operators can also be used, with the offspring chromosomes composed by alternating chromosome segments from the parents.

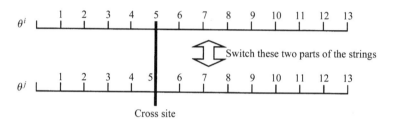

Figure 5.10 Crossover operation example.

Crossover perturbs the parameters near good positions to try to find better solutions to the optimization problem. It tends to perform a localized search around the more fit individuals (i.e., children are interpolations of their parents that may be more or less fit to survive).

Next, in the reproduction phase, after crossover, we have mutation. The biological analog of our mutation operation is the random mutation of genetic material. To do this, with probability p_m change (mutate) each gene location on each chromosome (in the mating pool) randomly to a member of the number system being used. Mutation tries to make sure that we do not get stuck at a local maximum of the fitness function and that we seek to explore other areas of the search space to help find a global maximum for $J(\theta)$. Since mutation is pure random search, p_m is usually near zero.

Finally, we produce the next generation by letting

$$P(k+1) = M(k).$$

Evolution is the repetition of the above process. For more details on GAs, see [10, 20, 22, 28].

5.2.3.3 Design Concerns

You can encounter many design concerns when using GAs to solve optimization problems. First, it is important to fully understand the optimization problem, and to know what you want to optimize and what you can change to achieve the optimization. You also must have an idea of what you will accept as an optimal solution. Second, choice of representation (e.g., the number of digits in a base-10 representation) is important. Too detailed a representation increases computational complexity, while too coarse a representation means you may not be able to achieve enough accuracy in your solution. Third, there are a wide range of other genetic operators (e.g., "elitism" where the most fit individual is passed to the next generation without being perturbed by crossover or mutation) and choosing the appropriate ones is important since they can affect convergence significantly. Fourth, just as for gradient optimization methods, it is important to pick a good termination method (even if it is simply a test on how much improvement has been made on J over the last several generations). Finally, for practical problems it is difficult to guarantee that you will achieve convergence owing to the presence of local maxima. Moreover, it can be difficult to select the best solution from the many candidate solutions that exist. (Most often you pick the parameters that resulted in the highest value of the fitness function, and these may have been generated in a past generation, not at the final one.)

In summary, the main advantage of genetic algorithms is that they offer an evolution-based stochastic search that can be useful in finding good solutions to practical complex optimization problems, especially when gradient information is not conveniently available.

5.2.4 Expert and Planning Systems

In this section, we briefly overview the expert and planning systems [27] approaches to control. We keep the discussion particularly brief because the use of expert systems for control (expert control) is conceptually similar to fuzzy control and because general planning operations often fall outside the area of traditional control problems (although they probably should not).

5.2.4.1 Expert Control

For the sake of our discussion, we will simply view the expert system that is used here as a controller for a dynamic system, as is shown in Figure 5.11. Here, we have an expert system serving as feedback controller with reference input r and feedback variable y. It uses the information in its knowledge base and its inference mechanism to decide what command input u to generate for the plant. Conceptually, we see that the expert controller is closely related to the fuzzy controller. There are, however, several differences. First, the knowledge base in the expert controller could be a rule base, but is not necessarily so. It could be developed using other knowledge-representation structures, such as frames, semantic nets, causal diagrams, and so on. Second, the inference mechanism in the expert controller is more general than that of the fuzzy controller. It

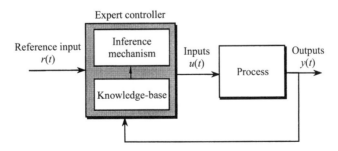

Figure 5.11 Expert control system.

can use more sophisticated matching strategies to determine which rules should be allowed to fire. It can also use more elaborate inference strategies such as refraction, recency, and various other priority schemes. Next, we should note that Figure 5.11 shows a direct expert controller. It is also possible to use an expert system as a supervisor for conventional or intelligent controllers.

5.2.4.2 Planning Systems for Control

Artificially intelligent planning systems (computer programs that are often designed to emulate the way experts plan) have been used for several problems, including path planning and high-level decisions about control tasks for robots [6, 27]. A generic planning system can be configured in the architecture of a standard control system, as shown in Figure 5.12. Here, the "problem domain" (the plant) is the environment in which the planner operates. There are measured outputs y_k at step k (variables of the problem domain that can be sensed in real time), control actions u_k (the ways in which we can affect the problem domain), disturbances d_k (which represent random events that can affect the problem domain and hence the measured variable y_k), and goals g_k (what we would like to achieve in the problem domain). There are closed-loop specifications that quantify performance and stability requirements.

The planner's task in Figure 5.12 is to monitor the measured outputs and goals and generate control actions that will counteract the effects of the disturbances and result in the goals and the closed-loop specifications being achieved. To do this, the planner performs "plan generation," where it projects into the future (usually a finite number of steps, and often using a model of the problem domain) and tries to determine a set of candidate plans. Next, this set of plans is pruned to one plan that is the best one to apply at the current time (where "best" can be determined based on, e.g., consumption of resources). The plan is then executed, and during execution the performance resulting from the plan is monitored and evaluated. Often, because of disturbances, plans will fail, and hence the planner must generate a new set of candidate plans, select one, and then execute that one. While not pictured in Figure 5.12, some planning systems use situation assessment to try to estimate the state of the problem domain. (This can be useful in execution monitoring and plan generation.) Others perform world modeling, in which a model of the problem domain is developed in an on-line fashion (similarly to on-line system identification); planner design uses information from the world modeler to tune the planner (so that it makes the right plans for the current problem domain).

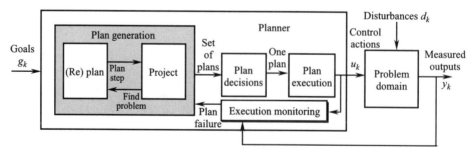

Figure 5.12 Closed-loop planning system.

The reader will perhaps think of such a planning system as a general adaptive (model predictive) controller.

5.2.5 Intelligent and Autonomous Control

Autonomous systems have the capability to independently (and successfully) perform complex tasks. Consumer and governmental demands for such systems are frequently forcing engineers to push many functions normally performed by humans into machines. For instance, in the emerging area of intelligent vehicle and highway systems (IVHS), engineers are designing vehicles and highways that can fully automate vehicle route selection, steering, braking, and throttle control to reduce congestion and improve safety. In avionic systems, a pilot's associate computer program has been designed to emulate the functions of mission and tactical planning that in the past may have been performed by the copilot. In manufacturing systems, efficiency optimization and flow control are being automated, and robots are replacing humans in performing relatively complex tasks. From a broad historical perspective, each of these applications began at a low level of automation, and through the years each has evolved into a more autonomous system. For example, automotive cruise controllers are the ancestors of the (research prototype) controllers that achieve coordinated control of steering, braking, and throttle for autonomous vehicle driving. And the terrain following and terrain avoidance control systems for low-altitude flight are ancestors of an artificial pilot's associate that can integrate mission and tactical planning activities. The general trend has been for engineers to incrementally "add more intelligence" in response to consumer, industrial, and government demands and thereby create systems with increased levels of autonomy.

In this process of enhancing autonomy by adding intelligence, engineers often study how humans solve problems and then try to directly automate their knowledge and techniques to achieve high levels of automation. Other times, engineers study how intelligent biological systems perform complex tasks and then seek to automate "nature's approach" in a computer algorithm or circuit implementation to solve a practical technological problem (e.g., in certain vision systems). Such approaches where we seek to emulate the functionality of an intelligent biological system (e.g., the human) to solve a technological problem can be collectively named intelligent systems and control techniques. By using these techniques, some engineers are trying to create highly autonomous systems such as those listed above.

Figure 5.13 shows a functional architecture for an intelligent autonomous controller with an interface to the process involving sensing (e.g., via conventional sensing technology, vision, touch, smell, etc.), actuation (e.g., via hydraulics, robotics, motors, etc.), and an interface to humans (e.g., a driver, pilot, crew, etc.) and other systems. The *execution level* has low-level numeric signal processing and control algorithms (e.g., PID, optimal, adaptive, or intelligent control; parameter estimators, failure detection, and identification [FDI] algorithms). The *coordination level* provides for tuning, scheduling, supervision, and redesign of the execution-level algorithms, crisis management, planning and learning capabilities for the coordination of execution-level tasks, and higher-level symbolic decision making for FDI and control algorithm management. The *management level* provides for supervising lower-level functions and for managing the interface to the human(s) and other systems. In particular, the management level will interact with the users in generating goals for the controller and in assessing the

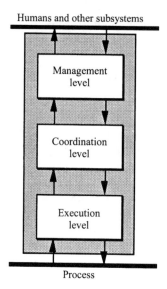

Figure 5.13 Intelligent autonomous controller.

capabilities of the system. The management level also monitors performance of the lower-level systems, plans activities at the highest level (and in cooperation with humans), and performs high-level learning about the user and the lower-level algorithms. Conventional or intelligent systems methods can be used at each level. For more information on these types of control systems see [1, 2, 11, 29, 30].

5.3 APPLICATIONS

This section outlines some of the main characteristics of the intelligent system methods that have proven useful in industrial applications and gives examples of the use of the methods.

5.3.1 Heuristic Construction of Nonlinear Controllers

Intelligent control has had a clear impact in industry in the area of heuristic construction of nonlinear controllers. Two areas in intelligent control have made most of the contributions to this area: fuzzy control and expert systems for control. (Here we will focus on fuzzy control, one type of rule-based controller, since the ideas extend directly to the expert control case.) The methods are heuristic because they normally do not rely on the development and use of a mathematical model of the process to be controlled.

5.3.1.1 Model-Free Control?

To begin with, it is important to critically examine the claim that fuzzy control is model-free control. So, is a model used in the fuzzy control design methodology? It is possible that a mathematical model is not used and that the entire process simply relies on the ad hoc specification of rules about how to control a process (in an analogous manner to the way PID controllers are often designed and implemented in industry).

Section 5.3 Applications 123

However, often a model is used in simulation to redesign a fuzzy controller. (Consider the earlier ship-steering controller design problem.) Others argue that a model is always used: even if it is not written down, some type of model is used "in your head" (even though it might not be a formal mathematical model).

Since most people claim that no formal model is used in the fuzzy control design methodology, the following questions arise:

1. Is it not true that there are few, if any, assumptions to be violated by fuzzy control and that the technique can be indiscriminately applied? Yes, and sometimes it is applied to systems where it is clear that a PID controller or lookup table would be just as effective. So, if this is the case, then why not use fuzzy control? Because it is more computationally complex than a PID controller and the PID controller is much more widely understood.
2. Are heuristics all that are available to perform fuzzy controller design? No. Any good models that can be used probably should be.
3. By ignoring a formal model, if it is available, is it not the case that a significant amount of information about how to control the plant is ignored? Yes. If, for example, you have a model of a complex process, we often use simulations to gain an understanding of how best to control the plant—and this knowledge can be used to design a fuzzy controller.

Nonetheless, at times it is either difficult or virtually impossible to develop a useful mathematical model. In such instances, heuristic constructive methods for controllers can be very useful. (Of course, we often do the same thing with PID controllers).

In the next section, we give an example of where fuzzy controllers were developed and proved to be very effective, and no mathematical model was used.

5.3.1.2 Example: Vibration Damping in a Flexible-Link Robot

For nearly a decade, control engineers and roboticists alike have been investigating the problem of controlling robotic mechanisms that have very flexible links. Such mechanisms are important in space structure applications where large, lightweight robots are to be utilized in a variety of tasks, including deployment, spacecraft servicing, space-station maintenance, and so on. Flexibility is not designed into the mechanism; it is usually an undesirable characteristic that results from trading off mass and length requirements in optimizing the effectiveness and "deployability" of the robot. These requirements and limitations of mass and rigidity give rise to many interesting issues from a control perspective. Why turn to fuzzy control for this application?

The modeling complexity of multilink flexible robots is well documented, and numerous researchers have investigated a variety of techniques for representing flexible and rigid dynamics of such mechanisms. Equally numerous are the works addressing the control problem in simulation studies based on mathematical models, under assumptions of perfect modeling. Even in simulation, however, a challenging control problem exists; it is well known that vibration suppression in slewing mechanical structures whose parameters depend on the configuration (i.e., are time varying) can be extremely difficult to achieve. Compounding the problem, numerous experimental studies have shown that when implementation issues are taken into consideration, modeling uncertainties either render the simulation-based control designs useless or demand extensive tuning of controller parameters (often in an ad hoc manner).

Hence, even if a relatively accurate model of the flexible robot can be developed, it is often too complex to use in controller development, especially for many control design procedures that require restrictive assumptions for the plant (e.g., linearity). It is for this reason that conventional controllers for flexible robots are developed either (1) via simple crude models of the plant behavior that satisfy the necessary assumptions (e.g., either from first principles or using system identification methods) or (2) via the ad hoc tuning of linear or nonlinear controllers. Regardless, heuristics enter the design process when the conventional control design process is used.

It is important to emphasize, however, that conventional control-engineering approaches that use appropriate heuristics to tune the design have been relatively successful. For a process such as a flexible robot, one is left with the following question: How much of the success can be attributed to use of the mathematical model and conventional control design approach, and how much should be attributed to the clever heuristic tuning that the control engineer uses upon implementation? Why not simply acknowledge that much of the problem must be solved with heuristic ideas and avoid all the work that is needed to develop the mathematical models? Fuzzy control provides such an opportunity and has in fact been shown to be quite successful for this application [23] compared to conventional control approaches, especially if one takes into account the efforts to develop a mathematical model that are needed for the conventional approaches.

5.3.2 Data-Based Nonlinear Estimation

The second major area where methods from intelligent control have had an impact in industry is in the use of neural networks to construct mappings from data. In particular, neural network methods have been found to be quite useful in pattern recognition and estimation. Here we explain how to construct neural network-based estimators and give an example of where such a method was used.

5.3.2.1 Estimator Construction Methodology

In conventional system identification, you gather plant input-output data and construct a model (mapping) between the inputs and outputs. In this case, model construction is often done by tuning the parameters of a model (e.g., the parameters of a linear mapping can be tuned using linear least-squares methods or gradient methods). To validate this model, you gather novel plant input-output data and pass the inputs into your constructed model and compare its outputs to the ones that were generated by the model. If some measure of the difference between the plant and model outputs is small, then we accept that the model is a good representation of the system.

Neural networks or fuzzy systems are also tunable functions that can be used for this system identification task. Fuzzy and neural systems are nonlinear and are parameterized by membership function parameters or weights (and biases), respectively. Gradient methods can be used to tune them to match mappings that are characterized with data. Validation of the models proceeds along the same lines as with conventional system identification.

In certain situations, you can also gather data that relates the inputs and outputs of the system to parameters within the system. To do this, you must be able to vary system parameters and gather data for each value of the system parameter. (The gathered data should change each time the parameter changes, and it is gathered via either a sophisticated simulation model or actual experiments with the plant.) Then, using a gradient method, you can adjust the neural or fuzzy system parameters to minimize the estimation error. The resulting system can serve as a parameter estimator (i.e., after it is tuned—normally it cannot be tuned on-line because actual values of the parameters are not known on-line, and they are what you are trying to estimate).

5.3.2.2 Example: Automotive Engine Failure Estimation

In recent years, significant attention has been given to reducing exhaust gas emissions produced by internal combustion engines. In addition to overall engine and emission system design, correct or fault-free engine operation is a major factor determining the amount of exhaust gas emissions produced in internal combustion engines. Hence, there has been a recent focus on the development of on-board diagnostic systems that monitor relative engine health. Although on-board vehicle diagnostics can often detect and isolate some major engine faults, because of widely varying driving environments they may be unable to detect minor faults, which may nonetheless affect engine performance. Minor engine faults warrant special attention because they do not noticeably hinder engine performance but may increase exhaust gas emissions for a long period of time without the problem being corrected. The minor faults we consider in this case study include calibration faults (here, the occurrence of a calibration fault means that a sensed or commanded signal is multiplied by a gain factor not equal to one, while in the no-fault case the sensed or commanded signal is multiplied by one) in the throttle and mass fuel actuators, and in the engine speed and mass air sensors. The reliability of these actuators and sensors is particularly important to the engine controller since their failure can affect the performance of the emissions control system. Here, we simply discuss how to formulate the problem so that it can be solved with neural or fuzzy estimation schemes. The key to this problem is to understand how data are generated for the training of neural or fuzzy system estimators.

The experimental setup in the engine test cell consists of a Ford 3.0 L V-6 engine coupled to an electric dynamometer through an automatic transmission. An air charge temperature sensor (ACT), a throttle position sensor (TPS), and a mass airflow sensor (MAF) are installed in the engine to measure the air charge temperature, throttle position, and air mass flow rate. Two heated exhaust gas oxygen sensors (HEGO) are located in the exhaust pipes upstream of the catalytic converter. The resultant airflow information and input from the various engine sensors are used to compute the required fuel flow rate necessary to maintain a prescribed air-to-fuel ratio for the given engine operation. The central processing unit (EEC-IV) determines the needed injector pulse width and spark timing, and outputs a command to the injector to meter the exact quantity of fuel. An ECM (electronic control module) breakout box is used to provide external connections to the EEC-IV controller and the data acquisition system.

The angular velocity sensor system consists of a digital magnetic zero-speed sensor and a specially designed frequency-to-voltage converter, which converts frequency signals proportional to the rotational speed into an analog voltage.

Data are sampled in every engine revolution. A variable load is produced through the dynamometer, which is controlled by a DYN-LOC IV speed/torque controller in conjunction with a DTC-1 throttle controller installed by DyneSystems Company. The load torque and dynamometer speed are obtained through a load cell and a tachometer, respectively. The throttle and the dynamometer load reference inputs are generated through a computer program and are sent through an RS-232 serial communication line to the controller. Physical quantities of interest are digitized and acquired utilizing a National Instruments AT-MIO-16F-5 A/D timing board for a personal computer. Because of government mandates, periodic inspections and maintenance for engines are becoming more common. One such test developed by the Environmental Protection Agency (EPA) is the Inspection and Maintenance (IM) 240 cycle. The EPA IM240 cycle represents a driving scenario developed for the purpose of testing compliance of vehicle emissions systems for contents of carbon monoxide (CO), unburned hydrocarbons (HC), and nitrogen oxides (NO_x). A modified version of this cycle was used in all the tests.

Using the engine test cell, we take measurements of engine inputs and outputs for various calibration faults (i.e., we gather sequences of data for each fault). Then, we induce faults over the whole range of possible values of calibration faults. Data from all these experiments become our training data set (the set G described in the neural network section). This allows us to construct neural or fuzzy estimators for calibration faults that can be tested in the actual experimental testbed. Additional details on this application are given in [18].

5.3.3 Intelligent Adaptive Control Strategies

In this section we overview how intelligent systems methods can be used to achieve adaptive control. Rather than providing a detailed tutorial on of all the (many) strategies that have been investigated and reported in the literature, an overview will be provided in the first subsection of this section that will show how all the methods broadly relate to each other. The reader should keep in mind that all of these methods bear very close relationships to the work in conventional adaptive control [15].

5.3.3.1 Fuzzy, Neural, and Genetic Adaptive Control

There are two general approaches to adaptive control. In the first one, depicted in Figure 5.14, we use an on-line system identification method to estimate the parameters of the plant (by estimating the parameters of an identifier model) and a controller designer module to subsequently specify the parameters of the controller. If the plant parameters change, the identifier will provide estimates of these and the controller designer will subsequently tune the controller. It is inherently assumed that we are certain that the estimated plant parameters are equivalent to the actual ones at all times. (This is called the certainty equivalence principle.) Then if the controller designer can specify a controller for each set of plant parameter estimates, it will succeed in controlling the plant. The overall approach is called *indirect adaptive control* since we tune the controller indirectly by first estimating the plant parameters.

Section 5.3 Applications

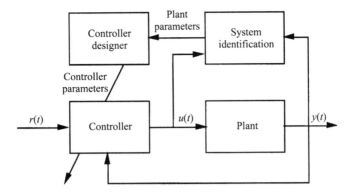

Figure 5.14 Indirect adaptivie control.

The model structure used for the identifier model could be linear with adjustable coefficients. Alternatively, it could be a neural or fuzzy system with tunable parameters (e.g., membership function parameters or weights and biases). In this case, the model that is being tuned is a nonlinear function. Since the plant is assumed to be unknown but constant, the nonlinear mapping it implements is unknown. In adjusting the nonlinear mapping implemented by the neural or fuzzy system to match the unknown nonlinear mapping of the plant, we are solving an on-line function approximation problem. Normally, gradient or least-squares methods are used to tune neural or fuzzy systems for indirect adaptive control (although problem-dependent heuristics can sometimes be useful for practical applications). The stability of these methods has been studied by several researchers (including Farrell and Polycarpou who provide an overview of this research in Chapter 6 [9]). Other times, a genetic algorithm has been employed for such on-line model tuning, and in this case it may also be possible to tune the model structure.

In the second general approach to adaptive control, which is shown in Figure 5.15, the adaptation mechanism observes the signals from the control system and adapts the parameters of the controller to maintain performance even if there are changes in the plant. Sometimes, in either the direct or indirect adaptive controllers, the desired per-

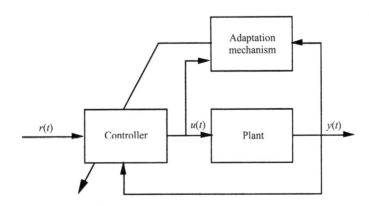

Figure 5.15 Direct adaptive control.

formance is characterized with a reference model, and the controller then seeks to make the closed-loop system behave as the reference model would, even if the plant changes. This is called model reference adaptive control (MRAC).

In neural control or adaptive fuzzy control, the controller is implemented with a neural or fuzzy system, respectively. Normally, gradient or least-squares methods are used to tune the controller (although sometimes problem-dependent heuristics have been found to be quite useful for practical applications, such as in the fuzzy model reference learning controller discussed later). The stability of direct adaptive neural or fuzzy methods has been studied by several researchers. (Again, for an overview of the research, see Chapter 6 by Farrell and Polycarpou.) Clearly, since the genetic algorithm is also an optimization method, it can be used to tune neural or fuzzy system mappings when they are also used as controllers. The key to making such a controller work is to provide a way to define a fitness function for evaluating the quality of a population of controllers. (In one approach a model of the plant is used to predict into the future how each controller in the population will perform.) Then, the most fit controller in the population is used at each step to control the plant. This is a type of adaptive model predictive control (MPC) method.

In practical applications it is sometimes found that a supervisory controller can be very useful. Such a controller takes as inputs data from the plant and the reference input (and any other information available, e.g., from the user) and tunes the underlying control strategy. For example, in the flexible-link robot application discussed earlier, such a strategy was found to be very useful in tuning a fuzzy controller. In an aircraft application, it was found useful in tuning an adaptive fuzzy controller to try to ensure that the controller was maximally sensitive to plant failures in the sense that it would quickly respond to them, but it still maintained stable high-performance operation.

5.3.3.2 Example: Adaptive Fuzzy Control for Ship Steering

How good is the fuzzy controller that we designed for the ship-steering problem earlier in this chapter? Between trips, let there be a change from ballast to full conditions on the ship (a weight change). In this case, using the controller we had developed earlier, we get the response in Figure 5.16.

Clearly there has been a significant degradation in performance. It is possible to tune the fuzzy controller to reduce the effect of this disturbance, but then other disturbances may occur and may have adverse effects on performance. This presents a fundamental challenge to fuzzy control and motivates the need to develop a method that can automatically tune the fuzzy controller if there are changes in the plant.

Fuzzy model reference learning control (FMRLC) is one *heuristic* approach to adaptive fuzzy control, and the overall scheme is shown in Figure 5.17. Here, at the lower level in the figure is a plant that is controlled by a fuzzy controller. (As an example, this one simply has inputs of the error and change in error.) The reference model is a user-specified dynamical system that is used to quantify how we would like the system to behave between the reference input and the plant output. For example, we may request a first-order response with a specified time constant between the reference input and plant output. The learning mechanism observes the performance of the low-level fuzzy controller loop and decides when to update the fuzzy controller. For this

Section 5.3 Applications

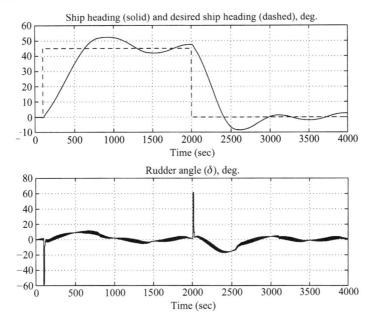

Figure 5.16 Response of fuzzy control system for tanker heading regulation, weight change.

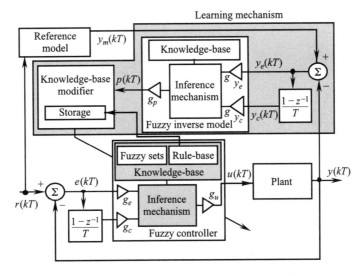

Figure 5.17 Fuzzy model reference learning controller.

example, when the error between the reference model output and the plant output is large, the learning mechanism will make large changes to the fuzzy controller (by tuning its output membership function centers). When this error is small, then it will make small changes. For more details, see [19].

How does the FMRLC work for the tanker ship? Assume that we initialize the controller with the one that was developed via manual tuning earlier. To see that it can tune a rule base see the response in Figure 5.18. (We use a first-order reference model.) Here, at $t = 9000$ sec the ship weight is suddenly changed from ballast to full, and we see that while initially the weight change causes poor transient performance, it quickly recovers to provide good tracking. Compare this response to the direct fuzzy controller results shown in Figure 5.16. You can see that it does very well at tuning the fuzzy controller (although it may not be done tuning at the end of the simulation). The tuned controller surface (at the end of the simulation) is shown in Figure 5.19, and we see that it produced some shape changes relative to the manually constructed one in Figure 5.6 since it is trying to compensate for the effects of the weight change.

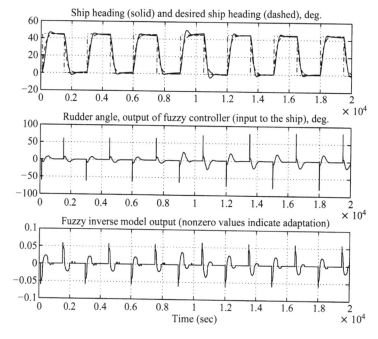

Figure 5.18 FMRLC response.

5.4 CONCLUDING REMARKS: OUTLOOK ON INTELLIGENT CONTROL

In this section we briefly note some of the current and future research directions in intelligent control. Current *theoretical research* in intelligent control is focusing on:

- Mathematical stability/convergence/robustness analysis for learning systems.
- Mathematical comparative analysis with nonlinear adaptive methods.

Section 5.4 Concluding Remarks: Outlook on Intelligent Control   131

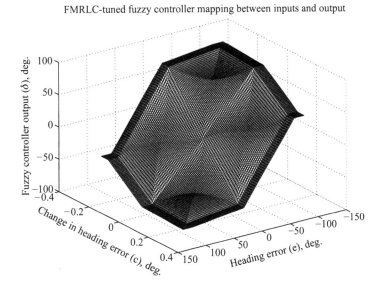

Figure 5.19 FMRLC, tuned controller surface.

However, as Albert Einstein once said: "So far as the laws of mathematics refer to reality, they are not certain. And so far as they are certain, they do not refer to reality." Or stated another way, your approaches developed with mathematical analysis are only as good as the model you use to develop them.

Current research on the development of new *techniques* in intelligent control focuses on the following:

- Complex heuristic learning strategies.
- Memory and computational savings.
- Coping with "hybrid" discrete-event/differential equation models.

Current research in *applications* and implementations is focusing on a wide variety of problems. It is important to note the following:

- There is a definite need for experimental research (especially in comparative analysis and new nontraditional applications).
- There have been definite successes in industry (though we are certainly not providing a complete overview of these successes).
- For researchers in universities, working with industry is challenging and important.

In summary, intelligent control tries to borrow ideas not only from physics and mathematics to help develop control systems, but also from biology, neuroscience, artificial intelligence, and others. It has proven useful in some applications, as we discussed in the previous section, and it may offer useful solutions to the challenging problems that you encounter.

ACKNOWLEDGMENTS

The author would like to thank J. Spooner who had worked with the author on writing an earlier version of Section 5.2.2.2. The author would also like to thank the editor T. Samad for his helpful edits and for organizing the writing of this book.

Related Chapters

- Chapter 6 provides a detailed technical introduction to neural networks and nonlinear approximation, including a discussion of stability properties of adaptive approximators.
- Other intelligent control techniques include agent-based complex adaptive systems. Some applications of these are outlined in Chapters 10 and 13.
- Some building control applications of neural networks, fuzzy logic, and expert systems can be found in Chapter 16.

REFERENCES

[1] J. S. Albus, "Outline for a theory of intelligence." *IEEE Trans. on Systems, Man, and Cybernetics*, Vol. 21, no. 3, pp. 473–509, May/June 1991.

[2] P. J. Antsaklis, and K. M. Passino (eds.), *An Introduction to Intelligent and Autonomous Control*. Norwell, MA: Kluwer Academic Press, 1993.

[3] K. J. Åström and B. Wittenmark, *Adaptive Control*. Reading, MA: Addison-Wesley, 1995.

[4] D. P. Bertsekas, *Nonlinear Programming*. Belmont, MA: Athena Scientific Press, 1995.

[5] M. Brown and C. Harris, *Neurofuzzy Adaptive Modeling and Control*. Englewood Cliffs, NJ: Prentice Hall, 1994.

[6] T. Dean and M. P. Wellman, *Planning and Control*. San Mateo, CA: Morgan Kaufman, 1991.

[7] D. Driankov, H. Hellendoorn, and M. Reinfrank, *An Introduction to Fuzzy Control*. New York: Springer-Verlag, 1993.

[8] J. Farrell, "Neural control." In W. Levine (ed.), *The Control Handbook*, pp. 1017–1030. Boca Raton, FL: CRC Press, 1996.

[9] J. Farrell and M. Polycarpou. "On-line approximation based control with neural networks and fuzzy systems." In T. Samad (ed.), *Perspectives in Control Engineering: Technologies, Applications, and New Directions*, pp. 134–164. New York: IEEE Press, 2001.

[10] D. Goldberg, *Genetic Algorithms in Search, Optimization and Machine Learning*. Reading, MA: Addison-Wesley, 1989.

[11] M. Gupta and N. Sinha (eds.), *Intelligent Control: Theory and Practice*. New York: IEEE Press, 1995.

[12] M. Hagan, H. Demuth, and M. Beale, *Neural Network Design*. Boston: PWS Publishing, 1996.

[13] J. Hertz, A. Krogh, and R. G. Palmer, *Introduction to the Theory of Neural Computation*. Reading, MA: Addison-Wesley, 1991.

[14] K. J. Hunt, D. Sbarbaro, R. Zbikowski, and P. J. Gawthrop, "Neural networks for control systems: A survey." In M. M. Gupta and D. H. Rao (eds.), *Neuro-Control Systems: Theory and Applications*, pp. 171–200. New York: IEEE Press, 1994.

[15] P. A. Ioannou and J. Sun, *Robust Adaptive Control*. Englewood Cliffs, NJ: Prentice Hall, 1996.

References

[16] J.-S. R. Jang, C.-T. Sun, and E. Mizutani, *Neuro-Fuzzy and Soft Computing: A Computational Approach to Learning and Machine Intelligence*. Englewood Cliffs, NJ: Prentice Hall, 1997.

[17] B. Kosko, *Neural Networks and Fuzzy Systems*. Englewood Cliffs, NJ: Prentice Hall, 1992.

[18] E. G. Laukonen, K. M. Passino, V. Krishnaswami, G.-C. Luh, and G. Rizzoni, "Fault detection and isolation for an experimental internal combustion engine via fuzzy identification." *IEEE Trans. on Control Systems Technology*, Vol. 3, no. 3, pp. 347–355, September 1995.

[19] J. R. Layne and K. M. Passino, "Fuzzy model reference learning control for cargo ship steering." *IEEE Control Systems Magazine*, Vol. 13, no. 6, pp. 23–34, December 1993.

[20] Z. Michalewicz, *Genetic Algorithms + Data Structure = Evolution Programs*. Berlin: Springer-Verlag, 1992.

[21] W. T. Miller, R. S. Sutton, and P. J. Werbos (eds.), *Neural Networks for Control*. Cambridge, MA: MIT Press, 1991.

[22] M. Mitchell, *An Introduction to Genetic Algorithms*. Cambridge, MA: MIT Press, 1996.

[23] V. G. Moudgal, K. M. Passino, and S. Yurkovich, "Rule-based control for a flexible-link robot." *IEEE Trans. on Control Systems Technology*, Vol. 2, no. 4, pp. 392–405, December 1994.

[24] R. Palm, D. Driankov, and H. Hellendoorn, *Model Based Fuzzy Control*. New York: Springer-Verlag, 1997.

[25] Kevin M. Passino and Stephen Yurkovich, *Fuzzy Control*. Menlo Park, CA: Addison-Wesley Longman, 1998.

[26] T. Ross. *Fuzzy Logic in Engineering Applications*. New York: McGraw-Hill, 1995.

[27] S. Russell and P. Norvig. *Artificial Intelligence: A Modern Approach*. Englewood Cliffs, NJ: Prentice Hall, 1995.

[28] M. Srinivas and L. M. Patnaik, "Genetic algorithms: A survey." *IEEE Computer Magazine*, pp. 17–26, June 1994.

[29] R. F. Stengel, "Toward intelligent flight control." *IEEE Trans. on Systems, Man, and Cybernetics*, Vol. 23, no. 6, pp. 1699–1717, November/December 1993.

[30] K. Valavanis and G. Saridis, *Intelligent Robotic Systems: Theory, Design, and Applications*. Norwell, MA: Kluwer Academic Press, 1992.

[31] L.-X. Wang, *A Course in Fuzzy Systems and Control*. Englewood Cliffs, NJ: Prentice Hall, 1997.

[32] D. White and D. Sofge (eds.), *Handbook of Intelligent Control: Neural, Fuzzy and Adaptive Approaches*. New York: Van Nostrand Reinhold, 1992.

Chapter 6
NEURAL, FUZZY, AND APPROXIMATION-BASED CONTROL

Jay A. Farrell and Marios M. Polycarpou

Editor's Summary

The assumption of linearity must be given due credit for the tremendous practical impact that control systems have had over the last several decades. However, as the original challenges have been encountered and overcome, and as the control and automation of complex, large-scale problems are being sought, effective methods for dealing with nonlinear systems have become essential.

One key component of nonlinear controls technology is representations or models of nonlinear systems that are derived from operational data. Such models, referred to as *approximators*, are the focus of this chapter. Specific attention is paid to neural networks and fuzzy models. These topics are discussed within a general formulation that emphasizes their close relationships with other approximator structures. In this chapter, several associated properties are noted and defined, including universal approximation, linear and nonlinear parameterizations, generalization, and approximator transparency. Compared to most other chapters in this volume, this one is relatively theoretical. Less formal introductions to neural networks and fuzzy logic can be found in Chapter 5; some applications are discussed therein and in Chapter 16.

An important problem in approximator development is the estimation of the approximator parameters. This chapter discusses some algorithms—specifically steepest descent, least-squares, and Lyapunov-based algorithms—that can be used for this purpose. Some degree of modeling error is inescapable, and this realization has motivated the development of extensions to parameter estimation algorithms.

Readers interested in additional nonlinear control methods may also find Chapter 8 of interest, which provides a readable technical introduction to a popular nonlinear control design technique, sliding-mode control.

Jay Farrell is an associate professor in the Department of Electrical Engineering at the University of California at Riverside and a former IEEE-CSS liaison representative to the IEEE Neural Networks Council. Marios Polycarpou is an associate professor in the Department of Electrical and Computer Engineering and Computer Science at the University of Cincinnati, and a current CSS representative to IEEE-NNC

6.1 INTRODUCTION

Introductory control courses focus on the design of linear control systems. However, many control applications involve significant nonlinearities. Although linear control design methods can sometimes be applied to nonlinear systems over limited operating regions through the process of linearization, the level of performance desired in other applications requires that the nonlinearities be directly addressed in the control system

Section 6.1 Introduction

design. The challenge of addressing nonlinearities during the control design process is further complicated when the description of the nonlinearities involves significant uncertainty. In such applications, the level of achievable performance may be enhanced by using on-line function approximation techniques to increase the accuracy of the model of the nonlinearities. Such on-line approximation-based control methods include the popular areas of adaptive fuzzy and neural control. This chapter discusses various issues related to on-line approximation-based control using a unifying framework and notation.

6.1.1 Components of Approximation-Based Control

Implementation or analysis of an on-line approximation-based control system requires that the designer properly specify the problem and solution. This section discusses major aspects of the problem specification.

6.1.1.1 Control Architecture

Specification of the control architecture is application dependent and has various aspects. The designer must determine how the nonlinear function affects the system dynamics and specify a control methodology capable of using the approximated nonlinear function to improve the system performance. Two examples will clarify these issues.

Consider a dynamic system that can be described as

$$\dot{x}_i(t) = x_{i+1}(t), \quad \text{for } i = 1, \ldots, n-1$$
$$\dot{x}_n(t) = f(\mathbf{x}(t)) + g(\mathbf{x}(t))h(u(t)),$$
$$\mathbf{y}(t) = \mathbf{x}(t)$$

where $\mathbf{x} = (x_1, \ldots, x_n)$ is the state of the system, $u(t)$ is the control input, f and g are accurately known functions, and the actuator function h involves significant nonlinearity. The actuator nonlinearity may, for example, represent dead-zone and saturation effects. If a satisfactory control system can be designed for the system

$$\dot{x}_i(t) = x_{i+1}(t), \quad \text{for } i = 1, \ldots, n-1$$
$$\dot{x}_n(t) = f(\mathbf{x}(t)) + g(\mathbf{x}(t))v(t)$$
$$\mathbf{y}(t) = \mathbf{x}(t)$$

and the function h can be approximated and inverted, then defining $u(t) = \hat{h}^{-1}(v(t))$ will solve the original control problem (see Figure 6.1).

Consider a dynamic system that can be described as

$$\dot{x}_i(t) = x_{i+1}(t), \quad \text{for } i = 1, \ldots, n-1$$
$$\dot{x}_n(t) = f(\mathbf{x}(t)) + g(\mathbf{x}(t))u(t),$$
$$\mathbf{y}(t) = \mathbf{x}(t)$$

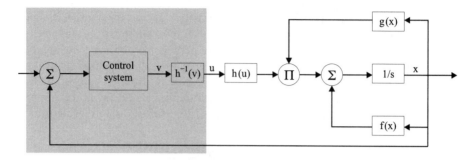

Figure 6.1 Actuator nonlinearity compensation. The shaded area contains the control system.

where **x** is the state of the system, $u(t)$ is the control input, and f and g are unknown nonlinearities. Let $\hat{f}$ and $\hat{g}$ represent approximations to the unknown functions f and g. Then, a control law can be defined as

$$u(t) = \frac{1}{\hat{g}(\mathbf{x}(t))}(v(t) - \hat{f}(\mathbf{x}(t))) \qquad (6.1)$$

when $\hat{g}(\mathbf{x}(t)) \neq 0$ where $v(t)$ can be specified as a function of the tracking error to meet the performance specification [23, 29, 30]. If the approximations were exact, then this control law would cancel the plant dynamics resulting in

$$\dot{x}_n(t) = v(t).$$

When the approximators involve error, it can be shown that the tracking error is directly related to the error in the function approximation [10]. Therefore, the designer will be interested in ensuring the convergence of the on-line approximator to the unknown function.

For generalizations of the control law of Eq. (6.1), see [30]. For a discussion of alternative control laws used with approximation-based control methods, see, for example, Figure 17.4 in [13, 15], or Section 6.2.

6.1.1.2 Approximator

Having analyzed the control problem and specified a control architecture capable of using an approximated function to improve the system control performance, the designer must specify the form of the approximating function. This specification includes the definition of the inputs and outputs of the function, the domain D over which the inputs can range, and the structure of the approximating function.

For the discussion that follows, the approximating function will be denoted as $\hat{f}(\mathbf{x}; \theta, \sigma)$ where

$$\hat{f}(\mathbf{x}; \theta, \sigma) = \theta^T \phi(\mathbf{x}, \sigma) \qquad (6.2)$$

where in this notation **x** is a dummy variable. The actual function inputs may include elements of the state, control input, or outputs. The notation $\hat{f}(\mathbf{x}; \theta, \sigma)$ implies that $\hat{f}$ is evaluated as a function of **x** when θ and σ are considered fixed for the purposes of function evaluation. In applications, the approximator parameters θ and σ will be adapted on-line to improve the accuracy of the approximating function.[1] The (neural network) literature refers to the parameters θ as the output layer parameters. The parameters σ are referred to as the input layer parameters. Note that the approximation of Eq. (6.2) is linear with respect to θ. The vector of basis functions ϕ will be referred to as the regressor vector. For the applications of interest in this chapter, the regressor vector is typically a nonlinear function of **x** and the parameter vector σ. Specification of the structure of the approximating function includes selection of the basis elements of the regressor ϕ, the dimension of θ, and the dimension of σ. The values of θ and σ are determined through parameter estimation methods based on the on-line data.

The approximator structure defined in Eq. (6.2) is sufficient to describe the various approximators used in the neural and fuzzy control literature, as well as many other approximators. In this chapter, we will not discuss specific types of approximators. Instead, we will analyze approximation-based control from a unifying perspective. Section 6.3 analyzes the properties of approximators as they relate to approximation-based control methods. References to publications discussing specific approximator structures are, for example, B-splines [7], CMAC [1], fuzzy logic [34, 39, 40], radial basis functions [5, 27], sigmoidal neural networks [28], and wavelets [33].

Regardless of the choice of the function approximator and its structure, normally perfect approximation is not possible. The approximation error will be denoted $e(\mathbf{x}; \theta, \sigma)$ where

$$e(\mathbf{x}; \theta, \sigma) = f(\mathbf{x}) - \hat{f}(\mathbf{x}; \theta, \sigma). \qquad (6.3)$$

If θ^* and σ^* denote the parameters that minimize the norm of the approximating error, then

$$e(\mathbf{x}) = e(\mathbf{x}; \theta^*, \sigma^*) = f(\mathbf{x}) - \hat{f}(\mathbf{x}; \theta^*, \sigma^*).$$

In applications, the quantities $e(\mathbf{x}), \theta^*$ and σ^* are not known but are useful for the purposes of analysis.

6.1.1.3 Stable Training Algorithm

Given that the control architecture and approximator structure have been selected, the designer must specify the algorithm for adapting the adjustable parameters θ and σ of the approximating function based on the on-line data and control performance.

Section 6.4 presents parameter estimation techniques and analyzes the related theoretical issues. The main issue to be considered in the development of the parameter estimation algorithm is the overall stability of the closed-loop control system. The stability of the closed-loop system requires guarantees of the convergence of the system state and of (at least) the boundedness of the error in the approximator parameter

---

[1] This is referred to as *training* in the neural network literature.

vector. This analysis must be completed with caution, as it is possible to design a system for which the system state is asymptotically stable while

1. even when perfect approximation is possible (i.e., $e(\mathbf{x}) = 0$), the error in the estimated approximator parameters is bounded but not convergent;
2. when perfect approximation is not possible, the error in the estimated approximator parameters may become unbounded.

In the first case, the lack of approximator convergence is due to lack of persistent excitation, which is discussed further in Section 6.4. This lack of approximator convergence may be acceptable, if the approximator is not needed for any other purpose, since the control performance is still achieved. However, control performance will improve as approximator accuracy increases. Also, the designer of a control system involving on-line approximation usually has interest in the approximated function and is therefore interested in its accuracy. In such cases, the designer must ensure the convergence of the control state and approximator parameters. In the second case, the fact that $e(\mathbf{x})$ cannot be forced to zero (the typical situation) must be addressed in the design of the parameter estimation algorithm. Such algorithms are discussed in Section 6.4.5.

6.1.2 Problem Statement

Given the discussion of the previous subsections, the approximation-based control problem can be summarized as follows.

Approximation-Based Control Problem. Given plant input-output data $\mathbf{z}(t) = (\mathbf{u}(t), \mathbf{y}(t))$ in compact set D

1. specify a control architecture utilizing an approximated function $f(\mathbf{z}(t))$;
2. find a positive integer M, vectors $\theta \in R^{M_1}$ and $\sigma \in R^{M_2}$ ($M = M_1 + M_2$), and a family of approximators $\hat{f}(\mathbf{z}; \theta, \sigma)$ such that for a cost function of the form

$$J(\theta, \sigma) = \int_D \|f(\mathbf{z}) - \hat{f}(\mathbf{z}; \theta, \sigma)\|^2 d\mathbf{z} \qquad (6.4)$$

there exists $(\theta^*, \sigma^*) \in R^M$ such that $(\theta^*, \sigma^*) = \mathrm{argmin}_{(\theta,\sigma)} J(\theta, \sigma)$, and the closed-loop system achieves the specified level of performance;
3. find an estimation algorithm $(\hat{\theta}(t), \hat{\sigma}(t)) = A(\mathbf{z}(\tau)), \tau \in [0, t]$ such that $(\hat{\theta}(t), \hat{\sigma}(t))$ approaches (θ^*, σ^*) and the closed loop system is stable.

Therefore, the designer has to select a family of approximators, an estimation algorithm, and a control methodology. The designer should be interested in proving (under reasonable assumptions) that

1. the tracking error $x(t) - x_d(t)$ is bounded and asymptotically approaches zero (or a small neighborhood of the origin); and

Section 6.1 Introduction

2. the function approximation error $f(\mathbf{z}) - \hat{f}(\mathbf{z})$ is bounded over D and asymptotically approaches zero (or is asymptotically less than some ϵ over D).

6.1.3 Discussion

The objective of on-line approximation-based control methods is to achieve a higher level of control system performance than could be achieved based on the *a priori* model information. Such methods can be significantly more complicated (computationally and theoretically) than nonadaptive or even linear adaptive control methods. This extra complication can result in unexpected behavior (e.g., instability) if the design is not rigorously analyzed under reasonable assumptions.

On-line function approximation has an important role to play in the development of advanced control systems. On-line function approximation-based control, including neural and fuzzy approaches, has become feasible in recent decades as a result of the rapid advances that have occurred in computing technologies. These advances have also spurred the reemergence of neural network research. Various motivations have been cited in the literature for the use of neural control. A few of the motivations are as follows:

- Neural networks are universal approximators. As discussed in Section 6.3.1, numerous families of approximators have this or related properties. Therefore, the fact that neural networks are universal approximators is not a motivation for using them over any other approximator with the same property.
- Neural networks are popular, convenient, or easy to compute. All of these are weak motivations.
- Neural networks are trainable by backpropagation (gradient descent). Gradient descent parameter adjustment applies to many families of approximators as long as the resultant approximator is a continuous function of the parameters. However, gradient descent is not a strong motivation for using a given approximator owing to the lack of robustness to residual approximation error as discussed in Section 6.4.5.
- Neural networks use distributed information processing. Distributed information processing refers to knowledge stored over many parameters and computations completed over many nodes. The claim is that this produces fault tolerance. This claim is justified by the analogy to biological systems. However, the neural networks that are typically implemented are much smaller and simpler than such biological systems, resulting in a weak analogy. In fact, additional parameter adjustment should be expected after node failures before performance might be recovered. Several other approximators can make the same distributed information processing claims. In addition, if the approximator is implemented on a traditional single-processor computer (as is the case in the vast majority of applications), then it is not possible for a "single nodal processor" to fail.
- Neural networks offer the inherent potential for parallel computation. Any approximation structure that can be written in vector product form is suitable for parallel implementation. Interesting questions are whether any particular application is worth special hardware, or more generally, whether any particular

approximation structure merits additional research funding to develop special hardware.

This list questions several typical motivations for the use of neural networks in approximation-based control applications. The intent is not to show that neural networks should not be used, but to encourage more careful consideration of the motivations before choosing a particular function approximator. Alternative motivations are discussed in greater depth in Section 6.3.

6.2 CONTROL ARCHITECTURES

An approximation-based controller is formed by combining one or more on-line approximators, which provide estimates of the unknown functions at each instant, with a control law, whose objective is to use the known components of the plant and the on-line estimates of the unknown components in order to achieve a desired control performance.

There are two approaches for combining the control law and the on-line approximation functions. In the first approach, referred to as *indirect control*, the on-line approximator is used to estimate the unknown nonlinearities of the plant. Based on these functional estimates, the control law is computed by treating the estimates as if they were the true functions, based on the *certainty equivalence principle* [2]. In the second approach, referred to as *direct control*, the on-line approximator is used to estimate directly the unknown nonlinear controller functions.

To illustrate the concepts of indirect and direct control, consider the problem of controlling an nth order single-input system of the form

$$\dot{\mathbf{x}} = \mathbf{f}(\mathbf{x}) + \mathbf{g}(\mathbf{x})u,$$

where the vector functions $\mathbf{f}$ and $\mathbf{g}$ are assumed to be unknown. According to the indirect control approach, two on-line approximation functions, denoted by $\hat{\mathbf{f}}(\mathbf{x})$ and $\hat{\mathbf{g}}(\mathbf{x})$, will be employed to estimate the unknown functions $\mathbf{f}(\mathbf{x})$ and $\mathbf{g}(\mathbf{x})$, respectively. By processing the input $u(t)$ and state variables $\mathbf{x}(t)$ in real time, on-line parameter estimation methods are designed for updating the parameters associated with each approximation function $\hat{\mathbf{f}}(\mathbf{x})$ and $\hat{\mathbf{g}}(\mathbf{x})$, as shown in Figure 6.2. These functional estimates are then used in place of the unknown functions in the control law. For example, for feedback linearizing control, the idea is to cancel the nonlinearities in the feedback loop and then employ standard linear control design methods in order to achieve a desired control performance. Alternatively, in direct control the approach is to approximate the controller functions directly without approximating $\mathbf{f}(\mathbf{x})$ and $\mathbf{g}(\mathbf{x})$. Therefore, for feedback linearization the control law

$$u = \alpha(\mathbf{x}) + \beta(\mathbf{x})v,$$

which is used to linearize the system from v to $\mathbf{x}$, is approximated by

$$u = \hat{\alpha}(\mathbf{x}) + \hat{\beta}(\mathbf{x})v,$$

where $\hat{\alpha}$ and $\hat{\beta}$ are on-line approximation functions (see Figure 6.3).

Section 6.2 Control Architectures

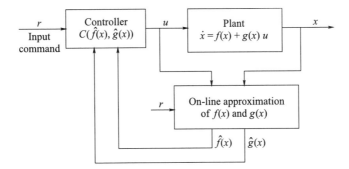

Figure 6.2 Indirect control architecture.

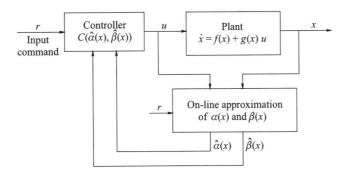

Figure 6.3 Direct control architecture.

Both the direct and indirect approaches present formidable challenges in developing provably stable on-line approximation control systems.

6.2.1 Indirect Methods

In indirect control approaches, the approximation function is used to estimate the unknown nonlinearities of the system. Therefore, the plant model $\mathcal{P}(\mathbf{f})$ is characterized in terms of all the unknown nonlinearities $\mathbf{f} = \{f_1, f_2, \ldots f_s\}$. For example, in a linearized system, the linear model may need to be augmented by unknown nonlinearities that represent higher-order terms left out during small-signal linearization. Approximation functions with on-line parameter estimation generate estimates $\hat{\mathbf{f}} = \{\hat{f}_1, \hat{f}_2, \ldots \hat{f}_s\}$, corresponding to each unknown function f_i. Each of these function estimates is generated in real time (or in almost real time) by processing the plant input $u(t)$ and output $y(t)$. The on-line approximators given by $\hat{f}$ yield an estimated plant model $\widehat{\mathcal{P}}(\hat{f})$, which is updated continuously. According to the indirect control approach, the estimated plant model $\widehat{\mathcal{P}}(\hat{f})$ is treated as the "true" plant, and a control law is designed for it.

An indirect control design consists of two separate tasks: (1) the on-line approximation of the unknown nonlinearities and (2) the nonlinear feedback control design. In principle, any on-line approximation algorithm can be combined with any nonlinear feedback control law in constructing an indirect on-line approximation control scheme. Indeed, one of the key appealing features of indirect control is the versatility to unify various on-line approximation schemes and feedback control laws. However, combining a stable estimation scheme with a stable control law does *not* necessarily imply that the overall scheme will be stable. Unlike linear systems, where separation of identification and control can be achieved using adaptive linear control techniques, for nonlinear systems the problem is more difficult. A cause of this difficulty is the difference in instability behavior between linear and nonlinear systems. Although the states of an unstable linear system remain bounded over any finite interval, in the case of nonlinear systems the states may become unbounded in finite time. Therefore, even small approximation errors may cause the state of the system to become unbounded in finite time—before the on-line approximation is able to "learn" the unknown nonlinearity. In general, because of difficulties in establishing stability of indirect control schemes for on-line approximation, care needs to be taken in their use.

In the practical implementation of indirect control schemes, both on-line approximation and control can be carried out synchronously at every instant of time, or asynchronously after processing some of the data over some period of time. For example, in the presence of noise and/or external disturbances, it is common to perform updates on the on-line approximation function at every instant of time but to update the control law over a slower time-scale.

6.2.2 Direct Methods

In direct control approaches, on-line approximation is performed directly on unknown functions in the control law. In order to design a direct control scheme, the plant model $\mathcal{P}(\mathbf{f})$ needs to be converted into a controller structure $\mathcal{C}(\alpha)$ that meets the performance requirements. The controller structure is characterized in terms of unknown control functions $\alpha = \{\alpha_1, \alpha_2, \ldots \alpha_q\}$, which appear because of the unknown plant functions $\mathbf{f}$. In the direct control approach, the on-line approximators are designed to approximate each unknown function α_i. Therefore, processing the plant input $u(t)$ and output $y(t)$ yields the estimated controller functions directly.

One appealing feature of direct control schemes is the ability to aggregate several unknown nonlinearities into one collective term. This can be especially useful for complex nonlinear systems where tracing the propagation of unknown nonlinearities into the design of a feedback control law can be intractable. The major source of difficulty in designing direct control schemes appears in the selection of the controller structure $\mathcal{C}(\alpha)$ and especially in the derivation of adaptive laws for updating the estimated parameters of the approximation functions. A useful tool in the design of direct control schemes is Lyapunov's stability theory [18], which is discussed in Section 6.4.4.

6.3 APPROXIMATOR PROPERTIES

This section focuses on the properties of families of function approximators. For each property, the technical meaning of the property will first be presented. Then the property will be interpreted in the context of approximation-based control applications.

6.3.1 Universal Approximator

For approximation-based control applications, a fundamental question is whether a particular family of approximators is capable of providing a close approximation to the function $f(\mathbf{x})$. There are at least three interesting aspects of this question:

1. Is there some subset of a family of approximators that is capable of providing an accurate approximation to $f(\mathbf{x})$?
2. If there exists some subset of the family of approximators that is capable of providing an accurate approximation, can the designer specify an approximation structure in this subset *a priori*?
3. Given that an approximation structure can be specified, can appropriate parameter vectors θ and σ be estimated using data obtained, while ensuring stable operation, during on-line system operation?

To answer these questions satisfactorily, various technical issues must be addressed. This section seeks to present a readable, yet rigorous, combination of the results of [14, 32] to analyze the first question. To enhance readability, some technical detail has been removed. The reader interested in a more complete discussion should consult the numerous articles on *universal approximation* (e.g., [6, 11, 14, 32]).

Definition 6.3.1 (Affine Functions) *For any $r \in \{1, 2, 3 \ldots\}$, $A^r : \Re^r \to \Re$ denotes the set of affine functions of the form*

$$A(\mathbf{x}) = \mathbf{w}^T \mathbf{x} + b$$

where $\mathbf{w}, \mathbf{x} \in \Re^r$ and $b \in \Re$.

Definition 6.3.2 (Single Hidden Layer Networks) *The family of r input, N node, single hidden layer network approximators associated with nodal processor $g(\cdot)$ is defined by*

$$S_{r,N} = \left\{ f : \Re^r \to \Re \,\middle|\, f(\mathbf{x}) = \sum_{i=1}^{N} \theta_i g(A_i(\mathbf{x})), \mathbf{x} \in \Re^r, \theta \in \Re^N, and A_i \in A^r \right\}.$$

The fact that various approximators, including sigmoidal networks and radial basis functions, can be coerced into this form is demonstrated in [32]. Any single hidden layer network can be written in the form of Eq. (6.2) by defining $\phi_i(\mathbf{x}, \sigma)$ to be $g(A_i(\mathbf{x}))$, where σ is a vector composed of the elements of $\mathbf{w}$ and b. Specification of a unique single hidden layer network approximator requires definition of the following 5-tuple $\mathcal{F} = (r, N, g, \theta, \sigma)$. If all parameters except for θ are specified, then we have a linear-in-the-parameters estimation problem (see Section 6.4).

Definition 6.3.2 explicitly defines single-output network functions. The definition of vector output network functions is a direct extension of the definition, where each vector component is defined as in Definition 6.3.2 and θ is a matrix. With the definition of vector output single hidden layer networks, multi-hidden layer networks can be defined by using the vector output from one network as the vector input to another network. The discussion that follows focuses on single hidden layer networks. Similar results apply to multi-hidden layer networks [11, 32].

To state the theorem that follows requires that two classes of nodal processors be specified.

Definition 6.3.3 (Squashing functions) *The nodal processor $g(\cdot)$ is a squashing function if $g(\cdot)$ is a non-constant, continuous, bounded, and monotone increasing function of its scalar argument.*

Definition 6.3.4 (Local functions) *The nodal processor $g(\cdot)$ is a local function if $g(\cdot)$ is continuous, $g(\cdot) \in \mathcal{L}_1 \cap \mathcal{L}_p, 1 \leq p < \infty$ and $\int g d\mu \neq 0$ for Lebesgue measure μ.*

Combining the results of [11, 14, 32],[2] the following theorem results.

Theorem 6.3.1 (Universal Approximation) *If $g(\cdot)$ satisfies either Definition 6.3.3 or 6.3.4, f is continuous on the compact set $D \in \Re^r$, and S is the family of approximators defined in Definition 6.3.2, then for a given ϵ there exist $\bar{N}(\epsilon)$ such that for $N > \bar{N}(\epsilon)$ there exist $\hat{f} \in S_{r,N}$ such that*

$$\rho(f,\hat{f}) < \epsilon$$

for an appropriately defined metric ρ for functions on D.

Results such as Theorem 6.3.1 are referred to as *universal approximation* results. Approximators that satisfy such theorems are referred to as *universal approximators*. Universal approximation theorems such as this state that under reasonable assumptions on the nodal processor and the function to be approximated, if the (single hidden layer) network approximator has enough nodes, then an accurate network approximation can be constructed by selection of θ and σ. Such theorems do not provide constructive methods for determining appropriate values of $\bar{N}, \theta$, or σ.

Universal approximation results are one of the most typically cited reasons for applying neural or fuzzy techniques in control applications involving significant unmodeled nonlinear effects. The reasoning is along the following lines: The dynamics involve a function $f(\mathbf{x}) = f_0(\mathbf{x}) + \Delta f(\mathbf{x})$ where $\Delta f(\mathbf{x})$ has a significant effect on the system performance and is known to have properties satisfying a Universal Approximation Theorem, but $\Delta f(\mathbf{x})$ cannot be accurately modeled *a priori*. Based on universal approximation results, the designer knows that there exists some subset of $\mathcal{F}$ that approximates $\Delta f(\mathbf{x})$ to an accuracy ϵ for which the control specification can be achieved. Therefore, the approximation-based control problem reduces to finding $\hat{f} \in \mathcal{F}$ that satisfies the ϵ accuracy specification. Most articles in the literature address the third question stated at the beginning of this section: selection of θ or (θ, σ) given that the remaining elements of $\mathcal{F}$ have been specified. However, selection of N for a given choice of g and σ (or (N, σ) for a specified g) is the step in the design process that limits the approximation accuracy that can ultimately be achieved. To cite universal approx-

---

[2] The results of these articles are more general than the theorem that follows but require a more technical discussion.

Section 6.3 Approximator Properties

imation results as a motivation and then select N as some arbitrary, small number is essentially contradictory.

Starting with the motivation stated in the previous paragraph, it is reasonable to derive stable algorithms for adaptive estimation of θ (or (θ, σ)) if N is specified large enough that it can be assumed larger than the unknown $\bar{N}$. Specification of too small a value for N defeats the purpose of using a universal approximation-based technique. When N is selected too small but a provably stable parameter estimation algorithm is used, stable (even satisfactory) control performance is still achievable; however, accurate approximation will not be achievable. Unfortunately, the parameter $\bar{N}$ is typically unknown, since $\Delta f(\mathbf{x})$ is not known. Therefore, the selection of N must be made overly large to ensure accurate approximation. The tradeoff for overestimating the value of N is the larger memory and computation time requirements of the implementation. In addition, if N is selected too large, then the approximator will be capable of fitting the measurement noise as well as the function. Fourier-analysis-based methods for selecting N are discussed in [29]. Online adjustment of N is an interesting area of research which tries to minimize the computational requirements while minimizing ϵ and ensuring stability.

Results such as Theorem 6.3.1 provide sufficient conditions for the approximation of continuous functions over compact domains. Other approximation schemes exist which do not satisfy the conditions of these particular theorems but are capable of achieving ϵ approximation accuracy. For example, the Stone-Weierstrass Theorem shows this property for polynomial series. In addition, some classical approximation methods can be coerced into the form necessary to apply the universal approximation results. Therefore, there exist numerous approximators capable of achieving ϵ approximation accuracy. The decision among them should be made by considering other approximator properties and carefully weighing their relative advantages and disadvantages.

6.3.2 Parameter (Non)Linearity

An initial decision that the designer must make is whether σ will be fixed *a priori* (i.e., $\sigma(t) = \sigma(0)$ and $\dot{\sigma} = 0$) or adapted on-line (i.e., $\sigma(t)$ is a function of the on-line data and control performance). If σ is fixed during on-line operation, then the function approximator is linear in the remaining adjustable parameters θ, so that the designer has a linear-in-the-parameter on-line function approximation problem. Proving theoretical issues, such as closed-loop system stability, is easier in the linear-in-parameter (LIP) case. In the case where the approximating parameters σ are fixed, these parameters will be dropped from the approximation notation, yielding

$$\hat{f}(\mathbf{x}, \theta) = \theta^T \phi(\mathbf{x}). \tag{6.5}$$

Fixing σ is beneficial in terms of simplifying the analysis but is limiting in terms of the functions that can be accurately approximated. To obtain a linear in the parameters function approximation problem, the designer must specify *a priori* (r, N, G, σ). If these parameters are not specified judiciously, then the desired ϵ accuracy may not be achievable for any value of θ. For later use, we define:

Definition 6.3.5 (Linear-in-Parameter Approximators) *The family of r input, N node, LIP approximators associated with nodal processor $g(\cdot)$ is defined by*

$$S_{r,N,g,\sigma} = \left\{ f : \Re^r \to \Re \,\middle|\, f(\mathbf{x}) = \sum_{i=1}^{N} \theta_i g_i(\mathbf{x}), \mathbf{x} \in \Re^r, \text{ and } \theta \in \Re^N \right\}. \tag{6.6}$$

In addition to simplifying the theoretical analysis, an additional motivation for the desire to use LIP approximations is that such approximators have a single global minimizing parameter vector (i.e., there are no local minima).

Theorem 6.3.2 (Unique Minimum) [8] *Given an approximator of the form Eq. (6.5), for any N, there exists a unique $\theta^* \in R^N$ such that $f(x) = (\theta^*)^T \phi(x) + e_f^*(x)$ where*

$$\theta^* = \operatorname{argmin}_\theta \int_D \|f(x) - \hat{f}(x : \theta)\|^2 dx. \tag{6.7}$$

In addition, there are no (non-global) local minima of the cost function.

Given that there exists a minimizing parameter vector, the uniqueness of θ^* can be proven by expanding Eq. (6.7) and noticing that it is quadratic in θ. Therefore, a major advantage of LIP approximators is that there exists a single global minimizing parameter vector. When a nonlinear in the parameter approximator is selected, there may be several local minima in the space of possible parameters. If the estimated parameter vector starts out in the basin of attraction of a local minimum, the estimated parameters will converge to the local minimum. In this case, it is immaterial that the global minimizing parameter vector achieves ϵ approximation accuracy if the parameter vector at the local minimum does not. An additional motivation for the use of LIP approximators is discussed in Section 6.3.3.

The relative drawbacks of approximators that are linear in the adjustable parameters are discussed, for example, by Barron in [3]. Barron shows that given certain technical assumptions, approximators that are nonlinear in their parameters have squared approximation errors of order $O(\frac{1}{N})$, while approximators that are linear in their parameters cannot have squared approximation errors smaller than order $\frac{1}{N^{2/d}}$ (N is the number of parameters, and d is the dimension of domain D). In spite of these disadvantageous order of approximation comparisons for high-dimension input domains, there is still significant interest in linear-in-parameter approximators. First, the theoretical performance guarantees necessary in dynamic applications prior to implementation may not be available for approximators with nonlinear parameter dependence. Second, when the approximator is linearly parameterized and the basis elements are local, significant computational advantages result [10]. Third, in low-input-dimension applications, more detailed analysis than order of approximation arguments is required to determine the relative merits of linear- or nonlinear-in-parameter approximators.

6.3.3 Best Approximator Property

Universal approximation theorems of the type discussed in Section 6.3.1 analyze the problem of whether for a family of function approximators $S_{r,N}$, there exists $a \in S_{r,N}$ that approximates a given function with at most ϵ error over a region D. This section considers an interesting related question: Given a convergent sequence of approximators $\{a_i\}$, $a_i \in S_{r,N}$, is the limit point of the sequence in the set $S_{r,N}$? If the limit point is guaranteed to be in $S_{r,N}$, then the family of approximators is said to have the "best approximator" property.

Let f be a continuous function on D (i.e., $f \in C(D)$). Let $S_{r,N}$ be a family of approximators defined on D such that $S_{r,N} \subset C(D)$. If the norm for functions in $C(D)$ is denoted by $\|\cdot\|$, then the distance between two elements of $C(D)$ will be defined as $\rho(f, g) = \|f - g\|$. The distance from f to $S_{r,N}$ is defined as $\rho(f, S_{r,N}) = \inf_{a \in S_{r,N}} \rho(f, a)$.

The best approximation problem [12] can be stated as: Given $f \in C(D)$ and $S_{r,N} \subset C(D)$, find $a^* \in S_{r,N}$ such that $\rho(f, a^*) = \rho(f, S_{r,N})$. Universal approximation theorems do not seek a best approximator, but rather an ϵ-accuracy approximator. However, a sequence of approximators can be conceived that achieve ϵ_i-accuracy approximation, where $\{\epsilon_i\}$ is a sequence that converges to zero. Depending on the properties of the set $S_{r,N}$, the limit point of such a sequence may or may not exist in $S_{r,N}$.

A set $S_{r,N}$ is called an *existence set* if for any $f \in C(D)$ there is at least one best approximation to f in $S_{r,N}$. A set $S_{r,N}$ is called a *uniqueness set* if for any $f \in C(D)$ there is at most one best approximation to f in $S_{r,N}$.

Proposition 4.2 of [12] shows that LIP approximators yield families of approximators (i.e., $S_{r,N,g,\sigma}$) that are existence sets. Nonlinear-in-parameter approximators may not have the best approximation property. In particular, [12] shows that radial basis functions with adaptive centers and sigmoidal neural networks with an adaptive input layer (or multiple adaptive layers) do not have the best approximator property.

6.3.4 Generalization

The term *generalization* is often used to motivate the use of neural network/fuzzy methods. The motivational phrase is typically of the form "neural networks have the ability to generalize from the training data." Analysis of such statements requires definition of the term *generalization*.

In [31] neural network applications are classified as either recognition or generalization. *Recognition* applications attempt to classify noisy inputs into one of a variety of categories that were deduced by the network during training (e.g., classify a handwritten character as one of the letters of a given alphabet). Fault identification applications could fall into this recognition category of applications. *Generalization* applications try to estimate the output value of a continuous function for given input values to the function. The estimated output value depends on the previous set of training data that was used to construct an approximating function to fit the training data in some well-defined sense. Most neural and fuzzy adaptive control applications fall into this generalization category.

The above categorization is not completely satisfying, since useful pattern recognition requires classification of input patterns outside the original training set. Therefore, recognition also incorporates the concept of generalization. Recognition can be interpreted as a mapping from a continuous set of real vectors to a set of *m* output values, where the output value indicates the appropriate classification. In this chapter, generalization will only be considered in the context of function approximation, as specified in Section 6.1.2.

As motivated in Section 6.1.2, the approximation-based control problem theoretically involves a cost function of the form:

$$J(\theta) = \int_D \|f(\mathbf{z}) - \hat{f}(\mathbf{z}; \theta)\|^2 d\mathbf{z}. \tag{6.8}$$

This cost function implies that the approximation error should be minimized by selection of θ over the region D. Unfortunately, the above approximation problem (as stated) can only be solved if $f(\mathbf{z})$ is known.

When $f(\mathbf{z})$ is not known, practical solutions to approximation-based control problems address the minimization of a cost function defined as a summation of sample errors

$$J_N(\theta) = \frac{1}{N} \sum_{i=1}^{N} \|y_i - \hat{f}(\mathbf{z}_i; \theta)\|^2 \tag{6.9}$$

where $y_i = f(\mathbf{z}_i)$ is known (or able to be estimated from noisy measurements) from variables sensed in the control application. Generalization refers to the capability of an approximator that minimizes the scattered data approximation cost function of Eq. (6.9) to also minimize the function approximation cost function of Eq. (6.8). This capability depends on (1) the degree of continuity of f and $\hat{f}$, (2) the available training data, and (3) the method of evaluation of the generalization results. Analysis of generalization claims should be split into analysis of the ability of approximators to *interpolate* and to *extrapolate*.

Interpolation is the process of providing an estimate of $f(\mathbf{z})$ at a point $\mathbf{z}$, where $\mathbf{z} - \mathbf{z}_i$ is small for some $1 \le i \le N$. Conceptually, interpolation averages appropriately weighted training points in the vicinity of the evaluation point. Therefore, interpolation is desirable as both a noise filtering and data reduction process. The capability of the function approximator to interpolate between training samples is necessary if the approximator is to make efficient use of memory and the training data.

Extrapolation is the process of providing an estimate of $f(\mathbf{z})$ at a point $\mathbf{z}$, where $\mathbf{z} - \mathbf{z}_i$ is large for all $1 \le i \le N$. Therefore, extrapolation attempts to predict the value of the function in a region far from the available training data. In off-line (batch) training scenarios, the set of training samples can be designed to be representative of the region D, so that extrapolation does not occur. In on-line control applications, operating conditions may force the designer to use whatever data the system generates, even if the training data do not representatively cover all of D. Since the class of functions to be approximated is large (i.e., all continuous functions on D) and the training data will include measurement noise, accurate extrapolation should not be expected. In fact, the control methodology should include provisions to accommodate regions of the state space for which adequate training has not occurred. Instead, the system should slowly move from regions for which accurate approximation has been achieved into regions still requiring exploration. In addition, it is desirable that exploration of new regions does not destroy approximation accuracy previously attained in other regions, which is one of the motivations for function approximators with locally supported influence functions.

6.3.5 Extent of Influence Function Support

In the specification of the approximators of Eqs. (6.2) or (6.5), a major factor in determining the ultimate performance that can be achieved is the selection of the influence functions $\phi(\mathbf{x})$. An important characteristic in the selection of ϕ is the extent of the support of the elements of ϕ which is defined to be $\text{Supp}_{\phi_i} = \{\mathbf{x} \in D | \phi_i(\mathbf{x}) \neq 0\}$. Let $\mu(A)$ be a function that measures the area of the set A. Then, the influence functions ϕ_i will be referred to as *global* influence functions if $\mu(\text{Supp}_{\phi_i}) = \mu(D)$. The influence functions ϕ_i will be referred to as *local* influence functions if $\mu(\text{Supp}_{\phi_i}) \ll \mu(D)$.

Based on the discussion in Section 6.3.4, the designer should not expect $\hat{f}$ to accurately extrapolate training data from one region into other (unexplored) regions. In addition, it is desirable that training data in new regions not affect the previously achieved approximation accuracy in distant regions. Both of these issues motivate the selection of local influence functions.

The on-line parameter estimation algorithms of Section 6.4 will adapt the parameter vector estimate $\hat{\theta}(t)$ based on the current tracking error $e(t)$. If the influence function ϕ_i has global support, then changing the estimated parameter $\hat{\theta}_i$ affects the approximation accuracy throughout D. Alternatively, if ϕ_i has local support, then changing the estimated parameter $\hat{\theta}_i$ affects the approximation accuracy only on Supp_{ϕ_i} which by assumption is a small region of D.

6.3.5.1 Approximators with Local Influence Functions

Several approximators with local influence functions have been proposed in the literature. This section analyzes such approximators in a general framework [9].
Local and global approximation structures can be distinguished as follows [9].

Definition 6.3.6 (Local Approximation Structure) *A function $\hat{f}(x, \hat{\theta})$ is a local approximation to $f(x)$ at x_0 if for any ϵ there exist $\hat{\theta}$ and δ such that $\| f(x) - \hat{f}(x, \hat{\theta}) \| \leq \epsilon$ for all $x \in B(x_0, \delta) = \{x | \|x - x_0\| < \delta\}$.*

Two common examples of local approximation structures are constant and linear functions. The constant, linear, or higher order polynomial function can be used to accurately approximate an arbitrary continuous function if the region of validity of the approximation is small enough.

Definition 6.3.7 (Global Approximation Structure) *A parametric model $\hat{f}(x, \hat{\theta})$ is an ϵ-accurate global approximation to $f(x)$ over domain D if for the given ϵ there exists $\hat{\theta}$ such that $\| f(x) - \hat{f}(x, \hat{\theta}) \| \leq \epsilon$ for all $x \in D$.*

The main objective of this section is to appropriately piece together a (large) set of local approximation structures to achieve a global approximation structure. The following definition of the class of *Basis-Influence Functions* [19] presents one means of achieving this objective.

Definition 6.3.8 (Basis-Influence Functions) *A function approximator is of the BI Class if and only if it can be written as*

$$\hat{f}(x, \hat{\theta}) = \sum_i f_i(x, \hat{\theta}; x_i) \Gamma_i(x; x_i) \qquad (6.10)$$

where each $f_i(x, \hat{\theta}; x_i)$ is a local approximation to $f(x)$ for all $x \in B(x_i, \delta)$, and $\Gamma_i(x; x_i)$ has local support S_i which is a subset of $B(x_i, \delta)$ such that $D \subseteq \bigcup_i S_i$.

Examples of basis-influence approximators include Boxes, CMAC [1], Radial Basis Functions [26], splines, and several versions of fuzzy systems [21, 34]. In the traditional implementation of each of these approximators, the basis functions are constant functions. If more capable basis functions (e.g., linear functions) were implemented, then the designer could expect a decrease in the number of required local approximation structures. Figure 6.4 illustrates basis-influence function approximation using linear approximations locally with normalized Gaussian influence functions. For clarity, the influence functions are plotted at a 10% scale and only a portion of each linear approximation is plotted. An alternative definition of local influence, which also provides a measure of the degree of localization based on the learning algorithm, is given in [35].

The partition of unity is defined as follows [37].

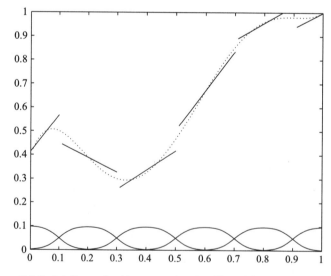

Figure 6.4 Basis-influence function approximation. The original function is shown as a dashed line. The local approximations (basis functions) are shown as solid lines. The influence functions (drawn at 10% scale) are shown as solid lines at the bottom of the figure.

Definition 6.3.9 (Partition of Unity) *The set of positive semi-definite influence functions $\{\Gamma_i\}(\mathbf{x})$ form a Partition of Unity on D if for any $\mathbf{x} \in D$, $\sum_{i=1}^{N} \Gamma_i(\mathbf{x}) = 1$.*

If a set of positive semi-definite influence functions $\{\bar{\Gamma}_i\}$ do not form a partition of unity, but have the coverage property (i.e., for any $\mathbf{x} \in D$ there exists at least one i such that $\bar{\Gamma}_i(\mathbf{x}) \neq 0$), then a partition of unity can be formed from $\{\bar{\Gamma}_i\}$ as

$$\Gamma_i(\mathbf{x}) = \frac{\bar{\Gamma}_i(\mathbf{x})}{\sum_{i=1}^{N} \bar{\Gamma}_i(\mathbf{x})}. \tag{6.11}$$

Function approximators with the partition of unity property, if well designed, are capable of accurate interpolation.

If the influence functions form a Partition of Unity, it can be shown that under the assumptions of Definitions 6.3.6 and 6.3.8, the basis-influence approximation achieves global ϵ approximation accuracy on D.

6.3.5.2 Lattice-Based Approximators

Specification of locally supported basis functions requires specification of the support of each basis element. Typically, this is implemented through the specification of center and width parameters of the basis elements. This specification includes the choice as to whether the center and width parameters are fixed *a priori* or estimated based on the acquired data.

Adaptive estimation of the center and width parameters is a nonlinear estimation problem. Therefore, the resulting approximator would not have the best approximator property but would have the beneficial order of approximation behavior discussed in Section 6.3.2.

Prior specification of the centers on a grid of points results in a *lattice-based approximator* [4]. Lattice-based approximators result in significant computational simplification over adaptive center-based approximators for two reasons. First, the center adaptation calculations are not required. Second, the nonzero elements of the vector ϕ can be determined without direct calculation of ϕ [10]. If the width parameters are also fixed *a priori*, then a linear parameter estimation problem results with the corresponding benefits.

6.3.5.3 Curse of Dimensionality

The main drawback of using locally supported basis elements is the fact that the required number of basis elements increases exponentially with the dimension of D. If D is d dimensional and m basis elements are allocated per dimension, then the total number of basis elements is m^d. This drawback is referred to as the *curse of dimensionality*.

6.3.6 Approximator Transparency

Approximator transparency refers to the designer's ability to preload *a priori* information into the function approximator and to interpret the approximated function as it evolves. Applications using fuzzy systems typically cite approximator transparency as a motivation. The fuzzy system can be interpreted as a rule base stating either the control value or control law applicable at a given system state [21, 34].

In any application, *a priori* information can be preloaded by at least two approaches. First, the function to be approximated can always be decomposed as

$$f(\mathbf{x}) = f_o(\mathbf{x}) + \Delta f(\mathbf{x}) \tag{6.12}$$

where $f_o(\mathbf{x})$ represents the known portion of the function and $\Delta f(\mathbf{x})$ represents the unknown portion for which an approximation will be developed on-line. In this case, the approximated function would be

$$\hat{f}(\mathbf{x}) = f_o(\mathbf{x}) + \Delta \hat{f}(\mathbf{x}).$$

Second, if for some reason, the approach described in Eq. (6.12) were not satisfactory, then $\hat{f}(\mathbf{x})$ could be initialized by off-line methods to accurately approximate the known portion of the function (i.e., $f_o(\mathbf{x})$). During on-line operation, the parameters of the approximator would be tuned to account for the unknown portion of the function (i.e., $\Delta \hat{f}(\mathbf{x})$).

Any approximator of the basis-influence class allows the user to interpret the approximated function. The influence functions dictate which of the basis functions are applicable (and the amount of applicability) at any given point.

6.4 PARAMETER ESTIMATION: ONLINE APPROXIMATION

This section examines the formulation of parametric models for the approximation problem and the derivation of parameter estimation algorithms with certain stability and robustness properties. Parameter estimation refers to the procedure for updating the parameters of the function approximator. For notational consistency and convenience, we focus on continuous-time parameter estimation methods. In general, there is no loss of generality in formulating the parameter estimation in continuous time since for all the algorithms discussed in this section, there are also corresponding discrete-time procedures [24].

6.4.1 Parametric Models

From a mathematical viewpoint, the selection of a function approximator provides a way to parameterize an unknown function. As discussed in Section 6.3, several approximator properties such as localization, generalization, and parametric linearity need to be considered. Once the structure of the approximation function has been selected, then the unknown function to be approximated is said to be parameterized and the problem reduces to one of parameter estimation. This problem falls within the domain of traditional adaptive control and on-line parameter estimation methods, provided that the structure of the on-line approximator remains fixed.

To further examine the construction of parametric models, let us focus on the on-line approximation problem of a nonlinear system represented by

$$\dot{\mathbf{x}}(t) = \mathbf{f}(\mathbf{x}(t), \mathbf{u}(t)), \tag{6.13}$$

where $\mathbf{u}(t) \in R^m$ is the control input vector, $\mathbf{x}(t) \in R^n$ is the state variable vector, and $\mathbf{f} : R^n \times R^m \mapsto R^n$ is a vector field representing the dynamics of the system. As discussed earlier, in most applications the vector field $\mathbf{f}$ is partially known either by analytical methods using first principles or by off-line identification methods. Therefore $\mathbf{f}$ can be decomposed as

$$\mathbf{f}(\mathbf{x}, \mathbf{u}) = \mathbf{f}_0(\mathbf{x}, \mathbf{u}) + \Delta\mathbf{f}(\mathbf{x}, \mathbf{u}),$$

where $\mathbf{f}_0$ represents the known system dynamics and $\Delta\mathbf{f}$ represents the discrepancy between the actual dynamics $\mathbf{f}$ and the nominal dynamics $\mathbf{f}_0$. The above decomposition is crucial because it allows the control designer to incorporate prior information; thereby the neural network (or other type of approximator) is needed to approximate only the uncertainty $\Delta\mathbf{f}$ (whose magnitude is typically small) instead of the overall function $\mathbf{f}$. Furthermore, if adaptation of the on-line approximator is disabled, then the residual controller is the one developed based on the nominal model, or a linear approximation of the nonlinear system in the case of linear control design methods.

The nonlinear system (6.13) can be rewritten as

$$\dot{\mathbf{x}} = \mathbf{f}_0(\mathbf{x}, \mathbf{u}) + \Delta\hat{\mathbf{f}}(\mathbf{x}, \mathbf{u}; \theta^*) + \mathbf{e}_f(\mathbf{x}, \mathbf{u}), \tag{6.14}$$

where $\Delta\hat{\mathbf{f}}$ is an approximating function of the type described in Section 6.3 and $\theta^* \in R^N$ is an "optimal" parameter vector that minimizes the cost function (6.4) between $\Delta\mathbf{f}$ and $\Delta\hat{\mathbf{f}}$ for all $(\mathbf{x}, \mathbf{u})$ belonging to a compact set D. The error term $\mathbf{e}_f$, defined as

$$\mathbf{e}_f(\mathbf{x}, \mathbf{u}) = \Delta\mathbf{f}(\mathbf{x}, \mathbf{u}) - \Delta\hat{\mathbf{f}}(\mathbf{x}, \mathbf{u}; \theta^*),$$

represents the *approximation error*, which is the minimum possible deviation between the unknown function $\Delta\mathbf{f}$ and the input-output function of the on-line approximator $\Delta\hat{\mathbf{f}}$. In general, increasing the number of adjustable parameters (denoted by N) reduces the function approximation error. Universal approximation results (discussed in Section 6.3.1) indicate that if N is sufficiently large, then $\mathbf{e}_f$ can be made arbitrarily small.

With a reasonably large number of parameters, the function approximation error, in general, is expected to be small (but not zero). The bound of the function approximation error is a critical quantity in approximation-based control, representing the optimal approximation capability of the selected function approximator within the compact region D. Linear modeling, which has dominated system theory and design during the last five decades, can be thought of as a special case of approximation-based control, where the approximators are linear models of the form $\hat{\mathbf{f}}(\mathbf{x}, \mathbf{u}) = \mathbf{Ax} + \mathbf{Bu}$. In the case of linear models, the approximation error $\mathbf{e}_f$ is zero at the point of linearization and may increase at state space regions farther away. The basic idea behind approximation-based control using nonlinear models is to expand the region where the approximation is valid from a small neighborhood around the linearizing point (in the case of linear models) to a larger region D, where D can be relatively large (i.e., defining the state space region of possible operation). It should be noted, however, that similar to linear control methods, if the state trajectories move outside the approximation region D, then the approximation-based controller may not be effective in achieving the desired control objectives.

In order to prevent the state trajectories from leaving the region D, some bound (possibly state-dependent) on the unknown function Δf is required. Otherwise, the state trajectories can move away from the desired trajectory faster than the feedback control can bring them back, possibly leading to instability. Unlike linear systems, where state trajectories can grow, at most, exponentially, in nonlinear systems the state trajectories can become unbounded in finite time. This is referred to as *finite escape time*. Therefore, in nonlinear systems the controller needs to be more aggressive in order to restrict the trajectories within a desired region. Several authors have designed control systems that assume known bounds on the unknown dynamics in order to restrict trajectories from leaving a specified region, and therefore obtaining global stability results (see, e.g., [22, 29]). These results employ the sliding-mode type of control methods to restrict the system within some desired region. Adaptive bounding methods have been used recently [23, 25] to relax some of the restrictive assumptions on the system uncertainty bounds.

If $\dot{\mathbf{x}}$ is available for measurement in Eq. (6.14), then the parameter estimation problem becomes a *static* nonlinear approximation problem of the general form

$$\bar{\mathbf{y}} = \Delta \hat{\mathbf{f}}(\mathbf{z}; \theta^*) + \mathbf{e}_f \tag{6.15}$$

where $\mathbf{z} = (\mathbf{x}, \mathbf{u})$ and $\bar{\mathbf{y}} = \dot{\mathbf{x}} - \mathbf{f}_0(\mathbf{x}, \mathbf{u})$ are measurable variables, $\mathbf{e}_f$ is the approximation error (or noise term), and θ^* is the unknown parameter vector to be estimated. Because in most applications only $\mathbf{x}$ is available for measurement and the use of differentiation is not desirable, the assumption of $\dot{\mathbf{x}}$ being available should be avoided. One way to avoid the use of differentiators is to use filtering techniques. By filtering each side of (6.14) with a first-order stable filter $\frac{1}{s+\lambda}$, where $\lambda > 0$, we obtain

$$\mathbf{y} = \frac{1}{s+\lambda}\left[\Delta \hat{\mathbf{f}}(\mathbf{z}, \theta^*)\right] + \delta \tag{6.16}$$

where[3]

$$\mathbf{y} = \frac{2}{s+\lambda}[\mathbf{x}] - \frac{1}{s+\lambda}[\mathbf{f}_0(\mathbf{x}, \mathbf{u})]$$

$$\delta = \frac{1}{s+\lambda}\left[\mathbf{e}_f(\mathbf{x}, \mathbf{u})\right]$$

In the special case where the approximation function is linearly parameterized (i.e., $\Delta \hat{\mathbf{f}}(\mathbf{z}; \theta^*) = (\theta^*)^T \phi(\mathbf{z})$), then (6.16) becomes a linear parametric model of the form

$$\mathbf{y} = (\theta^*)^T \zeta + \delta \tag{6.17}$$

where ζ is a vector of the filtered version of each basis; that is, $\zeta = \frac{1}{s+\lambda}[\phi(\mathbf{z})]$.

---

[3] The notation $y = H(s)[x]$, where $H(s)$ is a stable transfer function, is to be interpreted as $y(t)$ being the output of a linear system $H(s)$ with $x(t)$ as input.

Next we consider various on-line adaptive techniques for the estimation of θ^*. The gradient and least-squares methods are optimization-based methods, where the idea is to form an appropriate error function and minimize it using standard optimization techniques. Lyapunov-based methods, on the other hand, rely on the use of Lyapunov functions to derive a learning algorithm with inherent stability properties. In order to address the presence of the approximation error δ, in Section 6.4.5 we discuss the use of robust learning algorithms.

6.4.2 Gradient Algorithms

One of the most straightforward and widely used approaches for parameter estimation involves the use of the gradient (or steepest descent) method. The main idea behind the gradient method is to start with an initial estimate $\hat{\theta}(0)$ of the unknown parameter θ^* and to update at each time t the parameter estimate $\hat{\theta}(t)$ in the direction where the cost function $J(\hat{\theta})$ decreases the most. Several variations of the standard gradient algorithm have also been used in the parameter estimation literature. For example, the stochastic gradient approach leads to the well-known *least-mean-square* (LMS) algorithm, first developed by Widrow and Hoff [38]. Another useful modification of the gradient algorithm is the *gradient projection* algorithm, which restricts the parameter estimates in a specified region.

In this section, we focus on the deterministic, continuous-time version of the gradient learning algorithm. For continuous-time adaptive algorithms, infinitesimally small step lengths yield the following update law with respect to a specified cost function:

$$\dot{\hat{\theta}}(t) = -\nabla J(\hat{\theta}(t)),$$

where $\nabla J(\hat{\theta})$ denotes the gradient of the cost function J with respect to $\hat{\theta}$. Based on (6.17), if we minimize the cost function associated with the instantaneous error

$$J(\hat{\theta}) = \frac{1}{2}\Big(\mathbf{y}(t) - \hat{\theta}^T(t)\zeta(t)\Big)^T \Big(\mathbf{y}(t) - \hat{\theta}^T(t)\zeta(t)\Big) \tag{6.18}$$

we obtain the following gradient estimation algorithm:

$$\dot{\hat{\theta}}(t) = \Gamma \zeta(t)\Big(\mathbf{y}(t) - \hat{\theta}^T(t)\zeta(t)\Big), \tag{6.19}$$

where Γ is a positive-definite symmetric matrix representing the learning rate matrix and the initial condition is given by $\hat{\theta}(0) = \hat{\theta}_0$. In the special case where the same learning rate γ is used for each parameter estimate, then $\Gamma = \gamma \mathbf{I}$, where $\mathbf{I}$ is the identity matrix.

The *normalized gradient algorithm* is a variation of the gradient algorithm, which is sometimes used to improve the stability and convergence properties of the algorithm. The normalized gradient algorithm is described by

$$\dot{\hat{\theta}}(t) = \frac{\Gamma \zeta(t)\Big(\mathbf{y}(t) - \hat{\theta}^T(t)\zeta(t)\Big)}{1 + \beta \zeta^T(t)\zeta(t)},$$

where $\beta > 0$ is a design constant.

The backpropagation algorithm, which has been used extensively in the literature for training neural networks, is also a gradient-based algorithm. However, the extension of the backpropagation algorithm to dynamical systems (using learning algorithms such as *dynamic backpropagation* [20] and *backpropagation through time* [36]) yields adaptive laws that typically require the sensitivity $\frac{\partial \mathbf{x}}{\partial \theta^*}$ of the output $\mathbf{x}$ with respect to variations in the unknown parameters θ^*. Since these sensitivity functions are not available, implementation of such adaptive laws is not possible. In these cases, approximations of the sensitivity functions are used instead of the actual ones. One type of approximation used in dynamic backpropagation is to replace the gradient with respect to the unknown parameters by the gradient with respect to the estimated parameters. Such adaptive laws were used extensively in the early neural control literature, and simulations indicated that they performed well under certain conditions. Unfortunately, with approximate sensitivity functions, it is not possible, in general, to prove stability and convergence. It is interesting to note that approximate sensitivity function approaches also appeared in the early days of adaptive linear control, in the form of the so-called MIT rule [17].

One way to avoid the stability problems associated with approximate sensitivity functions is to reformulate the problem so that the cost function is convex with respect to the adjustable parameters. Based on the filtering techniques of Section 6.4.1, the cost function described by (6.18) with $\mathbf{y}$ as defined in (6.17) satisfies the convexity property for linearly parameterized approximators, and its gradient with respect to the estimated parameters is implementable ($\frac{\partial J}{\partial \theta}$ is calculable from available measurements). Therefore, the gradient algorithm described by (6.19) has some desirable stability properties, which are summarized as follows:

Theorem 6.4.1 (Stability of Gradient Algorithm) *Suppose the regressor vector is uniformly bounded (i.e., $\zeta \in \mathcal{L}_\infty$). If the on-line approximator is linearly parameterized (i.e., $\Delta \mathbf{f}(\mathbf{z}; \hat{\theta}) = \hat{\theta}^T \phi(\mathbf{z})$) and there is no approximation error (i.e., $\delta = 0$), then the gradient algorithm described by (6.17) and (6.19) has the following properties:*

(1) $(\mathbf{y}(t) - \hat{\theta}(t)\zeta(t)) \in \mathcal{L}_2 \cap \mathcal{L}_\infty,$ (2) $\hat{\theta}(t) \in \mathcal{L}_\infty,$

(3) $\lim_{t \to \infty}(\mathbf{y}(t) - \hat{\theta}(t)\zeta(t)) = 0,$ (4) $\lim_{t \to \infty} \dot{\hat{\theta}}(t) = 0.$

Even in the restrictive case of no approximation errors and a linearly parameterized approximator, it cannot be established that the parameter estimate vector $\hat{\theta}(t)$ will converge to the optimal vector θ^*. To guarantee that $\hat{\theta}(t)$ will converge to θ^*, the regressor vector $\zeta(t)$ needs to satisfy a so-called *persistency of excitation* condition. Intuitively, this implies that there should be sufficient variation in $\zeta(t)$ to allow the parameter estimates to converge to their optimal values. To get a basic idea of why persistency of excitation is important, consider the trivial case where $\zeta(t) = 0$. In this case, $y = (\theta^*)^T \zeta$ will be zero, and the parameter estimate will satisfy $\dot{\hat{\theta}} = 0$, which implies that $\hat{\theta}(t) = \hat{\theta}(0)$. Therefore, even though $y(t) - \hat{\theta}(t)\zeta(t) = 0$ for all $t \geq 0$, the estimated parameter vector $\hat{\theta}$ does not converge to θ^*, unless $\hat{\theta}(0)$ is incidentally selected to be the optimal parameter vector θ^*.

In the presence of approximation errors (i.e., $\delta(t) \neq 0$), the stability of the gradient algorithm (6.19) cannot be guaranteed. In fact, it is known from on-line parameter estimation of linear systems that even relatively small approximation errors are sufficient to make the adaptive system unstable. To address this problem, the standard update law described by (6.19) needs to be modified. Several modifications exist in the literature for enhancing the robustness of adaptive schemes. These modifications are discussed in Section 6.4.5.

6.4.3 Least-Squares Algorithms

Least-squares methods have been widely used in parameter estimation in both batch (nonrecursive) and recursive form [2, 16]. The basic idea behind the least-squares method is to fit a mathematical model to a sequence of observed data by minimizing the sum of the squares of the difference between the observed and computed data. To illustrate the least-squares method, consider the problem of computing the parameter vector $\hat{\theta}$ at time t that minimizes the cost function

$$J(\hat{\theta}) = \int_0^t \left(\mathbf{y}(\tau) - \hat{\theta}^T(t)\zeta(\tau) \right)^T \left(\mathbf{y}(\tau) - \hat{\theta}^T(t)\zeta(\tau) \right) d\tau,$$

where $\mathbf{y}(\tau)$ is the measured data at time τ, and $\zeta(\tau)$ is the regressor vector at time τ. The above cost function penalizes all the past errors $\mathbf{y}(\tau) - \zeta^T(\tau)\hat{\theta}(t)$ for $\tau = 0$ up to $\tau = t$, relative to the current parameter estimate $\hat{\theta}(t)$. By setting to zero the gradient (with respect to $\hat{\theta}$) of the cost function ($\nabla J(\hat{\theta}) = 0$), we obtain the least-squares estimate for $\hat{\theta}(t)$:

$$\hat{\theta}(t) = \left[\int_0^t \zeta(\tau)\zeta^T(\tau) d\tau \right]^{-1} \int_0^t \zeta(\tau) y^T(\tau) d\tau, \qquad (6.20)$$

provided that the inverse exists, which is a function of the level of regressor excitation.

The least-squares estimate given by (6.20) is derived for batch processing; in other words, all the data in the time interval $[0, t]$ is gathered before it is processed. In approximation-based control, the estimated parameter vector $\hat{\theta}(t)$ needs to be computed in real time, as new data becomes available. The recursive version of the least-squares algorithm is given by

$$\dot{\hat{\theta}}(t) = \mathbf{P}(t)\zeta(t)\left(\mathbf{y}^T(t) - \zeta^T(t)\hat{\theta}(t) \right) \qquad \hat{\theta}(0) = \hat{\theta}_0 \qquad (6.21)$$

$$\dot{\mathbf{P}}(t) = -\mathbf{P}(t)\zeta(t)\zeta^T(t)\mathbf{P}(t) \qquad \mathbf{P}(0) = \mathbf{P}_0 \qquad (6.22)$$

where $\mathbf{P}(t)$ is a square matrix of the same dimension as the parameter estimate $\hat{\theta}$. The initial condition $\mathbf{P}_0$ of the $\mathbf{P}$ matrix is chosen to be positive-definite. Because of the similarity of the recursive least-squares algorithm to the Kalman filter, when it is appropriately initialized the matrix $\mathbf{P}$ is called the *covariance matrix*.

The update law for $\hat{\theta}$, described by (6.21), is similar to the gradient learning algorithm (6.19), with $P(t)$ representing a time-varying learning rate. In practice, recursive least squares can converge considerably faster than the gradient algorithm at the expense of the increased computation required to compute $\mathbf{P}$. However, in its "pure" form, the recursive least squares may result in the covariance matrix $\mathbf{P}(t)$ becoming arbitrarily small. This problem, which is referred to as the *covariance wind-up* problem, can slow down adaptation in some directions and, as a result, critically dampen the ability of the algorithm to track time-varying parameters.

Several modifications to the "pure" least-squares algorithm have been considered. One such modification is *covariance resetting*, according to which the covariance matrix is reset to $\mathbf{P}(t_r) = \mathbf{P}_0$ at time t_r if the minimum eigenvalue of $\mathbf{P}(t_r)$ is less than a predefined small positive constant. This modification helps prevent the covariance matrix from becoming too small. Another commonly used modification to the least-squares algorithm leads to the *least-squares with forgetting factor*, which is given by

$$\dot{\hat{\theta}}(t) = \mathbf{P}(t)\zeta(t)\left(\mathbf{y}(t) - \zeta^T(t)\hat{\theta}(t)\right) \tag{6.23}$$

$$\dot{\mathbf{P}}(t) = -\mathbf{P}(t)\zeta(t)\zeta^T(t)\mathbf{P}(t) + \beta\mathbf{P}(t) \tag{6.24}$$

where $\beta > 0$ is typically a small positive constant. The extra term $\beta\mathbf{P}(t)$ in (6.24) prevents the covariance matrix from becoming too small; on the other hand, it may cause it to become too large. To avoid this complication, $\mathbf{P}(t)$ is either reset to $\mathbf{P}_0$ or adaptation is disabled (i.e., $\dot{\mathbf{P}}(t) = 0$) in the case that $\mathbf{P}(t)$ becomes too large. The literature on parameter estimation and adaptive control has several rules of thumb on how to choose the design variables that appear in the least-squares algorithm and its various modified versions.

The recursive least-squares algorithm described by (6.21) and (6.22) has similar stability properties as the gradient algorithm.

Theorem 6.4.2 (Stability of Recursive Least-Squares Algorithm) *Suppose the regressor vector is uniformly bounded (i.e., $\zeta \in \mathcal{L}_\infty$). If the on-line approximator is linearly parameterized (i.e., $\Delta\hat{\mathbf{f}}(\mathbf{z}; \hat{\theta}) = \hat{\theta}^T\phi(\mathbf{z})$) and there is no approximation error (i.e., $\delta = 0$), then the recursive least-squares algorithm described by (6.21) and (6.22) with $\mathbf{y}$ defined by (6.17) has the following properties:*

$(\mathbf{y}(t) - \hat{\theta}(t)\zeta(t)) \in \mathcal{L}_2 \cap \mathcal{L}_\infty$, $\quad\quad\hat{\theta}(t) \in \mathcal{L}_\infty$,
$\lim_{t\to\infty}(\mathbf{y}(t) - \hat{\theta}(t)\zeta(t)) = 0$, $\quad\quad\mathbf{P}(t) \in \mathcal{L}_\infty$,
$\lim_{t\to\infty}\hat{\theta}(t) = \bar{\theta}$, *where $\bar{\theta}$ is a constant vector.*

In comparing the stability properties of the gradient and least-squares algorithms, we notice that in addition to the other boundedness and convergence properties, the recursive least squares also guarantees that the parameter estimate $\hat{\theta}(t)$ converges to a constant vector $\bar{\theta}$. If the regressor vector ζ satisfies the persistency of excitation condition, then $\hat{\theta}(t)$ converges to the optimal parameter vector θ^*.

Despite its fast convergence properties, the recursive least-squares algorithm has not been widely used in problems involving large function approximation structures, mainly because of its heavy computational demands. Specifically, if the number of adjustable parameters is N, then updating of the covariance matrix $\mathbf{P}(t)$ requires adaptation of N^2 parameters.

6.4.4 Lyapunov-Based Algorithms

Lyapunov stability theory, and in particular Lyapunov's direct method, is one of the most celebrated methods for investigating the stability properties of nonlinear systems

[18]. The principal idea is that it enables one to determine whether or not the equilibrium state of a dynamical system is stable without explicitly solving the differential equation. The procedure for deriving such stability properties involves finding a suitable scalar function $V(\mathbf{x}, t)$, in terms of the state variables $\mathbf{x}$ and time t, and investigating its time derivative

$$\frac{d}{dt} V(\mathbf{x}, t) = \left(\frac{\partial V}{\partial \mathbf{x}}\right)^T \left(\frac{d\mathbf{x}}{dt}\right) + \frac{\partial V}{\partial t}$$

along the trajectories of the system. Based on the properties of $V(\mathbf{x}, t)$ (known as the Lyapunov function) and its derivative, various conclusions can be made regarding the stability of the system.

In general, there are no well-defined methods for selecting a Lyapunov function. However, in adaptive control problems, a standard class of Lyapunov function candidates is known to yield useful results. Furthermore, in some applications, such as mechanical systems, the Lyapunov function can be thought to represent a system's total energy, which provides an intuitive lead into selecting the Lyapunov function. In terms of energy considerations, the intuitive reasoning behind Lyapunov stability theory is that in a purely dissipative system the energy stored in the system is always positive and its time derivative is nonpositive.

The derivation of parameter estimation algorithms using the Lyapunov stability theory is crucial to the design of stable adaptive and learning systems. Historically, Lyapunov-based techniques provided the first algorithms for globally stable adaptive control systems in the early 1960s. In the recent history of neural control and adaptive fuzzy control methods, most of the results that deal with the stability of such schemes are based, to some extent, on Lyapunov-based algorithms. In many nonlinear control problems, Lyapunov stability theory is used not only for the derivation of learning algorithms but also for the design of the feedback control law.

In Lyapunov-based algorithms, the problem of designing an adaptive law is formulated as a stability problem in which the differential equation of the adaptive law is chosen so that certain stability properties can be established using Lyapunov theory. Because such algorithms are derived based on stability methods, by design they have some inherent stability and convergence properties.

According to the Lyapunov design method, an *estimation model* is derived based on Eq. (6.16). The estimation model is described by

$$\hat{\mathbf{y}} = \frac{1}{s + \lambda} \left[\Delta \hat{\mathbf{f}}(\mathbf{z}, \hat{\theta}) \right],$$

which in the case of linearly parameterized approximators becomes

$$\hat{\mathbf{y}} = \frac{1}{s + \lambda} \left[\hat{\theta}^T \phi(\mathbf{z}) \right].$$

A Lyapunov function is then selected, which is positive definite with respect to the estimation error $\mathbf{y} - \hat{\mathbf{y}}$ and the parameter estimation error $\theta^* - \hat{\theta}$. Taking the derivative of the Lyapunov function gives an expression in terms of $\hat{\theta}$. The idea is to select the righthand side of the adaptive law so that the derivative of the Lyapunov function is nonpositive, thus guaranteeing the boundedness of $\hat{\theta}(t)$ and $\hat{y}(t)$.

The Lyapunov design method generates the following adaptive law in the case of linearly parameterized approximators:

$$\dot{\hat{\theta}}(t) = \Gamma \phi(\mathbf{z}(t))(\mathbf{y}(t) - \hat{\mathbf{y}}(t)) \qquad (6.25)$$

The above parameter estimation algorithm derived using Lyapunov stability methods is essentially of the same form as the gradient algorithm (6.19) derived using optimization techniques. The stability and convergence properties of the two algorithms are also similar.

Theorem 6.4.3 (Stability of Lyapunov-Based Algorithm) *Suppose the regressor vector ϕ is uniformly bounded (i.e., $\phi \in \mathcal{L}_\infty$). If the on-line approximator is linearly parameterized (i.e., $\Delta\hat{\mathbf{f}}(\mathbf{z}; \hat{\theta}) = \hat{\theta}^T \phi(\mathbf{z})$) and there is no approximation error (i.e., $\mathbf{e}_f = 0$), then the Lyapunov-based algorithm described by (6.25) has the following properties:*

$$(\mathbf{y}(t) - \hat{\mathbf{y}}(t)) \in \mathcal{L}_2 \cap \mathcal{L}_\infty, \qquad \hat{\theta}(t) \in \mathcal{L}_\infty,$$
$$\lim_{t \to \infty}(\mathbf{y}(t) - \hat{\mathbf{y}}(t)) = 0, \qquad \hat{\mathbf{y}}(t) \in \mathcal{L}_\infty,$$
$$\lim_{t \to \infty} \dot{\hat{\theta}}(t) = 0.$$

The same remarks as in the gradient algorithm with regards to the persistency of excitation condition for parameter convergence are also valid here. Similarly, the Lyapunov-based algorithm (6.25) needs to be modified to handle approximation errors.

6.4.5 Robust Learning Algorithms

The learning algorithms described in Sections 6.4.2–6.4.4 are based on the assumption of no residual modeling errors. In other words, it was assumed that the only uncertainty in the dynamical system is due to $\Delta\mathbf{f}(\mathbf{x}, u)$, which can be represented *exactly* by an on-line approximation function $\Delta\hat{\mathbf{f}}(\mathbf{x}, \mathbf{u}; \theta^*)$ for some unknown parameter vector θ^*. In practice, the on-line approximation function $\Delta\hat{\mathbf{f}}(\mathbf{x}, \mathbf{u}; \hat{\theta})$ may not be able to match exactly the modeling uncertainty $\Delta\mathbf{f}(\mathbf{x}, \mathbf{u})$, even if θ was to be selected optimally. This discrepancy is usually called the approximation error, or the function reconstruction error. Furthermore, there may be unmodeled dynamics, and the measured input-output variables may be corrupted by noise and external disturbances. Unmodeled dynamics arise as a result of model reduction, which may be done either purposefully, in order to reduce the complexity of the model, or because of unknown dynamics of the full-order model. Indeed, in some cases (such as in the control of flexible structures) the full-order model may be of infinite dimension.

In this section, we consider modifications to the standard learning algorithms in order to provide stability and improve performance in the presence of modeling errors. These modifications lead to what is known as *robust learning algorithms*. The term *robust* is used to indicate that the learning algorithm is such that in the presence of modeling errors it retains its stability properties. It is well known from the adaptive control literature of linear systems [16] that in the presence of even small modeling errors, standard adaptive laws may cause the parameter vector $\hat{\theta}(t)$ to drift to infinity, a phenomenon usually referred to as *parameter drift*.

Intuitively, parameter drift occurs when the learning algorithm attempts to adjust the parameters in order to match a function for which an exact match does not exist for any value of the parameters (due either to approximation error or to other modeling errors such as external disturbances). Two approaches may be used to prevent parameter drift. In the first approach, the learning algorithm is modified so that it directly restricts the parameter estimates from drifting to infinity. The σ-modification, the ϵ-modification, and the projection algorithms belong to this category. In the second approach, the parameter estimates are prevented from drifting to infinity indirectly by not allowing the error, which is driving the learning algorithm, from becoming too small. The dead-zone algorithm has this characteristic.

To illustrate the various options for robustifying the adaptive laws discussed in Sections 6.4.2–6.4.4, we consider a generic adaptive law

$$\dot{\hat{\theta}}(t) = \Gamma \xi(t) \epsilon(t), \tag{6.26}$$

where Γ is the learning rate matrix, $\xi(t)$ is the regressor vector, and $\epsilon(t)$ is the estimation error. In the case of the gradient algorithm (6.19), the regressor is $\xi(t) = \zeta(t)$ and the estimation error is $\epsilon(t) = \mathbf{y}(t) - \hat{\theta}^T \zeta(t)$. For the Lyapunov-based algorithm given by (6.25), the regressor is $\xi(t) = \phi(\mathbf{z}(t))$, and the estimation error is $\epsilon(t) = \mathbf{y}(t) - \hat{\mathbf{y}}(t)$.

Based on (6.26), four different modifications for enhancing robustness are as follows.

Projection modification: One of the most straightforward and effective ways to prevent parameter drift is to restrain the parameter estimates within a predefined bounded and convex region $\mathcal{P}$. The projection modification implements this idea as follows: If the parameter estimate $\hat{\theta}$ is inside the desired region $\mathcal{P}$, or is on the boundary and directed inside the region $\mathcal{P}$, then the standard adaptive law (6.26) is implemented. In the case that $\hat{\theta}$ is on the boundary of $\mathcal{P}$ and its derivative is directed outside the region, then it is projected onto the tangent hyperplane. Therefore, the projection modification keeps the parameter estimation vector within the desired convex region $\mathcal{P}$ for all times. If $\mathcal{P}$ is selected to be sufficiently large so that it contains the optimal parameters θ^*, then it can be shown that in addition to the boundedness of the parameter estimates, the rest of the stability properties of the adaptive law are not affected.

σ-modification: In this approach, the adaptive law (6.26) is modified to

$$\dot{\hat{\theta}}(t) = \Gamma \xi(t) \epsilon(t) - \Gamma \sigma \hat{\theta}(t) \tag{6.27}$$

where σ is a small positive constant. The additional term $-\Gamma \sigma \hat{\theta}$ acts as a stabilizing component for the adaptive law. For example, if the parameter estimate $\hat{\theta}(t)$ starts drifting to large positive values, then $-\Gamma \sigma \hat{\theta}$ becomes large and negative, thus forcing the parameter estimate to decrease. Although the σ-modification does not require *a priori* information such as an upper bound on the approximation error, the robustness is achieved at the expense of destroying some of the convergence properties of the ideal (no approximation error) case. Therefore, several modifications have been suggested for addressing this issue, including the so-called switching σ-modification [16].

ϵ-modification: The ϵ-modification was motivated as an attempt to eliminate some of the drawbacks associated with the σ-modification. It is given by

$$\dot{\hat{\theta}}(t) = \mathbf{\Gamma}\xi(t)\epsilon(t) - \mathbf{\Gamma}|\epsilon(t)|\nu\hat{\theta}(t) \tag{6.28}$$

where $\nu > 0$ is a design constant. The idea behind this approach is to retain the convergence properties of the adaptive scheme by forcing the additional term $-\mathbf{\Gamma}|\epsilon|\nu\tilde{\theta}$ to be zero in the case that $\epsilon(t)$ is zero. In the case that the parameter estimate vector $\hat{\theta}(t)$ starts drifting to large values, then the ϵ-modification again acts as a stabilizing force.

Dead-zone modification: In the presence of approximation errors, the adaptive law (6.26) tries to minimize the estimation error ϵ, sometimes at the expense of increasing the magnitude of the parameter estimates. The idea behind the dead-zone modification is to enhance robustness by turning off adaptation when the estimation error becomes relatively small compared to the approximation error. The dead-zone modification is given by

$$\dot{\hat{\theta}}(t) = \begin{cases} \mathbf{\Gamma}\xi(t)\epsilon(t) & \text{if } |\epsilon| \geq \delta_0 \\ 0 & \text{if } |\epsilon| < \delta_0 \end{cases} \tag{6.29}$$

where δ_0 is a positive design constant that depends on the approximation error. One of the drawbacks of the dead-zone modification is that the designer needs an upper bound on the approximation error, which is usually not available.

In the presence of approximation errors (i.e., $\mathbf{e}_f \neq 0$), the above robust adaptive laws guarantee, under certain conditions, that the parameter estimates $\hat{\theta}(t)$ and the estimation error $\epsilon(t)$ remain bounded. Although, it cannot be established in the presence of approximation error that $\epsilon(t)$ will converge to zero, it can be shown that the estimation error is *small-in-the-mean* [16], in the sense that integral square error over a finite interval is proportional to the integral square approximation error.

6.5 CONCLUSIONS

On-line approximation-based control methods, including neural and fuzzy methods, offer a means to improve the performance of nonlinear control systems when the application involves functions that cannot be accurately modeled *a priori*. In addition, these methods provide the opportunity to develop a better understanding of the processes underlying the system to be controlled. Increased understanding is achieved through analysis of the approximated functions, if approximator convergence has been guaranteed in the control design.

Approximation-based control system design requires specification of a control architecture, an approximator structure, and a parameter estimation algorithm for which the stability of the overall system can be guaranteed under assumptions reflective of the application. This chapter has discussed each of these issues and provided references to articles that provide more in-depth discussion of the same issues.

> **Related Chapters**
>
> - For additional background material on neural networks and fuzzy logic see Ch. 5.
> - Applications of nonlinear approximation and approximation-based control are described in Chs. 5 and 16.
> - Ch. 4 reviews some other approaches to developing approximate models from data.

REFERENCES

[1] J. Albus, "A new approach to manipulator control: The cerebellar model articulation controller (CMAC)." *Trans. ASME, J. Dynamic Syst., Meas., Contr.*, Vol. 97, pp. 220–227, 1975.

[2] K. Åström and B. Wittenmark, *Adaptive Control*. Reading, MA: Addison-Wesley, 1995.

[3] A. Barron, "Universal approximation bounds for superpositions of a sigmoidal function." *IEEE Transactions on Information Theory*, Vol. 39, no. 3, pp. 930-945, 1993.

[4] M. Brown and C. Harris, *Neurofuzzy Adaptive Modelling and Control*. Englewood Cliffs, NJ: Prentice-Hall, 1994.

[5] D. Broomhead and D. Lowe, "Multivariable functional interpolation and adaptive networks," *Complex Systems*, pp. 321–355, 1988.

[6] G. Cybenko, "Approximation by superposition of a sigmoidal function." *Mathematics of Control, Signals, and Systems*, Vol. 2, no. 4, pp. 303–314, 1989.

[7] R. Eubank, *Spline Smoothing and Nonparametric Regression*. New York: Marcel Dekker, 1988.

[8] J. Farrell, "Motivations for local approximators in passive learning control." *Journal of Intelligent Control and Systems*, Vol. 1, no. 2, pp. 195–210, 1996.

[9] J. Farrell, "Neural control." In W. Levin (ed.), *The Control Handbook*, pp. 1017–1030. Boca Raton, FL: CRC Press, 1996.

[10] J. Farrell, "Stability and approximator convergence in nonparametric nonlinear adaptive control." *IEEE Transactions on Neural Networks*, Vol. 9, no. 5, pp. 1008–1029, 1998.

[11] K. Funahashi, "On the approximate realization of continuous mappings by neural networks." *Neural Networks*, Vol. 2, pp. 183–192, 1989.

[12] F. Girosi and T. Poggio, "Networks and the best approximation property." *MIT A.I. Memo No. 1164*, October 1989.

[13] M. Gupta and N. Sinha (eds.). *Intelligent Control Systems: Theory and Applications*. New York: IEEE Press, 1996.

[14] K. Hornik, M. Stinchcombe, and H. White, "Multilayer feedforward networks are universal approximators." *Neural Networks*, Vol. 2, pp. 359–366, 1989.

[15] L. Hunt and G. Meyer, "Stable inversion for nonlinear systems." *Automatica*, Vol. 33, no. 8, pp. 1549–1554, August 1997.

[16] P. Ioannou and J. Sun, *Robust Adaptive Control*. Englewood Cliffs, NJ: Prentice Hall, 1996.

[17] D. James, "Stability of a model reference control system." *AIAA Journal*, Vol. 9, no. 5, 1971.

[18] H. Khalil, *Nonlinear Systems*, 2nd ed. Englewood Cliffs, NJ: Prentice Hall, 1996.

[19] P. Millington, "Associative reinforcement learning for optimal control." S. M. Thesis: Department of Aeronautics and Astronautics, MIT, 1991.

[20] K. Narendra and K. Parthasarathy, "Identification and control of dynamical systems using neural networks." *IEEE Trans. Neural Networks*, Vol. 1, pp. 4–27, 1990.

[21] K. Passino and S. Yurkovich, *Fuzzy Control*. Menlo Park, CA: Addison-Wesley, 1998.

[22] M. Polycarpou and P. Ioannou, "Identification and control of nonlinear systems using neural network models: design and stability analysis." *Technical Report 91-09-01*, University of Southern California, Los Angeles, September 1991.

[23] M. Polycarpou, "Stable adaptive neural control scheme for nonlinear systems." *IEEE Transactions on Automatic Control*, Vol. 41, no. 3, pp. 447–451, March 1996.

[24] M. Polycarpou, "On-line approximators for nonlinear system identification: a unified approach." In C. Leondes (ed.), *Control and Dynamic Systems: Neural Network Systems Techniques and Applications*, Vol. 7, pp. 191–230. New York: Academic Press, 1998.

[25] M. Polycarpou and M. Mears, "Stable adaptive tracking of uncertain systems using non-linearly parametrized on-line approximators." *International Journal of Control*, Vol. 70, no. 3, pp. 363–384, May 1998.

[26] T. Poggio and F. Girosi, "Networks for approximation and learning." *Proceedings of the IEEE*, Vol. 78, no. 9, pp. 1481–1497, 1990.

[27] M. Powell, "Radial basis functions for multivariable interpolation: A review." In J. Mason and M. Cox (eds.), *Algorithms for Approximation of Functions and Data*, pp. 143–167. New York: Oxford University Press, 1987.

[28] D. Rumelhart, J. McClelland, et al. *Parallel Distributed Processing—Explorations in the Microstructure of Cognition, Volume 1: Foundations*. Cambridge, MA: MIT Press, 1986.

[29] R. Sanner and J. Slotine, "Gaussian networks for direct adaptive control." *IEEE Trans. on Neural Networks*, Vol. 3, pp. 837–863, 1992.

[30] S. Sastry and A. Isidori, "Adaptive control of linearizable systems." *IEEE Transactions on Automatic Control*, Vol. 34, no. 11, November 1989.

[31] S. Shekhar and M. Amin, "Generalization by neural networks." *IEEE Transactions on Knowledge and Data Engineering*, Vol. 4, no. 2, pp. 177–185, 1992.

[32] M. Stinchcombe and H. White, "Universal approximation using feedforward networks with non-sigmoid hidden layer activation functions." *Proceedings of the International Joint Conference on Neural Networks*, Vol. 1, pp. 613–617, 1989.

[33] G. Walter, *Wavelets and Other Orthogonal Systems with Applications*. Boca Raton, FL: CRC Press, 1994.

[34] L. Wang, *Adaptive Fuzzy Systems and Control: Design and Stability Analysis*. Englewood Cliffs, NJ: Prentice Hall, 1994.

[35] S. Weaver, L. Baird, and M. Polycarpou, "An analytical framework for local feedforward networks." *IEEE Transactions on Neural Networks*, Vol. 9, no. 3, pp. 473–482, 1998.

[36] P. Werbos, "Backpropagation through time: What it does and how to do it." *Proc. of the IEEE*, Vol. 78, no. 9, pp. 1550–1560, 1990.

[37] H. Werntges, "Partitions of unity neural function approximation," *Proc. IEEE Int. Conf. Neural Networks*, pp. 914–918, 1993.

[38] B. Widrow and M. Hoff, "Adaptive switching circuits." *IRE WESCON Convention Record*, pp. 96–104, 1960.

[39] L. Zadeh, "Fuzzy sets." *Information and Control*, Vol. 8, pp. 338–353, 1965.

[40] L. Zadeh, "Outline of a new approach to the analysis of complex systems and decision processes." *IEEE Transactions on Systems, Man, and Cybernetics*, Vol. 3, no. 1, pp. 28–44, 1973.

Chapter 7 | SUPERVISORY HYBRID CONTROL SYSTEMS

Michael D. Lemmon

Editor's Summary

The exploitation of well-honed techniques and the exploration of new challenges need not be mutually exclusive strategies for research. This maxim is illustrated by an emerging technology in control: hybrid dynamical systems. These systems combine, within a unified framework and formulation, discrete-event systems (for more on these systems, see Chapter 2) and continuous-time dynamics. Hybrid systems represent a broadening of the scope of control, with infusions of ideas and theories from other fields, especially formal methods in computer science.

This chapter discusses hybrid systems in some depth, with particular emphasis on supervisory applications. In this case, the discrete events are viewed as supervisory decisions affecting the qualitative behavior of a system, with different "modes" of behavior exhibiting different continuous dynamics. An example of a two-arm robotic platform is used to motivate the technical discussion, and other applications are also noted. Other chapters in this volume also outline applications of hybrid dynamical systems (e.g., Chapter 14). Variable structure control, the topic of Chapter 8, can also be considered a hybrid system approach to control.

This chapter introduces and explains several new concepts, borrowed in some cases from computer science, that are important for the analysis and synthesis of supervisory hybrid systems. Hybrid automata, an extension of finite state machines, are a popular representational formalism. Together with temporal logics, which can be used to formulate specifications for hybrid control systems, these representations allow the safety and performance of the system to be automatically determined, under some assumptions.

Michael Lemmon is an associate professor in the Electrical Engineering Department at the University of Notre Dame. He chairs the Technical Committee on Hybrid Systems of the IEEE Control Systems Society.

7.1 INTRODUCTION

Supervisory hybrid systems are systems that integrate high-level decision making with traditional regulatory control functions. Such systems are referred to as *hybrid* because they generate a mixture of continuous-valued (i.e., measurements of the physical process variables) and discrete-valued signals (i.e., discrete decisions that supervise the plant's behavior).

On the basis of the preceding description, it should be apparent that any complex engineering system employing some sort of decision making can be viewed as a supervisory hybrid system. To appreciate the potential benefits of this viewpoint, it is important to recall that most complex engineering systems are developed in an iterative manner. The system is decomposed into subsystems that separate decision-making functions from lower-level regulatory functions. These subsystems are designed independently of each other, and simulation testing evaluates the performance of the reintegrated system. If the results of simulation testing are unsatisfactory, then the subsystems may be redesigned and another cycle of simulation testing commences. In cases where a loose coupling exists between decision making and regulation subsystems, this iterative approach can converge relatively quickly to an acceptable design. Advances in computer and networking technology, however, make it possible to develop systems in which decision making and regulation are strongly coupled. When this happens, simulation-based testing can become expensive, time consuming, and still provide no provable guarantees of acceptable system performance. It is in these situations that supervisory hybrid systems theory provides a powerful new approach to system analysis and development.

Supervisory hybrid systems theory treats both the decision-making and regulation functions of the overall system at the same time. As a result, hybrid systems theory allows us to consider the impact of strong subsystem coupling much earlier in the design process. In recent years, there has been considerable interest in the development of a formal mathematical framework for the study of supervisory hybrid systems. Two different application areas have driven this interest. Computer scientists have found supervisory hybrid system methodologies a useful means for solving timing and safety problems in asynchronous digital circuits. Control systems engineers have found that hybrid systems provide a convenient and potentially powerful method for the analysis and synthesis of large-scale supervisory control systems. In both cases, the goal is to provide nothing less than a new framework for the analysis of complex systems in which previously disparate approaches to decision making and control are united in a single systematic framework.

This chapter provides a tutorial introduction to supervisory hybrid systems. The chapter opens by highlighting the distinction between the discrete and continuous parts of a supervisory hybrid system. Examples of hybrid systems are presented, and a popular modeling paradigm known as the *hybrid automaton* is described. The chapter overviews recent progress in the analysis and synthesis of systems modeled by hybrid automata and closes with a summary of open issues in the field.

7.2 EXAMPLES OF SUPERVISORY HYBRID SYSTEMS

Systems science is concerned with the use of formal mathematical methods in the modeling and analysis of engineering systems. We generally view a system as some sort of physical process whose behavior can be monitored by taking measurements of important process variables. If the system in question is a chemical reaction, for instance, then measurements of process temperature and pressure may be used to characterize the current state of the reaction. A robotic system, for example, may have its current state characterized by measuring the joint angles representing the robot's current spatial configuration. When such measurements of the physical process are indexed with respect to another independent variable (such as time), then we obtain a *signal*.

A systems scientist represents the signal x as an abstract mathematical function, $x : I \to M$, mapping elements of the index set I onto the measurement set M. A categorization of signals can be based on the type of measurement set. A discrete measurement set is a set whose elements can be placed in a one-to-one correspondence with integers. A continuous measurement set is a set that can be transformed in a continuous manner to Euclidean n-space $\Re^n$. Signals whose measurement sets are discrete (continuous) are referred to as discrete-valued (continuous-valued) signals. Discrete-valued signals are sometimes called *discrete-event* signals. Systems generating such signals are called discrete-event systems (DES). Any function taking values in $\Re^n$ can be viewed as a continuous signal. Discrete-event signals are often generated by decision-making systems.

Formal methods for dealing with signals and systems have traditionally assumed that the signals are either discrete- or continuous-valued. In practice, however, engineering systems consist of mixtures of continuous and discrete systems. A computer controlling a physical process generates just such a mixture of signals. The state of the physical process is represented by continuous-valued signals, whereas the state of the computer program controlling the physical process has a finite number of discrete states. Systems that generate signals containing a mixture of continuous and discrete valued signals are often referred to as *supervisory hybrid systems*. A large number of practical engineering systems can be represented as hybrid dynamical systems. The remainder of this section presents two examples. One example is based on robotic systems, and the other concerns digital circuits.

7.2.1 Switched Dynamical Systems

A common type of hybrid system arises when the system's differential equation has a discontinuous righthand side. Such systems possess continuous-valued and discrete-valued states. The continuous-valued state trajectory (denoted as $x(t)$) is governed by ordinary differential equations. The discrete-valued state trajectory (denoted as $i(t)$) takes values over a finite set of symbols, and its evolution is generated by a switching function q. Individual discrete states are sometimes referred to as system *modes*. Formally, we write the system equations for this switched system as

$$\dot{x}(t) = f(x(t), i(t)) \quad (7.1)$$
$$i(t) = q(x(t), i(t^-)), \quad (7.2)$$

where $i : \Re \to \Omega$ is a discrete-valued continuous-time signal representing the time history of switching modes. The continuous state trajectory $x(t)$ is generated by the function $f : \Re^n \times \Omega \to \Re^n$, and the discrete state's trajectory $i(t)$ is generated by the switching function $q : \Re \times \Omega \to \Omega$. $i(t^-)$ denotes the righthand limit of the signal i as time approaches t. It is customary to define q in such a way that there are well-defined switching sets between the various discrete states. In particular, we let the switching set Ω_{ij} between mode i and mode j be defined as

$$\Omega_{ij} = \{x \in \Re^n \ : \ j = q(x, i)\}$$

The switching set, therefore, represents a subset of the continuous state space in which a discrete mode switch from mode i to mode j can occur.

We now turn our attention to the type of signals generated by the system in Eqs. (7.1) and (7.2). Define the system's state space $\mathcal{H} = \Omega \times \Re^n$ as the Cartesian product of the discrete set Ω and the continuous state $x \in \Re^n$. A *hybrid system trajectory* is a continuous-time signal $\sigma : \Re \to \mathcal{H}$ taking values in the hybrid state space $\mathcal{H}$. Given a specific hybrid trajectory σ, the time τ is said to be a switching time if $i(\tau) = q(x(\tau), i(\tau^-)) \neq i(\tau^-)$. In other words, the switching time τ represents that instant in time when the discrete-valued component, i, changes value. We say that a hybrid system trajectory σ is *generated* by the system in Eqs. (7.1) and (7.2) if

- for any interval (τ_1, τ_2) that does not contain a switching time, the hybrid state $\sigma(\tau)$ (for all $\tau \in (\tau_1, \tau_2)$) satisfies Eq. (7.1); and
- for any switching time τ, the hybrid state $\sigma(\tau)$ satisfies the switching Eq. (7.2).

A hybrid trajectory σ that is generated by Eqs. (7.1)–(7.2) is also called a *solution* to the equations.

An important issue concerns the existence and uniqueness of solutions to equations in (7.1)–(7.2). We cannot expect the hybrid system trajectories to be continuous because of the discontinuity of the righthand side of Eq. (7.1). It is possible, however, to identify conditions guaranteeing the existence of piecewise continuous solutions to the system equations. These existence conditions [1] require the semi-continuity of set-valued mappings associated with Eq. (7.1).

Although it is usually easy to ensure the existence of piecewise continuous hybrid trajectories, it is not always possible to guarantee the uniqueness of these solutions. Switched systems often generate *nondeterministic* trajectories. This means that for a given initial condition, there may be many different trajectories that satisfy the system equations. In addition to nondeterministic piecewise continuous trajectories, it is possible for the system to generate *chattering* solutions. Chattering hybrid system trajectories arise when the system switches infinitely fast between various modes. In the variable structure control literature [2], these solutions are referred to as *sliding modes*. In general, it is often considered undesirable for a supervisory hybrid system to exhibit chattering behavior. Computer scientists also have an interesting term for this behavior. Systems capable of exhibiting such chattering solutions are sometimes referred to as *Zeno* systems. The name refers to the classical Zeno's paradox in which the concept of a limit is first informally introduced.

A concrete example of a switching system will be found in Figure 7.1. This figure shows a free-floating robotic vehicle with two articulated arms. The system is required to obtain components from a *parts bin* and to move these components to a *work area* where an assembly operation is to be performed. The tasks of fetching the workpiece, transporting it to the work area, and then returning to the parts bin are performed repeatedly. The equations of motion for the arms are expressed by the following differential equations

$$\frac{d^2\theta_1}{dt^2} = -\frac{d\theta_1}{dt} + k(\theta_1 + \theta_b - i_1(t)) \qquad (7.3)$$

$$\frac{d^2\theta_2}{dt^2} = -\frac{d\theta_2}{dt} + k(\theta_2 + \theta_b - i_2(t)) \qquad (7.4)$$

$$\frac{d\theta_b}{dt} = -\frac{J_a}{J_b + 2J_a}\left(\frac{d\theta_1}{dt} + \frac{d\theta_2}{dt}\right), \qquad (7.5)$$

Section 7.2 Examples of Supervisory Hybrid Systems

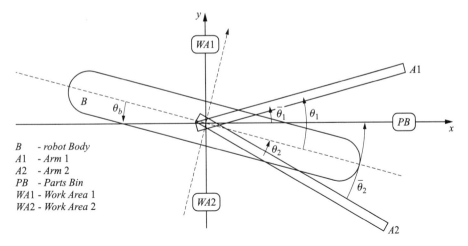

B   - robot Body
A1 - Arm 1
A2 - Arm 2
PB - Parts Bin
WA1 - Work Area 1
WA2 - Work Area 2

Figure 7.1 Free-floating robotic system.

where θ_1 and θ_2 are the arm angles with respect to the body axes (see Figure 7.1), θ_b is the angle of the body with respect to an inertial reference frame, J_a and J_b are the moments of inertia for the arm and body, respectively, i_1 and i_2 are reference inputs, and k is a proportional feedback gain. The arm angles with respect to an inertial frame are $\bar{\theta}_1 = \theta_1 + \theta_b$ and $\bar{\theta}_2 = \theta_2 + \theta_b$. Equations (7.3) and (7.4) represent the controlled behavior of the robotic arms in body coordinates. Equation (7.5) requires that the system's total angular momentum be conserved.

Equations (7.3)–(7.5) characterize the continuous-valued state trajectory of this system. The discrete state trajectory is represented by the reference inputs $i_1(t)$ and $i_2(t)$. The reference inputs for arm 1 have the form

$$i_1(t) = \begin{cases} i_1(t^-) + \pi/2 & \text{if } |\theta_1 + \theta_b| < .1 \\ i_1(t^-) - \pi/2 & \text{if } |\theta_1 + \theta_b - \pi/2| < .1 \\ i_1(t^-) & \text{otherwise} \end{cases} \quad (7.6)$$

under the assumption that $i_1(0) = \pi/2$. A similar equation characterizing the reference input for arm 2 can also be defined. This discrete switching equation corresponds to Eq. (7.2). It directs robot arm 1 to move back and forth between the parts bin and work area in an alternating manner.

7.2.2 Asynchronous Sequential Circuits

The switched system of Eqs. (7.1)–(7.2) is a classical example of a hybrid system. These equations provide an *equational* representation of the system which is familiar to most systems engineers. Another important type of hybrid systems, however, arises in the analysis of sequential digital circuits. A great deal of the research in hybrid systems has been driven by this particular application. In this case, we start with a system that was originally treated using formal methods from discrete mathematics and then was *hybridized* when real-world considerations began to play an important role. The resulting system models are not generally represented as equations but rather as *directed graphs*.

A digital circuit is a circuit (system) taking binary-valued inputs and producing binary-valued outputs. An AND gate, for instance, represents a simple example of a digital system with two binary inputs and a single binary output. The AND gate is an example of a *combinational* circuit, a circuit whose output is completely specified by the present inputs. In many applications, we are more interested in the behavior of *sequential circuits*. A sequential circuit is a digital circuit whose current output is dependent on the current and previous inputs to the system. Sequential circuits that change their internal states in step with a global clock tick are referred to as synchronous sequential circuits. Essentially, we can view such circuits as discrete-time, discrete state systems. Synchronous sequential digital circuits provide convenient models for digital integrated circuits. They can model the behavior of simple circuits such as flip-flops. Synchronous sequential circuits can also model the behavior of very large scale integrated (VLSI) chips such as microprocessors.

It is of practical importance to be able to check whether or not VLSI chips behave correctly. Because of the large size of these chips, a great deal of effort has been devoted to the development of computationally efficient methods for checking circuit correctness. Circuit *verification* refers to the activity of checking circuit correctness. *Symbolic model checking* (SMC) [3] is a very efficient means of checking the correctness of VLSI chips that can be modeled as synchronous sequential circuits. This algorithmic approach to circuit verification makes use of a graph theoretic model for the system that is known as the finite automaton. Checking the safety of the circuit involves computing a collection of discrete states that can be reached from a specified set of target states.

Although SMC methods work well for synchronous sequential circuits, it should be noted that many digital systems cannot be modeled this way. Synchronous sequential models assume that all machine states change in step with a global clock. In chips that need to respond in a reactive way to the outside world, or in extremely large circuits, synchronous operation may not be a realistic assumption. In such systems, the discrete states of the circuit may change at times between contiguous clock ticks. As a result, these systems generate signals that may be discrete-valued *and* continuous-valued. We sometimes refer to this type of sequential circuit as an *asynchronous sequential circuit*. Asynchronous circuits are clearly hybrid systems. Asynchronous circuits are found with increasing frequency, particularly in the context of real-time or embedded control. For these real-time systems, traditional SMC methodologies cannot provide provable guarantees of circuit correctness. The recent advances in supervisory hybrid systems theory have been driven by this need to extend traditional SMC methods to asynchronous digital circuits. Many of these advances make use of a specific hybrid system modeling paradigm known as the *hybrid automaton* [4].

7.3 HYBRID AUTOMATON

Early system theoretic models for hybrid systems tended to focus on *switched system* representations of the form found in Eqs. (7.1)–(7.2). A reference to these early models may be found in [5]. Although familiar to most systems scientists, these equational models do not provide a convenient means of representing discrete behaviors. Other modeling paradigms have emerged which provide computationally tractable frameworks for dealing with discrete and continuous dynamics. These models include logical DES models [6], hybrid automata [4], and continuous Petri nets [7]. Of all these models, the hybrid automaton has been the most influential.

Computer scientists have long used formal graph-theoretic models to represent concurrent computer processes. Finite state machines and Petri nets represent two well-known examples of such formal methods. Although powerful computational tools were developed for the manipulation of such formal methods, it was apparent in dealing with multiprocessors, real-time systems, and asynchronous digital circuits that these tools were inadequate. It was realized that existing graph-theoretic formalisms would need to incorporate continuous dynamics, and this realization led to the development of the hybrid automaton.

7.3.1 Definition of the Hybrid Automaton

The hybrid automaton is a modeling framework for hybrid systems that combines the graph-theoretic formalisms of traditional computer science with the equational formalisms found in traditional systems science. It can be defined as a three-tuple $(\mathcal{N}, \Delta, \mathcal{L})$ where $\mathcal{N}$ is a marked network representing the discrete-event behavior of the system. Δ is a set of mappings, $f_i : \Re^n \to \Re^n$ ($i = 1, \ldots, p$), that map vectors in $\Re^n$ back onto vectors in $\Re^n$. The mappings in Δ represent the various continuous dynamical subsystems that the hybrid automaton can switch between. Finally, $\mathcal{L}$ maps the arcs and vertices of networks $\mathcal{N}$ onto predicates in a propositional logic. This mapping characterizes the interactions between the hybrid automaton's continuous and discrete dynamics. These three components—the network $\mathcal{N}$, the set Δ, and the labels $\mathcal{L}$—are described in greater detail below.

The network $\mathcal{N}$ is represented by the ordered pair (V, A) where V is a set of vertices and $A \subset V \times V$ is a set of directed arcs between vertices. The vertex set is finite, with its cardinality denoted as $|V|$. Networks are often represented graphically. An open circle is used to represent each vertex of the network. An arrow starting at vertex v_i and terminating with an arrowhead at node v_j is used to represent the arc (v_i, v_j). Arcs and vertices are frequently labeled as a means of binding the network with a real-life process. As a specific example of a network, let's consider the set of vertices $V = \{v_1, v_2, v_3, v_4, v_5\}$ and the set of arcs

$$A = \{(v_1, v_2), (v_2, v_3), (v_3, v_4), (v_4, v_1), (v_2, v_5), (v_4, v_5)\}.$$

Figure 7.2 shows the graphical representation of this network. This network is labeled, the specific labels referring to the robotic system shown in Figure 7.1.

Network $\mathcal{N} = (V, A)$ enumerates all possible states that a discrete-event system might occupy. The current state of the system is denoted by *marking* the network. A marked network is the triple, (V, A, μ) where V and A are the network's vertices and arcs, respectively. The final element of the triple is a function $\mu : V \to \{0, 1\}$ that associates either zero or one with each vertex of the network. If $\mu(v) = 1$, then vertex v is marked. Otherwise the vertex is unmarked. Graphically, a marked vertex is denoted by placing a small solid circle (also called a *token*) in the marked vertex. In Figure 7.2, the vertex v_2 is marked. It is common to think of μ as a vector $\bar{\mu}$ in which the value of the *i*th element of this vector is the marking of the *i*th vertex. This *marking vector*

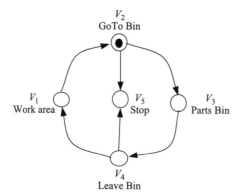

Figure 7.2 Network for a discrete-event system.

constitutes the discrete state of the hybrid automaton. The value that this vector takes at time τ is denoted as $\bar{\mu}(\tau)$.

The set Δ is a finite set of vector fields over $\Re^n$ characterized as

$$\Delta = \{f_1, f_2, \cdots, f_p\}.$$

Each $f_i : \Re^n \to \Re^n$ (for $i = 1, \ldots, p$) maps the continuous state space $\Re^n$ back onto itself. The elements of Δ represent continuous dynamical systems that generate state trajectories $x : \Re \to \Re^n$ satisfying the differential equation $\dot{x}(t) = f_j(x(t))$ for some $f_j \in \Delta$. The *continuous state* of the hybrid system is characterized by the following four-tuple, $z = (\dot{x}, x, \tau_0, x_0)$ where $\dot{x}$ is one of the mappings (say f_i) in Δ, $x : \Re \to \Re^n$ is a continuous function of $\Re$ taking values in $\Re^n$, $\tau_0 \in \Re$, and $x_0 \in \Re^n$. The objects in this continuous state $z = (\dot{x}, x, \tau_0, x_0)$ are referred to as the continuous state's rate, value, initial time, and initial value, respectively. Together these objects form an initial value problem

$$\dot{x}(t) = f_i(x(t)) \qquad (7.7)$$
$$x(\tau_0) = x_0 \qquad (7.8)$$

that is satisfied by the function $x : [\tau_0, \infty) \to \Re^n$.

Combining the four-tuple z with the marking vector $\bar{\mu}$ yields the system's *hybrid state*. The hybrid automaton's state, therefore, is represented by an ordered pair $\sigma = (z, \bar{\mu})$ consisting of the continuous state z and the discrete state $\bar{\mu}$ marking the network. The set of all ordered pairs, $(z, \bar{\mu})$, is denoted as $\mathcal{H}$ and will be called the *hybrid state space*. We will be interested in functions $\sigma : \Re \to \mathcal{H}$ of time that take values in the hybrid state space. These hybrid-valued continuous-time signals will be called *hybrid trajectories*. Note that this definition parallels the notion of hybrid trajectory introduced in the context of switched dynamical systems (see Eqs. (7.1)–(7.2)).

The event labels $\mathcal{L}$ form the third component of the hybrid automaton. $\mathcal{L}$ is the interface between the discrete subsystem $\mathcal{N}$ and the continuous subsystems in Δ. The event labeling is represented as a mapping $\mathcal{L} : V \cup A \to \mathcal{P}$ from the arcs and vertices of the network onto formulas in a propositional logic $\mathcal{P}$. The atomic formulas of this logic are defined with respect to the continuous state z. Although the vertices and arcs can be

Section 7.3 Hybrid Automaton

labeled in a variety of ways, one convenient labeling is introduced below. We first introduce the following set of *atomic equations*:

- *Invariant equations* are equations of one of three forms. The first form of the invariant is $[\dot{x} = f_j]$ where $\dot{x}$ is the rate of the continuous state and f_j is in Δ. Letting τ be a switching time, then the second type of invariant equation has the form $[h(x_0(\tau)) = 0]$ where $h : \Re^n \to \Re$ is a function of the initial condition x_0 of the hybrid state. The third type of invariant has the form $[\tau_0 = \tau]$ and acts to reset the initial time in the continuous state z to the switching time τ.
- *Guard equations* have the form $[g(x) > 0]$ where $g : \Re^n \to \Re$ is a functional defined over the continuous state space, $\Re^n$.

An atomic formula p is said to be satisfied by hybrid state σ if and only if the equation is true when the hybrid state σ is substituted into the equation. We denote the satisfaction of p by σ as $\sigma \models p$. Other legal fomulas in $\mathcal{P}$ are generated recursively by the conjunction or negation of other formulas in $\mathcal{P}$. For instance, if both p and q are in $\mathcal{P}$, then the conjunction $p \wedge q$ is in $\mathcal{P}$. We say that the hybrid state σ satisfies $p \wedge q$ (denoted as $\sigma \models p \wedge q$) if and only if $\sigma \models p$ and $\sigma \models q$. In a similar way if $p \in \mathcal{P}$, then the negation $\neg p$ is also in $\mathcal{P}$. Moreover, we say that the hybrid state σ satisfies $\neg p$ (i.e., $\sigma \models \neg p$) if and only if σ does not satisfy p. All formulas in $\mathcal{P}$ can be generated by the recursive application of conjunction ($\wedge$) and negation ($\neg$). For example, the disjunctive formula $p \vee q = \neg(\neg p \wedge \neg q)$.

The labeling function $\mathcal{L}$ associates each vertex and arc of the network with a proposition in $\mathcal{P}$. The bindings implied by $\mathcal{L}$ determine how the continuous and discrete parts of our hybrid system interact. For the hybrid automata considered in this chapter, we assume that network arcs are labeled with guard equations and that network vertices are labeled with invariant equations.

We now define the *dynamics of the hybrid automaton* by characterizing all hybrid trajectories that can be generated by a given automaton $(\mathcal{N}, \Delta, \mathcal{L})$. A time $\tau \in \Re$ is said to be a *switching time* if and only if $\bar{\mu}(\tau^-) \neq \bar{\mu}(\tau^+)$. In other words, a switching time is the instant when the marking of the network $\mathcal{N}$ changes. Given an arbitrary hybrid trajectory $\sigma : \Re \to \mathcal{H}$, we say that this trajectory is generated by the hybrid automaton $(\mathcal{N}, \Delta, \mathcal{L})$ if and only if the trajectory σ satisfies the following conditions:

- For all $\tau \in (\tau_1, \tau_2)$ that are not switching times and where τ_1 is a switching time, there exists a marked vertex v such that the hybrid state $\sigma(\tau) \models L(v)$.

    This condition requires, essentially, that between switching times the continuous part of the hybrid system state must satisfy the differential equations implied by the vertex predicate $L(v)$. Recall that the vertex predicates are invariant equations of the form $[\dot{x} = f_i]$, $[h(x_0) = 0]$, and $[\tau_0 = \tau]$. For these predicates to be satisfied, the continuous state z must be reset so that the rate $\dot{x}$, the initial time τ_0, and the initial state x_0 satisfy the equations in $L(v)$. These objects characterize the initial value problem in Eqs. (7.7) and (7.8) generating the hybrid system's continuous state value x. Therefore the switch at time τ_1 causes the hybrid system to switch the underlying continuous-time dynamics of the hybrid system.

- If τ is a switching time, then there is an arc (w, v) such that $\sigma(\tau^-) \models L((w, v))$, $\mu(w(\tau^+)) = \mu(w(\tau^-)) - 1 = 0$ and $\mu(v(\tau^+)) = \mu(v(\tau^-)) + 1 = 1$.

  This condition states that at the switching time τ, there is an arc (w, v) in the network which *fires*. The arc fires when the continuous state trajectory $x(\tau)$ satisfies the guard equation $L((w, v))$ on the arc and when the discrete enabling conditions of the marking vector are satisfied. Recall that the label $L((w, v))$ is a guard equation representing an inequality constraint on the continuous state's value x. This label, therefore, represents a necessary condition that the continuous state x has to satisfy before the arc (w, v) can fire. In addition to this continuous enabling condition, there are discrete enabling conditions. The final two conditions state that the vertex w must be marked just prior to the switch. These conditions also tell us what must happen to the marking vector after the switch. The firing of arc (w, v) will modify the network's marking vector by removing a token from w and placing a token in vertex v.

Two important classes of hybrid automata are obtained by restricting the nature of the objects in the triple, $(\mathcal{N}, \Delta, \mathcal{L})$. If the elements of Δ are all unity (i.e., $\dot{x} = 1$), then the hybrid automaton is called a *timed automaton*. The class of *rectangular* hybrid automata occurs when elements of Δ are set valued mappings in $\Re^n$ characterizing rectangular regions and when the guard and invariant equations form rectangles in the continuous state space $\Re^n$.

7.3.2 Robotic System Example: Revisited

As a concrete example of a hybrid automaton, let's reexamine the robotic system in Figure 7.1. Recall that this system is a robotic vehicle with two articulated arms repeatedly moving between the *parts bin* and *work area*. The dynamics of the continuous-valued variables (the arm angles) were given in Eqs. (7.3)–(7.5). Let's assume that each arm is controlled by a computer process (an instantiation of the arm control program). The processes execute on the same machine. The network in Figure 7.2 can be used to represent this program. In Figure 7.2, we've labeled the vertices of our network with the names GoToBin, WorkArea, Stop, PartsBin, and LeaveBin. These labels characterize which discrete state the arm is in when the labeled vertex is marked. For instance, if the vertex labeled GoToBin is marked, then the arm is moving toward the parts bin. If the vertex labeled PartsBin is marked, then the arm is in the parts bin. Note that Figure 7.2 also contains an additional error state (Stop) that the control program can enter in the event of a fault.

What type of faults might this program encounter? One type of fault occurs if the robotic arms collide. As shown in Figure 7.1, it can be seen that the parts bin is common to both robotic arms. If both arms enter the parts bin at the same time, there is a high probability that the arms might collide. We therefore need to impose a *mutual exclusion* requirement on the system. Not only must the arms complete their respective tasks, but they must also execute the tasks so that both arms don't enter the parts bin at the same time. In other words, the robotic system needs to treat the parts bin as a *critical section* that both arms access in a mutually exclusive manner.

A candidate solution to the mutual exclusion problem can be readily developed. Assuming that both processes execute on the same computer under the direction of a multitasking operating system, then we can use operating system (O/S) control struc-

Section 7.3 Hybrid Automaton

tures such as *semaphores* or *mutexes* [8] to ensure mutually exclusive execution of that section of program code requesting access to the parts bin. In other words, by requiring that the virtual (i.e., computer) processes respect the mutual exclusion requirement, we expect the robotic arms (i.e., the physical system) to respect the requirement as well.

The use of semaphores ensuring mutual exclusion is illustrated in the following pseudocode.

```
ENTRY: if(mutex==1) goto ENTRY;
       mutex=1;
CRIT1: if(arm_not_in_partsbin) goto_partsbin();
EXIT:  mutex=0;
ERR:   if(arm_locked) STOP;
REM:   if(arm_not_in_workarea) goto_workarea();
       goto ENTRY;
```

This code segment has four distinct parts. There is an entry section (ENTRY) which tests the lock variable mutex to see if the other arm is moving toward the parts bin. As a practical matter, the lock variable could be implemented as an O/S semaphore. If the lock variable mutex is 0, then the program sets the lock variable to alert the other process that the arm is heading to the parts bin. This process then enters its critical section (CRIT1), which represents that code which must be executed mutually exclusively. In other words, both computer processes cannot be executing their critical sections at the same time. While in the critical section, the program checks to see if the arm is in the parts bin (the function call arm_not_in_partsbin) and outputs the command signal to the arm's motor (the function call goto_partsbin()). Upon leaving the parts bin, the process releases the lock variable and then enters its remainder (REM) section from which it commands the arm to move back to the work area. This remainder section checks to see if the arm is in the work area (the function call arm_not_in_workarea) and outputs the command signal to the arm (the function call goto_workarea()), which moves the arm toward the work area. We've also included an *error* state (ERR) that aborts the program's execution if the arm hits its mechanical limits (arm_locked evaluates to true).

A hybrid automaton capturing the dynamics of our robotic system along with the supervisory control logic represented in the above pseudocode is shown in Figure 7.3. This figure shows two concurrent networks representing the control programs for both robotic arms. The set Δ consists of the vector fields

$$\Delta = \begin{cases} f_{11} = -\dot{\theta}_1 + k(\theta_1 + \theta_b), & f_{21} = -\dot{\theta}_2 + k(\theta_2 + \theta_b), \\ f_{12} = -\dot{\theta}_1 + k(\theta_1 + \theta_b - \pi/2), & f_{22} = -\dot{\theta}_2 + k(\theta_2 + \theta_b - \pi/2), \\ f_{13} = -\dot{\theta}_1, & f_{23} = -\dot{\theta}_2 \end{cases} \quad (7.9)$$

The labels for the network are shown in Figure 7.3.

We now step through the discrete transitions of one of the automatons in Figure 7.3 and show how network labels affect the transitions. The mutex is represented by a specific label on the arc between the vertices labeled GoToBin and WorkArea. The conditional predicate $\neg[\text{mutex} > 0]$ asserts that if the lock variable is zero, then the arc may fire and the system will command the arm to move toward the parts bin. The

176　Chapter 7 Supervisory Hybrid Control Systems

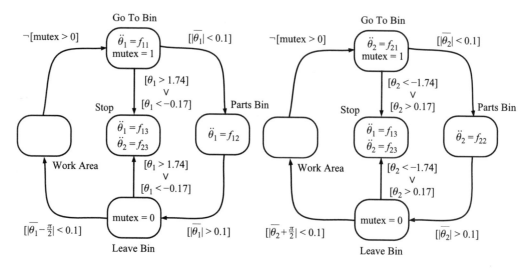

Figure 7.3 Hybrid automaton for robotic system.

discrete state GoToBin is labeled with the predicate [mutex = 1] $\wedge$ [$\ddot{\theta}_1 = f_{11}$]. This predicate sets the lock variable, which signals that arm 1 is moving toward the parts bin. The second part of the predicate switches the continuous dynamics actually controlling the arm's movement toward the parts bin. The arc connecting GoToBin to the discrete state PartsBin is labeled with the predicate [$|\bar{\theta}_1| < .1$]. This guard condition represents a strip in the continuous state space characterizing the extent of the parts bin for robot arm 1. Once the arm is in the parts bin, the system begins moving the arm out of the bin; therefore the state PartsBin is labeled with the predicate [$\ddot{\theta}_1 = f_{12}$]. Once the arm is out of the bin, we allow the system's discrete state to transition to the state LeaveBin. The predicate guarding this transition is [$|\bar{\theta}_1| > 0.1$]. Upon leaving the bin, the system resets the lock variable so that the other arm can access the parts bin, hence the predicate on LeaveBin is [mutex = 0]. Finally, the system returns to the WorkArea state if the appropriate conditions on the angle are satisfied.

Note that the variable mutex has a mixed interpretation. It is actually a discrete variable of the control program. But it is used in the guard predicate as if it were a continuous-valued variable. This usage highlights one common convention in hybrid automaton modeling. Discrete variables used in process synchronization are frequently treated as continuous-valued states with a zero rate. Since the rate is zero, these states are constant until the satisfaction of an invariant predicate resets the initial value of the state at a switching instant. Also note that Figure 7.3 does not explicitly note the predicates [$\tau_0 = \tau$] and [$x_0 = x(\tau)$]. We have dropped these predicates from the figure since we assume that they label every vertex.

The preceding discussion stepped through the various discrete states of the automaton controlling the first arm of the vehicle and related this automaton back to the control program introduced above. Mutexes represent a standard synchronization mechanism in computer systems, but it is not apparent whether ensuring mutually exclusive access to the process's critical sections is sufficient to guarantee the safe operation of the physical system. This particular system exhibits a subtle coupling between the arm and body dynamics. Angular momentum is conserved in this system so that the movement of the arms will induce a body rotation. This conservation relation is found in Eq. (7.3). The coupling between arm and body can lead to system failures that are not predicted by an analysis of the system's control program. It may be possible for commanded arm movements to rotate the body to a position from which one or both of the arms cannot reach the parts bin. If this were to occur, then the system would *deadlock* (i.e., the program would be stuck in one of its discrete states). This deadlock represents the other way in which our hybrid system might fail.

The possibility of system deadlock is also captured in the hybrid automaton model of Figure 7.3. Note that the transition out of discrete state GoToBin has a nondeterministic next state in the sense that we can either transition to PartsBin or Stopped. The condition for transitioning to the Stopped state is $[\theta_1 > 1.74] \vee [\theta_1 < -0.17]$. This is a safety condition that is triggered if the arm moves too far (i.e., hits its physical stops). In this case, we transition to the Stopped state. The Stopped state is a deadlocked state from which all forward progress in the system ceases. It is labeled with the predicate $[\ddot{\theta}_1 = f_{13}] \wedge [\ddot{\theta}_2 = f_{23}]$, which causes both arms to eventually stop their motion.

7.4 HYBRID SPECIFICATIONS

In traditional control systems, design specifications are often captured through performance measures such as integrated squared error (ISE) and maximum overshoot. These performance measures, however, are often inadequate in completely characterizing what a system designer wants the system to do. It may, for instance, be necessary to condition system performance on an applied reference signal. A gain-scheduled system may need to satisfy one bound on its ISE at one setpoint, but it may be acceptable to relax this performance bound at another operating point. Finally, these traditional control performance measures do not capture discrete-event system specifications such as deadlock freedom. The conclusion that must be drawn from these observations is that traditional control measures, by themselves, are inadequate to completely represent the performance requirements that occur in many complex engineering systems.

An important aspect of hybrid systems theory involves the development of a specification framework which more faithfully captures the design requirements for a complex system. These specification frameworks are also *hybrid* objects since they combine traditional control theoretic performance measures with logical constraints on the system's desired discrete-event behavior. Moreover, the specification framework must be compatible with the chosen modeling framework (i.e., the hybrid automaton) in the sense that verifying whether or not a given hybrid system satisfies a hybrid specification is tractable (i.e., can be determined after a finite number of calculations). For hybrid automata, several logical specification languages have been proposed. These languages include the duration calculus [9], μ-temporal logics [10], and timed computation tree logics [11]. This section shows the reader how hybrid system specifications can be posed using a timed temporal logic.

A logic may be characterized by its atomic formulas, its syntax, and its semantics. The atomic formulas are a set of elementary formulas or equations. The logic's syntax is the set of rules defining how atomic formulas may be combined to form legal formulas or predicates in the logic. The semantics characterize the meaning of a logical predicate with respect to a specified *frame*. The frame, F, is a set of states through which a system might evolve (i.e., the hybrid states in our hybrid automaton). The meaning of logical formulas is determined by defining the truth values of all logical equations with respect to the frame states. In particular, if a logical formula p is true with respect to frame state $s \in F$, then we say that s satisfies p, and we denote this as $s \models_F p$ where F is the frame over which s exists. In cases where the frame is clear, we drop the subscript F.

We will consider posing complex system specifications as predicates in a temporal logic that is an extension of the computation tree logic (CTL) [12]. We refer to this logic as CTL1. It is an extension of CTL because it simply allows atomic formulas whose truth values are determined with respect to equations or inequalities on the continuous state of the hybrid system. Let $\sigma(t)$ be a hybrid system trajectory, the *atomic* formulas will take the form $[g(x(t)) > 0]$ or $[\bar{\mu}(t) = \bar{\mu}_0]$. The first atomic formula is an inequality constraint on the continuous value of the hybrid state and was used earlier as a guard condition in the hybrid automaton. The current hybrid state $\sigma(t)$ is said to satisfy this formula if the inequality is true for the given state at time t. The second atomic formula is a specific marking of the network. In this case, the hybrid state at time t satisfies the predicate if and only if the network's marking at time t equals $\bar{\mu}_0$.

The frame is a hybrid automaton, so no explicit mention of the frame is made below. The syntax and semantics of CTL1 are defined as follows:

$s \models p \Leftrightarrow p$ is satisfied by state s

$s \models \neg p \Leftrightarrow p$ is not satisfied by state s

$s \models p \wedge q \Leftrightarrow p$ and q are satisfied by state s

$s \models p \exists \mathcal{U} q \Leftrightarrow$ *there exists* a hybrid trajectory $\sigma(t)$ such that

$\sigma(0) = s$ and a time t_1 such that

$\sigma(t) \models p \vee q$ for $t < t_1$ and $\sigma(t_1) \models q$.

$s \models p \forall \mathcal{U} q \Leftrightarrow$ *for all* hybrid trajectories $\sigma(t)$ such that $\sigma(0) = s$,

there exists a time t_1 such that

$\sigma(t) \models p \vee q$ for $t < t_1$ and $\sigma(t_1) \models q$.

Thus the formula $p \wedge q$ represents our usual notion of logical conjunction, $p \vee q$ represents logical disjunction and $\neg p$ represents the logical not operation. The other two formulas, $p \forall \mathcal{U} q$ and $p \exists \mathcal{U} q$, have a special meaning that is specific to temporal logics. These operators provide a way of describing temporal relationships between predicates. The formula $p \forall \mathcal{U} q$ can be seen as saying that *for all* hybrid trajectories, predicate p is true *until* predicate q is true. The formula $p \exists \mathcal{U} q$ is the existential formula, meaning that there exists a trajectory in which p is true *until* q is true. The formulas $\forall \mathcal{U} p$ and $\exists \mathcal{U} p$ are equivalent to $[\text{true}] \forall \mathcal{U} p$ and $[\text{true}] \exists \mathcal{U} p$, respectively.

We now present an example illustrating the use of CTL1. In referring to the example in Figure 7.1, the first requirement is that the system must satisfy a mutual exclusion requirement. A temporal logic specification capturing this desired constraint is

$$\forall \mathcal{U} \neg \big[[|\bar{\theta}_1| < .1] \wedge [|\bar{\theta}_2| < .1] \big].$$

This particular specification equation says that for all possible hybrid trajectories, the computer programs controlling both arms will not enter their critical sections at the same time. The critical sections are defined by inequality constraints on the absolute value of the arm angles $\bar{\theta}_1$ and $\bar{\theta}_2$ in the inertial frame.

Not all solutions to the mutual exclusion problem are equally desirable. An easy way to guarantee mutual exclusion is to require that the system deadlocks in a safe state. In other words, if one of the arms stops moving, then we can always ensure that the other arm accesses the parts bin in a mutually exclusive manner. For this reason, it is also essential to require that the system be *deadlock-free*. A system attempting to enforce a mutual exclusion constraint is deadlock-free if each process in its entry section is guaranteed of eventually transitioning into its critical section after a finite waiting time. A weak version of deadlock-freedom may be expressed by the CTL1 formula,

$$[\text{WorkArea}] \forall \mathcal{U} [\text{PartsBin}]$$

This requirement is weak because no constraints have been imposed on the amount of time before deadlock is broken. A time limit on the duration of deadlock might be imposed by introducing a clock into the system that measures how long the arm has been deadlocked. Let x_1 denote the state of such a clock, and let's assume the clock is reset and restarted when the system first marks the vertex WorkArea. In this case, the following equation provides a useful characterization of the deadlock-freedom requirement,

$$[\text{WorkArea}] \forall \mathcal{U} [[\text{PartsBin}] \wedge [x_1 < c]]$$

The specification is requiring that all trajectories starting in WorkArea enter PartsBin in less than c time units.

The particular logic used here, however, is extremely simple, and no attempt has been made to formulate a complete logic. Considerable work is still being done to investigate specification logics for hybrid systems. This chapter has only presented some of the basic principles and ideas behind using logics to formally specify hybrid system behavior. The interested reader is referred to [9–11].

7.5 HYBRID SYSTEM ANALYSIS

The analysis problem asks whether or not the model satisfies the specification. Solving this problem involves identifying *sufficient* or *necessary and sufficient* tests for satisfiability of the specification with respect to the assumed model (a hybrid automaton). Necessary and sufficient tests are often referred to as *verification* tests, whereas merely sufficient conditions are often referred to as *validation* tests. Verification methods have been studied extensively by computer scientists interested in extending symbolic model checking to real-time systems.

Validation methods are frequently used in the control systems community where it is often impractical from a computational standpoint to verify system properties such as stability and robust performance. In both cases, we are concerned with determining whether there exists a set of initial conditions from which there emanate trajectories satisfying the formal specification. This section overviews recent progress in our understanding of verification and validation problems in hybrid systems analysis.

The greatest progress appears to have been made in the verification of temporal logic specifications for restricted classes of hybrid automata. These verification methods are an extension of symbolic model checking (SMC), a commercially successful approach used in checking the correctness of large-scale synchronous digital circuits. This work [11, 13] attempts to extend symbolic model checking to real-time systems.

Model checking in hybrid automata is based on an extension of symbolic model checking for digital circuits. A full discussion of symbolic model checking is beyond the scope of this chapter, but a simple example will serve to illustrate the basic principle.

Let's consider the finite state machine shown in Figure 7.2 and the CTL predicate,

$$p = \exists \mathcal{U}[\texttt{PartsBin}]$$

This CTL specification asks us to identify all discrete states from which there exists a state trajectory eventually ending up in the parts bin, PartsBin.

Now consider a sequence of sets, Ω_i, for $i = 0, 1, 2, \ldots$. The first set, Ω_0, consists of all those discrete states for which the predicate p in the CTL formula $\exists \mathcal{U} p$ is true. In this case, we see that $\Omega_0 = \{\texttt{PartsBin}\}$. The next set, Ω_1, is generated by the relation

$$\Omega_1 = \Omega_0 \cup \Pi$$

where the set Π consists of the preset of all vertices in Ω_0. This preset represents those discrete states from which there exists at least one trajectory reaching Ω_0. In this case, we see that

$$\Omega_1 = \{\texttt{GoToBin}, \texttt{PartsBin}\}.$$

We repeat the above iteration until, in the ith iteration, we compute all of those discrete states that can reach Ω_0 after i steps. The first observation that can be made about this iteration is that it is monotonic, since $\Omega_i \subseteq \Omega_{i+1}$. The second observation is that because the state machine has a finite number of vertices, we are guaranteed that there exists some j such that $\Omega_j = \Omega_k$ for all $k \geq j$. In other words, the iteration has a *fixed point*, which we denote as Ω. This fixed point represents all the discrete states of the system satisfying the CTL formula $\exists \mathcal{U} p$. Moreover, this fixed point can be identified after a finite number of iterations, so the fixed point is computable. In this example, the fixed point is the set

$$\Omega = \{\texttt{PartsBin}, \texttt{WorkArea}, \texttt{GoToBin}, \texttt{LeaveBin}\}.$$

Figure 7.4 illustrates the basic steps in this iteration leading to the final determination of the fixed point. This figure shows each set of states in the sequence Ω_i. This set consists of all discrete states that can reach a discrete state satisfying the predicate p in CTL formula $\exists \mathcal{U} p$. Thus the iterative procedures used in symbolic model checking are essentially solving reachability problems over the discrete-event system's state space. The specification $\exists \mathcal{U} p$ is then verified by comparing this fixed point against the initial

Section 7.5 Hybrid System Analysis

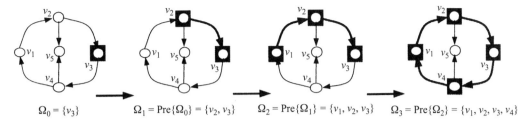

Figure 7.4 Model checking iteration.

starting states for our system. If the starting states are contained within this fixed point, then the specification can be considered to be verified.

Extending SMC methods to hybrid systems involves solving the reachability problem for both continuous and discrete system states. As before, let's consider the verification of the CTL formula $\exists \mathcal{U} p$ where $p = $ [PartsBin]. The SMC method described earlier identifies those discrete states that can reach the parts bin solely on the basis of the connectivity between logical states in the network. The enabling and firing of arcs in hybrid automata, however, are also conditioned on satisfaction of the guard equation labeling the arc in question. This implies that although connectivity between discrete states is certainly necessary for reachability, it is by no means sufficient. To fire the arc between the discrete states GoToBin and PartsBin, we must also ensure that the continuous state $\bar{\theta}_1$ satisfies the guard condition, $|\bar{\theta}_1| < 0.1$. Extensions of SMC methods to hybrid systems must therefore determine methods for computing subsets, Ξ, of continuous states that allow the firing of the arc.

These subsets can be computed using a recursive procedure similar to that used in traditional SMC methods. This recursive procedure computes a sequence of discrete sets Ω_i and continuous sets Ξ_i. The initial set Ξ_0 consists of all continuous states satisfying the guard conditions on the vertices marked by Ω_0. Unlike traditional model checking, however, there is no guarantee that the sequence of continuous state subsets Ξ_i will ever converge to a fixed point after a finite number of steps. This last point concerning the nonfinite nature of the computation highlights one of the weaknesses of model-checking methods as applied to hybrid systems. Since the computation may not terminate in a finite number of steps, the computation of these reachable sets is not decidable [14].

The decidability of the verification problem for hybrid systems has been an important issue driving a great deal of current work. In general, verification problems for hybrid automata are undecidable. Restricted classes of rectangular hybrid automata, however, have been shown to be decidable. Yet for minor perturbations of these restricted classes, decidability can be lost. The primary obstacle in establishing the decidability of hybrid systems rests with the fact that the precursor operation for determining Ξ may not converge. In particular, it was implied in [14] that the decidability boundary for hybrid systems may well rest with rectangular hybrid automata.

The preceding discussion has focused on algorithmic verification methods. These methods search through the state space of the system in order to verify a given specification. In concurrent systems, however, the number of states grows exponentially with the number of concurrent processes. As a result, algorithmic model-checking methods are frequently impractical for highly concurrent systems. One way around this limitation is to use *deductive verification* tools [15–16]. Deductive verification uses automated theorem-proving techniques to deduce the satisfiability of the specification. The symbolic model-checking methods are often referred to as automatic verification methods because they require little user intervention. Deductive verification, however, often requires some set of heuristics or user input to help guide the proof process.

In view of the undecidability of verification problems for many hybrid systems, it is natural to ask whether or not we should relax our demands and settle for *validation* tests. Recall that validation only requires finding sufficient conditions for a specification's satisfiability. The hope, of course, is that the sufficient condition is easier to compute and yet is sufficiently tight to be useful. The use of sufficient conditions in control theory has a long history. A number of fundamental control problems can be shown to be undecidable, but this fact has not prevented the development of sufficient methods that are still very useful.

An example of a useful sufficient test can be found in Lyapunov's second method. This method provides a sufficient test for system stability and serves as the basis for a number of analysis and synthesis methods in control theory. Given a dynamical system $\dot{x} = f(x)$ with state trajectories $x(t)$, we say that x_0 is an *equilibrium* point if and only if $f(x_0) = 0$. The equilibrium point is stable in the sense of Lyapunov if for all $\epsilon > 0$ there is a $\delta > 0$ such that $\|x(0)\| < \delta$ implies $\|x(t)\| < \epsilon$ for all $t \geq 0$. Lyapunov's method states that if there exists a positive definite functional $V : \Re^n \to \Re$ such that $V(x_0) = 0$ and $\dot{V}(x(t)) < 0$, then the equilibrium point is Lyapunov stable. Lyapunov methods are well known to only provide sufficient tests for system stability (though converse results exist for linear systems). Despite this shortcoming, Lyapunov methods are still an extremely useful tool in the study of nonlinear dynamical systems.

Given the importance of Lyapunov methods, it is not surprising to find a variety of results on the Lyapunov stability of hybrid systems. Recall that a switched system consists of a collection of continuous systems $\dot{x} = f_i(x)$. The hybrid system switches between these continuous subsystems on the basis of a supervisory control logic. Assuming that system switching is non-Zeno, for the *j*th mode, we can identify a collection of closed bounded time intervals over which that system is active. Figure 7.5 illustrates one such hybrid system trajectory and identifies the set of disjoint time intervals over which the first mode is active. Assuming there are N systems to switch among, we associate a function V_j ($j = 1, \ldots, N$) with the *j*th subsystem. We say that this family of functionals is *Lyapunov-like* if $V_j(x(t))$ is decreasing over the intervals in which the *j*th mode is active. Figure 7.5 illustrates a set of Lyapunov-like functionals for this particular system. The result in [17] states that if there exists such a family of Lyapunov-like functions, then the switched system is stable in the sense of Lyapunov. Note that in this result, there can be discontinuous jumps in the value of $V_j(t)$ between different modes. Extensions of this approach will be found in [18], and methods for constructing such Lyapunov functions will be found in [19–22].

Although extensions of Lyapunov methods provide considerable insight into validating hybrid system stability, these methods can also be used to validate temporal logic specifications regarding a hybrid system's deadlock freedom [23]. If we look at the level curves of the Lyapunov function, it should be apparent from the negative definite nature of $\dot{V}$ that these sets are *invariant* with respect to a cycle of discrete events. By invariance, we mean that if the continuous state starts in that set, then repeated application of the cycle of events will always return to the same set. These sets are sometimes referred to as *viability kernels* [24]. Recall that the extension of the SMC iteration to

Section 7.6 Hybrid Control System Synthesis

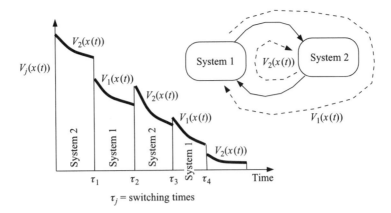

Figure 7.5 Switched Lyapunov system.

hybrid automata required computing a set Ξ which is a fixed point of a recursion. The recursion is computing the sets that can be reached from an initial state, and the fixed point of this recursion (if it exists) is precisely the invariant set which the Lyapunov analysis attempts to approximate. In this regard, therefore, Lyapunov methods provide a useful tool for hybrid system validation as well as traditional stability analysis.

7.6 HYBRID CONTROL SYSTEM SYNTHESIS

Verification/validation methods are analysis methods concerned with determining whether a system is capable of meeting a logical specification. When the verification method shows that the specification is infeasible, then we need to consider the use of controllers enforcing the specified behavior. In this context, we need to develop synthesis methods for hybrid control systems.

Probably the first issue to be dealt with concerns precisely what is a hybrid control system. As introduced earlier, a hybrid system is a system generating a mixture of discrete- and continuous-valued signals. A controller for this system may be viewed in a number of ways. Consider a hybrid plant with two sets of inputs and outputs. The inputs are categorized as continuous and discrete ($u(t)$ and $\tilde{u}(t)$, respectively), and the outputs also possess a discrete and continuous nature. A hybrid control system is formed when another system (which may or may not be hybrid) is connected to the plant in a way that regulates the plant's output behavior in a specified manner. In particular, we want our hybrid control system to enforce performance specifications that are posed as a formula in a hybridized temporal logic.

As shown in Figure 7.6, we can consider three distinct types of hybrid control system topologies. These topologies represent three different viewpoints of hybrid control. The controller, for instance, can be viewed as a discrete-event supervisor, a continuous system controller, or a true hybrid feedback controller. An example of the first viewpoint will be found in [6]. In this case, the controller is a logical discrete-event system that is designed to prevent the discrete state from transitioning to some illegal value. In the second case, we adopt the robust control viewpoint in which discrete-event switching leads to perturbations of the continuous-time plant. The objective is then to

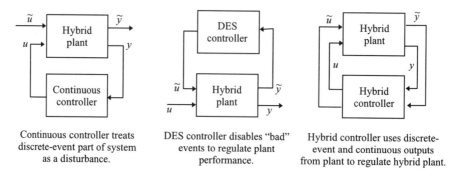

Figure 7.6 Hybrid control system topologies.

design a controller that ensures robust performance of the continuous variable subsystem over all possible switching events. Some details on this viewpoint to hybrid control can be found in [25]. The third viewpoint attempts to treat the continuous and discrete characters of the plant with an even hand and therefore requires the development of a truly hybrid controller. This approach to hybrid control system synthesis, however, is still relatively novel, and we will briefly outline it in more detail.

Hybrid controllers accept hybrid-valued inputs and produce hybrid-valued outputs. Because the controller is hybrid, it allows the designer greater flexibility in coordinating the behavior of the discrete and continuous parts of the system. The topology of this hybrid control system is shown in Figure 7.6 (right). As a concrete example, let's consider the robotic system of Figure 7.1. The plant in this case is a dynamical system whose continuous states θ_1 and θ_2 are governed by a pair of differential equations in Eqs. (7.3)–(7.4). The constraint on arm movement (Eq. 7.5) introduces a hybrid nature into the system. In addition, we have also included mechanical stops that prevent the arms from moving beyond a certain range. These constraints are of the form

$$-0.17 < \theta_1 < 1.74$$
$$-1.74 < \theta_2 < 0.17$$

This plant, therefore, generates two types of outputs; a discrete-valued output indicating when the system's mechanical stops have been reached and a continuous-valued output consisting of the angles that each robotic arm makes with respect to the parts bin. Our hybrid controller has access to these two types of signals.

The controller for this system is represented by the automaton in Figure 7.3, and since this is a hybrid automaton, the controller is itself a hybrid system. The discrete-valued events indicating when the arms have encountered their mechanical limits is used to transition the controller's discrete state to the Stop vertex representing the error condition, ERR. The continuous-valued signals θ_1 and θ_2 are used in the automaton's guard and switching conditions. As used in the guard conditions, these continuous-valued signals determine the type of feedback that guides the robotic arms to their desired positions. We therefore see that hybrid automata provide a convenient structure for the modeling of both the hybrid plant and the hybrid controller. In this particular example, the discrete control involves the determination of those control structures (i.e., the locking variable `mutex`), which help ensure mutual exclusion as well as the controller parameters (i.e., the control gains k) that control the commanded arm's movements. Hybrid control system synthesis is concerned with developing automatic methods for the simultaneous determination of both the control switching logic and the continuous-valued controllers.

The preceding example outlines the hybrid controller framework. The development of a systematic design methodology for such controllers is an intensive area of research. There is insufficient space in this introductory chapter to discuss hybrid controller synthesis in any detail. Recent approaches include gain-scheduling [26, 27], formal power series methods [28], game-theoretic formulations [29], the use of verification tools [30], and Horn clause logical control [31]. The interested reader is referred to the References for further study.

7.7 SUMMARY

This chapter has provided an introduction to many of the concepts and trends in hybrid systems science. Hybrid systems science is an interdisciplinary field requiring familiarity with methods and concepts from computer science and traditional systems science. Because of the introductory nature of this chapter, it was impossible to itemize all of the important work being performed. Much of the work outlined here will be found in a series of workshop proceedings published by Springer-Verlag [32–38] as well as numerous other workshops and various special issues of technical journals (*Theoretical Computer Science*, *IEEE Transactions on Automatic Control*, *Discrete Event Dynamic Systems*, *International Journal of Control*, *System and Control Letters*, *Automatica*). There have recently been significant applications of these methods in traffic control [39], automotive systems [40], and chemical process control [41]. All of these accomplishments point to a science that shows excellent potential for having a profound impact on the way we design and develop the engineering applications of the future.

Hybrid systems science is far from a mature field. This chapter summarizes a recent synthesis of computer science and systems science methods and a number of open questions remain. One of the predominant issues concerns computational complexity. As noted earlier, algorithmic model checking can be computationally intensive. Recursive computation of the viability kernel is not guaranteed to converge, and the finite state machine formalism suffers from state explosion problems when dealing with highly concurrent systems. Other open issues in hybrid systems concern synthesis and identification issues. Although a variety of frameworks have been proposed for hybrid control system synthesis, no universal agreement has formed concerning the best approach to follow. Additional work is needed in determining the role of Zeno-type or sliding-mode control in hybrid system supervision. Finally, it should be noted that very little attention has been paid to the issue of hybrid system identification and event detection.

Finally, hybrid systems theory is a highly interdisciplinary field requiring a familiarity with methods and concepts from computer science and traditional systems science. Current engineering curricula, however, often emphasize one or the other set of mathematical tools, thereby handicapping many engineers in their study of this field. A crucial prerequisite for the application of hybrid system methods will require an engineering education that embraces both discrete and continuous mathematics. The need for such a shift in engineering education has already been recognized by some of the more progressive elements of the academic community, and it can be expected that future graduates from these engineering schools will be well versed in hybrid aspects of the systems sciences.

ACKNOWLEDGMENTS

The partial financial support of the Army Research Office (DAAH04-96-10285 and DAAG5-98-1-0199) is gratefully acknowledged.

Related Chapters

- Chapter 1 includes a discussion of real-time programming for control systems, one application area for hybrid automata.
- An introduction to discrete-event systems can be found in Chapter 2.
- Variable structure control, the topic of Chapter 8, is another prominent type of hybrid control scheme.
- Additional types of system models, including other compositional models, can be found in Chapter 4.

REFERENCES

[1] J. P. Aubin and A. Cellina, *Differential Inclusions*. Berlin: Springer-Verlag, 1984.
[2] R. A. DeCarlo, S. H. Zak, and G. P. Matthews, "Variable structure control of nonlinear multivariable systems: a tutorial." *Proceedings of the IEEE*, Vol. 76, no. 3, March 1988.
[3] K. McMillan, *Symbolic Model Checking*. Norwell, MA: Kluwer Academic, 1993.
[4] R. Alur, C. Courcoubetis, T. A. Henzinger, and P.-H. Po, "Hybrid automata: An algorithmic approach to the specification and verification of hybrid systems." In R. L. Grossman, A. Nerode, A. P. Ravn, and H. Rischel (eds.), *Hybrid Systems*, Lecture Notes in Computer Science, Vol. 736, pp. 209–229. New York: Springer-Verlag, 1993.
[5] M. S. Branicky, "Studies in Hybrid systems: Modeling, analysis, and control." LIDS-TH-2304, Ph.D. Dissertation, Massachusetts Institute of Technology, LIDS, 1995.
[6] J. A. Stiver, P. J. Antsaklis, and M. D. Lemmon, "A logical DES approach to the design of hybrid control systems." *Mathematical Computer Modeling*, Vol. 23, nos. 11/12, pp. 55–76, 1996.
[7] J. LeBail, H. Alla, and R. David, "Hybrid petri nets." *Proceedings 1st European Control Conference*, Grenoble, France, 1991.
[8] R. Gallmeister, *POSIX.4, Programming for the Real World*. O'Reilly and Associates, 1995.
[9] C. Zhou, "Duration calculi: An overview." In Bjorner, Broy and Pottosin (eds.), *Proc. Formal Methods in Programming and Their Application*. Lecture Notes in Computer Science, Vol. 735, pp. 256–266. New York: Springer-Verlag, 1993.
[10] R. Alur, C. Courcoubetis, and D. Dill, "Model checking in dense real time." *Information and Computation*, Vol. 104, pp. 2–34, 1993.
[11] T. A. Henzinger, X. Nicollin, J. Sifakis, and S. Yovine, "Symbolic model checking for real-time systems." *Information and Computation*, Vol. 111, pp. 193–244, 1994.

References

[12] E. M. Clarke and E. A. Emerson, "Synthesis of synchronization skeletons for branching time temporal logic." *Logic of Programs*. Lecture Notes in Computer Science, Vol. 131, New York: Springer-Verlag, 1981.

[13] T. A. Henzinger, P.-H. Ho, and H. Wong-Toi, "A user's guide to HyTech." *First Workshop on Tools and Algorithms for the Construction and Analysis of Systems: TACAS94*. Lecture Notes in Computer Science, Vol. 1019, pp. 41–71. New York: Springer-Verlag, 1995.

[14] T. A. Henzinger, P. Kopke, A. Puri, and P. Varaiya, "What's decidable about hybrid automata?" *Proc. of the 27th Annual ACM Symposium on the Theory of Computing*, 1995.

[15] Z. Manna and A. Pnueli, *Temporal Verification of Reactive Systems: Safety*. New York: Springer-Verlag, 1995.

[16] Z. Manna and H. Sipma, "Deductive verification of hybrid systems using STeP." In *Hybrid Systems: Computation and Control*, Vol. 1386, LNCS, New York: Springer-Verlag, 1998.

[17] M. S. Branicky, "Multiple Lyapunov functions and other analysis tools for switched and hybrid systems." *IEEE Trans. on Automatic Control*, Vol. 43, no. 4, pp. 475–482, April 1998.

[18] L. Hou, A. N. Michel, and H. Ye, "Stability analysis of switched systems." *Proceedings of the 35th IEEE Conference on Decision and Control*, Kobe, Japan, 1996.

[19] M. Johansson and A. Rantzer, "Computation of piecewise quadratic Lyapunov functions for hybrid systems." *IEEE Transactions on Automatic Control*, Vol. 43, no. 4, pp. 555–559, 1998.

[20] C. A. Yfoulis, A. Muir, N. B. O. L. Pettit, and P. E. Wellstead, "Stabilization of orthogonal piecewise linear Lyapunov-like functions." *Proc. of the IEEE Conference on Decision and Control*, December 1998.

[21] A. Hassibi and S. Boyd, "A class of Lyapunov functionals for analyzing hybrid dynamical systems." *Proc. of the American Control Conference*, February 1999.

[22] K. X. He and M. D. Lemmon, "Lyapunov stability of continuous valued systems under the supervision of discrete event transition systems." *Hybrid Systems: Control and Computation*. Lecture Notes in Computer Science Vol. 1386, New York: Springer-Verlag, 1998.

[23] K. X. He and M. D. Lemmon, "Using dynamical invariants in the analysis of hybrid dynamical systems." *Proceedings of the IFAC World Congress*, Beijing, 1999.

[24] A. Deshpande and P. Varaiya, "Viable control of hybrid systems." In P. J. Antsaklis, W. Kohn, A. N. Nerode, and S. Sastry (eds.), *Hybrid Systems II*, Lecture Notes in Computer Science, Vol. 999, pp. 128–147. New York: Springer-Verlag, 1995.

[25] A. S. Morse, *Control Using Logic Based Switching*, Lecture Notes in Control and Information Sciences, Vol. 222. New York: Springer-Verlag, 1997.

[26] C. J. Bett and M. D. Lemmon, "Bounded amplitude performance of switched LPV systems with applications to hybrid systems." *Automatica*, Vol. 35, pp. 491–503, 1999.

[27] M. D. Lemmon and C. J. Bett, "Safe implementations of supervisory commands." *International Journal of Control*, Vol. 70, no. 2, pp. 271–288, 1998.

[28] A. Nerode and W. Kohn, "Multiple agent hybrid control architecture." In R. L. Grossman, A. N. Nerode, A. P. Ravn, and H. Rischel (eds.), *Hybrid Systems*, Lecture Notes in Computer Science, Vol. 736, pp. 297–316. New York: Springer-Verlag, 1993.

[29] J. Lygeros, C. Tomlin, and S. Sastry, "Multiobjective hybrid controller synthesis." *Proc. Hart'97*, 1997.

[30] H. Wong-Toi, "Synthesis of controllers for linear hybrid automata." *Proceedings of the 36th IEEE Conference on Decision and Control*, San Diego, California, 1997.

[31] A. Bemporad and M. Morari, "Control of systems integrating logic, dynamics, and constraints." *IFA Technical Report AUT-98-04*, Institut für Automatik, Swiss Federal Institute of Technology. Also in *Automatica*, Vol. 35, pp. 407–428, 1999.

[32] R. L. Grossman, A. N. Nerode, A. P. Ravn, and H. Rischel (eds.), *Hybrid Systems*. Lecture Notes in Computer Science, Vol. 736. New York: Springer-Verlag, 1993.

[33] P. J. Antsaklis, W. Kohn, A. N. Nerode, and S. Sastry (eds.), *Hybrid Systems II*. Lecture Notes in Computer Science, Vol. 999. New York: Springer-Verlag, 1995.

[34] R. Alur, T. A. Henzinger, and E. D. Sontag (eds.), *Hybrid Systems III; Verification and Control.* Lecture Notes in Computer Science, Vol. 1066. New York: Springer-Verlag, 1996.

[35] P. J. Antsaklis, W. Kohn, A. N. Nerode, and S. Sastry (eds.), *Hybrid Systems IV.* Lecture Notes in Computer Science, Vol. 1273. New York: Springer-Verlag, 1997.

[36] P. J. Antsaklis, W. Kohn, M. D. Lemmon, A. N. Nerode, and S. Sastry (eds.), *Hybrid Systems V.* Lecture Notes in Computer Science, Vol. 1567. New York: Springer-Verlag, 1999.

[37] T. A. Henzinger and S. Sastry (eds.), *Hybrid Systems: Control and Computation,* Lecture Notes in Computer Science, Vol. 1386. New York: Springer-Verlag, 1998.

[38] O. Maler (ed.), *Hybrid and real-time systems: Hart '97.* Lecture Notes in Computer Science, Vol. 1201. New York: Springer-Verlag, 1997.

[39] C. Tomlin, G. Pappas, and S. Sastry, "Conflict resolution for air traffic management: a study in multiagent hybrid systems." *IEEE Trans. of Automatic Control,* Vol. 43, no. 4, 1998.

[40] B. Lennartson, M. Tittus, B. Egardt, and S. Petterson, "Hybrid systems in process control." *Control Systems Magazine,* Vol. 16, no. 5, pp. 45–56, 1996.

[41] R. Balluchi, M. De Benedetto, C. Pinello, C. Rossi, and A. Sangiovanni-Vincentelli, "Hybrid control for automotive engine management: The cut-off case." In T. A. Henzinger and S. Sastry (eds.), *Hybrid Systems: computation and control,* Lecture Notes in Computer Science, Vol. 1386. New York: Springer-Verlag, 1998.

Chapter 8
VARIABLE STRUCTURE AND SLIDING-MODE CONTROL

Fumio Hamano and Younchan Kim

Editor's Summary

Control design is often construed as an activity directed toward the development of a single controller for a given system. This is a limiting perspective, however. In many control applications, changes in controller structure, including discrete jumps in parameter values, are necessary: manual-to-automatic startups and gain scheduling are common examples. Furthermore, switching between two controllers can lead to higher performing control loops than if either of the controllers is used exclusively, and such switching can stabilize a loop that is unstable with either controller.

Chapter 7 discussed one important class of variable structure control: supervisory hybrid control systems. This chapter focuses on another: sliding-mode control. An important difference between the two is that hybrid control often assumes that the system to be controlled is subject to uncontrolled structural variations, whereas sliding-mode control uses controlled structural variation (i.e., switching) as an integral part of a control mechanism. Here, different control laws are used depending on whether the state is on one side of a (hyper-)surface ("switching surface") or on the other. In either case, the control causes the state to move toward the surface. Once the surface is reached, the same pair of control laws attempts to keep the state on it—the surface is chosen so that the state then naturally slides toward the target point. The resulting control is robust and invariant; that is, the desired performance is maintained in the presence of a class of uncertainties and disturbances.

This chapter provides a technically and mathematically detailed, yet accessible, tutorial on sliding-mode control, starting with some motivating illustrations. For ease of exposition, the initial discussion is limited to single-input single-output systems. Modifications to the basic algorithms are then noted that can handle uncertainty in the system model and that can prevent chattering in the control signal. The general multivariable case is also treated in detail, and issues related to sampled data control systems are discussed.

Fumio Hamano is both a professor and chair of the Department of Electrical Engineering at California State University, Long Beach, and the chair of the Technology Review Subcommittee of the IEEE Control Systems Society. Younchan Kim is a Ph.D. candidate in the Department of Electrical Engineering at the same institution.

8.1 INTRODUCTION

This chapter deals with control systems for which the structure of the control law may change (e.g., jump of controller parameter values, change of the form of the function) during the course of action in accordance with the state, output, or error measurement. Such systems are generally referred to as variable structure control systems. The primary focus is on a specific type of variable structure control called sliding-mode control. The chapter serves as a tutorial introduction to these topics.

Control system performance can be improved significantly by allowing the controller to switch from one mode to another. For instance, for certain linear systems switching from a proportional controller to integral controller in a feedback loop may provide a fast response, small overshoot, and no offset by resolving conflicting requirements, and turning on and off a feedback loop may produce a fast response without overshoot [4]. Switching between two values of a parameter may result in a stable linear system even if the system with either of the parameter values is unstable. For instance, Figures 8.1(a) and (b) show sample trajectories of unstable systems represented, respectively, by $\dot{x}_1 = x_2$, $\dot{x}_2 = x_2 - 9.99x_1$ and $\dot{x}_1 = x_2$, $\dot{x}_2 = x_2 + 9.99x_1$. But the system described by $\dot{x}_1 = x_2$, $\dot{x}_2 = x_2 - 9.99|x_1|\text{sgn}(x_2 + 1.5x_1)$ shows stable behavior as depicted in Figure 8.1(c). An important type of variable structure control is called *sliding-mode control*, and this method of control has been well studied in the literature. Under this control, the state of the system first approaches a (hyper-)surface containing a target point and then slides along the surface (thus the surface is called a *sliding surface*, or *sliding manifold*) toward the target point. (See Figure 8.1(c).) The surface must be chosen properly so that any trajectory on the surface tends toward the target point, which is normally the origin of the state space with a proper formulation of the problem. The state is forced to remain on the sliding surface by switching between two different controls. Thus the sliding surface is also called *switching surface*. An advantage of the sliding-mode controller design is that the design process is decoupled into two stages, each of which involves a lower order system, and the controller is designed based

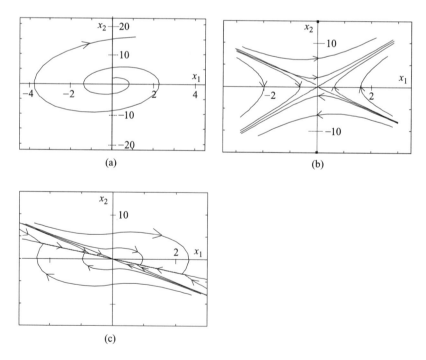

Figure 8.1

Section 8.1 Introduction

on a simple measure, that is, the distance of the state from the sliding surface. As a result, it is relatively easy to design the control law to compensate for modeling uncertainties and disturbances; that is, the control can be made robust without much difficulty. In addition, since the sliding surface does not change with the presence of modeling errors and disturbances, the system behavior does not change (at least ideally) once the state is on the surface (i.e., the control is invariant). Robustness and invariance are important properties of sliding-mode control.

Sliding mode control may require infinitely fast switching. Therefore, in practice, we can only use the method approximately. The approximation should, however, be used with care as demonstrated in the following example [27]. Figure 8.2 depicts the responses of the system described by $\dot{x}_1 = 0.3x_2 + ux_1$, $\dot{x}_2 = -0.7x_1 + 4u^3 x_1$, and $u = -\text{sgn}\{x_1(x_1 + x_2)\}$ when the sgn function (Figure 8.3(a)) is approximated by (i) hysteresis with $\varepsilon = 0.01$ (Figure 8.3(b)), and by (ii) saturation with $\varepsilon = 0.01$ (Figure 8.3(c)). The trajectory converges to the neighborhood of the origin for case (i), and it diverges for case (ii).

The theoretical issues of existence and uniqueness of a solution for a differential equation describing a system under sliding-mode control cannot be treated within the conventional framework based on piecewise continuity and Lipschitz condition, and will not be discussed here. Interested readers should refer to the theory of differential equations with a discontinuous righthand side due to Filippov [13], which approximates the function at the discontinuity by "averaging." For brief explanations, the reader should see [4, Chapter 1], [7], [8], [27], for instance.

Because of the "nonideal" aspects of physical systems such as hysteresis and delay, the fast oscillation (but at a finite frequency) called chattering may occur in practice. (See the trajectory for the system with the hysteresis in Figure 8.2.) How to eliminate chattering will be discussed at the ends of Sections 8.2 and 8.3. The use of computers in control systems results in sampled-data (or hybrid) systems in which both discrete- and continuous-time elements are present. Such systems will be discussed in Section 8.4.

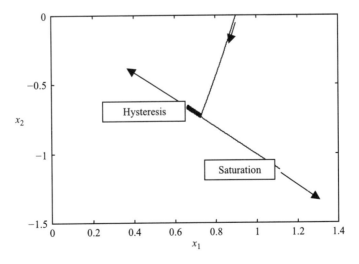

Figure 8.2 System responses with approximations by hysteresis and saturation.

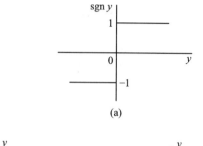
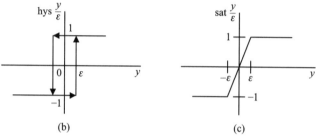

Figure 8.3 (a) Signum sgn y. (b) Hysteresis hys $\frac{y}{\varepsilon}$. (c) Saturation sat $\frac{y}{\varepsilon}$

Sliding-mode control was pioneered in the Soviet Union in the 1950s, initially for second-order continuous-time linear systems [4, Chapter 1], [12]. The method has been extended to deal with various types of systems, such as multi-input/multi-output, discrete-time, nonlinear, distributed, linear parameter-varying, and stochastic systems [7], [19], [24]. A variety of applications of sliding-mode control have also appeared in the literature. (See, for instance, [6, Chapter 7], [7], [19], [23], [26], [27], and [30].) Sliding-mode control has been implemented in real systems (experimental, prototype, and factory systems), and its practicality and effectiveness have been reported. The reader interested in practical implementations of sliding-mode control may refer to [7, Chapter 18] for controlling electric drives used for machine tools, vehicle control, and process control; [17], [21], and [22] for position servo systems; [29] for underwater vehicle control; [18] and [25] for robot control; [24] for controlling an active magnetic bearing system; [16] for magnetic servo levitation; and [28] for mobile robot control.

Notation: The small letters x, x_i, a, b, f, λ, etc., denote vectors, scalars, or vector- or scalar-valued functions. The capital letters A, B, G, K, etc., possibly with subscripts, denote matrices, matrix-valued functions, and sets (with the exceptions given below). If A is a matrix, A' denotes the transpose of A. To simplify the notation, the letters x, x_i, etc., are used in place of $x(t)$, $x_i(t)$, etc., with a slight abuse of notation except when the latter notation is desirable for clarity. The letters V, R, R_+, and t are reserved for the following quantities: V is used for a particular scalar valued function (Lyapunov function), R denotes the set of all real numbers, R_+ stands for the set of nonnegative numbers, R^k is the set of all real vectors of dimension k (interpreted as column vectors unless otherwise specified), and t denotes time. In addition, M and J are used for mass and moment of inertia. "Expression 1 := expression 2" (or "expression 1 =: expression 2") stands for "expression 1 is defined by expression 2" (or, respectively, "expression 2 is defined by expression 1.") If x is a function of t, then $\dot{x} := \frac{d}{dt}x$, the derivative of x with respect to t. In case x is a vector-valued function, the derivative operates on each element of x. The function sgn is the signum function, that is, sgn $s = 1$ if $s > 0$ and sgn $s = -1$ if $s < 0$. The symbol $\in$ means "belongs to" or "an element of"; for example, $x \in R^k$ indicates that x is an element of R^k (i.e., x is a real vector with dimension k). If $x \in R^k$, then $\|x\|$ stands for the Euclidean norm of x, and $\|x\|_\infty$ is the sup norm of x, i.e., $\|x\|_\infty := \max_i |x_i|$ where x_i is the ith element of x.

8.2 BASIC IDEA OF SLIDING-MODE CONTROL

In this section we examine a single-input second-order system with sliding-mode control to illustrate the basic mechanism of sliding-mode control.

8.2.1 Tracking Problem and Tracking Error Dynamics

Let us consider the following tracking problem. The plant to be controlled is described by the equations

$$\dot{x}_{p1} = x_{p2}$$
$$\dot{x}_{p2} = \bar{f}_{p2}(x_{p1}, x_{p2}) + u + d$$

where x_{p1}, x_{p2}, u, and d are real-valued functions of time $t \in R_+$, and $\bar{f}_{p2}$ is a real-valued function of two variables. The functions x_{p1} and x_{p2} are state variables, and u and d represent, respectively, the control input and unknown external disturbances. We wish to design a control law u so that x_{p1} and x_{p2} closely follow (or converge to) a given reference trajectory (x_{r1}, x_{r2}) where $\dot{x}_{r1} = x_{r2}$. Defining the tracking errors x_i by

$$x_i := x_{pi} - x_{ri}, \qquad i = 1, 2,$$

and using the plant equations, we obtain the state equations for the tracking errors

$$\dot{x}_i = x_2 \tag{8.1}$$

$$\dot{x}_2 = \bar{f}_{p2}(x_{p1}, x_{p2}) - \dot{x}_{r2} + u + d$$
$$= \bar{f}_{p2}(x_1 + x_{r1}, x_2 + x_{r2}) - \dot{x}_{r2} + u + d. \tag{8.2}$$

Since x_{ri}'s and $\dot{x}_{ri}$'s are given functions of t, we will write

$$\bar{f}_2(x_1, x_2, t) := \bar{f}_{p2}(x_1 + x_{r1}, x_2 + x_{r2}) - \dot{x}_{r2}.$$

Then, Eq. (8.2) becomes

$$\dot{x}_2 = \bar{f}_2(x_1, x_2, t) + u + d. \tag{8.3}$$

The problem therefore reduces to that of finding a proper control law so that x_1 converges to zero (for $i = 1, 2$) for the system described by Eqs. (8.1) and (8.3). In practice, $\bar{f}_{p2}$, and so $\bar{f}_2$, may not be accurately modeled. Therefore, we should write $\bar{f}_{p2} = f_{p2} + \Delta f_{p2}$ and $\bar{f}_2 = f_2 + \Delta f_2$ where f_{p2} and $f_2 = f_{p2} - \dot{x}_{r2}$ are known functions (i.e., the estimates of $\bar{f}_{p2}$ and $\bar{f}_2$) and Δf_{p2} and Δf_2 represent modeling uncertainties. For readability, we defer the consideration of the uncertain terms until Section 8.2.5, and we will first discuss the above problem with the assumption that the uncertainties Δf_{p2}, Δf_2, and d are zero. Thus we consider the system described by Eqs. (8.1) and (8.3) with the assumptions $\bar{f}_2 = f_2$ (known) and $d = 0$. To solve the above problem, we first select a line S in the state space of the system given by Eqs. (8.1) and (8.3). (In higher-order systems, we deal with a hypersurface instead of a line. This line or surface will become a sliding surface with a proper control law.) The line S must be well-behaved so that state trajectories confined in S are convergent to zero. We then design a controller to bring the state $(x_1(t), x_2(t))$ to S in finite time if the initial state $(x_1(t_0), x_2(t_0))$ is off the line S and to force the state to stay on S once it is on S.

8.2.2 Choosing a Sliding Surface (or Line)

To be specific, let s be a function defined by

$$s(x_1, x_2) := x_2 + \lambda x_1 \tag{8.4}$$

where λ is a positive number. We define a line S in the (x_1, x_2)-space by the equation

$$s = 0. \tag{8.5}$$

That is, we define $S := \{x \in R^2 | s(x_1, x_2) = 0\}$. Notice that it is important to choose $\lambda > 0$. In fact, since $\dot{x}_1 = x_2$, $s = 0$ means

$$\dot{x}_1 = -\lambda x_1.$$

Therefore, λ must be positive for x_1 to converge to zero if the trajectory is confined to S. The function x_2 then converges also to zero by Eqs. (8.4) and (8.5).

8.2.3 Control Law to Confine the State on the Sliding Surface

Suppose that the state $(x_1(t), x_2(t))$ is on S at time t. In order for the state to remain on S, the time derivative of s evaluated along the trajectory of (x_1, x_2) must be zero, that is, $\dot{s} = \frac{d}{dt}s(x_1, x_2) = 0$. By Eqs. (8.1), (8.3), and (8.4),

$$\dot{s} = \dot{x}_2 + \lambda \dot{x}_1 = f_2(x_1, x_2, t) + \lambda x_2 + u. \tag{8.6}$$

Let

$$u_{eq} := -f_2(x_1, x_2, t) - \lambda x_2 \tag{8.7}$$

and select u by

$$u = u_{eq}.$$

Then it is trivial to see $\dot{s} = 0$. The control law u_{eq} is called *equivalent control* (meaning that without uncertainties and disturbances it leads to a motion along the sliding surface S, and thus is equivalent to the robust controller to be designed later).

8.2.4 Control Law for Reaching the Sliding Surface (and Staying on It)

The state space is divided into two half spaces by S: One half of the space is characterized by $s > 0$, and the other half corresponds to $s < 0$. The magnitude $|s| = |s(x_1, x_2)|$ is a measure of the distance of (x_1, x_2) from S. Suppose that (x_1, x_2) is not on S. To move the state (x_1, x_2) toward the sliding surface, we choose the control u so that the value of $|s|$ decreases and will vanish in a finite time, more specifically,

$$\dot{s} \leq -\alpha < 0 \quad \text{if } s > 0, \tag{8.8}$$

$$\dot{s} \geq \alpha > 0 \quad \text{if } s < 0 \tag{8.9}$$

for all $t \geq 0$ where α is a positive number. Let t_0 be the initial time, and also let t_1 be the time at which the state reaches the sliding surface. Then, clearly,

$$t_1 - t_0 \leq \left|s(x_1(t_0), x_2(t_0))\right|/\alpha. \tag{8.10}$$

To find a desired control law, suppose $s > 0$. From Eqs. (8.6) and (8.7), it is easy to see that the control law

$$u = u_{eq} - \alpha$$

leads to $\dot{s} = -\alpha$. Thus Eq. (8.8) is satisfied. Similarly, for $s < 0$, the control law

$$u = u_{eq} + \alpha$$

results in $\dot{s} = \alpha$, and Eq. (8.9) holds. Combining the above two control laws, we obtain

$$u = u_{eq} - \alpha \operatorname{sgn} s. \tag{8.11}$$

Remark. The above control law applies not only to reaching the sliding surface but also to staying on the surface since the switching term in Eq. (8.11) forces the state to go back to the sliding surface if it deviates from the surface.

8.2.5 Robust Sliding-Mode Control

In this section we deal with the case where the uncertain functions Δf_{p2}, Δf_2, and d may be nonzero. We assume that these functions are bounded by known functions. Thus let

$$\left|\Delta f_2(x_1, x_2, t) + d(t)\right| \leq \rho(x_1, x_2, t)$$

for all $x_1, x_2 \in R$, and $t \in R_+$ where ρ is a known continuous function. (The condition will be generalized in Section 8.3.) It is easy to modify the control law discussed in Section 8.2.4 to accommodate the above uncertainties. For this, recalling Eqs. (8.7) and (8.11), let

$$u = u_{eq} - \alpha \operatorname{sgn} s + \nu \tag{8.12}$$

Note that the first two terms in the righthand side of this equation produce a proper control when the uncertainties Δf_2 and d are not present. We will choose v to overcome the effect of the uncertainties. Recalling Eqs. (8.1) and (8.3), the system equations (i.e., the tracking error equations) are given by

$$\dot{x}_1 = x_2 \qquad (8.1)$$

$$\dot{x}_2 = f_2 + \Delta f_2 + u + d. \qquad (8.13)$$

From Eqs. (8.1), (8.4), and (8.13),

$$\dot{s} = \dot{x}_2 + \lambda \dot{x}_1 = f_2 + \Delta f_2 + u + d + \lambda x_2.$$

Using Eqs. (8.7) and (8.12), we get

$$\dot{s} = \Delta f_2 + d - \alpha \,\mathrm{sgn}\, s + v.$$

Similarly to the previous section, for the state satisfying $s > 0$, we have

$$\dot{s} = -\alpha + \Delta f_2 + d + v \leq -\alpha + \rho + v.$$

Thus, if we choose v such that $v \leq -\rho$, then $\dot{s} \leq -\alpha$. On the other hand, for $s < 0$, we have

$$\dot{s} = \alpha + \Delta f_2 + d + v \geq \alpha - \rho + v.$$

So, choosing v satisfying $v \geq \rho$ results in $\dot{s} \geq \alpha$. Thus, combining the above two cases, v may be chosen as

$$v = -\rho \,\mathrm{sgn}\, s. \qquad (8.14)$$

Thus, by Eqs. (8.12) and (8.14), the overall control law is given by

$$u = u_{eq} - (\alpha + \rho)\,\mathrm{sgn}\, s. \qquad (8.15)$$

Here, ρ is a function of x_1, x_2, and t defined above, and α is a positive number (though it can be chosen as a function of x_1, x_2, and t).

8.2.6 Generalized Lyapunov Function

The control law obtained in the previous section can also be found by using the idea of Lyapunov functions. (Background materials on Lyapunov stability theory as well as extensive discussion of its application can be found in [5], [6], and so on. But this section may be read without the background.)

Define a function V (which is a generalized Lyapunov function) by

$$V(s) := \frac{1}{2}s^2.$$

Note that $V(s(x_1, x_2))$ represents a measure of distance (or squared distance) of a point (x_1, x_2) from the sliding surface S (on which the value of V, and therefore the value of s, is zero). Thus $\dot{V} < 0$ means that the state moves toward the surface S. Using the control law defined by Eqs. (8.12) and (8.7)

$$\dot{V} = \frac{\partial}{\partial s} V \cdot \dot{s} = s\dot{s} = s(\Delta f_2 + d - \alpha \operatorname{sgn} s + v)$$
$$\leq -\alpha|s| + |s|\rho + sv.$$

It is easy to see that the choice of the control

$$v = -\rho \operatorname{sgn} s$$

results in

$$\dot{V} \leq -\alpha|s|. \tag{8.16}$$

Note that the overall control law u obtained above is identical to the one given by Eq. (8.15).

Remark. Equation (8.16) implies $(d/dt)|s| \leq -\alpha$ for $s \neq 0$, which, in turn, implies Eq. (8.10); that is, V will vanish in at most $|s(x(t_0), x(t_0))|/\alpha$. Equation (8.16) also assures that the state will remain on S once it has reached S.

8.2.7 Preventing Chattering by Continuous Approximation

As stated in Section 8.1, the control law designed above forces the state trajectory to remain on S possibly by way of infinitely fast switching. However, in practice, the switching may occur with a slight delay when the trajectory reaches the switching surface, which causes the state trajectory to overshoot from one side of the surface to the other. The process repeats itself, causing the trajectory to cross the surface back and forth at a fast pace. This phenomenon is called *chattering* (because of the noise that it may create). The chattering is generally not desirable since it may excite unmodeled high-frequency dynamics resulting in unexpected instability. (For more discussion on chattering, refer to [6, Chapter 7], [8], and [27] for instance.) The chattering can be avoided by "smoothing" the switching controller. Use of continuous approximation of the switching function (i.e., use of a saturation function in place of the switching function) [5, Chapter 13], [6, Chapter 7], use of an asymptotic observer [7, Chapter 14], [27], and use of a sliding sector in place of a sliding surface [15] have been reported to be effective. Here, we will discuss the first approach. Let ε be a (small) positive number. In place of Eq. (8.15), we use

$$u = u_{eq} - (\alpha + \rho) \operatorname{sat}\left(\frac{s}{\varepsilon}\right), \tag{8.17}$$

where the saturation function sat is defined by

$$\text{sat}(y) = \begin{cases} y, & \text{if } |y| \leq 1 \\ \text{sgn } y, & \text{if } |y| > 1. \end{cases}$$

The functions sgn s and sat(s/ε) are depicted in Figure 8.3(a) and (c). Clearly, as $\varepsilon \to 0$, sat(s/ε) approaches sgn s. Instead of a switching (or sliding) surface S, we have a *boundary layer* $S_{BL} := \{x \in R^2 | \ |s(x_1, x_2)| \leq \varepsilon\}$ separating the space into two regions $s > \varepsilon$ and $s < -\varepsilon$. Within the boundary layer, the value of u transitions continuously. The above control with saturation clearly forces the trajectory to within the layer once the state reaches the layer. (*Note*: $(d/dt)|s| \leq -\alpha$ for $|s| \geq \varepsilon$.) Furthermore, it can be shown that, when the trajectory is confined in the layer, the trajectory (i.e., the tracking error) is bounded by the quantity dependent on ε. Hence we have the following.

Theorem 8.2.7.1. If the initial state $(x_1(t_0), x_2(t_0)) \in S_{BL}$, then with the control law given by Eq. (8.17) we have

$$|x_1(t)| \leq \frac{\varepsilon}{\lambda} + |x_1(t_0)| e^{-\lambda(t-t_0)}$$

$$|x_2(t)| \leq 2\varepsilon + \lambda |x_1(t_0)| e^{-\lambda(t-t_0)}$$

for all $t \geq t_0$.

Proof. By the definition of s,

$$\dot{x}_1 + \lambda x_1 = s.$$

Solving this for x_1, we get

$$x_1(t) = e^{-\lambda(t-t_0)} x_1(t_0) + \int_{t_0}^{t} e^{-\lambda(t-\tau)} s(x_1(\tau), x_2(\tau)) d\tau.$$

Since $|s| \leq \varepsilon$,

$$|x_1(t)| \leq \left| e^{-\lambda(t-t_0)} x_1(t_0) \right| + \left| \int_{t_0}^{t} e^{-\lambda(t-\tau)} s(x_1(\tau), x_2(\tau)) d\tau \right| \leq e^{-\lambda(t-t_0)} |x_1(t_0)| + \frac{\varepsilon}{\lambda}.$$

Then, since $x_2 = -\lambda x_1 + s$,

$$|x_2(t)| \leq \lambda |x_1(t)| + |s| \leq \lambda |x_1(t)| + \varepsilon \leq 2\varepsilon + \lambda |x_1(t_0)| e^{-\lambda(t-t_0)} \qquad \blacksquare$$

Remark. $(x_1(t), x_2(t))$ approaches an arbitrary small neighborhood of the zero state as $t \to \infty$ if $\varepsilon > 0$ is chosen arbitrarily small.

Remark. An extension of Theorem 8.2.7.1 to higher order systems can be found in [6, Chapter 7].

8.3. SLIDING-MODE CONTROL: GENERAL CASE

In this section we discuss the sliding-mode control in a more general context.

Section 8.3. Sliding-Mode Control: General Case

8.3.1 Problem Formulation

We will be concerned with the system described (in the so-called *regular form*) by

$$\dot{x}_1 = f_1(x_1, x_2, t) + \Delta f_1(x_1, x_2, t) \tag{8.18}$$

$$\dot{x}_2 = f_2(x_1, x_2, t) + G_2(x_1, x_2, t)\{u + \Delta g_2(x_1, x_2, u, t)\} \tag{8.19}$$

where $x_1(t) \in R^{n-m}$, $x_2(t) \in R^m$, $u(t) \in R^m$ for each time $t \in R_+$, and $f_1(x_1, x_2, t) \in R^{n-m}$, $\Delta f_1(x_1, x_2, t) \in R^{n-m}$, $f_2(x_1, x_2, t) \in R^m$, $G_2(x_1, x_2, t) \in R^{m \times m}$, $\Delta g_2(x_1, x_2, u, t) \in R^m$ for each $x_1 \in R^{n-m}$, $x_2 \in R^m$, $u \in R^m$, and $t \in R_+$. The column vectors $x(t) := [x_1'(t)\ x_2'(t)]'$ and $u(t)$ are, respectively, the state and control input at time t. The functions $f_1, f_2,$ and G_2 are assumed known, and Δf_1 and Δg_2 represent uncertain terms. Thus $f_1, f_2,$ and G_2 can be used in the control law, while Δf_1 and Δg_2 cannot. We assume that $G_2(x_1, x_2, t)$ is nonsingular for each $x_1 \in R^{n-m}$, $x_2 \in R^m$, and $t \in R_+$, and $f_1, \Delta f_1,$ and f_2 vanish at the origin of the state space (i.e., they are zero whenever x_1 and x_2 are zero). The uncertainty Δg_2 is called a *matched* uncertainty in the sense that this uncertainty affects the same channel as u (i.e., Δg_2 can be eliminated by an appropriate u if we know what Δg_2 is), and Δf_1 is called an *unmatched* uncertainty.

Remark. Equation (8.19) is equivalent to the form

$$\dot{x}_2 = f_2(x_1, x_2, t) + \tilde{\Delta} f_2(x_1, x_2, t) + G_2(x_1, x_2, t)\{u + \tilde{\Delta} g_2(x_1, x_2, u, t)\} \tag{8.20}$$

where $\tilde{\Delta} f_2(x_1, x_2, t) \in R^m$, and $\tilde{\Delta} g_2(x_1, x_2, u, t) \in R^m$, via $\Delta g_2 = G_2^{-1} \tilde{\Delta} f_2 + \tilde{\Delta} g_2$.

Remark. A control system equation

$$\dot{\xi} = \bar{f}(\xi) + \delta_1(\xi) + \bar{G}(\xi)\{u + \delta_2(\xi, u)\}$$

where $\xi \in R^n$ and $u \in R^m$ can be transformed into the form described by Eqs. (8.18) and (8.20) using a nonlinear coordinate transformation

$$\begin{bmatrix} x_1 \\ x_2 \end{bmatrix} = T(\xi)$$

satisfying

$$\left(\frac{\partial}{\partial \xi} T\right) G = \begin{bmatrix} 0 \\ G_2 \end{bmatrix}.$$

(See [3, Chapter 4], [5, Chapter 12], [6, Chapter 6], or [8] for the details.) The time-varying representation such as Eqs. (8.18) and (8.20) may appear for tracking problems as described in Section 8.2.

Definition. The zero state is *asymptotically stable* if the following two conditions hold: (1) There exists a nontrivial region in the state space called a region of convergence and $x_1(t), x_2(t) \to 0$ as $t \to \infty$ when the initial state $(x_1(t_0), x_2(t_0))$ is within the region, and (2) (stability in the sense of Lyapunov) the trajectory can be kept arbitrarily close to zero if the initial state is chosen sufficiently close to zero;

that is, for any number $\varepsilon > 0$, there is a number $\delta > 0$ such that $\|x(t_0)\| \leq \delta$ implies $\|x(t)\| \leq \varepsilon$ for all $t \geq t_0$. If in addition the region of convergence is the entire state space, the above stability is *global*. (See [6].)

Noting that Eqs. (8.18) and (8.19) are generalizations of Eqs. (8.1) and (8.3), we will find u such that the zero of the system described by Eqs. (8.18) and (8.19) is globally asymptotically stable, in particular, $x_1(t), x_2(t) \to 0$ as $t \to \infty$. As before, we will solve this problem by using the sliding-mode control.

Remark. In this chapter we discuss the use of sliding-mode control to achieve global asymptotic stability for simplicity of technical presentation. For the problem of attaining local asymptotic stability, see the remark at the end of Section 8.3.3.

8.3.2 Sliding Surface

As we did in Section 8.2.2, we first define a continuously differentiable function s by

$$s(x_1, x_2, t) := x_2 - \phi(x_1, t) \tag{8.21}$$

where $x_1 \in R^{n-m}$, $x_2 \in R^m$, $\phi(x_1, t) \in R^m$, $t \in R_+$, and

$$\phi(0, t) = 0 \text{ for all } t \in R_+.$$

As before

$$s(x_1, x_2, t) = 0, \text{ or } x_2 = \phi(x_1, t) \tag{8.22}$$

defines the sliding surface $S(t)$, that is, $S(t) := \{x \in R^n | s(x_1, x_2, t) = 0\}$ (provided that we can find an appropriate u). To find the equation governing the system behavior on the sliding surface S, we substitute Eq. (8.22) into Eq. (8.18). Then, we have

$$\dot{x}_1 = f_1(x_1, \phi(x_1, t), t) + \Delta f_1(x_1, \phi(x_1, t), t). \tag{8.23}$$

The function ϕ is chosen so that the zero state of this reduced order system is (globally) asymptotically stable and so $x_1(t) \to 0$ as $t \to \infty$. Then, by Eq. (8.22), it follows also that $x_2(t) \to 0$ as $t \to \infty$.

Example 3.1

For the system described by Eqs. (8.1) and (8.3) in Section 8.2, $\Delta f_1(x_1, \phi(x_1), t) = 0$ and $\dot{x}_1 = f_1(x_1, \phi(x_1), t) = \phi(x_1) = -\lambda x_1$, $\lambda > 0$. Clearly, $x_1(t) \to 0$ as $t \to \infty$. Since $x_2 = \phi(x_1) = -\lambda x_1$, $x_2(t) \to 0$ as $t \to \infty$ as well.

Example 3.2

Suppose Eq. (8.18) is linear and time-invariant, say

$$\dot{x}_1 = A_{11}x_1 + A_{12}x_2 \tag{8.24}$$

where A_{11} and A_{12} are, respectively, known $(n-m) \times (n-m)$ and $(n-m) \times m$ real matrices. Applying

$$x_2 = \phi(x_1, t) = Kx_1,$$

we get

$$\dot{x}_1 = (A_{11} + A_{12}K)x_1$$

where K is an $m \times (n-m)$ real matrix. Choose K so that $A_{11} + A_{12}K$ is Hurwitz (i.e., has eigenvalues with negative real parts). Such a K exists if and only if the pair (A_{11}, A_{12}) is stabilizable, or if (A_{11}, A_{12}) is controllable. (See [1, Chapter 3] or [9].) A specific K may be found by solving a linear quadratic optimal control or regulator problem with the solution of a matrix Riccati equation, or by using a coordinate transformation to reduce Eq. (8.24) to a controllable canonical form. (These can be found in a standard linear control textbook, for example, [2]. Also see Eq. (8.32).)

Example 3.3

Stiction usually occurs when the velocity of the system (such as a positioning system) is low. To reduce the effect of the stiction, the following s may be chosen:

$$s = x_2 + \gamma(\operatorname{sgn} x_1)\sqrt{x_1}$$

where $x_1 \in R$ is the position, $x_2 = \dot{x}_1$ is the velocity, γ is a positive constant, and the reference trajectory is identically zero [21].

8.3.3 Robust Sliding-Mode Control

As before, we use the $\dot{s}$ equation for controller design. From Eqs. (8.18), (8.19), and (8.21), we have

$$\begin{aligned}
\dot{s} &= \dot{x}_2 - \frac{\partial \phi}{\partial x_1}\dot{x}_1 - \frac{\partial \phi}{\partial t} \\
&= f_2(x_1, x_2, t) + G_2(x_1, x_2, t)\{u + \Delta g_2(x_1, x_2, u, t)\} \\
&\quad - \frac{\partial \phi}{\partial x_1}\{f_1(x_1, x_2, t) + \Delta f_1(x_1, x_2, t)\} - \frac{\partial \phi}{\partial t}.
\end{aligned} \qquad (8.25)$$

Without uncertainties, that is, if Δg_2 and Δf_1 are zero, the equivalent control

$$u_{eq} = G_2^{-1}\left\{-f_2 + \frac{\partial \phi}{\partial x_1}f_1 + \frac{\partial \phi}{\partial t}\right\} \qquad (8.26)$$

results in $\dot{s} = 0$, and so the trajectory remains on the surface S if the state is already on S. But to compensate for uncertainties and to bring in or bring back to the surface S the state not on S, we will use the control of the form

$$u = u_{eq} + G_2^{-1}(x_1, x_2, t)v \qquad (8.27)$$

where $v(t) \in R^m$ is to be determined below. From Eqs. (8.25)–(8.27), we have

$$\dot{s} = v + \Delta h(v, x_1, x_2, t) \qquad (8.28)$$

where

$$\Delta h(v, x_1, x_2, t) := G_2(x_1, x_2, t)\Delta g_2(x_1, x_2, u, t) - \frac{\partial \phi}{\partial x_1}\Delta f_1(x_1, x_2, t).$$

We assume that the uncertainty Δh is bounded by the inequality

$$\|\Delta h(v, x_1, x_2, t)\|_\infty \leq \rho(x_1, x_2, t) + \kappa\|v\|_\infty$$

for all $x_1 \in R^{n-m}$, $x_2 \in R^m$, $v \in R^m$, and $t \in R_+$, where a continuous function $\rho(x_1, x_2, t) \geq 0$ and a number $\kappa \in [0, 1)$ are known. As in Section 8.2.6, using the above uncertainty bound, we will find v to move the (off-the-surface) state toward the surface S. Rewriting Eq. (8.28) element-wise, we have

$$\dot{s}_i = v_i + \Delta h_i(v, x_1, x_2, t), \quad i = 1, \ldots, m$$

where s_i, v_i, and Δh_i are, respectively, the ith elements of s, v, and Δh. Following the procedure used in Section 8.2.6, define for each $i = 1, \ldots, m$,

$$V_i := \tfrac{1}{2}s_i^2.$$

As indicated in Section 8.2, $V_i(s(x_1(t), x_2(t)), t)$ is a measure of (squared) distance of the state $x(t)$ from $S_i(t) := \{x \in R^n | s_i(x_1, x_2, t) = 0\}$, and we will find v_i such that

$$\dot{V}_i \leq -\alpha|s_i|$$

where α is a positive number (specified by the designer). We have

$$\dot{V}_i = s_i\dot{s}_i = s_iv_i + s_i\Delta h_i \leq s_iv_i + |s_i|\{\rho(x_1, x_2, t) + \kappa\|v\|_\infty\}.$$

Choose a function η satisfying

$$\eta(x_1, x_2, t) \geq \rho(x_1, x_2, t) + \alpha$$

for all $x_1 \in R^{n-m}$, $x_2 \in R^m$, and $v \in R^m$, and define

$$v_i = -\frac{\eta(x_1, x_2, t)}{1 - \kappa}\,\mathrm{sgn}\,s_i. \tag{8.29}$$

Then, we have

$$\dot{V}_i \leq -\frac{\eta}{1-\kappa}|s_i| + \rho|s_i| + \kappa\frac{\eta}{1-\kappa}|s_i| = -\eta|s_i| + \rho|s_i| \leq -\alpha|s_i|.$$

Remark. The above inequality assures that the state trajectory remains on the surface $S_i(t) := \{x \in R^n | s_i(x_1, x_2, t) = 0\}$ if it is already on the surface, and the trajectory initially off the surface S_i reaches it in finite time. (See the Remark after Eq. (8.16) in Section 8.2.6.) the sliding surface S is the intersection of $S_1, \ldots, S_m$.

Remark. For local asymptotic stability, the zero state of Eq. (8.23) needs to be asymptotically stable only locally with the region of convergence Ω_S near the zero. But,

Section 8.3. Sliding-Mode Control: General Case   203

then, for a trajectory of Eqs. (8.18) and (8.19) to converge to the zero state, the initial state must be within the region where Ω_S can be reached in finite time.

8.3.4 Continuous Approximation to Avoid Chattering

As in Section 8.2.7, let ε be a (small) positive number. In place of Eq. (8.29), we use

$$v_i = -\frac{\eta(x_1, x_2, t)}{1-\kappa} \operatorname{sat}\left(\frac{s_i}{\varepsilon}\right), \quad i = 1, \ldots, m. \tag{8.30}$$

As noted in Section 8.2.7, as $\varepsilon \to 0$, $\operatorname{sat}(s_i/\varepsilon)$ approaches $\operatorname{sgn} s_i$.

Remark. It can be shown that under some conditions the trajectories of the system of Eqs. (8.18) and (8.19) with application of Eqs. (8.26), (8.27), and (8.30) converge to a neighborhood $N(\varepsilon)$ of the zero state, and the size of the neighborhood can be specified by the choice of ε. (See [5, Chapter 13] for the details.)

8.3.5 Example: Single Degree of Freedom Robot

We consider a single degree of freedom (revolute) robot represented by

$$J\ddot{\theta} + Mg\ell \sin\theta + b_0\dot{\theta} = u$$

where θ is the joint angle of the robot (i.e., the angle between the directions of the link and gravity), J is the total moment of inertia about the joint axis, M is the total mass, ℓ is the distance from the joint axis to the center of mass of the system, b_0 is the friction constant, and u is the input torque. Let θ_r be the reference trajectory. We wish to design a sliding-mode controller u so that θ closely follow θ_r. For this let $x_{p1} := \theta$ and $x_{p2} := \dot{\theta}$. Then, we have the plant state equation

$$\dot{x}_{p1} = x_{p2}$$
$$\dot{x}_{p2} = -(Mg\ell/J)\sin x_{p1} - (b_0/J)x_{p2} + (1/J)u.$$

With the tracking errors defined by $x_1 := x_{p1} - \theta_r$ and $x_2 := x_{p2} - \dot{\theta}_r$, we have the state equations for the error

$$\dot{x}_1 = \dot{x}_{p1} - \dot{\theta}_r = x_{p2} - \dot{\theta}_r = x_2$$
$$\dot{x}_2 = \dot{x}_{p2} - \ddot{\theta}_r = -(Mg\ell/J)\sin(x_1 + \theta_r) - (b_0/J)(x_2 + \dot{\theta}_r) + (1/J)u.$$

The coefficients $Mg\ell/J$, b_0, and $1/J$ may not be known exactly. So, we write $Mg\ell/J = a + \Delta a$, $b_0/J = b + \Delta b$, and $1/J = c + \Delta c$ where a, b, and c are the estimates of $Mg\ell/J$, b_0/J, and $1/J$, and Δa, Δb, and Δc are parameter uncertainties (which may be due to an unknown payload for instance). Then, the above equations reduce to

$$\dot{x}_1 = x_2,$$
$$\dot{x}_2 = -(a + \Delta a)\sin(x_1 + \theta_r) - (b + \Delta b)(x_2 + \dot{\theta}_r) + (c + \Delta c)u$$
$$= -a\sin(x_1 + \theta_r) - b(x_2 + \dot{\theta}_r) + c\{u + \Delta g_2\}$$

where $\Delta g_2 := (1/c)\{-\Delta a \sin(x_1 + \theta_r) - \Delta b(x_2 + \dot{\theta}_r) + \Delta c u\}$.
Define $s := x_2 + \lambda x_1$. Then,

$$\dot{s} = \dot{x}_2 + \lambda \dot{x}_1 = -a\sin(x_1 + \theta_r) - b(x_2 + \dot{\theta}_r) + c\{u + \Delta g_2\} + \lambda x_2.$$

So, the equivalent control is given by

$$u_{eq} = (1/c)\{a\sin(x_1 + \theta_r) + b(x_2 + \dot{\theta}_r) - \lambda x_2\}.$$

Let $u = u_{eq} + (1/c)v$. Then, we have $\dot{s} = v + \Delta h$ where

$$\Delta h = c\Delta g_2 = -\Delta a \sin(x_1 + \theta_r) - \Delta b(x_2 + \dot{\theta}_r) + \Delta c\{u_{eq} + (1/c)v\}.$$

Substituting into this equation u_{eq} obtained above, it is easy to see that

$$|\Delta h| = \left|\left(-\Delta a + \frac{\Delta c}{c}a\right)\sin(x_1 + \theta_r) + \left(-\Delta b + \frac{\Delta c}{c}b - \frac{\Delta c}{c}\lambda\right)x_2 \right.$$
$$\left. + \left(-\Delta b + \frac{\Delta c}{c}b\right)\dot{\theta}_r + \frac{\Delta c}{c}v\right| \le \left|-\Delta a + \frac{\Delta c}{c}a\right| + \left|-\Delta b + \frac{\Delta c}{c}b\right||\dot{\theta}_r|$$
$$+ \left|-\Delta b + \frac{\Delta c}{c}b - \frac{\Delta c}{c}\lambda\right||x_2| + \left|\frac{\Delta c}{c}\right||v|.$$

Thus

$$\rho \ge \left|-\Delta a + \frac{\Delta c}{c}a\right| + \left|-\Delta b + \frac{\Delta c}{c}b\right||\dot{\theta}_r| + \left|-\Delta b + \frac{\Delta c}{c}b - \frac{\Delta c}{c}\lambda\right||x_2| \text{ and } \kappa \ge \left|\frac{\Delta c}{c}\right|,$$

and ρ may be chosen as

$$\rho = \left|-\Delta a + \frac{\Delta c}{c}a\right|_{\max} + \left|-\Delta b + \frac{\Delta c}{c}b\right|_{\max}|\dot{\theta}_r| + \left|-\Delta b + \frac{\Delta c}{c}b - \frac{\Delta c}{c}\lambda\right|_{\max}|x_2|,$$

$$\text{and } \kappa = \left|\frac{\Delta c}{c}\right|_{\max} < 1$$

where $|y|_{\max}$ denotes the maximum value of $|y|$ over the uncertainties. Then, the sliding-mode control and its continuous approximation are given by, respectively

$$u = u_{eq} - \frac{\rho + \alpha}{c(1 - \kappa)} \operatorname{sgn} s$$

Section 8.3. Sliding-Mode Control: General Case

and

$$u = u_{eq} - \frac{\rho + \alpha}{c(1 - \kappa)} \operatorname{sat}\left(\frac{s}{\varepsilon}\right)$$

where α is a positive number, ε is a small positive number, and ρ and κ are as given above. For specific numerical values of the system parameters ([5, Chapter 13])

$$a = b = c = \lambda = 1$$
$$-\tfrac{1}{2} \leq \Delta a \leq 1, \quad -1 \leq \Delta b \leq 2, \quad -\tfrac{1}{2} \leq \Delta c \leq \tfrac{1}{2},$$

we have

$$\rho = 1.5 + 2.5|\dot{\theta}_r| + 2|x_2|, \quad \kappa = 0.5.$$

If we choose $\alpha = 0.5 > 0$, the sliding-mode controller and its continuous approximation are given by

$$u = \sin(x_1 + \theta_r) + \dot{\theta}_r - 2(2 + 2.5|\dot{\theta}_r| + 2|x_2|)\operatorname{sgn} s$$
$$u = \sin(x_1 + \theta_r) + \dot{\theta}_r - 2(2 + 2.5|\dot{\theta}_r| + 2|x_2|)\operatorname{sat}\left(\frac{s}{\varepsilon}\right).$$

The simulation results for the continuous approximation are shown in Figures 8.4–8.6. The parameters and functions used for different simulations are tabulated in Table 8.1. Figure 8.6 depicts simulation results for the system with the matched random disturbance d with values between -0.5 and 0.5 as shown in Figure 8.6(e), that is, for the system with the second plant equation described by

$$\dot{x}_{p2} = -(Mgl/J)\sin x_{p1} - (b_0/J)x_{p2} + (1/J)(u + d).$$

Matlab Version 5.2 was used for the simulations.

TABLE 8.1 Parameters and Functions Used for Section 8.3.5 Simulations

Fig. No.	θ_r	ε	$(x_{p1}(0), x_{p2}(0))$	Disturbance
Figure 8.4	$\frac{\pi}{6}t$ for $0 \leq t \leq 3$	0.5	(0.2, 0)	None
	$\frac{\pi}{2}$ for $t > 3$			
Figure 8.5	Same as above	0.1	(0.2, 0)	None
Figure 8.6	Same as above	0.1	(0, 0)	Random matched. See Figure 8.6(e)

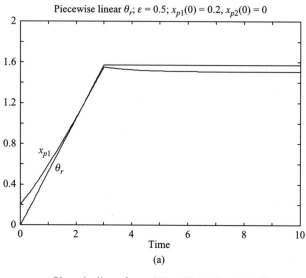

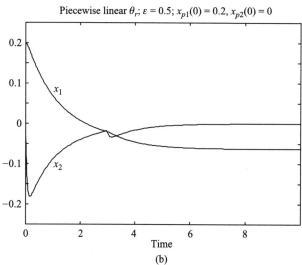

Figure 8.4

8.4. SLIDING-MODE-LIKE CONTROL FOR SAMPLED DATA CONTROL SYSTEMS

In this section, we consider a single-input linear sampled data system described by

$$\dot{x} = Ax + bu \tag{8.31}$$

where A and b are in the control canonical form, that is,

Section 8.4. Sliding-Mode-Like Control for Sampled Data Control Systems

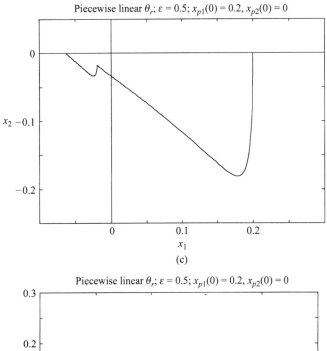

Figure 8.4 (continued)

$$A := \begin{bmatrix} 0 & 1 & 0 & \cdots & 0 \\ 0 & 0 & 1 & \cdots & 0 \\ \vdots & \vdots & \ddots & \ddots & 0 \\ 0 & 0 & \cdots & 0 & 1 \\ a_1 & a_2 & \cdots & a_{n-1} & a_n \end{bmatrix}, \quad b := \begin{bmatrix} 0 \\ 0 \\ \vdots \\ 0 \\ b_1 \end{bmatrix} \quad (8.32)$$

and $x = [x_1, \ldots, x_n]'$.

The control input u is assumed fixed over a sampling interval T, that is,

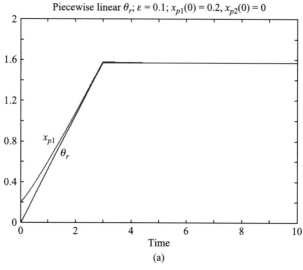

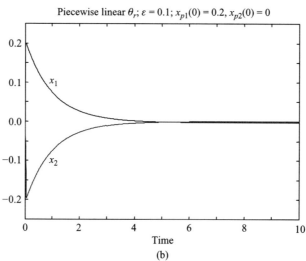

Figure 8.5

$$u(kT + \tau) = u(kT), \quad 0 \leq \tau < T. \tag{8.33}$$

Note that $x_2 = \dot{x}_1$, $x_3 = \dot{x}_2 = \ddot{x}_1, \cdots, x_n = \dot{x}_{n-1} = \cdots = x_1^{(n-1)}$. We seek a control law such that $\|x(t)\| \to 0$ as $t \to \infty$. Following the idea of the sliding-mode control design discussed in the previous sections, define

$$s(x) := c'x = c_1 x_1 + \cdots + c_{n-1} x_{n-1} + x_n.$$

Then, as before, $S := \{x \in R^n | s(x) = 0\}$ is a surface in the state space. We choose real constants c_i's so that $p^{n-1} + c_{n-1} p^{n-2} + \cdots + c_2 p + c_1$ is Hurwitz; that is, the zeros have negative real parts.

Section 8.4. Sliding-Mode-Like Control for Sampled Data Control Systems

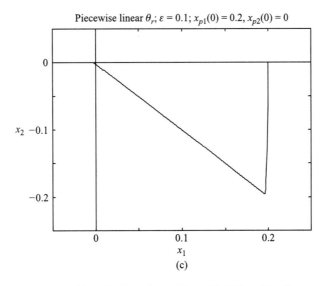

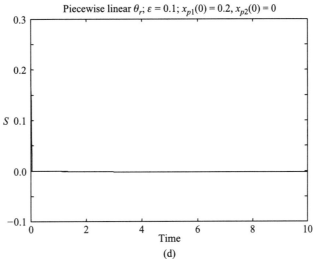

Figure 8.5 (continued)

Remark. The above choice of c_i's means that, since the motion of the state on S is characterized by

$$s(x) = c_1 x_1 + c_2 \dot{x}_1 \cdots + c_{n-1} x_1^{(n-2)} + x_1^{(n-1)}$$
$$= \left(\frac{d^{n-1}}{dt^{n-1}} + c_{n-1} \frac{d^{n-2}}{dt^{n-2}} + \cdots + c_2 \frac{d}{dt} + c_1 \right) x_1 = 0,$$

we have $x(t) \to 0$ as $t \to \infty$ provided the trajectory is confined to S.

210   Chapter 8 Variable Structure and Sliding-Mode Control

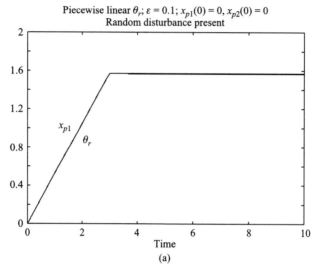

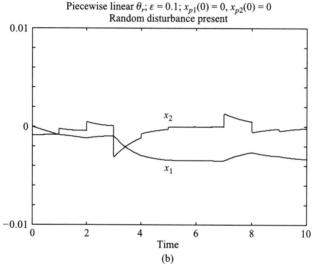

Figure 8.6

Since the input is fixed for a certain interval of time, it is not in general possible to confine the trajectory to S. But it is possible to asymptotically stabilize the system by introducing a cone-shaped region (also called a sector) surrounding S that plays a similar role to that of a sliding surface (see Figure 8.7). To capture the essence of the treatment for the hybrid situation (discrete-time controller in a continuous time system) without elaborate technicality, we assume that the system parameters a_i's and b_1 are known real numbers. With a slight abuse of notation, we will use $u(k)$, $x(k)$, and $s(k)$ in place of $u(kT)$, $x(kT)$, and $s(x(kT))$. In an attempt to bring the state to S (as in the previous sections), we will find u so that

$$|s(k+1)| < |s(k)|. \tag{8.34}$$

Section 8.4. Sliding-Mode-Like Control for Sampled Data Control Systems 211

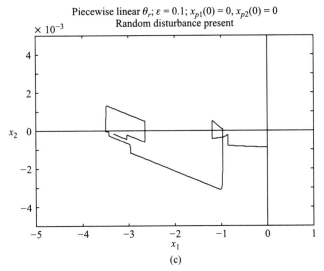

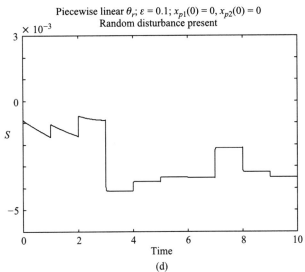

Figure 8.6 (continued)

For this, recall the variation of constants formula

$$x(t) = e^{A(t-t_0)}x(t_0) + \int_{t_0}^{t} e^{A(t-\tau)}bu(\tau)d\tau. \tag{8.35}$$

Using Eqs. (8.33) and (8.35), and a change of variables, we get the discrete-time state equation

$$x(k+1) = A_d x(k) + b_d u(k) \tag{8.36}$$

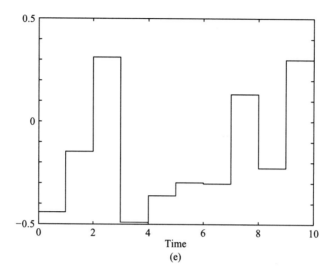

Figure 8.6 (continued) Disturbance with random amplitude.

where $A_d := e^{AT}$ and $b_d := \int_0^T e^{A\tau} b \, d\tau$. By the definition of s and Eq. (8.36), we have

$$s(k+1) = c'x(k+1) = c'A_d x(k) + c'b_d u(k).$$

But

$$A_d = e^{AT} = I + AT + \frac{1}{2!} A^2 T^2 + \cdots.$$

Therefore,

$$s(k+1) = c'x(k) + c'\left\{AT + \frac{1}{2!} A^2 T^2 + \cdots\right\} x(k) + c'b_d u(k).$$

Let γ be the largest singular value of A, that is, the largest (nonnegative) square root of the eigenvalues of $A'A$, and recall $\|A\| = \gamma$. Then,

$$s(k+1) \leq c'x(k) + \|c\| \left\{\gamma T + \frac{1}{2!} \gamma^2 T^2 + \cdots\right\} \|x(k)\| + c'b_d u(k)$$

$$= s(k) + \|c\| \|x(k)\| (e^{\gamma T} - 1) + c'b_d u(k).$$

Write $s_N(k) := \|c\| \|x(k)\| (e^{\gamma T} - 1)$. Then, we have

$$s(k+1) \leq s(k) + s_N(k) + c'b_d u(k).$$

Section 8.4. Sliding-Mode-Like Control for Sampled Data Control Systems

Similarly,

$$s(k+1) \geq s(k) - s_N(k) + c'b_d u(k)$$
$$= -s(k) + \{2s(k) - s_N(k)\} + c'b_d u(k).$$

Suppose $s(k) > 0$. Clearly, from the above inequalities, to achieve Eq. (8.34), that is, $-s(k) < s(k+1) < s(k)$, it is sufficient to choose u so that the following inequalities hold

$$s_N(k) + c'b_d u(k) < 0,$$
$$\{2s(k) - s_N(k)\} + c'b_d u(k) > 0,$$

which are equivalent to

$$s_N(k) - 2s(k) < c'b_d u(k) < -s_N(k).$$

Such u clearly exists provided $s_N(k) - 2s(k) < -s_N(k)$, that is, $s(k) > s_N(k)$, which means that γT must be sufficiently small. Assuming $c'b_d > 0$ (the other case can be treated similarly), u can be chosen, for instance, as

$$u(k) = \frac{1}{2} \frac{(s_N(k) - 2s(k)) + (-s_N(k))}{c'b_d} = -\frac{s(k)}{c'b_d}. \tag{8.37}$$

Similarly, for $s(k) < 0$, Eq. (8.34) is achieved by a control law satisfying

$$s_N(k) < c'b_d u(k) < -2s(k) - s_N(k),$$

and such a control law exists provided $-s(k) > s_N(k)$. Combining the above two cases, we conclude that if

$$|s(k)| > s_N(k), \tag{8.38}$$

a control law can be found to satisfy Eq. (8.34). An example of such a control law is given by Eq. (8.37). It can also be shown (see [20]) that

$$s(kT + \tau) \leq s(k), \quad 0 \leq \tau \leq T.$$

Remark. The region S_{cone} in which Eq. (8.38) is not valid, that is $S_{cone} := \{x \in R^n | |c'x| \leq \|x\| \|c\|(e^{\gamma T} - 1)\}$ defines a cone-shaped region. (See Figure 8.7.)

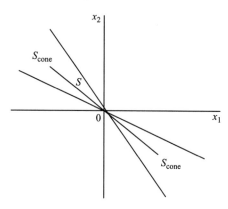

Figure 8.7 S_{cone} and S.

The following theorem is due to Jabbari and Tomizuka [20] tailored to the situation where all a_i's and b_1 are known.

Theorem 8.4.1. Given the system described by Eqs. (8.31)–(8.33) and a vector c defining a stable surface, the control u chosen in Eq. (8.37) results in

$$\begin{aligned} |s(k+1)| &< |s(k)| \text{ for } x(k) \notin S_{cone} \\ |s(k+1)| &< s_N(k) \text{ for } x(k) \in S_{cone}. \end{aligned} \quad (8.39)$$

Furthermore, $\|x(t)\| \to 0$ as $t \to \infty$ provided T is sufficiently small, or more specifically, if T satisfies

$$e^{\gamma T} - 1 < \frac{\alpha}{\rho_2 \sqrt{\left(\frac{\alpha}{\rho_2} + \frac{\|c_0\|}{\rho_1}\right) + \frac{1}{\rho_1^2}}}.$$

Here, $c_0 := [c_1 \ldots c_{n-1}]'$, and α, ρ_1, and ρ_2 are positive numbers satisfying the following inequalities:

Define a matrix

$$P := \begin{bmatrix} 0 & 1 & 0 & \cdots & 0 \\ 0 & 0 & 1 & \cdots & 0 \\ \vdots & \vdots & \vdots & \ddots & \vdots \\ 0 & 0 & 0 & \cdots & 1 \\ -c_1 & -c_2 & -c_3 & \cdots & -c_{n-1} \end{bmatrix}$$

which is Hurwitz by assumption. Denote by $_\beta\|\ \|$ the norm with respect to a basis β. Then, the solution of the differential equation $\dot{\zeta} = P\zeta$, $\zeta(0) = \zeta_0 \in R^{n-1}$ satisfies $_\beta\|e^{Pt}\zeta_0\| \le e^{-\alpha t} {_\beta}\|\zeta_0\|$ for some basis β. Furthermore, the norm in the basis β is related to that of the original basis by $\rho_1\|\cdot\| \le {_\beta}\|\cdot\| \le \rho_2\|\cdot\|$.

Remark. Equation (8.39) means that the trajectory starting in S_{cone} stays near or in S_{cone}.

Remark. The control law described by Eq. (8.37) is a fixed control law. This is due to the assumption that the system parameters are known. For systems with uncertain parameters (with known bounds), a discrete-time variable structure control law continuous in x can be designed. See [20] for the details.

Section 8.4. Sliding-Mode-Like Control for Sampled Data Control Systems

Example and Simulation.

Consider the system

$$\dot{x} = Ax + bu$$

where

$$A = \begin{bmatrix} 0 & 1 \\ 0 & 1 \end{bmatrix}, \quad \text{and } b = \begin{bmatrix} 0 \\ 1 \end{bmatrix}.$$

Let $T = 0.04$. Then,

$$u = -\frac{c'}{c'b_d}x(k) = -\frac{\begin{bmatrix} c_1 & 1 \end{bmatrix}}{\begin{bmatrix} c_1 & 1 \end{bmatrix}\begin{bmatrix} 0.000050167 \\ 0.0100502 \end{bmatrix}}x(k).$$

For $c' = [20 \ 1]$, we have

$$u = -\frac{\begin{bmatrix} 20 & 1 \end{bmatrix}}{0.0110535}x(k).$$

And for $c' = [1.5 \ 1]$, we have

$$u = -\frac{\begin{bmatrix} 1.5 & 1 \end{bmatrix}}{0.0101254}x(k).$$

The simulation results with $c' = [20 \ 1]$ and $[1.5 \ 1]$ are given in Figure 8.8. Both show convergence to the zero state.

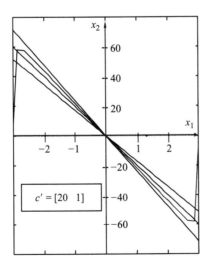
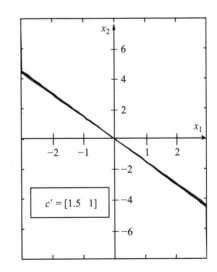

Figure 8.8 Sampled data control.

8.5 CONCLUDING REMARKS

In this chapter we have discussed robust variable structure control systems—more specifically, the sliding-mode control. Continuous-time sliding-mode control has been treated in a fairly general context. In practice, sliding-mode controls tend to create undesirable chattering. As a method to prevent chattering, modification of the sliding-mode control using saturation functions (replacing switching functions) has also been discussed. Use of computers in control systems leads to sampled-data systems. This chapter has illustrated a sliding-mode-like control for time-invariant linear systems with known coefficients. For further discussions on this topic, see [4, Chapter 5], [14], and [20]. There are various specific design methods for sliding-mode control, particularly for multi-input systems. The interested readers should refer to [5, Chapter 13], [6, Chapter 7], [8], [15], [19]. Sliding-mode control has found many practical applications as described in Section 8.1, and new applications are emerging [24], [26], [28]. As a final note, a design method called min-max design, or Lyapunov redesign, also leads to a variable structure control [5, Chapter 13], [10]. The resulting control law resembles that of the sliding-mode design with a boundary layer. But instead of controlling toward and within the boundary layer, it provides convergence of the trajectory toward the ball near the zero state and stability near the zero. Similar to the boundary layer, the ball can be made as small as the designer desires.

Related Chapters

- For a detailed discussion of hybrid dynamical systems, see Ch. 7.
- A technical tutorial on robot control can be found in Ch. 18.
- Lyapunov methods are also discussed in Ch. 6.

REFERENCES

[1] G. Basile and G. Marro, *Controlled and Conditioned Invariants in Linear System Theory*. Englewood Cliffs, NJ: Prentice Hall, 1992.

[2] K. Furuta, A. Sano, and D. Atherton, *State Variable Methods in Automatic Control*. Chichester, UK: John Wiley & Sons, 1988.

[3] A. Isidori, *Nonlinear Control Systems*, 2nd ed. New York: Springer, 1989.

[4] U. Itkis, *Control Systems with Variable Structure*. New York: Halsted Press, John Wiley & Sons, 1976.

[5] H. K. Khalil, *Nonlinear Systems*, 2nd ed. Englewood Cliffs, NJ: Prentice Hall, 1996.

[6] J.-J. E. Slotine and W. Li, *Applied Nonlinear Control*. Englewood Cliffs, NJ: Prentice Hall, 1991.

[7] V. I. Utkin, *Sliding Modes in Control Optimization*. Berlin, Heidelberg, New York: Springer, 1992.

[8] R. A. DeCarlo, S. H. Zak, and S. V. Drakunov, "Variable structure, sliding-mode controller design." In W. Levine (ed.), *The Control Handbook*. Boca Raton, FL: CRC and IEEE Press, 1996.

[9] F. Hamano, "Geometric theory of linear systems." In W. Levine (ed.), *The Control Handbook*. Boca Raton, FL: CRC and IEEE Press, 1996.

References

[10] M. Coreless and G. Leitmann, "Continuous state feedback guaranteeing uniform ultimate boundedness for uncertain dynamical systems." *IEEE Trans. Automat. Contr.*, Vol. AC-26, no. 5, pp. 1139–1144, 1981.

[11] R. A. DeCarlo, S. H. Zak, and G. P. Matthews, "Variable structure control of nonlinear multivariable systems: A tutorial." *Proc. IEEE*, Vol. 76, no. 3, pp. 212–232, March 1988.

[12] S. V. Emel'yanov. "Use of nonlinear correcting devices of switching type to improve the quality of second-order automatic control systems." *Avtomat. i Telemekh.*, Vol. 20, no. 7, 1959.

[13] A. F. Filippov, "Differential equations with discontinuous right-hand side." *Mathematicheskii Sbornik*, Vol. 51, no. 1, pp. 99–128, 1960.

[14] K. Furuta, "Sliding mode control of a discrete system." *Systems & Control Letters*, Vol. 14, pp. 145–152, 1990.

[15] K. Furuta and Y. Pan, "Variable structure control of continuous-time system with sliding sector," *Proc. IFAC*, World Congress, Beijing, 1999.

[16] M. M. Gutierrez and P. I. Ro, "Sliding mode control of a nonlinear input system: Application to a magnetically levitated fast tool servo." *IEEE Transactions on Industrial Electronics*, Vol. 45, no. 6, pp. 921–927, December 1998.

[17] F. Harashima, H. Hashimoto, and S. Kondo, "MOSFET converter-fed position servo system with sliding mode control." *IEEE Transactions on Industrial Electronics*, Vol. 32, no. 3, pp. 238–244, August 1985.

[18] H. Hashimoto, K. Maruyama, and F. Harashima, "A microprocessor-based robot manipulator control with sliding mode." *IEEE Transactions on Industrial Electronics*, Vol. 34, no. 1, pp. 11–18, February 1987.

[19] J. Y. Hung, W. Gao, and J. C. Hung, "Variable structure control: a survey." *IEEE Transactions on Industrial Electronics*, Vol. 40, no. 1, pp. 2–22, February 1993.

[20] A. Jabbari and M. Tomizuka, "Robust discrete-time control of continuous time plants in the presence of external disturbances." Japan/USA Symposium on Flexible Automation, Vol. 1, ASME, 1992.

[21] A. Jabbari, M. Tomizuka, and T. Sakaguchi, "Robust nonlinear control of positioning systems with stiction." *Proceedings of the American Control Conference*, San Diego, CA, 1990.

[22] O. Kaynak and F. Harashima, "Disturbance rejection by means of a sliding mode." *Proc. IEEE Transactions on Industrial Electronics*, Vol. 32, no. 1, February 1985.

[23] P. K. Nandam and P. C. Sen, "Industrial applications of sliding mode control." *Proc. IEEE/IAS International Conference on Industrial Automation and Control*, pp. 275–280, 1995.

[24] S. Sivrioglu and K. Nonami, "Sliding mode control with time-varying hyperplane for AMB systems." *IEEE/ASME Transactions on Mechatronics*, Vol. 3, no. 1, March 1998.

[25] C.-Y. Su, T.-P. Leung, and Y. Stepanenko, "Real-time implementation of regressor-based sliding mode control algorithm for robotic manipulators." *IEEE Transactions on Industrial Electronics*, Vol. 40, no. 1, February 1993.

[26] C. Unsal and P. Kachroo, "Sliding mode measurement feedback control for antilock braking systems." *IEEE Transactions on Control Systems Technology*, Vol. 7, no. 2, pp. 271–281, March 1999.

[27] V. I. Utkin, "Sliding mode control design principles and applications to electric drives." *IEEE Transactions on Industrial Electronics*, Vol. 40, no. 1, pp. 23–36, February 1993.

[28] J.-M. Yang and J.-H. Kim, "Sliding mode motion control of nonholonomic mobile robots." *IEEE Control Systems Magazine*, Vol. 19, no. 2, pp. 15–23 and 73, April 1999.

[29] D. R. Yoerger, J. B. Newman, and J.-J. E. Slotine, "Supervisory control system for the JASON ROV." *IEEE J. Oceanic Eng.*, Vol. 11, no. 3, pp. 392–400, 1986.

[30] K. D. Young, "Controller design for a manipulator using theory of variable structure systems." *IEEE Trans. Syst., Man, Cybernetics*, Vol. 8, pp. 101–109, 1978.

Chapter 9
CONTROL SYSTEMS FOR "COMPLEXITY MANAGEMENT"

Tariq Samad

Editor's Summary

At the midway point in this volume, this chapter attempts a "big picture" discussion on the trends and developments in advanced control systems. Previous chapters have described some of the new control technologies that are being pursued in research, and the second half of this book reviews the state of the art and future prospects in a variety of application arenas. Here, we view controls as a key discipline for "complexity management," a perspective that recognizes the substantially increased complexity of our engineering systems and the role that control systems can play in ensuring their safe, efficient, and profitable operation.

Talk of complexity is rife in virtually all technological (and many nontechnological) fields. However, it is in the area of control and automation that a practice of complexity is urgently needed. It is one matter to analyze the increasing complexity of systems or phenomena of interest; it is another matter entirely to "close the loop" on such systems. Indeed, the complexity of control solutions can increase disproportionately to the complexity of the target system.

But complexity is not just a problem; it is an opportunity too. Organizations (and people) that can harness the increasingly sophisticated technologies that are now at our disposal and leverage them into a new generation of automation and control solutions can expect to realize substantial economic (and intellectual) benefits.

This chapter reviews several objectives for control and automation systems, from human and environmental safety to increased autonomy; it briefly outlines some emerging control technologies that are not covered in depth elsewhere in this book; and it describes a few new, general application opportunities for control. Finally, a brief review of various schools of complexity management is provided, contrasting the diverse motivations that different intellectual communities are bringing to this intriguing new topic.

Tariq Samad is with Honeywell Technology Center and is a former vice president of technical activities for IEEE-CSS.

9.1 INTRODUCTION

Part 1 of this volume has focused, for the most part, on some of the algorithmic and theoretical machinery that control scientists and engineers have been recently developing. New specialized fields of control have emerged, such as discrete-event systems, hybrid control, intelligent control, and computer-aided control system design, and these are changing our notions of what control engineering is about. Linear systems and the PID will always be part of the control engineer's lexicon, but now our expertise and relevance also encompass large-scale, nonlinear, and decision-making problems.

Similar generalizations can also be made regarding the second half of this book. In some cases, the scope of our activities in traditional industries that have already reaped considerable benefit from control is expanding to higher levels of systems and enterprises. In other cases, entirely new domains and application areas are now looking to control, anticipating the same sorts of rewards that the process industries or building automation or aviation have historically gained.

Neither in technology nor in application is the trend in control toward one of "more of the same." The redefining and broadening that are underway can ultimately be seen as a response to the increasing complexity of our technological world. Effective technologies for "complexity management" are being sought by government, industry, and society. In this chapter, we discuss the implications of this for control. (A multi-disciplinary perspective can be found in [17] and [18]; some of the material here is taken from the first chapter of the latter.) New technological developments and new application opportunities are both highlighted, and this chapter thereby serves as a transition from the previous technology-oriented chapters to the subsequent application-oriented ones. Issues related to the complexity of an increasingly automated world are also discussed more generally, and some reactions within the scientific and engineering communities are reviewed. We begin by discussing the interplay between domain expertise and technique in control solutions.

9.1.1 Control Systems: Domain Knowledge and Solution Technologies

For devising effective control and automation solutions, two types of expertise are required. *Domain knowledge* is the first of these—we can control systems only if we understand them. This knowledge can exist in various forms, explicit and implicit: domain models for conceptual design, dynamic models for control algorithms, mental models of human operators, and many others. The degree and effectiveness of automation achievable are proportional to the quality and quantity of domain knowledge that is brought to bear. Thus controlling a distillation column requires that we capture an understanding of its construction and the chemistry of its operation; modern flight control would be impossible without knowledge of aircraft aerodynamics and jet propulsion; environmental control for buildings is contingent on knowing building layouts and pollutant diffusion mechanisms, among other factors.

The second component for control systems can be called the *solution technologies*. The automation and control of any complex engineering system is a multi-disciplinary undertaking. Sensing and actuation technologies are needed for measuring and moderating the physical world; control algorithms furnish the "intelligence" for mapping measurements and commands to actions; effective interaction between the automation system and its human users requires knowledge of human factors and cognitive science; software engineering is essentially the new manufacturing discipline for knowledge- and information-intensive systems; system health management technology is needed to help minimize and manage maintenance; and so on. Detailed treatments of most of these disciplines are beyond the scope of this book, the majority of which is focused on the "core" of control systems—control technologies in a narrower sense. Here, too, there is variety: Modeling, identification, optimization, estimation, and the other algorithmic methods of control engineering all need to be successfully mapped to the problem at hand if a practical solution is to be realized.

There is no gainsaying the critical importance of control systems in our technological society today. This is especially evident in industries such as oil refining, paper manufacturing, building management, robotics, and aviation—industries that, over the last few decades, have made substantial strides in efficiency, throughput, safety, and numerous other parameters. As an example of the state of the art in industrial process control, Figure 9.1 schematically shows the scale of controller that is now available for oil refining applications. In some other industries, such as automotive, civil construction, semiconductor manufacturing, and communication networks, control engineering has not historically had an equally substantial impact. Over the last few years, however, it has gained considerable prominence and is now seen as a key enabler for future progress. Then there are other industries in which the role of advanced controls is only just starting to be explored; biomedicine is one example of these industries.

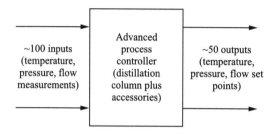

Figure 9.1 The state of the art in refinery control.

Even in the first tier of industrial domains, however, today's control systems are quite limited in comparison to the full range of problems for which automation is seen as a potential solution. A few of these limitations are as follows:

- Different controllers are developed for different operating conditions, and switchings are done manually.
- Automatic control may exist for several subsystems, but the overall supervisory and coordinating function is performed by people.
- Alarm management and handling of abnormal situations are largely unautomated.
- Little or no attempt is made to integrate the physical control of an engineered system with the business process.

We can view the current state of automation and control as a partial and incomplete connection between existing solution technologies and domain understanding, as depicted in Figure 9.2.

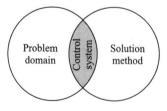

Figure 9.2 Current control systems represent a limited intersection of solution technologies and domain expertise.

9.2 CONTROL AND AUTOMATION TOMORROW: TOWARD COMPLEXITY MANAGEMENT

Much of our existing control system technology predates the profound, ongoing revolution in computational infrastructure. Hardware, software, displays, memory, communications are all orders of magnitude more powerful or capable today than they were even a decade ago. Progress has also been made in algorithms, theory, and domain understanding.

One implication of these developments is that it is now technologically feasible, in principle, to effect a much larger overlap than depicted in Figure 9.2. The foundation has been laid for a potentially revolutionary increase in the scale and impact of automation. But realizing the possibilities for dramatic advances in control systems requires equally dramatic changes to existing approaches and methodologies for their design and operation. If there is one word that captures the multifarious, interconnected opportunities and challenges for automation systems of tomorrow, it is complexity—and it is the management of this complexity that is a new research need (Figure 9.3).

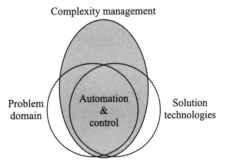

Figure 9.3 Control and automation can now play much larger roles in engineering systems. A science (and engineering) of complexity management is needed to exploit the opportunities and to avoid the accompanying pitfalls.

To illustrate this new vision, we can contrast the controller of Figure 9.1, which, despite its multivariable sophistication, is restricted to regulating one unit in a refinery under more-or-less steady-state conditions, with the enterprise-wide multiprocess diagram of Figure 9.4. The research needs in the process industries, as in all other domains that have in the past benefited substantially from controls, are increasingly related to problems at this scale. In this figure, liquor recovery loop around which most of the processes are structured can have a residence time on the order of weeks, during which large-scale fluctuations in ore quality, weather conditions, product requirements, and equipment health are not just possible but likely. Complex systems may never be in steady-state, even approximately.

9.3 OBJECTIVES FOR CONTROL AND AUTOMATION

In this section, we note some of the objectives that control systems must be designed to satisfy, and how these objectives have been getting increasingly more complex. It is not clear whether "performance at all costs" has ever been an acceptable dictum, but there is no doubt that criteria today are multifaceted. Human and environmental safety,

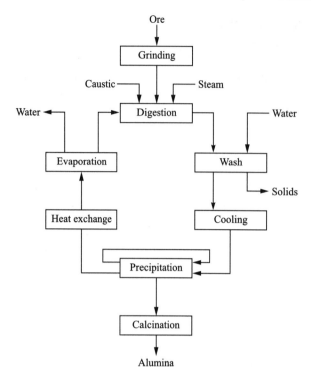

Figure 9.4 Tomorrow's need: enterprise-wide optimization and control. The picture shows the processes involved in alumina refining—turning mined bauxite into alumina. The liquor loop in this system can have a residence time of weeks. Industry is starting to look at the automation of systems of this scale.

design and development time and cost, regulatory compliance, and workforce reduction are some of the considerations that are driving automation initiatives.

Since commercial organizations comprise a significant proportion of both the suppliers and users of new technology, an underlying driver for automation is often economics. However, it is useful to examine the economic motivation in finer detail, decomposing it into different aspects that, to a first approximation, can be considered separable. Thus, while we do not explicitly break out profits and revenues as objectives for automation, these are often (but by no means always) the ends served by the means discussed below.

9.3.1 Human and Environmental Safety

Only in some cases is safety the primary motivation driving the development of an engineering system, but it is a near-universal concern that technology and automation are often called on to address. For some mature systems, safety issues can spur a large fraction of the associated innovation and research. We build cars primarily to transport people and their possessions, but a variety of devices are employed to ensure that this transportation does not endanger lives or the environment. Many of the recent controls-related technological developments in automobiles have been driven by the need to improve the safety of drivers and passengers—witness airbags and antilock brakes.

There are also systems in which safety is the primary objective and that are designed to protect us and our environments against natural forces or other engineering systems. Civil structures, such as dikes and bomb shelters, are obvious examples. This primary versus secondary distinction may be debated and in any case depends on the sometimes arbitrary delimitation of a system. Safety is certainly the primary consideration in a bicycle helmet, but it seems natural to consider seatbelts as a safety feature within the automobile system.

9.3.2 Regulatory Compliance

In the eyes of the public, human and environmental safety is an overriding concern with new technological developments. This concern directly influences the design, development, and operation of engineered systems, as alluded to earlier, but it also focuses attention on another facet of complexity management. At all political and governmental levels—from city councils to international organizations—there is increasing oversight on technology and its products. Regulations on environmental impact and human safety have resulted in new industries being established and have affected existing ones in profound ways.

In the automotive industry, regulatory compliance has been (arguably) the single most significant factor in increasing the complexity of engine control systems. Automobiles do not drive that much faster today than they did three decades ago, but their tailpipe emissions of nitrous oxides and hydrocarbons are orders of magnitude lower. New sensors; electronic fuel injection systems; catalytic converters; on-board diagnostic modules; advanced algorithms that regulate fuel flow, engine speed, and ignition timing in response to instantaneous conditions—these are some of the major enhancements in engines today whose primary purpose is to ensure that concentrations of regulated exhaust chemicals do not exceed mandated limits.

Legislation has been a major driver for new technological developments in the process industries as well. New product and service businesses have been formed to facilitate regulatory compliance and reporting. Emissions monitoring software for power plants and other process industries is one example, and many others exist in biomedicine, pharmaceuticals, consumer products, and food.

9.3.3 Time and Cost to Market

The design, development, and manufacturing of new technological products are complex processes, and performing them well requires time and effort. As the products themselves become more complex, these processes do so as well and in many respects disproportionately. Doubling the number of parts in a device quadruples the number of potential (binary) interactions, and validation and testing become that much more problematic.

Yet marketplace realities are demanding that companies develop new products, of increasing complexity, in less time. Product turnover and innovation indices are widely seen as indicators of quality in technology sectors. Good technical ideas are one ingredient, but their rapid commercialization is often seen as more important. (After all, a corporation can generally acquire the technical idea more readily than a marketable product.)

Fast time-to-market demands are not wishful thinking on the part of executive management; numerous recent instances can be cited in which product development times have been considerably shortened in comparison to previous developments. Computing technologies have been central to these achievements. Easy-to-use software packages have been developed to facilitate control design, implementation, and testing in all major control application domains.

9.3.4 Increased Autonomy

In many businesses, personnel costs constitute a large fraction of total expenses. Where human involvement is not considered critical, its replacement by suitably sophisticated automated substitutes can result in substantial financial savings. The benefits of automation are of course not limited to cost reduction. Autonomous systems are being sought for higher productivity, for operation in hostile or otherwise inhospitable environments, for miniaturization reasons, and for other purposes.

Complete automation of any significant system is not feasible now and will not be feasible in the foreseeable future. Increased autonomy thus implies essentially that the role of the human is shifted from lower-level to higher-level tasks. What used to be accomplished with a large team may now be done with a smaller team or with one person. With improvements in navigation systems and avionics, for example, the aviation industry has been able to reduce its cockpit flight crew by more than half. Fifty years ago, the state-of-the-art commercial airliner was the newly developed Lockheed Constellation. Its crew consisted of the pilot, copilot, navigator, flight engineer, and radio operator. Today's aircraft fly with just a pilot and copilot. In process industry sectors, automation solutions are being sought for some of the functions that plant operators perform. Preventive maintenance, prognostics, and diagnostics—aspects of system health management—are a particular focus of activity in this context.

Increased autonomy and the technological infrastructure that has enabled it also imply that larger-scale systems are now falling under the purview of automation. Sensors, actuators, processors, and displays for entire facilities can now be integrated through one distributed computing system. The control room in an oil refinery may provide access to or manipulate 20,000 or more "points" or variables. In possibly the largest integrated control system implemented to date, the Munich II international airport building management system can be used to control everything from heating and cooling to baggage transport. The system controls more than 100,000 points and integrates 13 major subsystems from nine different vendors, all distributed over a site that includes more than 120 buildings [1].

9.3.5 Other Criteria: Yield, Capacity, Efficiency, and More

Technological systems are developed for a primary purpose, whether it's production of gasoline in the case of oil refineries, transport of people and goods in the case of automobiles and airplanes, information processing in the case of computers, or organ stimulation in the case of implanted biomedical devices. Some of the objectives discussed earlier have illustrated how these primary purposes are not the only ones that matter and how in fact aspects of complex technological systems that might originally have been considered secondary from the point of view of system functionality can take on substantial importance.

Improvements in how well the system performs its primary function are also a continuing need—if not the only one. The specific nature of these performance improvements is system and function dependent, but it is useful to consider some examples:

- *Manufacturing yield.* An obvious performance metric for a manufacturing process is the within-specification product output per unit time. Improvements in yield relate directly to increased revenues (demand permitting) and so are always driving the development of automation and control systems. In many industries, plants are distinguished first on the basis of their production: megawatt output in power generation units, barrels per year for oil refineries, annual units for many discrete manufacturing lines.

- *Transportation capacity.* With vehicles, measures of speed, payload (number of passengers or cargo capacity), and distance are primary. In the case of automobiles and commercial aircraft, we have seen little or no improvement on the first two metrics over at least two decades. (In the latter case, however, the capacity of the air transportation system as a whole has seen considerable growth.)
- *Information processing power.* Processing speeds and memory capacity define performance in computers, and the continuing demand for growth on these counts has required increased complexity—in device physics, design methods, and manufacturing processes—to satisfy. At the same time, these improvements have themselves fueled the development of complex systems.
- *Energy efficiency.* This applies to all technological products, although its importance varies widely. Fuel efficiency in automobiles is of relatively less importance compared to fuel efficiency of aircraft. Even for a given system, energy efficiency can vary over time or location. Thus automobile fuel economy is much more important in most parts of the world than in the United States with its relatively low gasoline prices.
- *Miniaturization.* The utility of some functions is dependent on their encapsulation in small packages. Heart pacemakers, for example, have existed for over 40 years, but initially they were bulky, used external electrodes, and operated off line voltage. Compact, implantable pacemakers have been a revolutionary improvement; miniaturized sensors and actuators (electrodes) in particular have facilitated implantation. Miniaturization is also central to the development of effective substitutes for organs such as the heart [10].

9.4 EMERGING CONTROL TECHNOLOGIES FOR COMPLEX SYSTEMS

Given the qualitative tenor of the preceding remarks, it does not seem feasible to identify specific developments in control science and engineering that are directly and immediately motivated by specific drivers for automation. However, broader research trends can be correlated with these new challenges. In general, we can discern a movement away from narrowly specialized techniques that can give globally optimal, exact solutions for highly constrained problems; and toward techniques that are flexible, scalable, and broadly applicable. Most of the previous chapters in the book can be seen in this light. Here we highlight emerging control technologies that have not already been elaborated in detail.

9.4.1 Randomized Algorithms

As we attempt to address larger-scale, more complex problems, the limitations of our existing toolbox of techniques are an obstacle. The problem formulations of modern control theory, ideally suited for the applications of past decades, are inherently intractable in terms of their scaling properties. The question arises, Is there an alternative way of looking at the analysis and synthesis of models and controllers that is tractable, theoretically well-founded, and potentially useful for practical application?

One attractive property of conventional theories and derived algorithms for modeling, optimization, and control has been the "guarantees" on various counts—such as accuracy, speed, stability—that result. Unfortunately, these guarantees are gained at some cost, notably computational complexity. For example, determining whether the structured singular value is less than one is a certain indicator of stability, but the problem is NP-Hard and, hence, cannot be expected to be solvable for large-dimensional problems.

The area of randomized algorithms suggests an answer [19]. The concepts of Monte Carlo simulation and Vapnik-Chervonenkis (VC) dimension (neither of which is especially new—this emerging technology could more accurately be labeled "reemerging") lead to an entirely different way of looking at problems of interest. We eschew deterministic guarantees in heading in this direction, but there is nothing ad hoc about the approach. The theory can furnish probabilities of performance or stability and, furthermore, can precisely quantify the confidence in these estimated probabilities. This can be done for linear or nonlinear problems, under relatively weak assumptions. We assume, for example, that samples can be drawn at random and independently from a distribution of models or controllers and that the VC dimension of a model or controller (or an upper bound) can be computed. The numbers often work out surprisingly favorably, rebutting the usual "curse of dimensionality" arguments. Recent work also shows that Bayesian priors can be used to quantify expected performance in cases where the VC dimension of a concept space is infinite—a setup that would otherwise be considered uselessly underdetermined [11].

9.4.2 Biologically Motivated Control

Biological organisms represent a class of effective solutions to complexity management. At the higher levels of the biological world, vertebrates are able to survive, find sustenance, fend off predators, and raise kin, all over lifetimes that span decades or years and in the face of a physical environment that presents a host of challenges.

Little in our conventional control systems technology bears any substantive similarity to the design of biological control systems. Yet the success of the latter in dealing with complex systems (the "body" of the organism and the external environment are both examples of such) is unquestionable. This fact behooves us to gain a deeper understanding of the architecture and function of biocontrol. The central nervous system (CNS) of vertebrates is an ideal candidate for study in this context, and considerable literature is now available that illuminates such aspects of vertebrate CNS as its modular structure, the functions performed by different modules, the information processing pathways, and its integration with sensory organs and muscular tissue.

An especially notable characteristic of vertebrate CNS is the distinctive bidirectional bridging connections across parallel sensory and motor pathways (Figure 9.5). This structure allows actions to sensory stimuli to be initiated at various, more or less abstract, levels of processing, ranging from the presensory cortex to the prefrontal cortex. The inhibitory functions of the cortical areas enable this scheme.

Enough is now known about vertebrate CNS that we can start to outline CNS-based solution approaches to difficult problems in the automation of engineering sys-

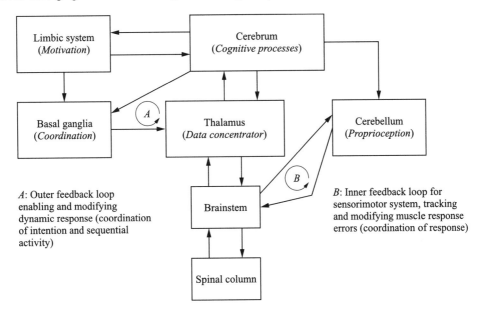

Figure 9.5 Control system architecture for the vertebrate central nervous system. (Figure courtesy of Blaise Morton)

tems. As one example, a control system for autonomous (uninhabited) fighter aircraft can be suggested. Of course, the biological analogy should only be carried so far as is useful; there are many important functions or subfunctions for which our existing technology already provides effective solutions. It is at the overall architectural level that the CNS analogy is likely to be most useful.

9.4.3 Complex Adaptive Systems

Complexity, as a phenomenon, is by no means limited to our artifacts. At virtually all scales of space and time, the natural world is replete with complex systems. A view that has gained considerable popularity over the last decade or so is that complexity in nature can be understood as emerging from the interaction of numerous, but individually simple, entities—often referred to as agents. This perspective has led to the establishment of complex adaptive systems (CAS) as a new interdisciplinary field of study that draws on research in such diverse disciplines as artificial intelligence, biology, physics, and nonlinear dynamical systems. Successful explorations with CAS have already been conducted in domains such as market economics, manufacturing logistics, chemistry, and ecology.

In CAS simulations, agents are object-oriented representations of domain entities. By endowing agents with adaptation capabilities, these simulations can address analytically intractable optimization problems. The adaptation mechanisms are generally based on algorithms inspired by biological evolution, such as genetic algorithms. Essentially, perturbations and modifications of parameters or structures within agents are essayed, and the effects of the variations are evaluated through the simulation. Criteria of interest, which may be agent-specific or global, can thus be optimized as the agents evolve.

9.4.4 Distributed Parameter Systems

The mathematics of distributed parameter systems has a long history, but only in relatively recent years has the control of such systems become recognized as a specialized and important topic. Cross-directional control of paper machines and other flat sheet systems, flexible structure control for space applications, the control of membranes and deformable mirrors—these are some of the applications that now are seen as having much in common.

Unlike conventional, lumped parameter systems, distributed parameter systems cannot be modeled purely by ordinary differential equations. The dynamics must be expressed in terms of partial differential equations to capture the spatio-temporal couplings. Distributed parameter system control requires large numbers of sensors and actuators to deal with spatial variation. A thousand actuators may be used in today's state-of-the-art paper machine controllers for ensuring that the thickness of the manufactured paper is within tolerance. (The tolerance limit may be a half micron out of a thickness of 20 or so microns.)

The development of microelectromechanical systems (MEMS) has also spurred research in the control of distributed parameter systems. For example, microvalves have now been manufactured in which the valve opening is regulated by a large number of individually actuated "gates." These are currently used simplistically to meet a desired "percent open" setpoint, but with more sophisticated control schemes it has been suggested that reliable laminar flow could be achieved over a wide operational range. Analogously, proposals have been proffered for drag reduction in aircraft and other vehicles by controlling large numbers of microflaps on exterior surfaces.

9.5 NEW APPLICATION OPPORTUNITIES FOR CONTROL AND AUTOMATION

In previous sections of this chapter, and elsewhere in this book, some of the "how's" and "why's" of complexity management for control systems are noted. Here we briefly outline a few of the associated "what's"—the new application possibilities that are now being explored. Although the opportunities discussed here and the earlier sections of this chapter will overlap somewhat, the additional specificity may help readers gain a deeper appreciation of the dramatic changes that new developments in control promise.

9.5.1 Large-Scale and Enterprisewide Optimization

Examinations of progress in control solutions in established application domains reveal some common themes. As lower-level problems are successfully managed, attention turns to the next higher level—and so on. In the control of process plants, the first targets for automation were single-input single-output loops for physical parameters such as the temperature in a vessel. Next, multivariable systems were automatically controlled—pressure, temperature, flow outputs in a reactor are dynamically coupled, and multiple inputs should be coordinated to realize desired operation. The current state of the practice allows entire units, such as distillation columns in a refinery, to be put under feedback control.

Analogously, commercial aviation has seen the successive automation of "inner-loop" processes that directly control elevators, ailerons, and rudders; "handling qualities" that ensure desired transient behavior to a given new vehicle state; and flight management systems that can automatically fly an aircraft from way-point to way-point.

The next step in the progression is seen as a quantum leap. In the process industries, the new target of research is enterprisewide optimization in which one automation umbrella is unfurled over all the processes and systems involved in turning raw material into product. The scale of such undertakings is truly daunting (recall Figure 9.4, for example). In the aviation industry, the interest has turned from controlling an aircraft to controlling the airspace. The hot topic today is "free flight," in which control and automation systems are used to ensure that aircraft can fly routes according to airline preferences rather than as mandated by air traffic control. In principle, both human safety and airline profitability can be significantly enhanced.

9.5.2 Integration of Business and Physical Systems

An important corollary of the interest in enterprisewide optimization is that control system objectives can no longer be narrowly constrained to the operation of the physical system. It is still important to ensure that temperatures stay within bounds, line speeds are maximized, product quality requirements are met, energy is efficiently expended, and so on. In addition—and this is a somewhat novel development in the application of control science—economic objectives have explicitly intruded into the picture. In closing the loop around a refinery, the control algorithm must factor in the cost of crude oil and the prices and demands of different products that the refinery can produce. Monetary costs of energy, of meeting (or failing) emissions constraints, even of maintenance shutdowns, are part of the picture.

One industry in which the integration of business and physical realms has recently taken on a new importance is electric power. Deregulation and competition in the power industry in several countries over the last few years has resulted in new business structures. At the same time, generation and transmission facilities impose hard physical constraints on the power system. For a utility to attempt to maximize its profit while ensuring that its power delivery commitments can be accommodated by the transmission system—which is simultaneously being used by many other utilities and power generators—economics and electricity must be jointly analyzed. An illustration is provided by a prototype modeling and optimization tool recently developed under the sponsorship of the Electric Power Research Institute. The tool integrates classical power flow calculations, corporate transaction models, and generation and consumption profiles within a complex adaptive systems framework. Figure 9.6 shows the user interface for the tool (named SEPIA for Simulators for Electric Power Industry Agents) [6].

9.5.3 Autonomous Vehicles

Vehicles that can fly, drive, or otherwise propel themselves have held an enduring fascination for people. Researchers in control systems have hardly been immune from such imaginings. Many of them, in fact, have attempted to turn these visions into real systems. Research in autonomous vehicles has made substantial progress recently, and there is now considerable excitement in the controls community that practical auton-

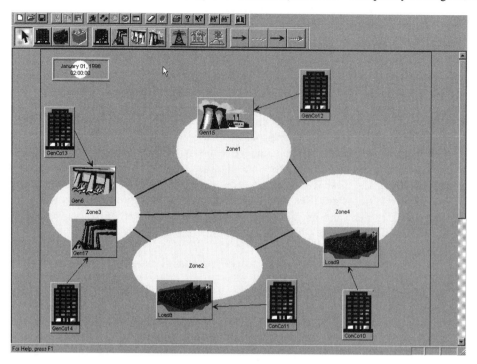

Figure 9.6 User interface for the SEPIA software tool showing a four-zone scenario.

omous vehicles may be a reality in the foreseeable (though not necessarily immediate) future.

Autonomous behavior in all types of vehicles has now been demonstrated. Videos of a Mercedes cruising the autobahn under automatic control can be seen on television shows. The car can adjust speed automatically on encountering speed limit signs and execute safe overtake maneuvers, among other capabilities. In the United States, several research groups demonstrated self-driving cars on a controlled section of the San Diego freeway in 1997.

Unpiloted aircraft are in use in warfare for precisely prescribed surveillance and reconnaissance missions today. Several research groups are actively working on the major extensions to this technology that are necessary before such vehicles can be autonomous in a meaningful way. Both autonomous fixed wing aircraft and autonomous helicopters are the objects of this research. Small remote-controlled helicopters are already available for civilian tasks; for example, rice farmers in Japan are using them for spraying paddies. Progressing from remote-control (where a human must directly control the rotor speeds and tilts) to autonomous control (where high-level tasks can be commanded) remains a challenge.

Numerous prototype underwater vehicles have also been developed. Because there are fewer safety issues in this case than for autonomous cars or aircraft, in some respects the field of autonomous underwater vehicles has seen more advances. Finally, recent breakthroughs in walking robots have shown that autonomous terrestrial vehicles need not be of the wheeled variety. Honda's humanoid robot (http://www.honda.co.jp/english/technology/robot/index.html) is capable of pushing carts and climbing stairs, among other behaviors that used to be well outside the realm of automated systems.

9.5.4 Data Mining and Intelligent Data Analysis

Virtually all engineering systems are sufficiently stationary so that their past behavior can help explicate current performance. Historical data can thus be invaluable for controlling complex systems—provided that it is available in large quantity, that it is readily accessible in electronic form, that appropriate statistical algorithms are available for deriving useful information from it, and that computational platforms can process it rapidly. These provisions were problematic even a decade ago, but not any longer, and data mining is on the "in" list of new technologies.

Although applications in areas such as direct marketing are already widespread, a fuller treatment of data mining for *dynamical* systems is necessary before the technology can have an impact on control systems. Research is well underway in this direction. As a hypothetical vision of the benefits that may lie in store, Figure 9.7 illustrates a potential application in an oil refinery.

9.5.5 Control Systems and the World Wide Web

Control engineering has not been at the center of the Internet and World Wide Web explosions, but we are starting to see some implications of the Web as an enabling technology for control. For example, Honeywell Hi-Spec Solutions has announced a series of Web-based products for the process industries [7]. One is called LoopScout™: a data collector can be downloaded free over the Web; the collected data are shipped back and automatically analyzed by statistical algorithms at the Hi-Spec Solutions server. Customers can then order a variety of reports that highlight problem control loops, allowing maintenance and asset management resources to be efficiently used. Similarly, IntelliScout™ focuses on plant heat exchangers. Site-resident software applets can be downloaded that monitor and analyze shell-and-tube heat exchanger performance in real time. Plant staff is alerted if abnormalities are uncovered.

In neither of these cases is closed-loop control being attempted over the Web. The Internet is not a real-time network, and no guarantees are provided for data rates or latencies. Real-time Web-based control is thus not a feasible proposition for any practical application. However, this situation could change if a global network with better real-time performance were to be developed. The next generation internet (NGI) could potentially be one such development.

9.6 SCHOOLS OF COMPLEXITY MANAGEMENT

Our focus in this chapter has been on examining the effects of the increasing complexity of automation on control science and engineering. The problem of complexity management, however, is of widespread interest within the technological community generally and even beyond it. In this section, we take a step back and review a selection of the discussion and proposals that have arisen out of a broad-based interest in managing complexity on the part of scientists, engineers, and others. The connection with previous sections is somewhat peripheral; the intent

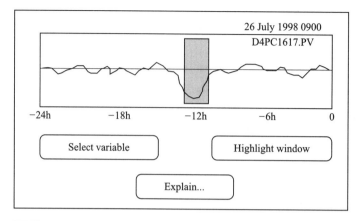

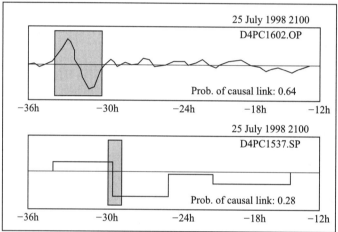

Figure 9.7 Hypothetical screens showing a possible data mining application for an oil refinery. The top screen allows a refinery supervisor to highlight an anomalous trajectory of a process variable and request the data mining system to identify likely causes for it. The bottom screen shows two possibilities: a disturbance (probably unmeasured) that affected a controller output and a setpoint change. Note the time-scales involved—data over days will need to be analysed for this application.

here is to provide a background within which the earlier control-specific discussion can be better appreciated.

Four different "schools" of complexity management are identified here, and significant differences exist within each. The schools are distinguished based on the particular aspect of complexity they foreground: safety, design efficiency, inspiration from nature, and societal connection.

9.6.1 Human and Environmental Safety: Forfeiture and Risk Assessment

In an influential book, Perrow [13] suggested that for many complex systems catastrophic accidents can never be made literally impossible, regardless of the diligence with which we design or operate the systems. Perrow considers three cases. For systems such as chemical plants and aircraft, the benefits are substantial (no feasible alternatives are known), and the freak worst-case accident will not be unthinkably catastrophic. With modest efforts we can further improve the safety of such systems and live with the remaining risk. His second category includes large marine transports and genetically engineered systems. Here benefits are again substantial, and risks, though higher, are outweighed. His recommendation is that we invest the substantial resources necessary for minimizing the likelihood of accidents such as the sinking of the Estonia ferry. The final category is reserved for systems that we must abandon because the scale of catastrophe that can result from their failures is absolutely intolerable, and, further, because less risky alternatives to these systems exist. The primary instance of this class is nuclear power.

Perrow's argument is not quantitative. Indeed, in some respects, it is a reaction to a mathematically sophisticated approach to complexity management known as risk assessment with which it shares the concern for safety. Given any system, risk assessment develops models that describe various ways that accidents can result, and it attempts to calculate the probability of each accident by considering the individual abnormal events that produce the accident. Probabilities for these individual events are needed in this calculation. Accidents are also often quantified in terms of the monetary losses or losses of lives they cause (and sometimes the two are explicitly related to each other). The end result is a number, a monetary amount, which can then be compared to the expense of engineering workarounds that will sufficiently reduce the probabilities of individual abnormal events. A variety of risks to human safety in this vein, with attendant statistics, are considered in [9].

9.6.2 Efficiency in Design: System Engineering and Virtuality

Whereas risk assessment and the "forfeiture" tack focus on the issue of safety, systems engineering (e.g., [16]) also addresses other aspects of complex technological systems, such as how to design systems cost-effectively, so that their reliability is maximized. Depending on the needs of specific projects, systems engineering can include systems analysis, requirements specification, system architecting, and other topics. Systems engineering (also sometimes referred to as systems management) often focuses on structured, formal process models for design and development. These models, at the highest level, specify the steps and tasks involved in the development of engineering systems, from the conceptual to product stage, and they can encompass the entire life cycle of products or technologies. Depending on the characteristics of the technologies and products, different models may be appropriate. When development is a sequential process, "waterfall" models can be considered. For more complex systems, with substantial couplings and interdependences, "spiral" models may be more suitable [15].

A related topic is virtual engineering, which is a term used to capture the increasing use of computational tools for facilitating system design [3]. These tools allow complex systems to be engineered "virtually," greatly reducing the need for manufacturing physical prototypes. The paradigmatic example is integrated circuit design. A technology that is fueling complexification requires complexity management itself, as might be expected. Putting millions of transistors on a few square centimeters is not humanly possible, and the electronics industry has been able to accomplish it only by relying extensively on libraries of largely automated tools for placement, layout, verification, simulation, synthesis, and virtually all tasks associated with the design and analysis of VLSI devices.

The use of conceptually similar tools is spreading to all technology-driven industries. The recent development of the Boeing 777 is a well-publicized example. At one time, 238 teams were working in concert to design the 777 using "design/build" computational tools. Full-scale physical mockups were not needed at every stage, correct part fits were ensured before manufacturing through digital preassembly, and multinational teams were able to work in coordination. The design/build virtual engineering teams were also considered essential for delivering "service-ready" aircraft to airlines.

Although design is a primary application of virtual engineering, there are many others, Simulation-based training, as another example, is commonplace in many industries. Aircraft pilots, power system dispatchers, and oil refinery operators routinely hone their skills on simulators, which allow appropriate human responses to dangerous incidents to be learned by trial and error, but without the potentially catastrophic consequences of trials on the real system.

9.6.3 Nature and Biology: Evolution, Emergence, and Power Laws

As already noted, every living system represents a successful solution to a complex problem—the environment within which the organism must survive. The biological world thus provides innumerable examples of complexity management. These successes have been achieved through an approach unlike any of those noted above. No formal, structured development process was followed, no specifications were written in advance, and no computer-aided design or analysis tools used!

This has led many researchers to propose that complex engineering systems can be developed and operated using biologically inspired methods. Computational implementations (not necessarily especially faithful) of biological evolution can "learn" the right answers to difficult problems through a process that simulates natural selection but (since this process is simulation and not reality) without the inefficiencies of the original. The algorithms that have been developed for this field, generally referred to as evolutionary computing [4], are all highly flexible and customizable. They make few assumptions about the characteristics of the problem compared to more traditional optimization algorithms, and evolutionary computing algorithms are thus applicable to problem formulations for which the latter cannot be used. Thus the design space need not be differentiable, convex, or even continuous; discrete and continuous variables can be simultaneously accommodated; arbitrary inequality and equality constraints can be included; and so on. Genetic algorithms are perhaps the best known instance of an evolutionary computing algorithm and well established as an important intelligent control technique.

In most of the preceding discussion, complexity can be seen, at least in part, as a consequence of scale. Systems become more complex as their components become more numerous, as more couplings arise, or as more behaviors are encapsulated. There is, however, another sense in which complexity is used within the technical community. In many small-scale systems, certain parametric regimes can cause a transition from well-behaved "simple" behavior to unpredictable "chaotic" dynamics. The output of such a system may appear to be random, but the system is entirely deterministic, and the apparent randomness is due to (often mild) nonlinearities.[1]

The key attribute of this chaos is a sensitive dependence on initial conditions, as memorialized in the Chinese butterfly apothegm: lepidopteran flittings in Beijing can produce showers in Minneapolis (or blizzards, depending on the time of year). A variety of mathematical tools are now available to detect chaos in systems by analyzing their outputs. Chaos has been uncovered in a wide range of systems, including manufacturing operations, road transportation networks, biological organisms, contagious disease epidemics, and financial markets. Systems of all sorts—small-scale and large-scale, abstract and real—can thus be analyzed through the language of nonlinear dynamical systems, the "science of complexity" [12]. Chaotic attractors, bifurcations, Lyapunov exponents, and other characteristics can provide insight into the complexity of a system and suggest approaches for managing the complexity.

The leading exponent of work in this area is the Santa Fe Institute (SFI), a private, nonprofit, multidisciplinary research and education center founded in 1984. SFI is described by one of its founders as "one of very few research centers in the world devoted exclusively to the study of simplicity and complexity across a wide variety of fields." [5, p. xiv]. Chaos theory and nonlinear dynamical systems are among the main themes pursued, but in an interdisciplinary environment that also includes international experts in biology, quantum physics, information theory, and economic theory. Theoretical research is coupled with application explorations in financial markets, manufacturing operations, cosmology, and social sciences, among others.

Alternative and intriguing explanations have recently been proposed for some putatively chaotic phenomena—and many others [2]. It turns out that a vast array of systems exhibit power law spectra: A quantity of interest (for example, the frequency of occurrence of an event) can be expressed as some power of another quantity (such as the magnitude or severity of the event). Thus a log-log plot of earthquake magnitude versus the number of earthquakes of at least that magnitude over some geographical region is a straight line (of negative slope).

The pervasiveness of power laws in human systems, including distributions of cities on the basis of their populations and the distribution of English words as a function of their usage rank, has been known since Zipf [20]. A new wealth of data from natural systems and observations in controlled experiments have further validated the power law model, and we now have the beginnings of a theory that may ultimately lead to a science of "self-organized criticality," the label coined for the new field. Unlike chaos theory, which shows how low-dimensional deterministic systems can exhibit seemingly random behavior, self-organized criticality is concerned with large-scale systems. Power law dynamics arise from the statistics of the interactions between system components.

9.6.4 Societal Connections

Although we have referred to market considerations and the economics of technology development and commercialization, the focus of the discussion has been on

---

[1] Perhaps the simplest example of chaos arises for the logistic difference equation, $x[t+1] = ax[t](1 - x[t])$. For $a = 2.5$, the system will eventually lead to a constant value for x, given any starting value $x[0]$ between 0 and 1. For $a = 3.25$, x will oscillate forever between two values. For $a = 3.5$, a cycle of period 4 results. Chaotic dynamics arise when $a = 4$ (for $x[0] \neq 0.5$).

technological issues and approaches. Another mark of increasing complexity, however, is that technology spills over into societal arenas. Managing complexity in such an environment requires an awareness of the interconnectivity between technology and society. As engineers and scientists, we are used to thinking that the primary influences are unidirectional. Technological and scientific achievements lead to societal change—through the telephone, television, electric power, the automobile, aviation, synthetic fibers, gasoline, the computer, and so on, our ways of life have changed in ways that were once inconceivable.

But technology development itself does not happen in a vacuum. Government funding of scientific research is ultimately under political and societal control; grassroots movements can derail major industries; "slick" marketing campaigns can sometimes overcome technical shortcomings; small, and not necessarily rational, advantages can snowball into industry domination. Technology certainly affects society, but it is "how society shapes technology" in the words of Pool [14] that technologists must also understand in today's complex world. Pool's emphasis is on nuclear power—a paradigmatic example, at least in the United States, of how society has influenced technology—but he also discusses numerous other case studies.

In a similar vein, Latour [8] coins the word "technoscience" to "describe all the elements tied to the scientific contents no matter how dirty, unexpected, or foreign they seem" (p. 174). Science and technology themselves are just elements of this broader, socially influenced enterprise, in which laboratory researchers, product engineers, and technology managers are a part of a network that includes consumers, heads of funding agencies, military strategists, legal professionals, and even metrologists. Figure 9.8 illustrates some of the complexity associated with the objects and processes of technology development in this view.

9.7 CONCLUSIONS

This chapter has argued that a dramatic change is underway in automation and control. Fueled in part by the inexorably exponential advancements in hardware and software

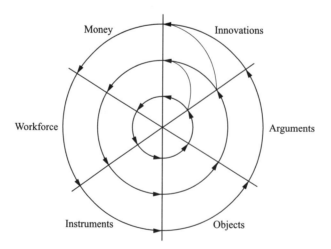

Figure 9.8 Spirals and circles of network-driven growth (from [8]).

technologies, and in part by economic and societal concerns—global competition and resource constraints, for example—our artifacts are becoming increasingly more complex.

Control systems are now being tasked to close the loop on substantially larger-scale problems than before. The complexity of these undertakings is truly daunting and is beyond the capabilities of the current state of the art. Many of the emerging technologies in control, such as hybrid and discrete-event systems, intelligent control, non-linear approximators, distributed parameter systems, and biologically inspired control architectures are attempting to correct this shortcoming. These and other research developments, if ultimately successful, will help furnish systematic, sound methods for dealing with tomorrow's control challenges.

Any impression that controls is a peripheral technology for meeting the automation challenges of the future would thus be misplaced. The portfolio of techniques and the research directions of control science and engineering will evolve, but the centrality of controls, as a unified discipline, for complexity management is assured. Indeed, there are good reasons to anticipate a revival of excitement and enthusiasm. As the scale of automation increases, the importance and visibility of modeling, identification, optimization, feedback, dynamics and so on, will inevitably be raised. The controls community has much to offer—and much to gain—by embracing complexity management.

REFERENCES

[1] M. Ancevic, "Intelligent building system for airport." *ASHRAE Journal*, pp. 31–35, November 1997.
[2] P. Bak, *How Nature Works*. New York: Copernicus Books, 1996.
[3] J. C. Doyle, "Theoretical foundations of virtual engineering for complex systems." http://www.cds.caltech.edu/vecs/, 1997.
[4] D. B. Fogel, *Evolutionary Computation: Toward a New Philosophy of Machine Intelligence*. New York: IEEE Press, 1995.
[5] M. Gell-Mann, *The Quark and the Jaguar: Explorations in the Simple and the Complex*. New York: W. H. Freeman, 1994.
[6] S. A. Harp, et al. *Complex Adaptive Strategies: Tools for the Power Industry*. Technical Report TR-112816, EPRI, 3412 Hillview Avenue, Palo Alto, CA, 1999.
[7] Hi-Spec Solutions. http://www.hispec.com/RecentNews/press/teleConReg.htm.
[8] B. Latour, *Science in Action*. Cambridge, MA: Harvard University Press, 1987.
[9] H. W. Lewis, *Technological Risk*. New York: W. W. Norton, 1990.
[10] E. H. Maslen, et al., "Artificial hearts." *IEEE Control Systems Magazine*, Vol. 18, no. 6, 1998.
[11] D. McAllester, "Some PAC-Bayesian theorems." *Proc. Computational Learning Theory (COLT '98)*, pp. 230–234, 1998.
[12] G. Nicholis and I. Prigogine, *Exploring Complexity*. New York: W. H. Freeman and Co., 1989.
[13] C. Perrow, *Normal Accidents*. New York: Basic Books, 1984.
[14] R. Pool, *Beyond Engineering: How Society Shapes Technology*. New York: Oxford University Press, 1997.
[15] E. Rechtin and M. Maier, *The Art of System Architecting*. Boca Raton, FL: CRC Press, 1997.
[16] A. P. Sage, *Systems Engineering*. New York: Wiley-Interscience, 1997.

[17] T. Samad (ed.), *Complexity Management: Multidisciplinary Perspectives in Automation and Control*. Technical Report CON-R98-001, Honeywell Technology Center, 3660 Technology Drive, Minneapolis, MN 55418, 1998.

[18] T. Samad and J. Weyrauch (eds.), *Automation, Control and Complexity: An Integrated Approach*. Chichester, UK: John Wiley and Sons, 2000.

[19] M. Vidyasagar, "Statistical learning theory and randomized algorithms for control." *IEEE Control Systems Magazine*, Vol. 18, no. 6, pp. 69–85, 1998.

[20] G. K. Zipf, *Human Behavior and the Principle of Least Effort*. Cambridge, MA: Addison-Wesley, 1949.

Chapter 10 | CONTROL OF MULTIVEHICLE AEROSPACE SYSTEMS

Jorge Tierno, Joseph Jackson, and Steven Green

Editor's Summary

Aircraft and spacecraft have been perhaps the most visible and awe-inspiring applications for control technology. The continuing march toward, and achievement of, ever higher performance in flight control have seemed inexorable and a perpetual source of problems as challenging as any that controls researchers could wish for.

Yet, while difficult problems in vehicle flight control remain outstanding, it is fair to say that algorithmic research in this area has encountered the law of diminishing returns. As discussed in Chapter 11, substantial challenges remain in single-vehicle control, related to the cost-effective development and deployment of controllers rather than new algorithms and theories. That aerospace control remains a vigorous, exciting field from algorithmic and theoretical perspectives as well can be attributed in large part to a broad new research direction: the control of multivehicle systems.

This chapter discusses three different and important multivehicle aerospace challenges. The first is in commercial aviation, where government and industry are seeking radical alternatives to today's air traffic control technology. Concepts such as "free flight" promise substantial improvements in safety and operational efficiency—under the proviso that the enabling control technology is available. The second example presented is formation flying for uninhabited combat air vehicles (UCAVs), with fleet coordination and autonomy requiring new fundamental research. Finally, the control of satellite clusters is discussed. This application is motivated by the cost and failure rates of large monolithic satellites and by the additional capabilities, such as synthetic aperture radar, that can be achieved by precise positioning of spatially distributed satellites.

Multivehicle systems are now a topic of general interest in control and automation; the interest is not limited to aerospace. Chapter 14 notes applications to road vehicles, including platooning automobiles.

Jorge Tierno is with the Dynamics and Control group at Honeywell Technology Center, Joseph Jackson is with Honeywell Air Transport Systems, and Steven Green is with NASA Ames Research Center.

10.1 INTRODUCTION

Automatic control technology for single-aerospace vehicles has seen significant progress in the past 30 years. It has achieved a level of sophistication that makes cost, and not theory, the principal limiting factor in performance. Although theoreticians and practicing engineers alike in this area still face significant challenges, it is no longer a major open question.

Automation is now extending beyond the single vehicle to a set of vehicles working in a coordinated fashion. Fleet or formation automation is being driven by many factors, the two most significant ones probably being the quests for increased safety and increased autonomy.

Coordinated control of a fleet of vehicles poses new challenges at all levels of the control hierarchy. On the one hand, theory and technology for the higher levels of the hierarchy, those dealing with the autonomy and coordination aspects, need to be developed from the ground up. On the other hand, the lower levels of the hierarchy, those dealing with the regulation aspect of control, need to be optimized in order to interact with the autonomy layer.

This chapter discusses some of these new challenges and briefly presents some ideas on how to solve them through three different examples derived from aerospace applications of multivehicle systems.

The first example is the commercial air traffic management (ATM) system. ATM is at the verge of revolutionary changes related to the introduction of "free flight." In future ATM systems, much more emphasis will be put on enabling users (i.e., airlines) to select their preferred routing, with minimal guidance from centralized facilities. Implementing such a system, while increasing overall capacity and maintaining or increasing current safety levels, will require the development of new automation both on-board the aircraft and on the ground.

The second example deals with the technological demands of uninhabited combat air vehicles (UCAVs) and autonomous and semi-autonomous UCAV formations. In a meaningful sense, UCAVs are a reality today. Success with cruise missiles has shown that it is feasible to inflict severe damage on enemy or terrorist nerve centers without putting pilots in harm's way. Although cruise missiles can be considered first-generation UCAVs, they are a far cry from the ultimate objective of autonomous UCAV fleets capable of executing complex missions in uncertain environments. We will describe some of the technology developments that need to come to fruition in order to build and field successful and safe UCAV formations.

The final example describes some of the recent developments and current research topics in the field of low Earth orbit satellite clusters. Smaller, simpler, cheaper satellites, operating as a unit, are more economical than large satellites. By being physically separated, the cluster is more capable (e.g., distributed aperture radar applications) and less vulnerable to being taken out by a single event such as a meteor or a collision with spacecraft debris. To achieve its potential in flexibility and cost reduction, it is required that all satellites be identical, that they all perform similar tasks (forming a network of peers and not a master/slave configuration), and that all distributed operations be readily reconfigurable in case of failure. These requirements pose exciting new challenges to the controls community. One of these challenges is to determine how simple identical control systems for individuals in the cluster can be chosen in order to achieve a given collective behavior. Some of the possibilities being considered will be discussed.

Control of fleets of aerospace vehicles is at the moment an open, quite active, and exciting research area. By presenting the examples mentioned above in this book, we hope to interest young engineers in this area.

10.2 FUTURE CONTROLS APPLICATIONS AND CHALLENGES IN ATM

On the road toward free flight, the designers of future ATM systems and components will face a number of interesting control challenges. Only a few key control challenges are described here, but we hope they will give the reader an appreciation of the role of control systems studies and contributions in the evolution of ATM. The examples described are as follows:

- Air traffic capacity management in the presence of disturbances (Section 10.2.2).
- Enabling user preferences in a safety-constrained air traffic system (Section 10.2.3).
- "Executing to plan" in constrained airspace—terminal area operations (Section 10.2.4).

10.2.1 Preliminaries: Airspace and Air Traffic Management

We will focus on flight operations within "controlled" airspace (i.e., airspace within which a flight must obtain an air traffic control, ATC, clearance to operate). The en route airspace within the continental United States (CONUS) is divided into 20 geographic regions (Figure 10.1), each of which is under the jurisdiction of an Air Route Traffic Control Center. In general Centers control the airspace from just above the surface on up to flight level (FL) 600 (60,000 ft). Each Center itself is subdivided into sectors (Figure 10.2), upwards of 20 or more, each operated by at least one

Figure 10.1 US Air Route Traffic Control Centers.

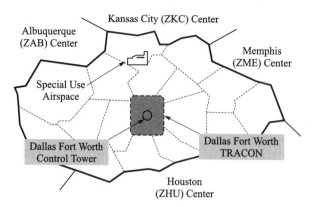

Figure 10.2 Ft. Worth Center airspace.

responsible human controller. Sectors are stratified into low altitude (e.g., surface up to FL240), high altitude (e.g., at and above FL240), and sometimes ultra-high altitude (e.g., above FL350). The geographic dimensions are typically designed based on nominal traffic conditions to facilitate traffic flow and distribute workload among the controllers. Each sector has its own unique radio frequency for air-ground voice communications. During sustained periods of low traffic density, multiple sectors are often combined into one sector. As density and workload increase within a sector, additional controllers may be brought in to assist.

Within a Center (Figure 10.2), many portions of the airspace may be reserved for uses other than civilian air transport. For reasons of national security or safety of flight, these regions of special use airspace (SUA) are off-limits to civilian aircraft. Depending on the application, SUAs vary in size, with areas ranging from several tens to hundreds of square miles and altitudes ranging from just above the surface to the upper flight levels. When active, SUAs may require significant detours from the most cost-efficient flight paths. Other regions of airspace may be delegated from a Center to a Terminal Radar Approach Control (TRACON) facility. TRACONs are also divided into sectors, like the Centers, but are designed around high-density terminal areas (often serving multiple airports). TRACON airspace generally extends from near the surface up to 15,000 ft or so above the ground. Although ATC radar coverage extends over most of the CONUS, TRACONs are generally served by shorter-range radar, with a higher update rate and accuracy that supports lower radar separation standards than en route (e.g., 3 vs. 5 nautical miles). The purpose of the TRACON is to facilitate the arrival and departure streams transitioning between the airport and Center airspace. A typical sequence of interactions between a flight and air traffic control is represented in Figure 10.3.

10.2.2 Air Traffic Capacity Management in the Presence of Disturbances

The goal of ATM is to maintain safe and efficient operations that meet the needs of airspace users. The U.S. ATM infrastructure has systematic procedures and constraints in place to robustly accommodate disturbances such as inclement weather, airport and runway closures, special-use airspace restrictions, terminal area congestion, aircraft

Section 10.2 Future Controls Applications and Challenges in ATM

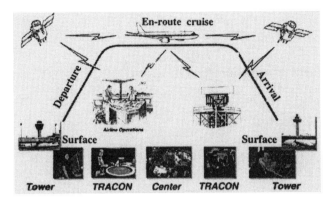

Figure 10.3 ATM phases of flight

equipment failures, and medical emergencies among others. These procedures will likely need to be modified as ATM progresses toward a free flight environment.

It is illustrative and potentially valuable for future ATM designers to analyse these procedures in the presence of disturbances, with the view of ATM as a regulator operating at multiple levels of control. Here, the term *regulator* denotes the classic control system with closed-loop control of a dynamic system. The "regulator" view of the ATM system components and procedures lends itself to the time-varying characteristics of the system disturbances, the high degree of dependency between the components of ATM, the "controls" that each procedure is intended to impose, and the feedback nature of the "system" to maintain safety of operations. The decomposition of the air traffic system attributes in this view are identified as follows:

10.2.2.1 Initial Conditions and Framework

Several parameters can be considered to describe the "nominal" state of an air traffic system. These include:

- *Airway structures and procedures.* Aircraft are normally constrained to fly along prescribed "highways in the sky." These are generally straight-line routes between way-points. Near terminals, however, the airway structures are more complex. In addition to the airways, procedural rules are also enforced. For example, minimum vertical and horizontal separation between aircraft must be maintained.
- *Airport configurations.* The placement and direction of runways in an airport affect air traffic. Parallel runways are common in major airports, but they may or may not be usable simultaneously, depending on conditions.
- *Airline published schedules.* The schedule of flights published by an airline represents an initial condition for air traffic management as well. Deviations are possible and not uncommon, but these can be viewed as perturbations from a nominal design.
- *Navigation aids.* The air traffic infrastructure provides several orientation and navigation signals to aid aircraft. Notable among these are ground-based navi-

gational beacons for en route guidance, inertial landing systems (ILS) near airport terminals, and the global positioning system (GPS).
- *Fleet equipage.* The number and configuration of aircraft also influence the role of ATM as regulator. Aircraft velocity, range, and on-board equipment must be considered.

10.2.2.2 Control Variables

Given the "problem structure" and initial conditions, the next issue is to identify what degrees of freedom are available to modify the state of the air traffic system:

- *Delay restrictions.* Air traffic management can impose delays to maintain a target level of safety (e.g., level of sector-traffic density). The ground delay procedure holds flights in order to reduce congestion or other limitation at the destination. Sector traffic is also controlled by adjusting delays of incoming and outbound traffic.
- *Traffic rerouting.* When faced with severe weather fronts, airport closures, or medical emergencies, aircraft can be rerouted or diverted.
- *Runway configurations.* The availability of runways for landing or takeoff is contingent on weather conditions. A good example of this "control" is at San Francisco International, where inclement weather can shut down one of a pair of runways/approaches to maintain safety of operations, thereby cutting capacity in half.
- *Traffic metering.* The metering and sequencing of traffic in the terminal area are used to avoid congestion of airport facilities. Note that the Center-TRACON Automation System (CTAS) uses route changes (vectoring, fanning, base leg extensions) as well as altitude and speed clearances to perform metering and sequencing operations.
- *Equipment upgrades.* Users may add new fleet equipment if the cost-benefit tradeoff justifies it.

10.2.2.3 State Variables

The next items to be considered are the key state variables that can be measured or estimated in air traffic systems and that are correlated with the performance of the system.

- *Sector-traffic density.* This refers to the number of aircraft under control by each ATM control center.
- *Airport state of operations.* Airports can operate under different conditions, with specific rules, that is, visual meteorological conditions (VMC) or instrument meteorological conditions (IMC). Other factors, such as the runway visual range (RVR), will also affect airport capacity and throughput.
- *Geographically distributed delay across the air traffic system.* Delays will propagate through the air traffic system.
- *Flight trajectories.* The flight plans of the aircraft in the air are also a fundamental and measurable state of the system.

10.2.2.4 Disturbances

Disturbances, that is, uncontrolled variables, can affect the behavior of the overall system. The most significant of these disturbances are:

- *Weather.* This is the single most significant disturbance to consider. Weather patterns can close areas to air traffic and degrade or impede operations at airports.
- *Medical emergencies.* On-board medical emergencies can alter flight plans and require emergency procedures in the terminal.
- *Aircraft equipment failures.* Equipment failures can delay departures, affecting gate assignments, connecting flights, and crew availability.
- *Change in special use airspace restrictions.*

10.2.2.5 Control System

After a model of the air traffic system has been determined, a control system can be devised. In order to do so, we first must identify the achievable performance. Some fundamental variables determine hard constraints on system performance, including the following:

- *Capacity versus safety of operations tradeoff.* Higher capacity can be achieved by lowering the separation requirements between aircraft in flight, but this also compromises the safety of the system. This is a fundamental tradeoff for the overall performance of the system.
- *Airport capacity.* Separation of arriving and departing vehicles is dictated by wake vortices behind the airplanes. This is a fundamental physical limit that sets an upper bound on the airport's capacity.
- *Economics of modernization.* Although more modern systems could increase performance, it may not be economically justifiable to implement them uniformly across the system.
- *Aircraft capability (and variation across fleet).* Related to the previous point, not all aircraft will be equally capable, and older airplanes may be in the system for many years yet.
- *Human-in-the-loop workload (controller, flight crew).* Any system relying on human controllers and pilots must respect their task load limits.

The second component of the control system is the measurement and estimation of the state variables described previously. Different state variables have different time constants associated with them and should be treated accordingly (see Figure 10.4).

- Short-term or tactical measurements such as airport runway visual range, airport ceiling, and the measurements of tactical hazard alerting systems such as TCAS (Traffic Collision Avoidance System, an on-board system that alerts pilots of other aircraft in the immediate vicinity and recommends corrective action).

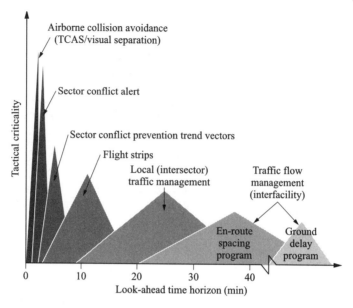

Figure 10.4 Key ATM decisions vs. time.

- Strategic measurements and estimates such as long-time-scale weather, AOC (Airline Operating Center) knowledge of ATM state, cleared flight plans, fleet position, and expected delays.

The final component is the desired performance or control objectives. These will be different for the different parties in the air traffic system. The airline's objective is to maintain safety and efficiency of operations, schedule integrity, and profit generation. The overall ATM objective is to maintain safety of operations and optimize capacity while supporting user needs. The ATM also will enforce a structured environment with deviations as exceptions and would arbitrate fairly across user's preferences. Airports strive to optimize throughput, while addressing environmental and noise restrictions in the communities they serve.

In order to assess the control or "regulation" performance of ATM, a system representation (model) and associated tool sets would be useful to simulate, analyze, design, and synthesize first-order changes to elements of the system. It would also be valuable for the design of new ATM functionality to have the ability to perform "what-if" scenario assessments and evaluate the feasibility of prototype procedures with decision support tools. These assessments should include the testing of nominal and off-nominal conditions; system performance and scheduling systems in the presence of delays; and real-time human-in-the-loop performance with the automation technology [5].

10.2.3 Enabling User Preferences in a Safety-Constrained ATM System

In a free flight environment, user (airline) preferences for flight trajectories will be accommodated by ATM unless the preferred flight path in some way compromises the safety of operations in a given airspace. User-preferred trajectories (UPTs) may be enabled by modifying the current system (ground-based responsibility for separation) to minimize deviations from the UPT [7, 19]. An alternative approach is to shift separation responsibility to the aircraft along with greater authority and flexibility in flight planning and maneuvering.

An operational mechanism to provide shared air and ground responsibility for separation assurance in the en route phase of flight is being investigated under the NASA Aeronautics Advanced Air Transportation Technology (AATT) program [10]. Early AATT studies propose methods to maximize user flexibility in relatively low-density airspace. Several controls applications and challenges associated with a new en route separation assurance system are described below.

10.2.3.1 Development of Distributed Separation Assurance Procedures

Shared separation in en route "unconstrained" airspace would require the following (necessary but not sufficient) air traffic system-level functionality:

- Mechanism to transfer position and intent data between aircraft and ATM.
- Unambiguous presentation of traffic data, conflict detections, and conflict resolutions that support user preferences.
- "Real-time" situation awareness of conflicts and resolution procedures by ATM.
- Determination of the role of controllers and flight crews under normal and nonnormal conditions to maintain the overall safety of ATM operations.
- Seamless reversion to positive ground separation control in the event of an unresolved conflict or abnormal event or upon the transition to constrained airspace.
- Intersector/facility coordination to orchestrate equitable, efficient, and stable metering of traffic flow management (TFM) affected flights.

Modeling and simulation of these functions in an air traffic system framework, as has been recommended in general by the RTCA Free Flight Select Committee and discussed in Section 10.2.2, will provide valuable insight into the safety, reliability, and robustness of candidate shared separation procedures under nominal, normal, and worst-case conditions. (The RTCA, organized as the Radio Technical Commission for Aeronautics in 1935, is a private, not-for-profit organization that addresses requirements and technical concepts for aviation.)

10.2.3.2 ATM Considerations

Independent of whether separation responsibility resides on the ground or is shared with the flight deck, ATM must fulfill a critical TFM role to enable user preferences. It may not be reasonable for the sophisticated aircraft of the future to plan "optimum" trajectories when the trajectory constraints (e.g., required time of arrival/metering) vary over airspace and time. It is not efficient for an aircraft to fly fast (to maintain schedule) into airspace with delays that could have been absorbed earlier, more efficiently. On the other hand, it may not be economically wise to slow down for a delay that is uncertain. A user's choice of preferences (such as speed profile or routing) is directly dependent on ATM constraints as well as the user's estimation of the state of the air traffic system (delays, weather). It is incumbent on the ATM system to provide users with the following two services: accurate, real-time updates of the status of the air traffic system (from which intelligent preferences may be selected); and stable, equitable, and minimal TFM constraints (required time-of-arrival, RTA) within which user preferences may be applied.

10.2.3.3 Flight Deck Considerations

The feasibility of shared separation will require the ability to detect and display conflicts, present resolution strategies to the flight crew for acceptance, and revert to ground separation control with a high degree of reliability. As it is likely that shared separation benefits will only be achieved if the presence of this flight deck functionality is the rule rather than the exception in the airspace where the procedure is exercised, this functionality will have to be implemented across a mixed fleet of aircraft and airlines, as discussed earlier. Integrating the separation task into those that the flight crews routinely execute in en route airspace might in itself be achievable, were it not for the limitations on the presentation of basic aircraft parameters, navigation, flight plan, weather radar, TCAS, wind-shear, terrain avoidance, cockpit display of traffic information (CDTI), and other demands for information presentation to the flight crew. This limitation is especially difficult to overcome in older aircraft with minimal display area.

An interesting controls-related problem here is to determine what information the flight crew needs to support the shared separation procedure, how to present the information that conveys the conflict and resolution strategy while maintaining situation awareness, and what control variables to exercise in order for the flight crew to perform the procedure. This process must account for the user preferences and equipage constraints of both aircraft and should not propagate the conflict to other pairs of aircraft. A related problem is to determine the least common denominator aircraft equipage that can support shared separation to a limited extent across a mixed fleet of aircraft with minimum cost to the users (airlines).

10.2.3.4 Airline Operating Center (AOC) Considerations

There are rare-normal conditions that will require a major redirection of en route traffic to maintain airspace safety, such as to avoid severe weather fronts or divert to an alternate destination in response to an airport closure. Under these conditions, it is unlikely that the flight deck will have the capability, in terms of information access and computational resources, to coordinate independently with ATM to derive a conflict-free resolution that is optimal from an airline fleet operation point of view. However, it may be beneficial to retain some measure of shared separation responsibility under these conditions to potentially reduce controller workload and sector traffic density. A direct involvement of AOCs with ATM and en route aircraft could facilitate shared separation responsibility and optimal conflict resolutions.

10.2.4 "Executing to Plan" in Constrained Airspace: Terminal Area Operations

Within the extended terminal area, that is, the airspace within 200 nm of a major airport, the degrees of freedom by which an individual aircraft can maneuver are significantly reduced from en route operations. In this high-density airspace (i.e., many constraints), positive controller intervention is required to maintain the safety of operations as arrival traffic converges. ATM automation tools have been developed to assist the controller and flight crews to operate more efficiently in the terminal area [1, 2].

A modern avionics system can automatically control a commercial aircraft from takeoff above 300 feet above ground level (AGL) through climb, cruise, and descent portions of the flight regime, with an automatic landing at a required time of arrival (RTA). However, because of today's controller intervention in the terminal area, or in the future ATM augmented as planned with the ground-based Traffic Management Advisor (TMA) and Passive Final Approach Spacing Tool (P-FAST), the flight path guidance portion of the flight deck automation is basically "turned off," and the flight crew operates the avionics using basic heading, speed, and altitude commands to the autoflight system. To enable a more efficient utilization of resources in a free flight environment, a more cooperative approach is needed between the evolving flight deck and ground system automation. Several efforts have investigated the use of two-way data communications (datalink) [7, 19] to facilitate the transfer of aircraft intent and state information to the ground automation, and traffic clearances to the flight deck. Further study is needed to determine what level of cooperation between the airborne and ground automation maximizes the closed-loop performance of terminal area operations, that is, capacity, while maintaining safety of operations with a mixed fleet of (possibly uncooperative) arrival traffic under varying environmental conditions. Assuming that user preferences and flexibility are supported by the ATM system, significant gains in capacity and efficiency can be achieved if aircraft fly more predictable trajectories. Airborne automation can help pilots fly more precise trajectories that will allow controllers to operate with less conservatism to protect against flight path uncertainties.

Research into closely spaced parallel approach procedures [1, 2] shows promise to maintain clear weather airport capacity in instrument meteorological conditions at airports with parallel runways spaced less than 3400 feet. This procedure requires precision navigation positioning (such as differential GPS, DGPS), aircraft-to-aircraft communications, display of the traffic position on the parallel runway, and escape maneuver alerting should aircraft position safety tolerances be exceeded. Although commercial aircraft are currently certificated with automatic approach and landing systems that are demonstrated to adequately reject gusts, turbulence, and other anomalies in the final approach phase of flight, this closely spaced parallel approach procedure may place new control demands on precision approach systems in order to maintain adequate safety margins.

10.3 EXAMPLE 2: UNINHABITED (COMBAT) AIR VEHICLES

Uninhabited vehicles will doubtless play a major role in air combat in the twenty-first century [14]. UCAVs can bring at least three significant advantages to combat missions: targets can be attacked without endangering human pilots; performance need not be constrained by human tolerance limits (e.g., of g forces) nor payload diminished because of human-centered vehicle–pilot interfaces; and vehicle and mission costs can be substantially reduced.

Uninhabited combat vehicles will be phased in gradually, performing first the missions for which they are most easily adapted (and for which human pilots are worst prepared). Suppression of air defenses is a likely candidate for initial use of UCAVs. Cruise missiles and other smart munitions are currently used to accomplish significant portions of this mission. This presents a clear path to the inclusion of increasing autonomy and nonexpendable uninhabited vehicles. More challenging missions will include the use of UCAVs for area access denial (as a substitute for land mines, for example), and eventually for aerial superiority combat.

Although UCAVs' flight performance will be higher, and thus so will be the requirements on the flight control systems, we expect the control of individual vehicles to remain fairly standard. However, fleet coordination and autonomy issues open large areas of research. In what follows we briefly describe these areas.

10.3.1 Inter-Fleet and Central-Command-to-Fleet Communications

A communication architecture and language has to be developed that can achieve three related but distinct objectives: coordination of the formation, robustness of the maneuvers carried out, and distributed optimization of the formation behavior. Besides having the ability and flexibility to express these tasks, the language has to be concise in order to reduce communication bandwidth, facilitate scaling of formation size, and simplify coordination of heterogeneous formations. Although the communications language is only part of this architecture, its careful design can have a significant impact on the performance and ease of implementation of the overall system. Coordinated control requires the exchange, first of all, of information about variables and parameters of interest. These can be considered the message objects and include the following:

- Trajectory modes and multivehicle trajectories
- Mission modes
- Vehicle states, including fuel, weapons, faults, damage
- Interpreted sensor data, for example target or threat detections

Figure 10.5 shows an example of possible trajectories that would constitute some of these message objects. These objects can be communicated with different intentions and for different communicative purposes. Thus we can distinguish between different types of messages—a simplified "speech act" portfolio for coordinated multivehicular control. Examples include:

- Queries
- Commands
- Warnings
- Assertions
- Acknowledgments

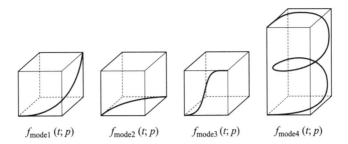

$f_{\text{mode1}}(t;p)$ $f_{\text{mode2}}(t;p)$ $f_{\text{mode3}}(t;p)$ $f_{\text{mode4}}(t;p)$

Figure 10.5 Modes as parameterized trajectory elements (simplified representations of, respectively, turn and climb, turn, altitude capture and hold, and dogfight climb and pursue).

Ultimately, messages are represented and exchanged through tokens and protocols. At the other extreme, a multipurpose autonomous vehicle communication capability would require attention to contextual message interpretation—the pragmatics of agent-based communication in general. More generally, an understanding of the semiotic aspects of coordinated control can help identify communication requirements and capabilities.

10.3.2 Safety Analysis and Conflict Resolution

To ensure formation safety, each UCAV in the fleet must remain within its aerodynamically safe flight envelope (this will be different for different vehicles), and it cannot collide with other vehicles. Game theory and optimal-control-based hybrid system design methods have been used in the past to solve similar problems in multi-vehicle applications. Through solving a dynamic game between the controller and a hostile disturbance generator, the UCAV safety boundary can be characterized. If the state of the system crosses the safety envelope, the controller can switch to a default safety envelope protection mode. Similar methods can be used for characterizing flight envelopes to avoid inter-UCAV collisions. In this case, if a collision is predicted, the controller will switch to a collision avoidance mode, communicating with the UCAVs involved in the conflict. Once the threat is successfully avoided, the controller switches back to the normal mode. Knowledge of current and future trajectory segments for other UCAVs plays an important role in the efficacy of this type of scheme.

Conflict detection under uncertainty can be viewed as a (somewhat) new application for robust control. The standard approaches to robust control—worst-case and stochastic analysis—can be applied, and similar techniques utilized.

A second area in safety analysis is that of conflict resolution; that is, once a conflict between two or more aircraft has been detected, how will we proceed to avoid the conflict and complete the ongoing tasks with minimal disruption? If conflict detection is a robustness analysis problem, conflict resolution is the equivalent of a robust control synthesis problem and as such is intrinsically more difficult.

The nominal conflict resolution problem can be solved using a variety of standard optimization techniques, adapted to the particular problem at hand. One technique that has been particularly fruitful for this problem is a variation of dynamic programming.

When significant uncertainties in the behavior of the vehicles are present, more elaborate methods can be used. One such set of techniques is referred to as robust optimization. A performance index and a set of constraints are derived from the problem specification (e.g., maintain separation, while keeping velocities, accelerations, and deviations from nominal routes within certain bounds). Next, a set of bounded structured uncertainty in the systems behavior is added. Finally, we determine the nominal trajectories that will minimize the worst-case cost while respecting the constraints. Although usually the exact solution cannot be computed, approximations can be computed using convex optimization algorithms that have guaranteed bounds on the robustness level.

10.3.3 Autonomy

To design an autonomous fighter, we need to enhance the functionality of current systems in many ways. Although a standard flight-management system with autopilot might be adequate for most of the flight, there will be alert modes and combat modes of operation when many complex, concurrent, coordinated actions and reactions may have to be planned and performed in real time. Figure 10.6 shows a schematic of a hierarchic architecture for a (semi)-autonomous vehicle.

One new problem associated with an autonomous vehicle arises from the lack of preprogrammed information about the situations in which the vehicle may find itself. Current sensor signal processing algorithms focus on detecting threats and identifying targets, but a more fundamental problem is trying to figure out where to look. Different sensor modalities (e.g., radar, video cameras, infrared sensors) provide different (though highly correlated) information—the decision of which sensor or group of sensors to use in identifying a potential target depends on the type of targets one is

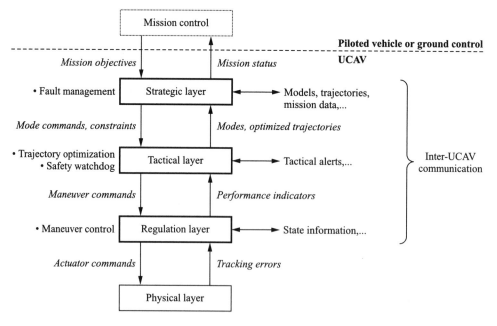

Figure 10.6 UCAV control system architecture for (semi-)autonomous multivehicle missions

expecting to encounter. In a human-operated system, this sensor allocation problem is solved by the operator, who makes a decision based on situational awareness. An autonomous vehicle must solve this problem by some other means, possibly by using an on-board situational-awareness module.

The situational-awareness function is another new problem that must be addressed for an autonomous vehicle. Needed is a module that builds and maintains a model of the environment in which the vehicle finds itself. The model includes state space models of the motion of both enemy and friendly forces, along with other relevant data (whether entities are friendly, armed, damaged, on a collision course, etc.). The model must be updated in real time, relying on predictive estimation for those dynamic elements of the environment that are not currently monitored.

Another new type of function is the temporal planning of actions. This is a high-level motor-control activity that continuously updates a time-parameterized set of commands to the highest level effector submodules. These commands are generated in order to realize system functions at the highest, most abstract level, there being a hierarchy of submodules determining commands for lower levels within the system. The motor-control hierarchy of command is another system function that will require new research to define.

Also needed is research in a strategy/tactical module. One approach here is to combine an extensive, situationally indexed set of memorized actions together with a faster-than-real-time simulation/analysis module to evaluate possible short-term/long-term alternatives. A difficult part of this module is the development of an evaluation function that can decide whether one hypothetical outcome is better than another.

There are other vehicle functions that will need modification for an autonomous system, but the four special areas mentioned here seem to need the most work. The final area of research we note is to determine a system architecture that will enable efficient implementation of all the vehicle functions. One approach to this problem is to examine how analogous functions are organized in the central nervous system of animals [3, 13], but many other possible approaches might be considered.

10.4 EXAMPLE 3: FORMATION FLYING AND SATELLITE CLUSTERS

Large, one-of-a-kind satellites are costly and very susceptible to single failures that can destroy the vehicle or substantially cripple its performance. Recent studies suggest that savings of cost and weight (which affects the launch cost) can be achieved by distributing the same functionality within the elements of a "cluster" of satellites. Such a cluster will be formed by small, fairly simple vehicles flying in formation. The cluster operates cooperatively to perform a function, in a sense as a "virtual" satellite. As an example, clusters currently being considered by the Air Force Research Laboratories (in the TechSat21 program) consist of 8 to 16 satellites operating within a radius of 100 to 1000 meters.

Many technologies need to be developed to make this concept fully operational. Relative position sensing is of particular importance, both for formation flying and for distributed sensor operations. The control of a relatively large number of closely spaced vehicles, however, also poses some interesting problems in distributed control. The cluster's distributed control algorithms and software must have the following properties:

- To achieve all the benefits, the cluster should have a large level of autonomy. Individual satellites should not require direct commands from ground control.
- All satellites and their software need to be identical; there cannot be a "keystone" satellite in the cluster.
- The cluster should be able to detect and resolve conflicts during normal operations (configuration, deconfiguration), and during failures.
- The cluster should minimize resource use (computing resources, communications resources, and fuel).

10.4.1 Multi-Agent Systems and Decentralized Distributed Control

A constellation of small satellites is a classic example of a *multi-agent system*, a system in which many independent agents interact in a common environment. In general, the agents in a multi-agent system may or may not cooperate with each other; the distinguishing characteristic of multi-agent systems is simply that the individual agents are autonomous rather than being directed by a centralized controller, so that the behavior of the whole system is controlled in a distributed fashion. The modeling of multi-agent systems is an important part of current artificial intelligence and artificial life research; tools such as SWARM [9] can simulate many different types of biological and economic interactions.

Decentralized, distributed control offers several important advantages, including:

1. *Simplicity*: Decentralized networks of agents may be able to work out reasonable solutions to problems without the need for complex and computationally intensive centralized control algorithms.
2. *Robustness*: Decentralized systems are less vulnerable to faults because there is no "command agent" whose malfunction would disable the entire system.
3. *Flexibility*: Decentralized systems can be easily reconfigured by making small changes in the individual agents' behavior.

In addition, decentralized systems of agents can be designed to exhibit "emergent behavior," behavior whose complexity is far beyond that of any individual agent's programming. The control protocols that enable constellations of satellites to exhibit useful types of emergent behavior include:

- Station-keeping within a particular configuration (e.g., a 2-D imaging array for a sparse aperture radar mission).
- Collision avoidance, both in the course of a normal orbit cycle and in response to fault conditions.
- Reconfiguration in response to changing mission requirements.
- Distributed processing and communications tasks (e.g., processing of distributed aperture radar, DAR, image data).

10.4.1.1 Emergent Behavior

Emergent behavior can be defined as "a global effect generated by local rules" [11]. More precisely, emergent behavior is complex behavior exhibited by a multi-agent system whose individual members operate according to simple rules and interact in simple ways. Numerous examples exist in the natural world, from the swimming patterns of fish schools to the cooperation of worker ants. See, for example, [4, 6].

Systems in which emergent behavior can arise are typically *subsumptive* systems. In a subsumptive system, the individual elements are interchangeable; have little or no usefulness when acting alone; cannot control the behavior of other elements; and work according to predefined rules without knowing the common goal they are trying to accomplish. Thus the behavior of a subsumptive system represents an extreme case of decentralized control.

10.4.1.2 Flocking

Flocking is a type of behavior in which agents moving in space are able to remain in a stable configuration—a "flock"—and respond properly to the presence of obstacles by following a few simple control rules. Reynolds [12] has simulated flocking by directing a set of agents called "boids" with three rules: avoid obstacles that you can see, maintain a constant distance from your neighbors, and move toward a locally approximated "center." The boids then "fly" in a simulated aerodynamic environment, accelerating in preferred directions determined by their obedience to the rules. The boids' resulting behavior is very similar to that of a real flock of birds.

Satellites keeping station in a constellation must obey similarly simple rules: avoid collisions and correct for drift relative to neighboring satellites.

10.4.1.3 Market-Oriented Programming

Market-oriented programming is, in its most general form, a method of allocating resources in a multi-agent system by simulating a bidding process carried out among agents producing and consuming the resources. This process may be carried out by trades among neighboring agents or by systemwide resource auctions. In an auctioneering model, consumer and producer agents submit bids specifying amounts of resources they are willing to buy or sell in return for other resources. A central "auction" receives these bids and sends back "price quotes" indicating current going rates for resources; it comes to a final allocation decision when there are no new bids. This decision reflects equilibrium prices for the various resources, set by the interplay between the consumer and producer bids.

10.4.2 Distributed Processing

In a satellite constellation performing a task such as DAR imaging, numerous processing tasks are required; image data must be correlated and analyzed, and results transmitted to ground stations. There are three important resource considerations for carrying out these tasks: computation time, intersatellite communications, and satellite-to-ground communications. Some of the tasks involved will only need a fixed set of resources that one satellite in the constellation can supply; others will require more than one satellite's capabilities (e.g., parallel computation); still others will require multiple copies to be run concurrently for redundancy purposes. We must establish a mechanism for allocating computing and communication resources that takes into account all of these possibilities.

Several groups of researchers have used auctioneering techniques to allocate computational tasks in parallel processing systems. In one such system, called Spawn [15], distributed processors act as producer agents, passively accepting bids for computational tasks. These tasks are consumers, endowed with a certain "budget" reflecting their criticality and the amount of time they require. Each processor runs an auction for each time slice it has available, soliciting bids from various tasks that might want to use the time slice; large applications may be divided into many subtasks bidding separately. Wellman et al. [17, 18], have implemented a similar, but more generally applicable, system called MARX; MARX is based on the WALRAS model [16] and allows allocations of multiple types of resources.

Neither of these systems, however, addresses the requirements of real-time functioning and fault-tolerance that are intrinsic to a satellite application. The establishment of guarantees—that a given task will be able to execute within a predetermined amount of time or that a failed satellite will have a strictly limited effect on processing—is not a feature of market models as currently implemented. One development that addresses these concerns is a real-time resource management system for distributed environments called RT-ARM [8]. This system uses a negotiation tree-based approach to allocate computing and other resources based on quality-of-service and criticality requirements; it supports dynamic adaptation to changes in system state, such as fault scenarios.

10.5 CONCLUSIONS

We have presented examples from three different domains: commercial flight operations, military combat formations, and space systems. All of them deal with the interactions of multiple vehicles sharing an operational space. Some of the problems that arise when considering these multivehicle systems are conflict avoidance, distributed resource allocation, distributed optimization, and communication and coordination.

Some of these problems can be recast in the framework of some theory well known to the controls community, including linear and nonlinear optimization, robustness analysis, and robust control synthesis. The new applications pose new challenges to these theories, and they will surely evolve and adapt to meet them.

Other problems, however, are newer to our community—such as autonomy and decision hierarchies, languages for command, control and coordination, and analysis and synthesis methods for hybrid dynamical systems.

As has happened many times in the past, control will draw from other fields to meet the new challenges. This much will never change.

ACKNOWLEDGMENTS

The authors wish to thank Dr. Blaise Morton, Nicholas Weininger, and Dr. Tariq Samad for their contributions and comments on this chapter.

Related Chapters

- Single-vehicle flight control, for both aircraft and missiles, is the focus of Chapter 11.
- See Chapter 7 for a tutorial on hybrid systems.
- Multi-agent systems are also being explored for power system applications (see Chapter 13)

REFERENCES

[1] Inter-agency Air Traffic Management (ATM) Integrated Product Team, "Integrated plan for air traffic management research and technology development." Technical report, FAA/NASA, December 1997.

[2] R. Ashford, "Technological developments in airspace optimization—a summary of NASA research." In *Developing Strategies for Effective Airport Capacity Management Conference*. London, February 1998.

[3] Per Brodal, *The Central Nervous System*. New York: Oxford University Press, 1998.

[4] J. L. Casti, *Cooperation and Conflict in General Evolutionary Processes*. New York: John Wiley & Sons, 1994.

[5] A.R. Odoni, et al., "Existing and required modeling capabilities for evaluating ATM systems and concepts." Technical Report, International Center for Air Transportation, Massachusetts Institute of Technology, March 1997.

[6] M. Gell-Mann, *The Quark and the Jaguar: Adventures in the Simple and the Complex*. New York: W. H. Freeman, 1994.

[7] S. M. Green, T. Goka, and D. H. Williams, "Enabling user preferences through data exchange." In *AIAA Guidance, Navigation, and Control Conference*, August 1997.

[8] J. Huang, R. Jha, W. Heimerdinger, M. Muhammad, S. Lauzac, B. Kannikeswaran, K. Schwan, W. Zhao, and R. Bettati, "RT-ARM: A real-time adaptive resource management system for distributed mission-critical applications." In *Proc. Workshop on Middleware for Distributed Real-Time Systems*, December 1997, New York, IEEE Press.

[9] C. Langton, The SWARM simulation system. http://www.santafe.edu/projects/swarm.

[10] AATT Program Office, "ATM concept definition and integrated evaluations." Technical Report, NASA, October 1997.

[11] R. Parent, "Emergent behavior: Particles and flocks." In *Computer Animation: Algorithms and Techniques*, book in progress.

[12] C.W. Reynolds, "Flocks, herds, and schools: A distributed behavioral model." *Computer Graphics*, Vol. 21, no. 4, pp. 25–34, 1987.

[13] Michael D. Rugg, *Cognitive Neuroscience*. Cambridge, MA: MIT Press, 1997.

[14] USAF Scientific Advisory Board, *New World Vistas. Air and Space Power for the 21st Century*. Aircraft and Propulsion Volume. Department of the Air Force, 1995.

[15] C. Waldspurger, T. Hogg, B. Huberman, J. Kephart, and S. Stometta, "Spawn: A distributed computational economy." *IEEE Transactions on Software Engineering*, Vol. 18, no. 2, pp. 103–117, 1992.

[16] M. Wellman, "The WALRAS algorithm: A convergent distributed implementation of general equilibrium outcomes." *Computational Economics*, Vol. 12, no. 1, pp. 1–24, 1998.
[17] M. Wellman, S. Jamin, and J. MacKie-Mason, Michigan Adaptive Resource eX-change. http://ai.eecs.umich.edu/MARX/.
[18] M. P. Wellman and P. R. Wurman. "Market-aware agents for a multiagent world." *Robotics and Autonomous Systems*, to appear.
[19] D. W. Williams, P. D. Arbuckle, S. M. Green, and W. den Braven. "Profile negotiation: An air/ground automation integration concept for managing arrival traffic." *AGARD Conference Proceedings 538, "Machine Intelligence in Air Traffic Management,"* Berlin, May 1993.

Chapter 11
AFFORDABLE FLIGHT CONTROL FOR AIRCRAFT AND MISSILES

Kevin A. Wise

Editor's Summary

Not so long ago, research in flight control was largely an algorithmic endeavor—the objective was to devise higher performance control laws. Such research still continues, but an additional imperative is now prominent. For many new flight control developments, the challenge is not so much to come up with the control algorithm; the existing literature describes how to do that sufficiently well. The new challenges are cost-efficiency and development time reduction.

This chapter revises some of the basic technology for flight control systems and discusses how computational tools are now being used to ensure that new control systems can be affordably developed and deployed. The target vehicles for the flight control technology described are military aircraft and missiles. There are notable differences between the two, such as the use of different sets of actuators (in missiles, for example, reaction jets can directly control roll, pitch, and yaw).

Flight control system design requirements also differ fundamentally for aircraft and missiles, due to the presence of a pilot in the former. As a consequence, control techniques can differ. Dynamic inversion—a form of feedback linearization—is a popular technique for aircraft flight control, whereas linear quadratic optimal control is commonly used for missiles. In either case, complications arise due to the fact that accelerometers and gyros are not located at the center of gravity of the vehicle. In some cases, performance and stability can be improved by appropriately locating the sensors.

Simulation and analysis software is now essential for flight control design. The chapter discusses tools within the MATRIXx toolset that permit performance and robustness evaluation over a range of operating conditions. Today's tools allow aerospace engineers to graphically design a flight control system, analyze its performance and stability, and generate autocode that can be ported directly to training simulators, hardware in the loop simulators, and the aircraft itself. Similar tools are in use in other industries too—see Ch. 15 for an automotive industry perspective. Further discussion on computer aided control system design can also be found in Ch. 3 which also outlines a flight control application.

What lies ahead for flight control? This chapter concludes by noting the interest in, and the challenges associated with, uninhabited aircraft. This topic is also discussed in Ch 10 where focus is on multi-vehicle aerospace challenges.

Kevin Wise is with Boeing Phantom Works.

11.1 INTRODUCTION

Aircraft and missile flight control engineers must design for stability, performance, robustness, and digital implementation. This is a very complex problem because of the highly nonlinear dynamics, aerodynamics, and large operating envelopes. In addition to these technical challenges, industry requires affordability in every step of the development process. This is of significant concern because the trends toward multiple control surfaces, more stringent performance requirements, and expanded flight envelopes all increase the time and cost of developing flight control systems.

The process for developing flight control systems typically has four phases. The first phase, the design phase, is the selection of the control architecture and design methodology. The second phase consists of analysis and simulation. During this phase, stability margins and flying qualities are assessed, and iterations to the architecture and design methods are made to improve the flight control system's performance and robustness. The third phase is the digital implementation of the flight control system and the development of flight software. The final phase is the validation and verification through flight simulation (manned or unmanned), hardware-in-the-loop simulation, detailed software test, and finally flight test.

The cost of making changes to the flight control system increases significantly as the process evolves into the later phases. Unfortunately, aerodynamic and hardware models of the aircraft and its subsystems come late in the design process, forcing changes to the architecture, design methodology, and control system requirements. New processes are needed for designing, analyzing, simulating, and generating real-time embedded flight control system software to streamline the development cycle and reduce costs, and allow the engineer to evolve the control system requirements later into the design process. By using new control system design tools and processes, engineers can significantly reduce the cost of change (cost, risk, and schedule) later in the implementation and validation and verification phases. This chapter addresses these issues and presents both aircraft and missile control architectures and tools and processes for achieving this goal.

11.2 AIRCRAFT AND MISSILE DYNAMICS AND LINEAR MODELS

The body axis six degree-of-freedom equations of motion (EOM) are used in the design of the flight control system. Figure 11.1 illustrates the body axis coordinate system and definition of important aerodynamic angles such as angle-of-attack α and sideslip β. Assuming a rigid body, constant mass, and inertia, we can write the standard six degree-of-freedom body equations of motion as

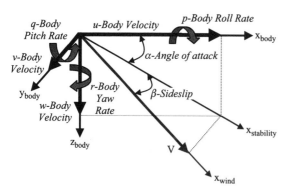

Figure 11.1 Body, stability, and wind axis coordinate systems.

Section 11.2 Aircraft and Missile Dynamics and Linear Models

$$\dot{u} = rv - qw + X + G_x + T_x$$
$$\dot{v} = pw - ru + Y + G_y + T_y$$
$$\dot{w} = qu - pv + Z + G_z + T_z$$
$$\dot{p} = -L_{pq}pq - L_{qr}qr + L + L_T \quad (11.1)$$
$$\dot{q} = -M_{pr}pr - M_{r^2p^2}(r^2 - p^2) + M + M_T$$
$$\dot{r} = -N_{pq}pq - N_{qr}qr + N + N_T$$

where G_i models gravity, (X, Y, Z) models the linear accelerations produced by the aerodynamic forces, (L, M, N) models the angular accelerations produced by the aerodynamic moments, (T_x, T_y, T_z) models propulsion system forces, and (L_T, M_T, N_T) models the moments produced by the propulsion system. Note that these variables have units of acceleration. The aerodynamic forces and moments are modeled as nondimensional quantities and are scaled to units of force. This scaling is described by

$$\begin{bmatrix} X \\ Y \\ Z \end{bmatrix} = \frac{\bar{q}S}{m} \begin{bmatrix} C_x \\ C_y \\ C_z \end{bmatrix}, \begin{bmatrix} L \\ M \\ N \end{bmatrix} = \begin{bmatrix} \frac{\bar{q}Sl}{I_{xx}I_{zz} - I_{xz}^2}(C_l I_{zz} + C_n I_{xz}) \\ \frac{\bar{q}Sl}{I_{yy}} C_m \\ \frac{\bar{q}Sl}{I_{xx}I_{zz} - I_{xz}^2}(C_n I_{xx} + C_l I_{xz}) \end{bmatrix} \quad (11.2)$$

where $\bar{q}$ (lb/ft^2) is the dynamic pressure, S (ft^2) is a reference area, m is the mass in slugs, (C_x, C_y, C_z) model nondimensional aerodynamic forces, (C_l, C_m, C_n) model nondimensional moments, l is a reference length, and $(I_{xx}, I_{yy}, I_{zz}, I_{xz})$ are moments of inertia. Note that the cross-axis inertia term I_{xz} couples the roll-yaw moment equations and can significantly impact handling and stability characteristics. The coefficients $L_{pq}, L_{qr}, M_{pr}, M_{r^2p^2}, N_{pq}$, and N_{qr} in Eq. (11.1) are functions of the moments of inertia.

The pitch-plane angle-of-attack α and yaw-plane sideslip angle β are defined in Figure 11.1. The total angle of attack, α_T, is the angle from the velocity vector to the x-body axis. The stability axis coordinates are a transformation of the body axes using α. The wind axis coordinates are a transformation from stability axes using β. The stability axis coordinate system will be used later in the design of the flight control system.

The aerodynamic forces (C_x, C_y, C_z) and moments (C_l, C_m, C_n) are typically modeled as functions of α, β, Mach, body rates (p, q, r), $\dot{\alpha}$, $\dot{\beta}$, the aerodynamic control surface deflections (δ_e, δ_a, and δ_r for elevator, aileron, and rudder surfaces, respectively), center-of-gravity (CG) changes, and propulsion system effects (plume effects). Also, the aerodynamic forces may depend on whether reaction jets are on or off (jet interaction effects). These complicated and highly nonlinear functions are used in the EOM to model the airframe's aerodynamics.

The gravitational forces are modeled as:

$$\begin{bmatrix} G_x \\ G_y \\ G_z \end{bmatrix} = g \begin{bmatrix} -\sin(\theta) \\ \cos(\theta)\sin(\phi) \\ \cos(\theta)\cos(\phi) \end{bmatrix} \quad (11.3)$$

where θ and ϕ are pitch attitude and roll angle, respectively.

For aircraft with thrust vector control (TVC), the flight control system is designed to command the TVC actuator angle δ_T (rad). The TVC forces and moments are modeled using a constant thrust T that is deflected by the actuator. It is assumed that the actuator can deflect the thrust vector only in the pitch (δ_{T_e}) and yaw (δ_{T_r}) planes, using separate actuators devoted to this task. The roll, pitch, and yaw moments (L_T, M_T, N_T) produced by the TVC will be the moment arm $l_T = x_{cg} - x_{TVC}$ times the above pitch and yaw forces, where x_{cg} and x_{TVC} are the x-distances from the CG and TVC actuator, respectively. The forces and moments used in Eq. (11.1) are:

$$\begin{bmatrix} T_x \\ T_y \\ T_z \end{bmatrix} = \frac{T}{m} \begin{bmatrix} \cos(\delta_{T_e})\cos\delta_{T_r} \\ -\cos(\delta_{T_r}) \\ -\sin(\delta_{T_e})\cos(\delta_{T_r}) \end{bmatrix}, \quad \begin{bmatrix} L_T \\ M_T \\ N_T \end{bmatrix} = \begin{bmatrix} \dfrac{-l_T I_{xz} T \sin(\delta_{T_r})}{I_{xx}I_{zz} - I_{xz}^2} \\ \dfrac{l_T T \sin(\delta_{T_e})\cos(\delta_{T_r})}{I_{yy}} \\ \dfrac{-l_T I_{xx} T \sin(\delta_{T_r})}{I_{xx}I_{zz} - I_{xz}^2} \end{bmatrix}. \quad (11.4)$$

For missile systems with a reaction control system (RCS), the flight control system is designed to command the RCS thrust level T_{RCS} (lb). The reaction jets are assumed to be positioned so that no axial force is generated. The RCS actuators are designed to provide roll, pitch, and yaw moment control. The forces produced by the pitch and yaw jets are modeled as $\bar{T}_y$ and $\bar{T}_z$. The moments produced by the thrusters are modeled by thrust forces multiplied by the moment arm $l_T = x_{cg} - x_{RCS}$. It is assumed here that the pitch and yaw jets are located at the same missile x-station x_{RCS}. Roll jets (with thrust $\bar{T}_{Roll}$ and moment arm l_{Roll}) may also be used to control missile roll. These jets are symmetrically placed so that only a rolling moment $L_T = l_{Roll} \bar{T}_{Roll}/I_{xx}$ is produced. The RCS forces and moments used in Eq. (11.1) are

$$\begin{bmatrix} 0 \\ T_y \\ T_z \end{bmatrix} = \frac{1}{m} \begin{bmatrix} 0 \\ \bar{T}_y \\ \bar{T}_z \end{bmatrix}, \quad \begin{bmatrix} L_T \\ M_T \\ N_T \end{bmatrix} = \begin{bmatrix} \dfrac{l_{Roll}\bar{T}_{Roll}}{I_{xx}} + \dfrac{-l_T I_{xz}\bar{T}_y}{I_{xx}I_{zz} - I_{xz}^2} \\ -\dfrac{l_T \bar{T}_z}{I_{yy}} \\ \dfrac{l_T I_{xx}\bar{T}_y}{I_{xx}I_{zz} - I_{xz}^2} \end{bmatrix}. \quad (11.5)$$

The following derivation will form a set of differential equations describing the dynamics for $\dot{V}$, $\dot{\alpha}$, and $\dot{\beta}$ valid for large α's and $\beta < 90°$. Consider the following definition of the body velocities from Figure 11.1:

$$u = V\cos(\alpha)\cos(\beta)$$
$$v = V\sin(\beta) \quad (11.6)$$
$$w = V\sin(\alpha)\cos(\beta)$$

where V is the magnitude of the velocity vector. This can be represented as a transformation of the wind-axis velocity vector to the body axes as follows:

Section 11.2 Aircraft and Missile Dynamics and Linear Models

$$\begin{bmatrix} u \\ v \\ w \end{bmatrix}_{Body} = \begin{bmatrix} c\alpha & 0 & -s\alpha \\ 0 & 1 & 0 \\ s\alpha & 0 & c\alpha \end{bmatrix} \begin{bmatrix} c\beta & -s\beta & 0 \\ s\beta & c\beta & 0 \\ 0 & 0 & 1 \end{bmatrix} \begin{bmatrix} V \\ 0 \\ 0 \end{bmatrix}_{Wind} \quad (11.7)$$

where $c(\bullet)$ and $s(\bullet)$ denote $\cos(\bullet)$ and $\sin(\bullet)$, respectively. The angular velocities in stability axes are given by

$$\begin{bmatrix} p_s \\ q \\ r_s \end{bmatrix} = \begin{bmatrix} c\alpha & 0 & s\alpha \\ 0 & 1 & 0 \\ -s\alpha & 0 & c\alpha \end{bmatrix} \begin{bmatrix} p \\ q \\ r \end{bmatrix}. \quad (11.8)$$

Differentiating Eq. (11.7) yields

$$\begin{bmatrix} \dot{u} \\ \dot{v} \\ \dot{w} \end{bmatrix} = \begin{bmatrix} c\alpha c\beta & -s\alpha c\beta & -c\alpha s\beta \\ s\beta & 0 & c\beta \\ s\alpha c\beta & c\alpha c\beta & -s\alpha s\beta \end{bmatrix} \begin{bmatrix} \dot{V} \\ V\dot{\alpha} \\ V\dot{\beta} \end{bmatrix}. \quad (11.9)$$

Inverting the coefficient matrix in the preceding equation yields

$$\begin{bmatrix} \dot{V} \\ V\dot{\alpha} \\ V\dot{\beta} \end{bmatrix} = \frac{-1}{c\beta} \underbrace{\begin{bmatrix} -c\alpha c^2\beta & -s\beta c\beta & -s\alpha c^2\beta \\ s\alpha & 0 & -c\alpha \\ s\beta c\beta c\alpha & -c^2\beta & s\alpha s\beta c\beta \end{bmatrix}}_{W(\alpha, \beta)} \begin{bmatrix} \dot{u} \\ \dot{v} \\ \dot{w} \end{bmatrix}. \quad (11.10)$$

Substituting from Eq. (11.1) yields

$$\begin{bmatrix} \dot{V} \\ V\dot{\alpha} \\ V\dot{\beta} \end{bmatrix} = W(\alpha, \beta) \left(-\begin{bmatrix} p \\ q \\ r \end{bmatrix} \times \begin{bmatrix} u \\ v \\ w \end{bmatrix} + \begin{bmatrix} X \\ Y \\ Z \end{bmatrix} + \begin{bmatrix} G_x \\ G_y \\ G_z \end{bmatrix} + \begin{bmatrix} T_x \\ T_y \\ T_z \end{bmatrix} \right). \quad (11.11)$$

Expanding Eq. (11.11) results in

$$\dot{V} = c\alpha c\beta (X + G_x + T_x) + s\beta (Y + G_y + T_y) + s\alpha c\beta (Z + G_z + T_z)$$

$$\dot{\alpha} = (1/Vc\beta)[-s\alpha(X + G_x + T_x) + c\alpha(Z + G_z + T_z)] + q - p_s \tan(\beta)$$

$$\dot{\beta} = (1/V)[-c\alpha s\beta(X + G_x + T_x) + c\beta(Y + G_y + T_y) - s\alpha s\beta(Z + G_z + T_z)] - r_s.$$

$$(11.12)$$

Chapter 11 Affordable Flight Control for Aircraft and Missiles

To develop an aircraft control law using feedback linearization, expressions for $\ddot{V}$, $\ddot{\alpha}$, and $\ddot{\beta}$ are needed:

$$\ddot{V} = \begin{bmatrix} s\beta G_z - s\alpha c\beta G_y \\ s\alpha c\beta G_x - c\alpha c\beta G_z \\ c\alpha c\beta G_y - s\beta G_x \end{bmatrix}^T \begin{bmatrix} p \\ q \\ r \end{bmatrix} + \begin{bmatrix} c\beta(c\alpha X_\alpha + s\alpha Z_\alpha) \\ c\alpha c\beta X_\beta + s\beta Y_\beta \end{bmatrix}^T \begin{bmatrix} \dot{\alpha} \\ \dot{\beta} \end{bmatrix}$$

$$+ \begin{bmatrix} c\alpha c\beta X_{\delta a} + s\beta Y_{\delta a} \\ c\alpha c\beta X_{\delta e} + s\alpha s\beta Z_{\delta e} \\ c\alpha c\beta X_{\delta r} + s\beta Y_{\delta r} \end{bmatrix}^T \begin{bmatrix} \dot{\delta}_a \\ \dot{\delta}_e \\ \dot{\delta}_r \end{bmatrix} + \begin{bmatrix} c\alpha c\beta \\ s\beta \\ s\alpha c\beta \end{bmatrix}^T \begin{bmatrix} \dot{T}_x \\ \dot{T}_y \\ \dot{T}_z \end{bmatrix}$$

$$+ \begin{bmatrix} -s\alpha c\beta \dot{\alpha} - c\alpha s\beta \dot{\beta} \\ c\beta \dot{\beta} \\ c\alpha c\beta \dot{\alpha} - s\alpha s\beta \dot{\beta} \end{bmatrix}^T \begin{bmatrix} G_x + X + T_x \\ G_y + Y + T_y \\ G_z + Z + T_z \end{bmatrix}$$

$$\ddot{\alpha} = \frac{1}{Vc\beta} \left\{ \begin{bmatrix} -c\alpha G_y \\ s\alpha G_z + c\alpha G_x \\ s\alpha G_y \end{bmatrix}^T \begin{bmatrix} p \\ q \\ r \end{bmatrix} + \begin{bmatrix} -s\alpha X_\alpha + c\alpha Z_\alpha \\ -s\alpha X_\beta \end{bmatrix}^T \begin{bmatrix} \dot{\alpha} \\ \dot{\beta} \end{bmatrix} \right.$$

$$+ \begin{bmatrix} -s\alpha X_{\delta a} \\ -s\alpha X_{\delta e} + c\alpha Z_{\delta e} \\ -s\alpha X_{\delta r} \end{bmatrix}^T \begin{bmatrix} \dot{\delta}_a \\ \dot{\delta}_e \\ \dot{\delta}_r \end{bmatrix} + \begin{bmatrix} -s\alpha \\ 0 \\ c\alpha \end{bmatrix}^T \begin{bmatrix} \dot{T}_x \\ \dot{T}_y \\ \dot{T}_z \end{bmatrix}$$

$$\left. + \begin{bmatrix} c\alpha \dot{\alpha} \\ -s\alpha \dot{\alpha} \end{bmatrix}^T \begin{bmatrix} G_x + X + T_x \\ G_z + Z + T_z \end{bmatrix} \right\}$$

$$+ \dot{q} - \dot{p}_s \tan\beta - p_s \dot{\beta} + (\dot{\alpha} - q)\tan\beta \dot{\beta} - \frac{\dot{V}}{V}(\dot{\alpha} - q + p_s \tan\beta)$$

$$\ddot{\beta} = \frac{1}{V} \left\{ \begin{bmatrix} c\beta G_z + s\alpha s\beta G_y \\ c\alpha s\beta G_z - s\alpha s\beta G_x \\ -c\beta G_x - c\alpha s\beta G_y \end{bmatrix}^T \begin{bmatrix} p \\ q \\ r \end{bmatrix} + \begin{bmatrix} -c\alpha s\beta X_\alpha - s\alpha s\beta Z_\alpha \\ -c\alpha s\beta X_\beta + c\beta Y_\beta \end{bmatrix}^T \begin{bmatrix} \dot{\alpha} \\ \dot{\beta} \end{bmatrix} \right.$$

$$+ \begin{bmatrix} -c\alpha s\beta X_{\delta a} + c\beta Y_{\delta a} \\ -c\alpha s\beta X_{\delta e} + s\alpha s\beta Z_{\delta e} \\ -c\alpha s\beta X_{\delta r} + c\beta Y_{\delta r} \end{bmatrix}^T \begin{bmatrix} \dot{\delta}_a \\ \dot{\delta}_e \\ \dot{\delta}_r \end{bmatrix} + \begin{bmatrix} -c\alpha s\beta \\ c\beta \\ -s\alpha s\beta \end{bmatrix}^T \begin{bmatrix} \dot{T}_x \\ \dot{T}_y \\ \dot{T}_2 \end{bmatrix}$$

$$\left. + \begin{bmatrix} s\alpha s\beta \dot{\alpha} - c\alpha c\beta \dot{\beta} \\ -s\beta \dot{\beta} \\ -c\alpha s\beta \dot{\alpha} - s\alpha c\beta \dot{\beta} \end{bmatrix}^T \begin{bmatrix} G_x + X + T_x \\ G_y + Y + T_y \\ G_z + Z + T_z \end{bmatrix} \right\} - \frac{\dot{V}}{V}(\dot{\beta} + r_s) - \dot{r}_s$$

Section 11.2 Aircraft and Missile Dynamics and Linear Models

where the subscripts on X, Y, and Z refer to partial derivatives with respect to that variable.

Equations relating the Euler angle rates to the stability axis rotational rates are also needed. These relationships are given by

$$\dot{\phi} = (c\alpha + \tan\theta c\phi s\alpha)p_s + \tan\theta s\phi q + (\tan\theta c\phi c\alpha - s\alpha)r_s$$
$$\dot{\theta} = -s\alpha s\phi p_s + c\phi q - c\alpha s\phi r_s.$$

To develop flight control laws for missiles using optimal control theory, linear models are needed. The pitch-plane nonlinear angle of attack and pitch rate dynamics are described in Eqs. (11.1) and (11.12). Neglecting the roll-yaw dynamics and linearizing about α_0 results in

$$\dot{\alpha} = \tfrac{1}{V}(Z_\alpha \alpha + q + Z_{\delta_e}\delta_e - \sin(\alpha_0)T_x + \cos(\alpha_0)T_z)$$
$$\dot{q} = M_\alpha \alpha + M_q q + M_{\delta_e}\delta_e + M_T \tag{11.13}$$

where

$$Z_\alpha = \frac{\partial \dot{\alpha}}{\partial \alpha}\bigg|_{\alpha=\alpha_0} = \left[\cos(\alpha)\left(\frac{\partial Z}{\partial \alpha} - G_x - T_x - X\right) - \sin(\alpha)\left(\frac{\partial X}{\partial \alpha} + G_z + T_z + Z\right)\right]\bigg|_{\alpha=\alpha_0}$$

$$Z_{\delta_e} = \frac{\partial \dot{\alpha}}{\partial \delta_e}\bigg|_{\alpha=\alpha_0} = \left[\frac{\partial Z}{\partial \delta_e}\cos(\alpha) - \frac{\partial X}{\partial \delta_e}\sin(\alpha)\right]\bigg|_{\alpha=\alpha_0}$$

$$M_\alpha = \frac{\partial M}{\partial \alpha}\bigg|_{\alpha=\alpha_0} \qquad M_q = \frac{\partial M}{\partial q}\bigg|_{\alpha=\alpha_0} \qquad M_{\delta_e} = \frac{\partial M}{\partial \delta_e}\bigg|_{\alpha=\alpha_0}.$$

Since most TVC actuators are limited to small deflection angles, $\sin(\delta_{T_e}) \approx \delta_{T_e}$ and $\cos(\delta_{T_e}) \approx 1$, resulting in

$$T_x = T/m \qquad T_z = -(T/m)\delta_{T_e} \qquad M_T = -(l_T T/I_{yy})\delta_{T_e}. \tag{11.14}$$

To model RCS thruster forces (axial thrust T is due to main engine) (see Eq. (11.5))

$$T_x = T/m \qquad T_z = T_{RCS}/m \qquad M_T = -(l_T/I_{yy})T_{RCS}. \tag{11.15}$$

Neglecting the influence of gravity on the α dynamics (since it is divided by V) and the $T\sin(\alpha_0)$ term (since it represents a constant), and combining these into a linear matrix model results in

$$\begin{bmatrix}\dot{\alpha}\\ \dot{q}\end{bmatrix} = \begin{bmatrix}\frac{Z_\alpha}{V} & 1\\ M_\alpha & M_q\end{bmatrix}\begin{bmatrix}\alpha\\ q\end{bmatrix} + \begin{bmatrix}\frac{Z_{\delta_e}}{V}\\ M_{\delta_e}\end{bmatrix}\delta_e + \begin{bmatrix}\frac{c\alpha_0}{mV}\\ \frac{l_T}{I_{yy}}\end{bmatrix}T_{TCS}$$
$$+ \begin{bmatrix}\frac{T(s\alpha_0 s\delta_{T_0} - c\alpha_0 c\delta_{T_0})}{mV}\\ \frac{-Tl_T}{I_{yy}}\end{bmatrix}\delta_{T_e}. \tag{11.16}$$

This state space model can be used to design a pitch autopilot at a specific flight condition (α_0, Mach, altitude, CG).

The lateral directional nonlinear dynamics are described in Eqs. (11.1) and (11.12). Zeroing the pitch dynamics and linearizing about α_0 (with $\beta = 0$) results in

$$\begin{bmatrix} \dot{\beta} \\ \dot{p} \\ \dot{r} \end{bmatrix} = \begin{bmatrix} \frac{Y_\beta}{V} & s\alpha_0 + \frac{Y_p}{V} & c\alpha_0 + \frac{Y_r}{V} \\ L_\beta & L_p & L_r \\ N_\beta & N_p & N_r \end{bmatrix} \begin{bmatrix} \beta \\ p \\ r \end{bmatrix} + \begin{bmatrix} \frac{Y_{\delta_a}}{V} & \frac{Y_{\delta_r}}{V} \\ L_{\delta_a} & L_{\delta_r} \\ N_{\delta_a} & N_{\delta_r} \end{bmatrix} \begin{bmatrix} \delta_a \\ \delta_r \end{bmatrix}$$

$$+ \begin{bmatrix} \frac{1}{V} & 0 & 0 \\ 0 & 1 & 0 \\ 0 & 0 & 1 \end{bmatrix} \begin{bmatrix} T_y \\ L_T \\ N_T \end{bmatrix} \quad (11.17)$$

where the elements of the matrices were obtained in a similar manner to Eq. (11.13).

For TVC (assuming a small TVC angle δ_{T_r}) results in

$$T_y = \frac{-T}{m}\delta_{T_r}, \quad L_T = \frac{-l_T I_{xz} T}{I_{xx}I_{zz} - I_{xz}^2}\delta_{T_r}, \quad N_T = \frac{l_T I_{xx} T}{I_{xx}I_{zz} - I_{xz}^2}\delta_{T_r}. \quad (11.18)$$

Modeling an RCS control system yields

$$T_y = \frac{\bar{T}_y}{m}, \quad L_T = \frac{l_{Roll}\bar{T}_{Roll}}{I_{xx}} + \frac{l_T I_{xz} \bar{T}_y}{I_{xx}I_{zz} - I_{xz}^2}, \quad N_T = \frac{-l_T I_{xx} \bar{T}_y}{I_{xx}I_{zz} - I_{xz}^2}. \quad (11.19)$$

Neglecting gravity results in the following linear autopilot design model

$$\begin{bmatrix} \dot{\beta} \\ \dot{p} \\ \dot{r} \end{bmatrix} = \begin{bmatrix} \frac{Y_\beta}{V} & s\alpha_0 + \frac{Y_p}{V} & c\alpha_0 + \frac{Y_r}{V} \\ L_\beta & L_p & L_r \\ N_\beta & N_p & N_r \end{bmatrix} \begin{bmatrix} \beta \\ p \\ r \end{bmatrix} + \begin{bmatrix} \frac{Y_{\delta_a}}{V} & \frac{Y_{\delta_r}}{V} \\ L_{\delta_a} & L_{\delta_r} \\ N_{\delta_a} & N_{\delta_r} \end{bmatrix} \begin{bmatrix} \delta_a \\ \delta_r \end{bmatrix} + \begin{bmatrix} \frac{-T}{m} \\ \frac{-l_T I_{xz} T}{I_{xx}I_{zz} - I_{xz}^2} \\ \frac{l_T I_{xx} T}{I_{xx}I_{zz} - I_{xz}^2} \end{bmatrix} \delta_{T_r}$$

$$+ \begin{bmatrix} \frac{1}{m} & 0 \\ \frac{l_T I_{xz}}{I_{xx}I_{zz} - I_{xz}^2} & \frac{l_{Roll}}{I_{xx}} \\ \frac{-l_T I_{xx}}{I_{xx}I_{zz} - I_{xz}^2} & 0 \end{bmatrix} \begin{bmatrix} \bar{T}_y \\ \bar{T}_{Roll} \end{bmatrix}. \quad (11.20)$$

This state space model can be used to design roll-yaw autopilots at a specific flight condition (α_0, Mach, altitude, CG). Note that β_0 was assumed to be zero assuming bank-to-turn.

Section 11.2 Aircraft and Missile Dynamics and Linear Models

In missile flight control systems, there are typically four tail fins, each driven by an electromechanical actuator. The fin actuator dynamics can usually be modeled with a second-order transfer function. The significant nonlinearities typically modeled include position and rate limits, as well as mechanical backlash.

The fin mixing logic that relates $\delta_e, \delta_u, \delta_a$, and δ_r commands to individual fin deflections is configuration specific and depends on whether the missile is flown with an "×" or "+"-tail. Here (for an ×-tail) the equations for the fin mixing logic are:

$$\begin{bmatrix} \delta_1 \\ \delta_2 \\ \delta_3 \\ \delta_4 \end{bmatrix} = \begin{bmatrix} 1 & -1 & -1 \\ 1 & -1 & 1 \\ 1 & 1 & -1 \\ 1 & 1 & 1 \end{bmatrix} \begin{bmatrix} \delta_e \\ \delta_a \\ \delta_r \end{bmatrix} \quad (11.21)$$

where δ_e, δ_a, and δ_r are the autopilot pitch, roll, and yaw fin commands, respectively, distributed to the four fins, and the δ_i, $i = 1, \ldots, 4$, are the actual fin deflections. Note that it is the δ_i that exhibit the nonlinearities (fin and rate limits, backlash).

In deriving the autopilot design models, it was assumed that the airframe was a rigid body. In fact, it is a flexible body, and these dynamics have a significant impact on the sensed accelerations and body rates. (This applies to aircraft control systems as well.) The discussion here is limited to the airframe's pitch plane. Also discussed is the tail-wags-the-dog effect due to fin mass imbalance and inertias, and TVC nozzle inertias. (See [1] for more detail on modeling these dynamics.) Consider the following rigid body model coupled with the first bending mode:

$$\begin{bmatrix} \dot{\alpha} \\ \dot{q} \\ \dot{b}_1 \\ \ddot{b}_1 \end{bmatrix} = \begin{bmatrix} Z_\alpha/V & 1 & Z_{b_1} & Z_{\dot{b}_1} \\ M_\alpha & M_q & M_{b_1} & M_{\dot{b}_1} \\ 0 & 0 & 0 & 1 \\ b_{1\alpha} & b_{1q} & b_{b1} & b_{\dot{b}1} \end{bmatrix} \begin{bmatrix} \alpha \\ q \\ b_1 \\ \dot{b}_1 \end{bmatrix} + \begin{bmatrix} Z_\delta/V & Z_{\ddot{\delta}}/V \\ M_\delta & M_{\ddot{\delta}} \\ 0 & 0 \\ b_{\delta 1} & b_{\ddot{\delta} 1} \end{bmatrix} \begin{bmatrix} \delta_e \\ \ddot{\delta}_e \end{bmatrix} \quad (11.22)$$

where the subscripts denote partial derivatives with respect to that variable. This analysis model describes the pitch plane rigid body dynamics (α, q) combined with the first bending mode (b_1), including the tail-wags-the-dog effects proportional to $\ddot{\delta}_e$. The pitch rate gyro measurement q_{Flex} and z-axis accelerometer measurement $A_{z_{Flex}}$ for this model are

$$q_{Flex} = q_{IMU} + F'_{A1} \dot{b}_1$$
$$A_{z_{Flex}} = A_{z_{IMU}} + F_{A1} \ddot{b}_1 \Big/ g \quad (11.23)$$

where q_{IMU} and $A_{z_{IMU}}$ are the rigid body pitch rate and acceleration from an inertial measurement unit (IMU), and F_{A1} and F'_{A1} are the mode displacement and slope, respectively. Partitioning the A-matrix in Eq. (11.22) into 2 × 2 blocks, the (1, 1) block is the same as in Eq. (11.16) and describes the rigid body dynamics. The (1, 2) block describes the changes in the aerodynamic forces and moments due to the body flexure. The (2,1) block describes how the rigid body states (α, q) excite the bending mode. The (2,2) block describes the first bending mode's second order dynamics.

268 Chapter 11 Affordable Flight Control for Aircraft and Missiles

In addition to the rigid body states (α, q), the control surface deflection δ_e also excites the bending dynamics. When the surface rotates, a bending torque is applied to the missile body that is proportional to both the surface's inertia and any mass imbalance (if the surface CG is located off the surface's rotational axis). This effect is called the tail-wags-the-dog effect (see [1] for more discussion) and can be significant. (When TVC is used, this effect is large because the nozzle is heavy and its CG is not located about its rotational axis.)

The flight control sensor measurements (angular velocities and accelerations) are corrupted by the flexible dynamics. Filters are designed to remove these signals from the sensed rates and accelerations. Unfortunately, these filters add gain attenuation and phase lag at the loop gain crossover frequency, thus impacting stability margins. It is very important when designing flight control systems that these filters be accounted for very early in the design process.

11.3 SIMULATION TOOLS

Simulation tools play a crucial role in the cost-effective development and analysis of flight control systems. High-fidelity nonlinear simulations are used to demonstrate, prior to flight, that flight control system requirements are satisfied and that the aircraft is safe to fly. These requirements (stability, performance, and robustness) must be evaluated over the entire flight envelope.

Computer-aided design tools are required to support the design, analysis, robustness, and simulation of flight control systems. By creating a graphical environment in which to work, manpower can be reduced in the development and use of simulations. An even greater cost savings can be realized if this same environment can be extended across other engineering disciplines that support other aspects of flight control and software engineering.

Figure 11.2 illustrates a simulation environment and its graphical user interface as provided by MATRIXx (a commercial software package from Integrated Systems, Inc.) that supports modeling, design, analysis, simulation, and autocode generation. This

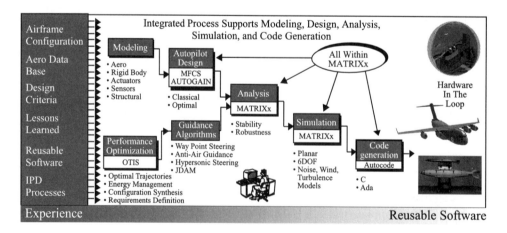

Figure 11.2 Integrated environment for flight control system design, analysis, simulation and code generation.

type of engineering environment offers significant productivity gains in terms of both schedule and cost for flight control system engineers and software developers. Figure 11.3 shows the top-level simulation architecture illustrating the modeling connection of the major air vehicle subsystems. Equations of motion (EOM), aerodynamics, propulsion, mass properties, actuation, and flight control are all represented at this top level and are connected through a configuration-controlled interface. This simulation framework allows the user to simulate aircraft, missiles, ejection seats, and unmanned air vehicles all with the same simulation toolset.

Using a simulation environment standard across project boundaries allows engineers to support and transition to various projects with moderate to little retraining. In addition, simulation reuse between different projects becomes much easier when the simulation is maintained in a graphical form. Block diagrams, or graphical objects, implementing algorithms and modeling subsystems are much easier to transition and reuse than actual software (like FORTRAN, C code, Ada, etc.) modeling those subsystems.

Central to any high-fidelity aircraft simulation is the flight control system. Using a graphical user interface to the simulation allows the engineer to perform detailed trade studies on key control system architecture attributes. At this level it becomes very easy to make changes to the control law architecture, gain tables, filter designs, as well as digital implementation rates. The productivity enhancements offered by this work environment allows the engineer to analyze and design a flight control system with increased performance and robustness at lower costs.

11.4 FLIGHT CONTROL SYSTEM DESIGN

Flight control system design requirements for aircraft and missiles are fundamentally different. In aircraft, a pilot guides the aircraft, thus closing outer loops around the aircraft's inner loop flight control system. The pilot adjusts his or her gain in these outer loops in real time to deal with the situation at hand. The inner loop flight control system is designed to respond to the pilot's commands in a way that provides the pilot with

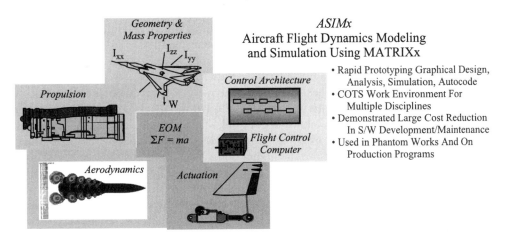

Figure 11.3 Simulation architecture connecting air vehicle subsystem models.

"flying qualities." In general, flying qualities are achieved by making the aircraft's dynamics from stick to commanded variable respond like a linear transfer function (whose dynamics vary or are scheduled with flight condition).

In contrast, missile control systems are designed to provide a fast response to commands generated by a guidance system. Like the pilot in an aircraft, the guidance system closes loops around the missile's inner loop flight control system. Unlike aircraft control systems, missile inner loop flight control systems are designed to have a high bandwidth (loop gain crossover frequency) so that the missile can respond very quickly.

Although the flight control design requirements are different for aircraft and missiles, both use successive loop closures to control their trajectories. A great deal of insight into flight control design can be gained by examining this principle. To stabilize an unstable air vehicle, angular rate feedback is used. Translational acceleration loops are often then closed around the inner rate loops.[1] The zero dynamics of the inner loop significantly affect the location of the resulting poles for the closed-loop system. In a classical sense, when the outer loops are closed and the gain in the loop is varied, the poles migrate towards the zeros. Thus, shaping the zero dynamics can be an important contributor to achieving a high-performance, robust feedback control system design. This is discussed further in Section 11.4.

11.4.1 Aircraft Control Law Design Using Dynamic Inversion

Feedback linearization, also known as dynamic inversion [2], has become a common method for the design of flight control systems for both aircraft and missiles. Control laws based on feedback linearization theory transform a nonlinear system into a linear system via nonlinear feedback control and a transformation of the state vector. Once the nonlinear system is in linear form, linear control methods can be used to design loop shapes that result in the desired response characteristics. The basic idea is that given a nonlinear plant $\dot{x} = f(x) = g(x)u$, a control law that achieves the desired response characteristics may be formulated as $u = g(x)^{-1}[v - f(x)]$, where v specifies the desired response. When the loops are closed, the resulting system is characterized by $\dot{x} = v$. This approach can be thought of as a deaugmentation-augmentation approach: Subtract (deaugment) off the aerodynamics and gyroscopic coupling effects and add (augment) the system with the desired dynamics (flying qualities). General feedback linearization theory requires that the nonlinear plant be minimum phase since the resulting control law effectively inverts the plant and would otherwise produce a closed-loop system that is not internally stable.

The typical approaches for applying dynamic inversion to the design of aircraft and missile control laws can be separated into two broad categories: those that consider the dynamics as a single coupled set of nonlinear differential equations, and those that model the dynamics as evolving in multiple time scales. The multiple time scale approach typically separates the inner loop rotational (rotational rates) dynamics and outer loop translational (angle of attack, sideslip, stability axis roll angle) dynamics by assuming that the rotational dynamics evolve much faster than the translational. This approach requires that the control effectors are primarily moment-producing

---

[1] This is true for aircraft at moderate to high speeds. At low speeds, it is common to close an angle-of-attack loop in the outer loop closure.

Section 11.4 Flight Control System Design

devices, which is reasonable for conventional aerodynamic control surfaces on winged aircraft.

The aircraft control architecture discussed here uses the two time scale assumption. The resulting control law structure is shown in Figure 11.4. The stick logic generates angle of attack, stability axis roll rate, and sideslip commands based on the pilot stick inputs, and is classically tuned to meet flying qualities objectives. Dynamic inversion is then applied to generate the effective control surface commands to produce the desired responses to the stick inputs. The propulsion system forces and moments are driven by the throttle setting controlled by the pilot and thus are not treated as command inputs to be generated by the feedback linearization controller. A control surface mixer is then used to generate individual control effector commands from the effective control surface commands.

The feedback control laws are obtained here as the solution of the noninteracting controls problem [2]. This produces a controller that completely decouples the system into independent control channels, where a given output is only influenced by its corresponding input. This type of controller directly supports aircraft flying qualities objectives providing decoupled responses in the longitudinal, lateral, and directional axes.

The dynamic inversion controller is derived by placing the dynamics of each time scale in the form $\dot{x} = f(x) + g(x)u$. In the fast time scale, the states are the stability axis rotational rates, $x = \begin{bmatrix} p_s & q & r_s \end{bmatrix}^T$, and the control variables are the effective aileron, elevator, and rudder surface deflections, $u = \begin{bmatrix} \delta_a & \delta_e & \delta_r \end{bmatrix}^T$. The objective of feedback linearization in the fast time scale is to use the surface deflections to control the rotational rates, and thus the outputs are chosen as $y = x = \begin{bmatrix} p_s & q & r_s \end{bmatrix}^T$. The dynamics governing this time scale can be expressed as

$$\begin{bmatrix} \dot{p}_s \\ \dot{q} \\ \dot{r}_s \end{bmatrix} = \begin{bmatrix} f_1(x) \\ f_2(x) \\ f_3(x) \end{bmatrix} + \begin{bmatrix} \bar{L}_{\delta_a} & 0 & \bar{L}_{\delta_r} \\ 0 & M_{\delta_e} & 0 \\ \bar{N}_{\delta_a} & 0 & \bar{N}_{\delta_r} \end{bmatrix} \begin{bmatrix} \delta_a \\ \delta_e \\ \delta_r \end{bmatrix}. \quad (11.24)$$

$$= f(x) + g(x)u$$

where the overbar on the aerodynamic derivatives represents a transformation to stability axes. Other variables such as α and β are treated as constants in the fast time scale since they evolve on a slower time scale. Each of the outputs of the fast time scale has a

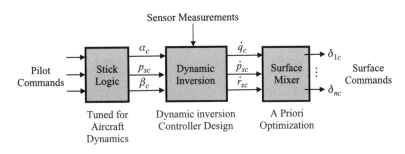

Figure 11.4 Dynamic inversion control law architecture.

relative degree of 1, and thus the total relative degree is 3. (See [2] for the definition of relative degree and its relationship to internal stability.) Since the total relative degree is equal to the number of states, this time scale will have no internal dynamics. The fast time scale control law is given by $u = g(x)^{-1}[v - f(x)]$ where the desired dynamics are chosen as commanded rotational accelerations, $v = [\dot{p}_{sc} \ \dot{q}_c \ \dot{r}_{sc}]^T$. The resulting closed-loop dynamics for the fast time scale are three decoupled integrators relating the rotational acceleration commands to their corresponding rotational rates. The left side of Eq. (11.24) can be derived by differentiating Eq. (11.8) and substituting from Eq.(11.1).

The translational dynamics of the aircraft are modeled and controlled in the slow time scale. The states of the slow dynamics are given by $x = [\alpha \ \dot{\alpha} \ \beta \ \dot{\beta} \ \phi \ \theta \ V \ \dot{V}]^T$. The control variables are the commands to the inner loop fast dynamics, which are the vehicle rotational accelerations, $u = [\dot{p}_s \ \dot{q} \ \dot{r}_s]^T$. The objective of feedback linearization in the slow time scale is to use the rotational accelerations to control α in response to the pilot's pitch stick input, β in response to the pilot's pedal input, and p_s in response to the pilot's lateral stick input. This defines the output vector to be $y = [\dot{p}_s \ \alpha \ \beta]^T$. The dynamics of the slow time scale can be expressed in the form

$$\begin{bmatrix} \dot{\alpha} \\ \ddot{\alpha} \\ \dot{\beta} \\ \ddot{\beta} \\ \dot{\phi} \\ \dot{\theta} \\ \dot{V} \\ \ddot{V} \end{bmatrix} = \begin{bmatrix} f_4(x) \\ f_5(x) \\ f_6(x) \\ f_7(x) \\ f_8(x) \\ f_9(x) \\ f_{10}(x) \\ f_{11}(x) \end{bmatrix} + \begin{bmatrix} 0 & 0 & 0 \\ -\tan(\beta) & 1 & 0 \\ 0 & 0 & 0 \\ 0 & 0 & -1 \\ 0 & 0 & 0 \\ 0 & 0 & 0 \\ 0 & 0 & 0 \\ 0 & 0 & 0 \end{bmatrix} \begin{bmatrix} \dot{p}_s \\ \dot{q} \\ \dot{r}_s \end{bmatrix} \quad (11.25)$$

where $f_5(x), f_7(x), f_8(x), f_9(x)$, and $f_{11}(x)$ are functions of the state vector, while each of the remaining functions equates a state derivative to the corresponding state that contains its derivative. It follows that this system has a vector relative degree of $\{0, 2, 2\}$ and thus a total relative degree of 4. Since the total relative degree is less than the number of states, internal dynamics will be present in this time scale. (These dynamics must be stable.)

The solution of the noninteracting controls problem yields a control law for the slow time scale as

$$\begin{bmatrix} \dot{p}_{sc} \\ \dot{q}_c \\ \dot{r}_{sc} \end{bmatrix} = \begin{bmatrix} 1 & 0 & 0 \\ -\tan(\beta) & 1 & 0 \\ 0 & 0 & -1 \end{bmatrix}^{-1} \left[v - \begin{bmatrix} 0 \\ f_5(x) \\ f_7(x) \end{bmatrix} \right]. \quad (11.26)$$

The desired dynamics, v, of the slow time scale (outer loop) are selected based on flying qualities requirements. Flying qualities requirements describe the aircraft response characteristics that pilots desire to complete various tasks. These guidelines have been developed through extensive research and are documented in the Military

Section 11.4 Flight Control System Design

Standard 1797A. Flying qualities are ranked as Level 1, 2, or 3. Level 1 indicates that flying qualities are clearly acceptable for the intended task; Level 2 indicates that flying qualities are adequate to complete a task but improvements are desirable; and Level 3 indicates that the aircraft is controllable but the pilot workload is excessive and/or the mission effectiveness is inadequate.

In the longitudinal axis, flying qualities requirements can be specified in terms of a second-order angle of attack to longitudinal stick δ_{LON} response, given by

$$\frac{\alpha}{\delta_{LON}} = \frac{K\omega_{sp}^2}{s^2 + 2\zeta_{sp}\omega_{sp}s + \omega_{sp}^2}. \tag{11.27}$$

Requirements for the short period frequency, ω_{sp}, and damping ratio, ζ_{sp}, are given in terms of the vehicle lift curve slope, n_z/α, where n_z is the normal acceleration in the negative z-body direction. Lateral axis flying qualities are specified by the stability axis roll rate response to a lateral stick δ_{LAT} input. The desired first-order response is characterized by

$$\frac{p_s}{\delta_{LAT}} = \frac{K}{\tau_R s + 1}. \tag{11.28}$$

Specifications for the roll mode time constant, τ_R, are given as a function of the maximum achievable steady-state roll rate. Directional axis flying qualities are specified in terms of the dutch roll frequency, ω_d, and damping ratio, ζ_d. The desired second-order response characterizing the sideslip response to a directional control (pedal, δ_{DIR}) input is given by

$$\frac{\beta}{\delta_{DIR}} = \frac{K\omega_d^2}{s^2 + 2\zeta_d\omega_d s + \omega_d^2}. \tag{11.29}$$

The control law of Eq. (11.26) linearizes the angle of attack and sideslip responses of the aircraft but produces internal dynamics whose stability must be evaluated. Since the slow time scale dynamics are in normal form, the states of the internal dynamics are directly identified as

$$\eta = \begin{bmatrix} \phi & \theta & V & \dot{V} \end{bmatrix}^T, \tag{11.30}$$

and thus consist of the phugoid and spiral modes of the aircraft. The phugoid mode is defined by changes in pitch attitude and velocity at a constant angle of attack, while the spiral mode is characterized by unconstrained roll motion at a constant sideslip angle. These modes are typically stabilized through feedback of attitudes, velocity, and accelerations. The stability of these modes is further augmented by the loop closure through the pilot.

11.4.2 Missile Control Law Design Using Linear Quadratic Optimal Control

Missile flight control design is different from aircraft flight control design. Typically, missiles are designed to have as fast a response as possible. This is different from designing for flying qualities as described in the last section. Maximizing overall missile performance requires choosing the appropriate autopilot command structure for each mission phase. This may include designing a different autopilot for separation (launch), an agile turn (high α turn), midcourse (long flyout), and endgame (terminal homing) maneuvers. The autopilot can command either body rates (p, q, r), wind angles (α, β), attitudes (ϕ, θ, ψ), or accelerations (A_z, A_y).

During launch, a body rate command system is typically used. Rate command autopilots are very robust to the uncertain proximity aerodynamics. During an agile turn, directional control of the missile's velocity vector relative to the missile body is desired. This equates to commanding α or β and regulating roll to zero. During midcourse and in the terminal phase, an acceleration command autopilot is typically used. At the end of terminal homing, during a guidance integrated fuse maneuver, the missile attitude may be commanded to improve the lethality of the warhead.

Separation, midcourse, and endgame autopilots have been designed and implemented in production missiles, and are in general well understood. Autopilot designs for agile turns (high α flight) are significantly less understood. Missile performance during the agile turn can be maximized by maximizing the missile's turn rate (higher turn rates lead to faster target intercepts). The missile's turn rate (for a pitch-plane maneuver) is given by $\dot{\gamma} = A_z \cos(\alpha) - A_x \sin(\alpha)/V$. High turn rates can be achieved by commanding a constant high α or by commanding large values of normal acceleration $(A_z \cos(\alpha) - A_x \sin(\alpha))$. Simulation studies have shown that because of the large changes in the missile's velocity (V) at high α's (due to the high drag), commanding body accelerations during an agile turn may not be desirable.

The nonlinear missile dynamics can be written as $\dot{x} = f(x, u)$. To form a linear model, partial derivatives of the f_i are needed with respect to each state variable and each control input. These partial derivatives are evaluated at a specific design point (flight condition). This would typically be at a trimmed equilibrium condition; however, at high α's the missile is generally not in what is considered an equilibrium condition.

The Robust Servomechanism Linear Quadratic Regulator (RSLQR) design incorporates integral control to track commands with zero steady-state error. The name "Robust" comes from its ability to track any magnitude command, without altering the structure or recalculating feedback gains. The mechanization of the RSLQR design algorithms allows for easy change of the commanded variable (i.e., attitude, rate, or acceleration) with no change in structure. This greatly simplifies the transition between different flight regimes where the commanded variable may change, and it allows for automation of calculating the feedback control gains.

Consider the following finite dimensional linear time-invariant model of the missile dynamics

$$\begin{aligned} \dot{x} &= Ax + Bu + Ew \\ y_c &= C_c x + D_c u + F_c w, \end{aligned} \quad (11.31)$$

with $w \in R^{n_w}$ an unmeasurable disturbance, and $x \in R^{n_x}$, $u \in R^{n_u}$, and $y_c \in R^{n_c}$.

The command input vector r to be tracked has the same dimension as the controlled outputs y_c. It is assumed that the kth differential equations for r and w are known and are given by

Section 11.4 Flight Control System Design

$$\overset{(k)}{r} = \sum_{i=1}^{k} \alpha_i \overset{(k-i)}{r}, \qquad \overset{(k)}{w} = \sum_{i=1}^{k} \alpha_i \overset{(k-i)}{w} \qquad (11.32)$$

where the α_i are known scalars and $\overset{(k-i)}{r}$ denotes the $(k-i)$th derivative of r. The polynomial formed by the Laplace transformation of Eq. (11.32) is

$$\alpha(s) = s^k - \sum_{i=1}^{k} \alpha_i s^{k-i} \qquad (11.33)$$

and describes a known class of inputs without knowledge of their magnitudes.

Define the error signal as $e = y_c - r$. Tracking in y_c is regulation in e; therefore, the objective is to make the error approach zero $e \to 0$ ($y_c \to r$) as $t \to \infty$, in the presence of unmeasurable disturbances w in a robust manner with respect to the plant description. The differential equation for the error may be written as

$$\overset{(k)}{e} - \sum_{i=1}^{k} \alpha_i \overset{(k-i)}{e} = \overset{(k)}{y_c} - \sum_{i=1}^{k} \alpha_i \overset{(k-i)}{y_c} \underbrace{- \overset{(k)}{r} + \sum_{i=1}^{k} \alpha_i \overset{(k-i)}{r}}_{=0}. \qquad (11.34)$$

Using the definition for r in Eq. (11.32), we find that the r variables of the right side of Eq. (11.34) sum to zero. Differentiating Eq. (11.31), we have

$$\overset{(k)}{y_c} = C_c \overset{(k)}{x} + D_c \overset{(k)}{u} + F_c \overset{(k)}{w}. \qquad (11.35)$$

Substituting this into Eq. (11.34) yields

$$\overset{(k)}{e} - \sum_{i=1}^{k} \alpha_i \overset{(k-i)}{e} = C_c \left[\overset{(k)}{x} - \sum_{i=1}^{k} \alpha_i \overset{(k-i)}{x} \right] + D_c \left[\overset{(k)}{u} - \sum_{i=1}^{k} \alpha_i \overset{(k-i)}{u} \right]$$
$$+ F_c \underbrace{\left[\overset{(k)}{w} - \sum_{i=1}^{k} \alpha_i \overset{(k-i)}{w} \right]}_{=0}. \qquad (11.36)$$

Using the definition for w, the third term of Eq. (11.36) is zero. This equation defines a set of simultaneous linear differential equations for the error. Let ξ and μ be defined as

$$\xi = \overset{(k)}{x} - \sum_{i=1}^{k} \alpha_i \overset{(k-i)}{x}, \qquad \mu = \overset{(k)}{u} - \sum_{i=1}^{k} \alpha_i \overset{(k-i)}{u}. \qquad (11.37)$$

The error dynamics in Eq. (11.36) then becomes

$$\overset{(k)}{e} - \sum_{i=1}^{k} \alpha_i \overset{(k-i)}{e} = C_c \xi + D_c \mu. \tag{11.38}$$

Differentiating ξ yields

$$\dot{\xi} = \overset{(k+1)}{x} - \sum_{i=1}^{k} \alpha_i \overset{(k-i+1)}{x}. \tag{11.39}$$

Using Eq. (13.31), we have

$$\dot{\xi} = A\left[\overset{(k)}{x} - \sum_{i=1}^{k} \alpha_i \overset{(k-i)}{x}\right] + B\left[\overset{(k)}{u} - \sum_{i=1}^{k} \alpha_i \overset{(k-i)}{u}\right] + E\underbrace{\left[\overset{(k)}{w} - \sum_{i=1}^{k} \alpha_i \overset{(k-i)}{w}\right]}_{=0} = A\xi + B\mu. \tag{1.40}$$

The error dynamics described in Eq. (11.38) is also a linear combination of ξ and μ. Using this definition, we define a new state vector z containing the $(k-1)$ derivatives of the error vector augmented with ξ, written as

$$z = \begin{bmatrix} e & \dot{e} & \cdots & \overset{(k-1)}{e} & \xi \end{bmatrix}^T. \tag{11.41}$$

This new state vector z has dimension $n_x + kn_c$. Differentiating Eq. (11.41) yields the "wiggle" system, defined as

$$\dot{z} = \tilde{A}z + \tilde{B}\mu \tag{11.42}$$

with $\tilde{A}$ and $\tilde{B}$ given by:

$$\tilde{A} = \begin{bmatrix} 0 & I & \cdots & 0 & 0 \\ 0 & 0 & \cdots & 0 & 0 \\ 0 & 0 & \cdots & I & 0 \\ \alpha_k I & \alpha_{k-1} I & \cdots & \alpha_1 I & C_c \\ 0 & 0 & \cdots & 0 & A \end{bmatrix}, \tilde{B} = \begin{bmatrix} 0 \\ 0 \\ \vdots \\ D_c \\ B \end{bmatrix}. \tag{11.43}$$

The RSLQR is obtained by applying linear quadratic regulator theory to the wiggle system in Eq. (11.42). By regulating z, we regulate to zero both e and ξ. In steady state, this allows the state vector x to be a nonzero constant vector in which $C_c x = r$.

Consider a constant input command r. This gives $\dot{r} = 0$ ($k = 1$) with $\alpha_1 = 0$ (Eq. (11.32)). The command error is $e = y_c - r$. The state space system described in Eq. (11.42) is given by

$$z = \begin{bmatrix} e \\ x \end{bmatrix}, \mu = \dot{u}, \tilde{A} = \begin{bmatrix} 0 & C_c \\ 0 & A \end{bmatrix}, \tilde{B} = \begin{bmatrix} D_c \\ B \end{bmatrix}. \tag{11.44}$$

LQR control theory is applied to Eq. (11.42) using the performance index (PI)

Section 11.4 Flight Control System Design

$$J = \int_0^\infty (z^T Q z + \mu^T R \mu) d\tau. \qquad (11.45)$$

where Q and R are weighting matrices. Solving this infinite time LQR problem yields a state feedback control law $\mu = -Kz$ given as

$$\begin{aligned} \mu &= -R^{-1} \tilde{B}^T P z \\ &= -Kz \\ &= -\begin{bmatrix} K_I & K_x \end{bmatrix} \begin{bmatrix} e \\ \dot{x} \end{bmatrix}. \end{aligned} \qquad (11.46)$$

where the real symmetric matrix P is the solution of the algebraic Riccati equation

$$0 = \tilde{A}^T P + P \tilde{A} - P \tilde{B} R^{-1} \tilde{B}^T P + Q. \qquad (11.47)$$

The optimal control $u(t)$ is obtained by integrating $\mu(t)$, that is,

$$u = \int \mu d\tau = -K \int z d\tau = -K_I \int e d\tau - K_x x. \qquad (11.48)$$

This controller mechanization yields integral control action on the error to provide zero steady-state error command following.

The Q in the LQR problem acts as a weight to penalize the states, that is, $z^T Q z$. The elements of Q are tuned to achieve performance and robustness. The controller gains are designed by tuning the LQR penalty matrices Q and R. Their "ratio" is an important factor in determining the size of the gains. Making Q large relative to R will increase the speed of response but may require large control effort. Making R large relative to Q will slow the speed of response while requiring less control effort. Since the "ratio" is important, the R matrix can be set to an identity (also since it is a scalar) and the elements of the Q matrix tuned to achieve the desired response. This reduces the number of parameters that have to be adjusted.

In using Eq. (11.48) the state vector x must be available for feedback. If the state vector is not available, then projective control theory can be used to project the optimal state feedback design to an output feedback architecture while preserving the dominant eigenstructure of the state feedback design. The number of closed-loop poles (eigenvalues) that can be retained from the state feedback design is equal to the number of outputs (n_y) available for feedback and the number of error states (n_e). Projective control theory retains both the eigenvalues and eigenvectors of the dominant poles, while the others are free to shift from their location in the state feedback design. This step in the control law design process can be used to eliminate the actuator state feedbacks from the control law.

The closed-loop system formed with the state feedback controller has $(n_e + n_x)$ poles, which are the eigenvalues of $\left(\tilde{A} - \tilde{B} K \right)$. Each eigenvalue (λ_i) has an associated eigenvector (v_i) defined by

$$\left(\tilde{A} - \tilde{B}K\right)v_i = \lambda_i v_i. \quad (11.49)$$

In the Robust Servo LQR technique, $n_r = n_e + n_y$ outputs are available for feedback, described by $y = \tilde{C}_z$. These n_r feedbacks can be used to retain n_r eigenvalues ($\lambda_i : i = 1, \ldots, n_r$) of the full state feedback design. The projective control output feedback gain matrix that accomplishes this task is computed as

$$\bar{K} = KX_{n_r}\left(\tilde{C}X_{n_r}\right)^{-1}.$$

The columns of the matrix X_{n_r} are those eigenvectors corresponding to the closed-loop eigenvalues (poles) to be retained. The matrix X_{n_r} must be real to guarantee real feedback gains. Each real eigenvalue ($\lambda_i = p_i$) has a corresponding real eigenvector ($v_i = \gamma_i$). Eigenvalues that appear in complex conjugate pairs ($\lambda_i = \sigma_i \pm j\omega_i$) have corresponding eigenvectors that also appear in complex conjugate pairs ($v_i = a_i \pm jb_i$). To retain a real pole and a complex conjugate pole pair, the matrix X_{n_r} is formed as $X_{n_r} = [\gamma_i \; a_i \; b_i]$. There is one column in X_{n_r} corresponding to each real eigenvalue to be retained, and two columns corresponding to each complex conjugate pair of eigenvalues to be retained.

The RSLQR combined with projective control has been used to design gain-scheduled control laws for both aircraft and missile inner loop flight control laws. Figure 11.5 shows a block diagram of a pitch loop autopilot that commands either normal acceleration or pitch rate. Sensor measurements from the inertial measurement unit (IMU) are filtered using third-order elliptic filters to remove noise and flexible body dynamics. The pitch rate signal is then fed into the inner pitch rate loop that is closed with proportional plus integral control. Pitch rate is also combined with normal acceleration to form normal acceleration at the center of percussion. This changes the zeros of the acceleration transfer function to a position that significantly improves stability margins. (This is discussed further in the next section.) When blending gyro and accelerometer signals, it is very important to keep the filtering of the signals identical. When the autopilot is commanding normal acceleration, the turn rate command q_{cmd} is zero, and similarly, when the turn rate command is to be tracked, A_{Zcmd} is set to zero.

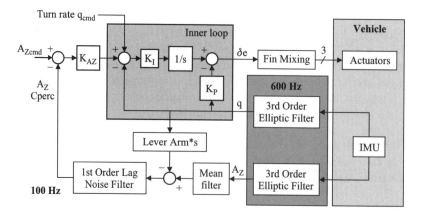

Figure 11.5 Pitch autopilot block diagram.

Figure 11.6 shows a roll-yaw autopilot. This autopilot can track lateral acceleration commands for skid-to-turn steering, bank angle error commands for bank-to-turn steering, or a turn rate command r_{cmd}. In this design fourth-order elliptic filters are used for filtering the IMU signals. The body rates are transformed to stability axes prior to closing the inner rate loops, thus requiring an estimate of angle of attack. Yaw rate is also combined with lateral acceleration to form lateral acceleration at the center of percussion.

The proportional plus integral autopilot gains in the inner loop shown in Figures 11.5 and 11.6 are designed using the RSLQR design approach. The design models used to compute the LQR state feedback gains include both the short period dynamics and second-order models of the actuators. Projective control is then used to eliminate the actuator state feedback gains from the control law. The outer loop proportional gains are designed classically. The design models are formed by linearizing the nonlinear dynamics at a specific flight condition (altitude, speed, angle of attack). The feedback gains are then scheduled over the flight envelope.

11.4.3 Zero Shaping to Improve Control System Design

Most aircraft and missile flight control sensor packages have three accelerometers and three gyros. Because of packaging considerations, the sensors are usually not located at the vehicle's center of gravity (CG). The location of the accelerometers relative to the CG greatly affects the measured accelerations and must be accounted for in the design of the flight control system. Figure 11.7 illustrates an acceleration measurement (from an accelerometer) from a fin deflection at different locations along the body. In the past, flight control designers compensated the measured acceleration signals to effectively move them to the center of gravity. Ideally, if the accelerometers

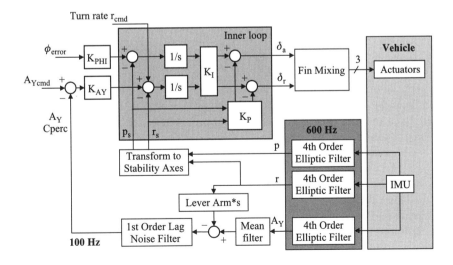

Figure 11.6 Roll-yaw autopilot block diagram.

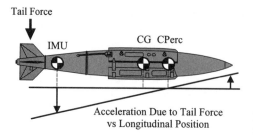

Figure 11.7 Fin force due to fin deflection measured along the x-axis of the body.

are located at the CG, then they measure just the translational accelerations. If the accelerometers are located off the CG, then they measure a combination of translational accelerations and rotational accelerations. This can be expressed as $a_{IMU} = a_{cg} + \dot{\omega} \times r_{Sensor} + \omega \times \omega \times r_{Sensor}$, where r_{Sensor} is a vector from the CG to the sensor package and $\omega = [p \ q \ r]^T$.

Consider the following expression for the z-axis accelerometer compensated for the x-axis CG offset as follows:

$$A_{z_{IMU}} = A_{z_{cg}} + (x_{cg} - x_{Sensor})\dot{q}. \tag{11.50}$$

This equation shows a blending of the rotational and translational dynamics. This changes the zeros of the transfer function from the control input to the sensor output and can have a dramatic impact on the flight control system design. The transfer function from elevator δ_e to acceleration $A_{z_{cg}}$ (from. [3], Eq. (4)) is given by

$$\frac{A_{z_{cg}}}{\delta_e} = \frac{\omega_a^2(Z_{\delta_e}s^2 + Z_\alpha M_{\delta_e} - Z_{\delta_e}M_\alpha)}{(s^2 - Z_\alpha s - M_\alpha)(s^2 + 2\zeta\omega_a s + \omega_a^2)}. \tag{11.51}$$

For tail-controlled vehicles, this transfer function is nonminimum phase (has a right half plane (RHP) zero). As the elevator δ_e deflects, the fin force $(Z_{\delta_e}\delta_e)$ accelerates the missile in the wrong direction. However, this fin force creates a pitching moment that rotates the missile. As the missile rotates, the body force builds $(Z_\alpha \alpha)$, accelerating the missile in the correct (commanded) direction. Aerodynamically unstable ($M_\alpha > 0$) tail-controlled missiles pose a considerable control challenge in that they have both RHP poles and zeros.

The transfer function from δ_e to $A_{z_{IMU}}$ does not have the same zeros as using $A_{z_{cg}}$. Figure 11.8 illustrates the location of the acceleration zeros as the sensor is moved along the body of the missile. When the IMU is aft of the CG, the two zeros are real with one in the RHP. As the IMU moves forward of the CG to the center of percussion, the zeros bifurcate and become complex, moving in along the $j\omega$ axis. The autopilot designer can shape the zeros in the acceleration transfer function by placing the sensor at a different location, thus creating a virtual IMU location. Depending on how the feedback gains are designed, this can be exploited to significantly improve stability and transient performance.

Section 11.5 Analysis Tools 281

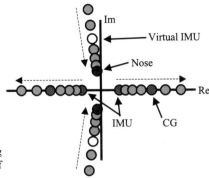

Figure 11.8 Acceleration zero locus varying accelerometer location along the *x*-axis of the body.

11.5 ANALYSIS TOOLS

Linear models are typically used to analyze the stability, performance, and robustness of flight control systems. In the past, linear models were developed and maintained in an environment separate from the nonlinear simulation. These models often did not characterize the flight control system's behavior as simulated/predicted via the nonlinear simulation. With the nonlinear simulation treated as the truth model, significant effort was spent analyzing these differences and tweaking model parameters until an adequate match was made. This process had to be repeated over the entire flight envelope.

Today, by using commercially available control system tools like **MATRIXx** with SystemBuild, linear models can be extracted directly from the nonlinear simulation. These models are created by trimming the air vehicle at a specific flight condition and linearizing about the operating point. This has proven to be a significant productivity enhancement for engineers, eliminating the nonlinear simulation versus linear model behavioral differences. It also greatly improves the accuracy of the analysis models and the ability to analyze the flight control system. Script files, or graphical user interfaces (GUIs), are created to automate the task over the flight envelope.

This modern design environment has led to a "what you test is what you fly" engineering practice. These tools give the engineer the ability to exactly simulate and analyze what will fly later in the aircraft. This eliminates any interpretation or language changes between the simulation environment, analysis tools, and the embedded flight software.

This environment is also very friendly for evaluating new control technologies and performing trade studies. By standardizing interfaces between subsystems (graphically in the simulation), different control laws or models can be easily "dropped in" and evaluated. By providing universities with the same simulation tools used in industry, new approaches can be easily incorporated and evaluated at low cost. In addition, new graduates arrive already trained in the same tools for high-fidelity simulations as used in industry.

11.5.1 Linear Analysis Models

Figure 11.9 illustrates the creation of a linear analysis model from the nonlinear simulation. Each block in the block diagram is represented using a state space quad-

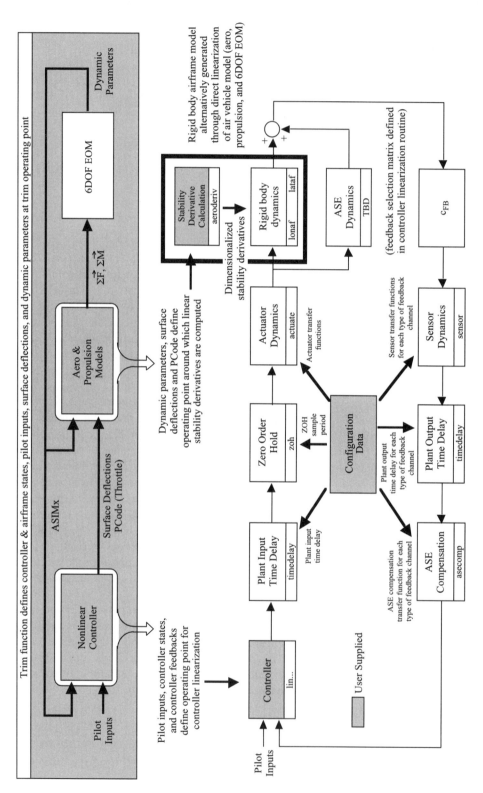

Figure 11.9 Linear analysis models numerically obtained from nonlinear simulation.

ruple (A, B, C, D). By using a trim capability provided by MATRIXx, the controller (which is also nonlinear), aerodynamic and propulsion models, and equations of motion are trimmed at a specific flight condition. After trim, linear models are extracted for each model. Both analytical linearizations and models from numerical perturbation of the nonlinearities are used. Included in the analysis model are time delays, zero order hold models modeling digital implementation, and aeroservoelastic compensation to filter flexible body dynamics. If information regarding a block is not available, the state space model for that block contains $(A = 0, B = 0, C = 0, D = I)$. As more information is available about a subsystem, like actuators, higher fidelity models can be easily inserted.

11.5.2 Performance Analysis

Figure 11.10 illustrates toolset modules that use the linear model for design and performance analysis. Performance analysis is always a very important aspect of the design process. This analysis is performed in both the frequency domain and the time domain. Toolset modules are used to extract performance data about the design to determine how well the design meets requirements.

Frequency domain tools perform classical and singular value frequency response analysis. In aircraft flight control systems, frequency response characteristics of the model are compared with the target flying qualities to assess the performance. Low-order equivalent system models are fitted to the frequency response and compared against the flying qualities model. Aeroservoelastic filter design tools are used to optimize filter coefficients. These filters are designed to provide a certain attenuation at modal frequencies while minimizing the phase lag introduced in the pilot frequencies. These filters must satisfy these requirements over a large operating envelope that depends on the aircraft configuration (what stores are on the wings). Aircraft flight control systems must be robust to (independent of) what weapons are loaded on the aircraft. (You do not want a different flight control system for each weapon loadout.) This makes robustness a key feature in the performance analysis of the flight control system.

11.5.3 Robustness Analysis

During the 1980s, many significant analysis capabilities were developed for linear time-invariant control systems. These robustness analysis methods gave engineers the ability to assess the control system's dependence on knowing model parameters (beyond a single parameter root locus), as well as the sensitivity to neglected or incorrectly modeled dynamics. In addition to assessing robust stability, these same methods can be used to develop and specify hardware requirements.

Robustness analysis methods typically transform the control system under study into a new block diagram in which the uncertainties have been isolated from known quantities. Stability of the control system is determined by examining the return difference matrix $I + L(s)$, where $L(s)$ is the loop transfer function matrix. It can be shown that $\det[I + L(s)] = \phi_{cl}(s)/\phi_{ol}(s)$, where $\phi_{cl}(s)$ is the closed-loop characteristic polynomial and $\phi_{ol}(s)$ is the open-loop polynomial. When the uncertainties in the control system are set to zero, representing the nominal system, the control system is stable ($\phi_{ol}(s)$ has no closed RHP zeros), and the $\det[I + L(s)]$ locus encircles the origin the same number of times as there are unstable open-loop poles. When the uncertainties are

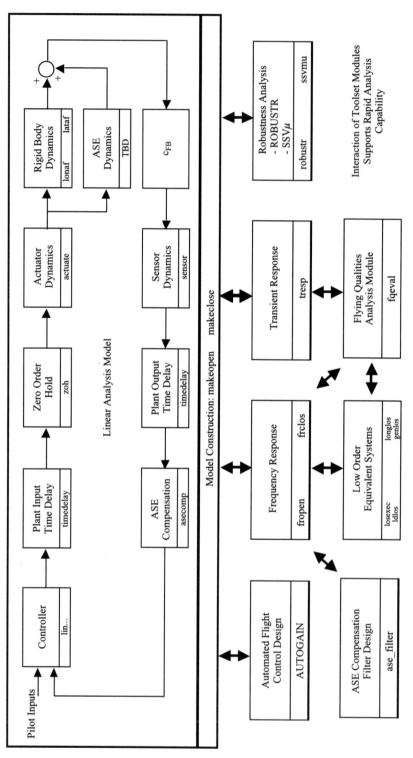

Figure 11.10 Analysis modules used to analysis control system stability, performance, and robustness properties.

introduced into the model, the destabilizing uncertainties must change the number of encirclements (multivariable Nyquist theorem). For the number of encirclements to change, the return difference matrix must be singular at some frequency. Thus, many of the robustness tests focus on the near singularity of the return difference matrix.

Figure 13.11 illustrates block diagrams for a nominal control system and the transformed system isolating the uncertainties into the Δ matrix. When $\Delta = 0$, the system is stable. The return difference matrix for the uncertain system is $I - M\Delta$. For the uncertain system to be unstable, viewed as a change in the number of encirclements, the return difference matrix must be singular at some frequency $s = j\omega$. Let $A = I$ and $B = -M\Delta$. The return difference matrix is then represented as $A + B$. The matrix A ($A = I$) is nonsingular. For the matrix $A + B$ to be singular, the matrix B, when added to A, must make $A + B$ singular.

If $A + B$ is singular, then $A + B$ is rank deficient, and there exists a vector $x \neq 0$ with unit magnitude such that $(A + B)x = 0$ (x is in the null space or kernel of $A + B$). This leads to $Ax = -Bx$, or $\|Ax\|_2 = \|Bx\|_2$. Using the definition of singular values, $\underline{\sigma}(A) \leq \|Ax\|_2 = \|Bx\|_2 \leq \bar{\sigma}(B)$, where $\underline{\sigma}(A)$ denotes the minimum singular value of A and $\bar{\sigma}(B)$ denotes the maximum singular value of B. To be singular, $\underline{\sigma}(A) \leq \bar{\sigma}(B)$. To be nonsingular, which means stable, $\underline{\sigma}(A) > \bar{\sigma}(B)$. Substituting for A and B, $1 > \bar{\sigma}(M\Delta)$. Substituting $\bar{\sigma}(M\Delta) \leq \bar{\sigma}(M)\bar{\sigma}(\Delta)$, we have $1/\bar{\sigma}(M) \leq \bar{\sigma}(\Delta)$, which is called the small gain theorem. This theorem bounds the uncertainties based on the gain of the nominally stable closed-loop system.

By modeling the actuator dynamics as an uncertainty and designing a stabilizing controller that meets performance requirements, the small gain theorem can be used to specify requirements for the actuator dynamics. It can also be used to assess how accurately these actuator dynamics must be known. In [4] both of these actuator analysis problems are solved for a missile flight control system. Also demonstrated is the assessment of how sensitive the flight control system is to unmodeled high-frequency flexible body dynamics. The small gain theorem as discussed here is conservative in many control system stability analysis problems. The structured singular value μ [5] can be used to reduce this conservatism.

For control system analysis problems where Δ models uncertain real parameters, the real-μ algorithm [6], as well as the real margin k_m [7], can be used. Figure 11.12 illustrates the real margin algorithm. These methods can be used to assess a flight control system's dependence on knowing the aerodynamics, mass properties, or any real parameters (as compared to complex uncertainties that result from dynamic elements) that are used in the control system. To use these methods, the analysis model must model the uncertainties in a form where the nominal control system can be isolated from the uncertainties (as shown in Figure 11.11).

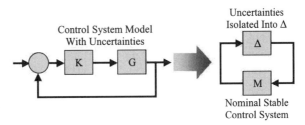

Figure 11.11 Robustness analysis block diagram model.

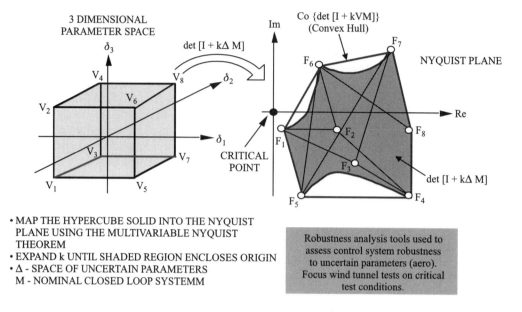

- MAP THE HYPERCUBE SOLID INTO THE NYQUIST PLANE USING THE MULTIVARIABLE NYQUIST THEOREM
- EXPAND k UNTIL SHADED REGION ENCLOSES ORIGIN
- Δ - SPACE OF UNCERTAIN PARAMETERS
  M - NOMINAL CLOSED LOOP SYSTEMM

Figure 11.12 The real margin robustness analysis approach.

In the early stages of an aircraft's design, these real parameter robustness test methods can be used to determine where the flight control system is most sensitive to uncertain aerodynamics. These analysis results can then be used to focus subsequent wind tunnel tests to gather more information about the aircraft. These tests have also proven to be very valuable in the design of future ejection seat control systems, such as the Navy/Air Force Fourth Generation Escape System (GEN4) program. On the GEN4 program a flight control system was designed and flight tested, demonstrating the capability to steer the pilot to a safe parachute deployment altitude regardless of orientation (inverted) as well as for supersonic ejections. In the ejection seat control problem, the size of the pilot, as well as his/her aerodynamic properties, are very uncertain. The control system must be designed to accommodate the 95% male pilot, who is quite large compared to the 5% female pilot, without knowing who is actually in the seat.

11.6 DIGITAL IMPLEMENTATION, REUSABLE SOFTWARE, AND AUTOCODE

In the past, aerospace engineers developed the requirements for a flight control system by developing a FORTRAN six degree-of-freedom nonlinear simulation for the application, designing the flight control system by linear and nonlinear analyses, defining the flight control system requirements (software requirements), and delivering the requirements to software engineers for software development. This process was expensive in time and manpower. In addition, the software development process precluded changes in the design (requirements) late in the development cycle as more accurate/higher fidelity models of hardware were obtained.

By using a tool like MATRIXx, the requirements definition process remains the same, but the analyses are all completed within the same tool. The big change is that the flight control system is designed graphically, and the autocode feature provided by MATRIXx allows for rapid development of the flight software. This autocoded software is used by the engineer; it is used in the manned simulators and in the hardware in the loop simulators (HIL); and it will fly in the aircraft.

This new process represents a new paradigm in developing flight software. Figure 11.13 illustrates the changes to the existing software development process. In the new process, there is no handoff of software requirements to a software development group. Also, instead of writing the flight software by hand, where errors could be introduced, the graphical block diagrams are autocoded. This software is then unit tested and transitioned to software engineers for integration testing and incorporation into the operation flight program (OFP). The vision behind this new process is to have the same engineer who is developing the requirements also create the software implementing those requirements, all within the same tool, thus minimizing errors and reducing costs.

Validation and verification of the software are shown in Figure 11.14. After autocode generation, the autocode software is unit tested. This is a very important step in the validation and verification of software and is often the most expensive. For each software module, test vectors are needed to test that the software has no errors and satisfies requirements. Figure 11.15 illustrates how this process can be automated using MATRIXx. In designing the flight control system (or algorithms) in MATRIXx, modules, or superblocks, are created implementing the functional requirements. The engineer in creating the superblock simulates the design, thus creating test vector inputs as well as expected results. The engineer reviews the results to make sure that the algorithms satisfy all functional requirements. Once the design is complete, the superblock becomes configuration controlled, and the test vectors and expected results are written to configuration-controlled directories for later use in integration and hardware testing. Using the autocode template from MATRIXx, unit test drivers are created that can automate the unit testing on workstations, in circuit emulators (ICE), or on flight control computer hardware. These test drivers execute the software using the test vectors and generate test results. They are then loaded back into MATRIXx and compared against the expected results.

In the past, software testing has required significantly more labor than developing the software. By using autocode, fewer software coding mistakes are made, thus reducing the number of errors uncovered during testing. Since engineers familiar with the functional requirements are performing the unit testing, anomalies in the requirements are uncovered early in the testing process and can be fixed. This new process allows the engineers to develop the requirements much later in the program without impacting schedule.

11.7 FLIGHT CONTROL CHALLENGES IN THE TWENTY-FIRST CENTURY: UNMANNED AIRCRAFT

Recently, several new programs have emerged that focus on the development of unmanned combat aircraft. The overall goal is to improve military effectiveness and the affordability of these systems. A new level of vehicle autonomy is needed to reduce the operating and support costs of these systems. These systems must possess the intelligence required to behave in a completely autonomous manner, to react to a changing environment or to changes in their own configuration, and to execute mis-

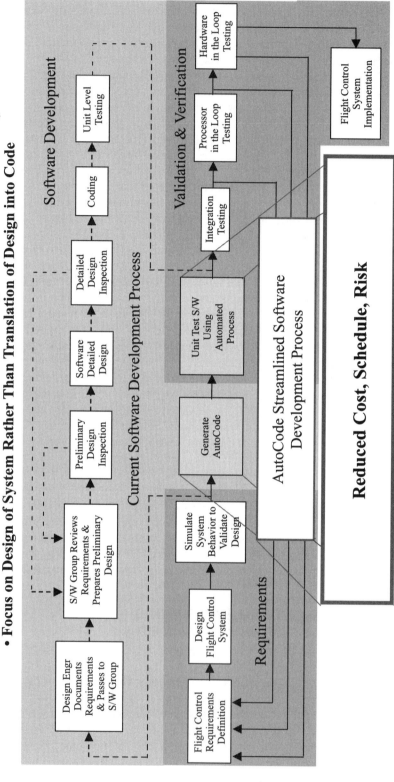

Figure 11.13 New software development process using autocode tools.

Section 11.7 Flight Control Challenges in the Twenty-First Century: Unmanned Aircraft

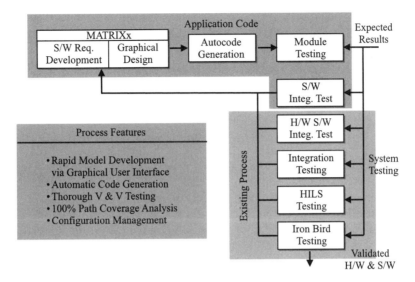

Figure 11.14 Embedded flight control software validation and verification.

sions collectively or in unison with other unmanned or manned aircraft, all without compromising their operational integrity. Intelligent control is needed to perform the operations once completed by a pilot. Autonomous vehicle guidance and flight control plays a major role in the development of these systems. For an overview of autonomous vehicle guidance and flight control see [8]; for a summary of ongoing research in this area, see [9]

Building the intelligence into these systems is the key challenge. Vehicle guidance on the ground and in the air will need to be autonomous to reduce operator workload. Both three-dimensional and four-dimensional guidance laws will be used at different

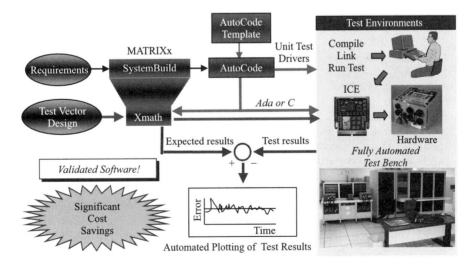

Figure 11.15 Automated software unit testing.

phases of the mission to coordinate position and time. In addition, when flying in formation, a follow-the-leader guidance mode will be required.

The flight control system will need to be robust and reconfigurable. The vehicle management system will have to be able to reconfigure the aircraft in case of failures or damages in order to make the system as safe to operate as manned systems. Robust damage-adaptive flight control will be required to provide safe and predictable control over the aircraft's trajectory. Building intelligence into the unmanned aircraft will require sensors to provide the information needed to diagnose vehicle health and reconfigure the system if needed. On-line system identification will play an important role in minimizing the costs of adding and instrumenting vehicle subsystems. On-board behavioral models of the vehicle subsystems for healthy and degraded mode operation will be needed to reduce/minimize the cost of vehicle autonomy.

The challenge that lies ahead is to build a new level of autonomous vehicle guidance and control technology but at the same time make it affordable. No answer is correct if it is too expensive to implement. Affordable design of the flight control laws as described in this chapter is just the first step. The next step will be to make these systems autonomous.

Related Chapters

- New control concepts for multiple piloted and uninhabited vehicles are described in Chapter 10.
- Computational tools for control design are discussed generally in Chapter 3, where flight control also serves as a motivating example.
- Chapter 15 discusses in some detail the use of computational tools for an automotive control application.

REFERENCES

[1] J. H. Blakelock, *Automatic Control of Aircraft and Missiles*, 2nd ed. New York: John Wiley & Sons, 1991.

[2] A. Isidori, *Nonlinear Control Systems*, 2nd ed. New York: Springer-Verlag, 1989.

[3] K. A. Wise, "A Comparison of six robustness tests evaluating missile autopilot robustness to uncertain aerodynamics." *Journal of Guidance, Control, and Dynamics*, Vol. 15, no. 4, pp. 861–870, 1992.

[4] K. A. Wise, "Singular value robustness tests for missile autopilot uncertainties." *Journal of Guidance, Control, and Dynamics*, Vol. 14, no. 3, pp. 597–606, 1991.

[5] J. C. Doyle, "Analysis of feedback systems with structured uncertainties." *IEEE Proceedings*, Vol. 129, Part D, no. 6, pp. 242–250, November 1982.

[6] P. Young, M. Newlin, and J. C. Doyle, "Practical computation of the mixed μ problem." *Proceedings of the American Control Conference*, Chicago, IL, 1992.

[7] K. A. Wise, "Missile autopilot robustness using the real multiloop stability margin." *Journal of Guidance, Control, and Dynamics*, Vol. 16, no. 2, pp. 354–362, 1993.

[8] M. Pachter and P. Chandler, "Challenges of autonomous control." *IEEE Control Systems Magazine*, Vol. 18, no. 4, August 1998.

[9] *Proceedings of the Association for Unmanned Vehicle Systems International—AUVSI'99*, Baltimore, MD, July, 1999.

Chapter 12 | INDUSTRIAL PROCESS CONTROL

Michael A. Johnson and Michael J. Grimble

Editor's Summary

In terms of economic impact, the process industries represent the industrial sector that has benefited the most from control technology. A large-scale process plant, such as an oil refinery, can contain over 10,000 individual control loops. A substantial hardware, software, and communications infrastructure, integrated in "distributed control systems," is required to support automation at this scale.

Process control is generally hierarchically organized. Sensors and actuators at the lowest level are interfaced to single-loop controllers, which are coordinated through multivariable control, with supervisory control, scheduling, and management information systems providing additional layers that ultimately connect physical equipment with the enterprise level. (The hierarchy can go even further—Chapter 13 notes applications at an industrywide level for one process industry, power generation.) There has been steady progress, driven in significant part by the information technology revolution, in expanding the domain of control systems to the higher levels of the hierarchy.

Several important technological research areas for process control are reviewed here. Performance monitoring, with classical techniques such as statistical process control and recent extensions thereof, can help capture the objectives of process operation and quantify the degree of success in their achievement. A number of tuning approaches for PID control are summarized, with tuning rules listed for developments spanning half a century. New developments in adaptive and robust control that appear to be well suited for process industry application are identified (see Chapter 5 for more on the topic of adaptive control). Model-based predictive control, virtually synonymous with advanced control in many segments of the process industries, is discussed. And, finally, the chapter notes the important new topic of plant-wide optimization.

Michael Johnson and Michael Grimble are professors associated with the Industrial Control Center, University of Strathclyde. Michael Grimble chairs the IEEE-CSS Technical Committee on Industrial Process Control.

12.1 INTRODUCTION

The process industries are a broad group of businesses comprising both the primary and secondary industrial sectors. The primary industries look after the first stages of processing raw materials, or producing the primary energy requirements for our society. These basic industries are the heavy industries such as oil refining, fossil power generation, pulp and paper producers, and metals refiners, or the utilities such as the water/wastewater sectors. The so-called secondary industries are those in which the first stage of product manufacturing is initiated. In this second group of industries would be found petrochemicals, pharmaceuticals, brewing, distillers, food producers, glass-making, and textile manufacturers. Manufacturing is the tertiary industrial layer which combines the primary outputs of energy with the partially refined raw materials to produce finished goods for our consumer-based society. Clearly, the whole process industrial sector is highly strategic to the support of civilized society, producing energy for light, warmth, and transportation, purified water and pharmaceuticals for healthy living, and the many materials (plastics, sheet metal, glass, and so on) for the convenience goods of our society.

A significant feature of the process industries is their sheer scale; plants are very large installations with substantial energy and/or material flows present. Thus, in these industrial sectors the emphasis is on using control technology and methods that produce commercial payback very quickly. For simple plants and processes, plant operators prefer automated procedures and minimal human operator interaction with the control system. However, in some complex and possibly dangerous plants, top-level manual operator control can still be seen today. This paradoxical situation is driven by the operational priority of producing maximum least cost throughput, with any form of control system experimentation being seen as having the potential to disrupt production and plant production schedules. The outcome of these constraints is at best slow incremental change to the plant control procedures and, at worst, no change at all, with operators maintaining "day one" performance only.

Even if process plant control strategies evolve slowly, however, the plant hardware, computer control systems, communications systems, and computer software change remorselessly. These new developments require that the installed plant be technologically upgraded, reengineered, and sometimes even rebuilt. It is at these major refurbishment or technological changeovers that new control systems and strategies can be considered and installed. Thus digital control capabilities are now quite widespread, and the payback on applying more sophisticated controls is excellent, granting a competitive advantage to those bold enough to exploit the potential of ideas available in the control research community.

In this chapter, industrial priorities are followed by a brief review of the technological framework of current process control hardware, software, and standards. This leads to a hierarchical paradigm consistent with industrial operational practice. Over the years, the process control industries have absorbed into this hierarchy those aspects of academic control concepts that have been deemed useful. Examples include automated loop tuning (the autotune culture) and predictive control algorithms. The chapter progresses through this hierarchy, describing some of the control concepts that have been used as well as those that might be used in future technological solutions to the problems of the process industries.

12.2 INDUSTRIAL PROCESS CONTROL TECHNOLOGY: STATE OF THE ART

The 1990s witnessed significant strides in the development of computer control technology and the related communications networks. In national groupings where co-ordinated and strategic control research is pursued, like the United States or the European Union (EU), the push to exploit information technology (IT) has been relentless. In the United States, the VISION 2020 strategic assessment of research priorities in the chemical engineering industries was underscored by strong IT themes. In the EU, the ESPRIT program was devoted to the many different interconnections between IT and industry. In the new EU Framework 5 programs, the emphasis on developments in information technology is extensive and far-reaching. Given this important background of IT innovation, it is useful to look first at the current status of process control technology as found in industry.

12.2.1 The Information Technology Infrastructure for Process Control

The basic structure of a distributed control system (DCS) is shown in Figure 12.1 [37]. As can be seen, the computing and communications in a DCS have a hierarchical structure. This structure is directly linked to the industrial tasks in the hierarchy. Thus at the process unit level a local area network (LAN) links the operator workstations and drives low-level programmable logic controllers (PLC's), sequential control, and simple unit controllers. At the top end of the hierarchy, the organization's enterprise management uses mainframe computer systems and long-distance telecommunications technology to monitor the economic well-being of the company. The geographical dimension of the company model can be captured by analogy with a spider's web, where the enterprise functions are at the very center of the web directing a network of production and sales operations.

Recent developments in computer networks have resulted in a new flexibility being added to the more rigid 1990s hierarchy of Figure 12.1. The importance of computer networking technology has increased greatly, so that companies have to react faster to customer demands. One effect of this change has been to focus attention on the

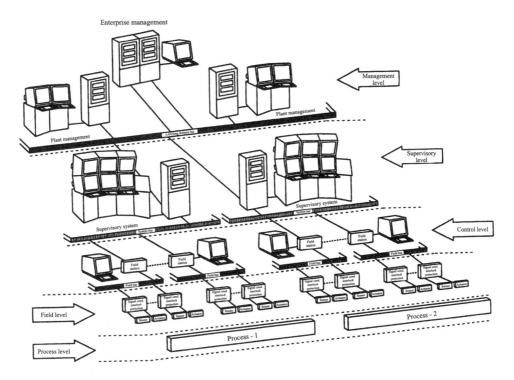

Figure 12.1 Basic structure of distributed control system [37].

dynamics of supply chains. A *supply chain* is the term given to the way feedstocks, components, and products flow through the organization in response to customer orders. The second important change has been the phenomenal development in personal computer hardware and software. Thus many new computer systems for process unit control are based on personal computer kits. This means that in addition to the control system being based on the PC, many supplementary software packages are available to the process operator for monitoring and administrative tasks like process performance data analysis, organization of maintenance schedules, or routine report writing. The company process or control engineer can exploit this extra process unit resource to achieve better control and more informed performance monitoring using novel multimedia methods. It is expected that the computer network and the use of PC process control hardware will flourish and grow substantially over the coming decade.

12.2.2 Process Control Applications Software

The recent dramatic changes in the man–machine interfaces of personal computers have enhanced the entry, manipulation, and presentation of data. For example, in the related field of systems simulation, tools like $MATRIX_X$, $EASY5_x$, and SIMULINK, which use graphical user interface (GUI) technology have changed the way process models are used and analyzed [35]. In the process controls area the presentation of process schematics and the way dynamic process data can be shown to the process unit operator have also changed over the last decade. The outcome of these changes is the use of far more sophisticated data processing routines in an attempt to provide the operator with much more meaningful representations and precise measures of process behavior. However an important question is whether the underlying process control techniques have also changed. Perhaps the best way to answer this question is to examine the market and review some of the products currently available.

12.2.2.1 Control Application Suite 1

For application in any DCS, Honeywell Hi-Spec Solutions has developed a family of control tools that goes by the name of Profit Suite. This is a suite of applications products whose emphasis is on optimizing the total process operation and which exploits the technical themes [24] of (1) being model based, (2) incorporating a prediction capability, and (3) using simple controllers that are robust to plant uncertainty.

Two key components are the use of model-based predictive control algorithms and a robust PID tuning tool. The model-based predictive control algorithm is promoted strongly from the viewpoint that the cost function can be directly related to system performance and hence profit [33]. The robust PID algorithm called R-PID uses the idea of choosing the PID coefficients to optimize a cost function across a family of process models [29].

An interesting feature of the Honeywell product is the attempt to push the optimization concepts further up the industrial operational hierarchy. Thus the idea is to consider the global optimization of a multiunit process plant to achieve additional economic gains. Hence the above technical themes are joined by two more, those of (4) performance monitoring and (5) global process optimization.

12.2.2.2 Control Application Suite 2

The second suite in the marketplace is essentially European in origin, and its controller development package is reviewed. The controller creation design process of this software is essentially a two-stage process:

Stage 1: Autotune based on a relay experiment [1,2] is used to establish some preliminary knowledge about the process. A PID rule base can then be used to produce an appropriate set of three term controller coefficients.

Stage 2: If the process engineer feels that the process is subject to change, or the model is insufficiently accurate, then an adaptation/identification stage can be used to complete the required design. In this second stage, the process is excited by a carefully designed pseudorandom-binary-sequence (PRBS) and a process identification is performed. The PID design is accomplished using dominant pole-placement design [2].

This process control design technology has the technical themes of

1. Nonparametric identification for the PID design.
2. Parametric identification of the process model for the enhanced controller design.

12.2.2.3 Control Application Suite 3

Virtual instrumentation (VI) seeks to emulate real hardware instrumentation on a personal computer. It exploits graphical user interface technology to enable the rapid creation of instruments and controllers from existing software libraries. Unlike C, BASIC, or FORTRAN, the software uses graphical symbols, or icons, to represent low-level operations or mathematical statements. These icons can be selected from existing libraries and connected together by virtual wires representing data links.

LabVIEW^{TM}[28] is an example of a virtual instrumental software that has enjoyed wide application. The software has three main sections:

1. *The front panel*. This is a template ready for embellishment as the front panel of the virtual instrument under construction.

2. *The block diagram*. This is the source code program constructed from the libraries of icons and connected together by wires.

3. *The icon and connector*. Once a VI has been constructed it can be captured as a single icon. Thus the VIs can be defined as separate units and then connected together. The connector specifies inputs and outputs and facilitates the construction of a hierarchical structure for an instrument or device. In this way, the construction of a very complicated instrument becomes modular and easy.

The facilities that make virtual instrumentation software like LabVIEW so successful include the ease of data acquisition and the simplicity by which it is possible to produce tailored data analysis instruments quickly. LabVIEW now comes with new supervisory software called BridgeVIEW^{TM}, which can be used to integrate instrumentation and controller units. The LabVIEW software also has some PID kits available. These modules contain sustained oscillation routines and open-loop and closed-loop tuning procedures. Other varieties of PID control algorithms are included, for example, PID control operating on setpoint error squared. Most recently, LabVIEW has released a setpoint relay experiment based on the work of Luo et al. [30].

In summary, the tools to construct different types of controller VIs are readily available in LabVIEW. The current PID kit uses the technical themes of:

1. Nonparametric identification procedures, sustained oscillation, and a set-point relay.
2. A PID coefficient rule base.

12.2.3 Data Communications and Standards

Data communications is the process of taking physical plant measurements and using them in a control system. In addition, the plant data can be used for monitoring and fault detection routines. Continuous variables are usually measured in analog form and then sampled for use in a digital process control system.

Field devices are often linked to a central computer using an analog point-to-point connection. In recent years, the use of microprocessor technology has created a generation of smart field devices that can perform self-diagnostic tests and calibration, and have enhanced numerical computation capability. However, it was the move to digital communications technology which enabled the development of much more interactive and responsive computer-based control systems.

Digital communications technology comprises hardware, software, and network aspects.

1. *Hardware aspects* include communications media (cables, fiber-optics, telephone networks, or radio links), data transfer architecture and electrical standards for equipment connections (RS-232, for example).
2. *Software aspects* include timing and the protocols for the software interfaces. The major development in this area was the Open Systems Interconnection (OSI) reference model which enabled communications to be organized into seven layers, each with designated functions.
3. *Network aspects* began from the need to have more than one device connected to more than one computer in a process system. A key outcome was the local area network (LAN), which, as the name suggests, is a geographically distributed communications network. The potential to effectively control, coordinate, and manage processes on a plantwide basis suddenly became possible.

To ensure the interoperability of different suppliers' equipment, the need for a worldwide communications standard was imperative. This standard was to be at the lowest functional level of the OSI model and would be a universal field bus standard. This goal of a single universal field bus standard has not yet been achieved, although there are several candidates.

12.2.4 Summary Conclusions

The information technology of process control is thriving. The last decade has seen rapid development in process control hardware, software, and data communications. Typical standard options in the applications products of many leading control system suppliers include the following:

- A library of traditional controllers with application guidelines, including three term controllers (PID family), lead/lag, dead-time compensators, ratio control-

lers, and signal selectors. These can support complex cascaded schemes encompassing constraint overrides, feedforward, and setpoint compensators.
- Advanced tools for PID, including robust designs, rule-based methods, and single-loop performance monitoring tools.
- Model-based tools for multivariable control, including predictive controllers and tools for difficult-to-control plants (long dead-time, inverse response plants, etc.).
- Steady-state optimizers using both linear and nonlinear model-based methods.
- Control systems to meet the S88 standard for batch processes.

These advances in system capability are enabling process control vendors *to globalize the control system to the whole plant*. Thus one of the most exciting goals for the next decade is to see the process industries exploit these new global system opportunities and create the applications experience of optimized and closed-loop production coordination.

However, control companies still have room to innovate in the low-level loop structures where robustness and adaptive theories are slowly being introduced in these regulators. The three-term PID controller remains as popular as ever with new methodologies being devised to ensure that the companies have new PID products to bring to the market.

12.3 ORGANIZING PROCESS CONTROL APPLICATIONS/PRODUCTION PROCESSES

Modern industrial plants range in size and scale from the relatively compact single-site conglomeration, comprising say an oil refinery or a steel works, to the geographically dispersed national electric power network or water distribution network. Whatever the size and geographical spread of such plants, their process operations tend to be characterized by high energy consumption and substantial feedstock and material usage. Supplementary utility support in terms of power and materials is often very high. The unit and global operation of these important industrial processes is almost always complex and demanding. Large-scale distributed computer systems are essential to provide the necessary plant synchronization and coordination needed to operate these processes efficiently and effectively. The drive for a competitive advantage often means that an older plant is refurbished with the installation of distributed computer control systems. Meanwhile, a newly installed plant will always be controlled using a high level of advanced computer technology. Such new on-line computer systems offer the potential to progress beyond simple control and coordination and to address the financially rewarding problems of global plant optimization.

12.3.1 The Industrial Operations Hierarchy: Strategy Issues

The technology of the process control operations hierarchy has been shown in Figure 12.1. In the lower layers of the hierarchy, the direct digital control (DDC), sensors, and actuators interface with the industrial process. In the supervisory control, scheduling, and management layers, the technological framework is one of interconnected computer networks. These provide the information links between the enterprise management (commercial/business divisions) and the plant management (production divisions). As discussed, the industrial communications networks follow a hierarchy: field bus at the DDC level, system bus at the supervisory level and local area network (LAN)/long-distance bus at the scheduling and management levels.

To discuss *the strategic aspects* and *the information flows* underpinning the operation of the large-scale industrial process, an extremely convenient generic framework is the standard process control hierarchy, as shown in Figure 12.2. This framework can be used to describe and conceptualize the tasks to be accomplished by state-of-the-art process control technology and data communications for large-scale industrial processes.

The strategy of global plant control usually has two components—one geographical, where the plant is subdivided into operational units, and one management control, where a plant control and command framework is needed. Figure 12.2 shows that a global plant optimization strategy must encompass every layer of the control hierarchy:

1. *Distributed Digital Control Level*: Control loops should be optimized and correctly tuned, with constant vigilance to retune the loops if operational conditions change.
2. *Supervisory Level*: Here the control centers on the process unit, comprising a number of items of plant equipment and a supervisor. In the case of feedstock changeover or new local conditions, setpoints supplied from the scheduling layer

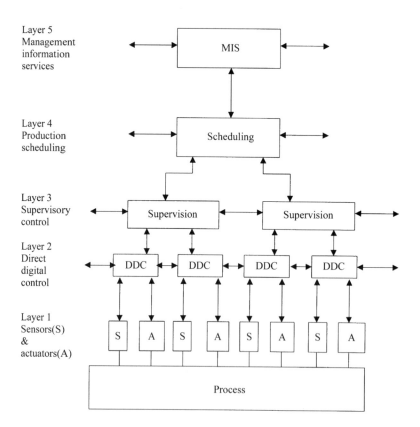

Figure 12.2 The industrial operations hierarchy.

have to be implemented. This requires routines and algorithms, first to automate the transition and second to optimize the changeover. Such operational setpoint maneuvers are usually subject to actuator and process output constraints. Consequently, constrained optimization and constrained control algorithms characterize the solution methods for this type of problem.

3. *Scheduling Level*: The response time of the processes through the hierarchy generally becomes slower at the higher levels, so that there is a longer time frame in which to provide control and commands to the lower layers. This feature is one reason why global plant optimization is able to function. At the scheduling level, plantwide monitoring, coordination, and planning occur. The problems at this level generally involve steady-state optimization algorithms concerned with longer-term plant changes and objectives. Production strategies to accommodate unplanned outages and planned maintenance and repair periods originate from this layer. At this level of the hierarchy, unit performance measures will be collated to give global plant performance figures.

4. *Management Information System (MIS)*: The management information system is the information network that facilitates the functioning of the enterprise. It comprises the interfaces between the different commercial activities of the organization: sales, production, finance and personnel. Ultimately, this system provides the appropriate facts and figures to the company boardroom to inform the top-level strategic decisions of the company.

12.3.2 The Industrial Operations Hierarchy: Information Issues

Real-time operating information of an appropriate quality is the key to global plant optimization. A diagram that uses the standard process control hierarchy to structure typical large-scale system information flows is shown in Figure 12.3. It is from this structure that three types of performance indices or diagnostics can be identified:

1. *Control Performance Indices*. These are associated with the direct digital control and the supervisory layer. They are the design metrics on which the low-level control design methods are based. Such measures can be classical, for example, process overshoot or settling time, or they can be modern and involve, say, robustness measures related to singular values, or alternatively use the optimization of mathematical cost indices.

2. *Fault Detection Indices*. Included in the sensor/actuator layer, the direct digital control layer, and the supervisory layer will be indices or diagnostics designed to detect and identify fault conditions. The outcome of these (often simple) algorithms will be operator alarms and subsequent actions.

3. *Plant Performance Indices*. These are invariably plant or process specific and are devised to capture some cost aspect of the plant/process/unit performance. Sometimes these indices have a direct financial interpretation. Consequently, they are usually associated with the scheduling and management layers of the operational hierarchy. Although plant performance indices have provided critical items of information for industrial personnel, only recently have academics begun to take an interest in this area and subject it to a systematic and formal analysis.

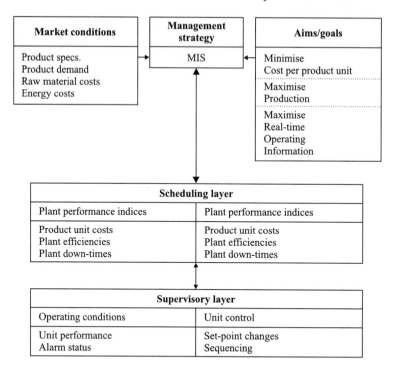

Figure 12.3 Hierarchical information structure.

The optimization of control performance indices has a long history as exemplified by Kalman's linear quadratic performance index [26,16,18]. However, the practical and systematic use of similar performance indices in supervisory control design and for controller performance assessment is a fairly new phenomenon. The availability of sensibly defined plant performance indices will also be important for the future development of global optimization tools. Although fault detection indices are not involved in plant optimization, these techniques are part of the portfolio of techniques of modern supervisory control. Hence it is useful to include them in this initial discussion of the operational hierarchy. In the next section, some of the methods that have been developed for performance monitoring, and subsequently extended to analyze performance quality, are discussed.

12.4 PERFORMANCE MONITORING

The procedures for *performance monitoring* provide an on-line means of quantifying and analyzing process output quality, plant efficiencies, and process unit performance. Once such an analysis is available, it is possible to consider whether a process operation needs to be optimized and how this might be achieved. In a range of process industries, the basic on-line tool for performance monitoring and optimization is *statistical process control*, or procedures based on this philosophy via the *magnificent seven*—the histogram, the checksheet, the Pareto chart, the cause-and-effect diagram, the defect concentration diagram, the scatter diagram, and the control chart [32].

Motivated by the success of the on-line methods of statistical process control [31,32], performance monitoring techniques seek to extend these principles to create entirely new facilities. These new methodologies include performance quality indices [4], function curve diagnostics [14] and controller performance indices [21,34]. As many of the names of these techniques indicate, a performance index or diagnostic often plays a fundamental role in these methods.

12.4.1 Statistical Process Control

Statistical process control (SPC) has its origins in the 1920s when Dr. W. A. Shewhart of the Bell Telephone Laboratories developed the control chart. The key result from statistics which is being exploited is the Central Limit Theorem. This indicates that if $\bar{x}_n$ is the sample mean of a random sample of size n, from any distribution with finite variance σ^2 and mean μ, then $\bar{x}_n$ is approximately a normal distribution of mean μ and variance σ^2/n. Consequently, it is possible to assign a probability to a sample mean occurring in a specific range. For example:

$$\Pr\{|\bar{x}_n - \mu| < \sigma/\sqrt{n}\} = 0.6820 \quad \text{or} \quad \Pr\{|\bar{x}_n - \mu| > \sigma/\sqrt{n}\} = 0.318$$

$$\Pr\{|\bar{x}_n - \mu| < 2\sigma/\sqrt{n}\} = 0.9546 \quad \text{or} \quad \Pr\{|\bar{x}_n - \mu| > 2\sigma/\sqrt{n}\} = 0.0454$$

$$\Pr\{|\bar{x}_n - \mu| < 3\sigma/\sqrt{n}\} = 0.9973 \quad \text{or} \quad \Pr\{|\bar{x}_n - \mu| > 3\sigma/\sqrt{n}\} = 0.0027$$

Thus, if a series of sample means occurs outside the $\pm 2\sigma/\sqrt{n}$ range, it is highly likely that the underlying process variable mean has changed. Shewhart's innovation was to translate this into a control chart, where sample means are plotted in real time, in the presence of upper and lower control limits, UCL and LCL, respectively:

$$\text{UCL} = \mu + k\sigma_v$$
$$\text{Center Line} = \mu_v$$
$$\text{LCL} = \mu_v - k\sigma_v$$

where the process variable mean is μ_v, the process variable standard deviation is σ_v, and the distance constant k is related to the number of standard deviations for the control limits ($2\sigma_v$ for $<5\%$ limits; $3\sigma_v$ for $<1\%$ limits). This theory can be developed for both discrete attributes and continuous variables, the latter being more appropriate for process and industrial control applications. A typical chart is shown in Figure 12.4.

Abnormal patterns on control charts then form the means by which deteriorating process performance is identified. For example:

1. A sequence of (eight) successive points in one direction or a trend is often indicative of component wear, a slow deterioration of a process component, or even operator fatigue.
2. A sequence of (seven) successive points on one side of the process mean indicates a shift in process variable levels. This may be due to a systematic change in feedstock or raw material quality, a change in shift, or a change in machine components.
3. A cyclic pattern can indicate systematic environmental changes, rotation of process operators or a component, cyclically affecting the variable.

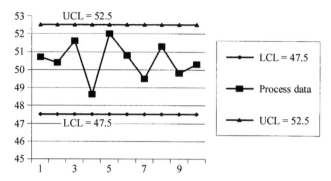

Figure 12.4 A control chart showing process data plotted against sample time with upper control limit (UCL) and lower control limit (LCL) shown.

The performance monitoring and diagnostic capabilities of statistical process control are extensive. The various types of charts are supported by the use of on-line statistical process control alarms and cause-and-effect diagnostic software. In addition, off-line studies using Pareto analysis, scatter diagrams, and so on ensure that adverse process effects are eradicated. A further component of the statistical process control philosophy is the involvement of the workforce in the need to achieve process quality objectives. The importance of this for process optimization cannot be stressed too highly. It is only after the process is *in control*, with repeatable behavior, that on-line optimization and fine-tuning of the process performance can begin. The types of adverse process effects that are identifiable by statistical process control include the following:

Material Variability: Variations in the quality of raw materials and utility supplies.

Personnel Problems: Some operators more vigilant than others, with some operators overworked by the excessive demands that arise.

Equipment Problems: Maintenance or servicing of equipment required; excessive wear identified; sensors inaccurate, improperly calibrated, or broken; actuators underrated and working at their limit; leaks or loss of materials identified.

Process Operational Problems: Disturbances due to load changes disrupting the process; process interactions causing instability; control loops incorrectly tuned; quality specifications too demanding for the installed process or production strategy being followed.

Environmental Factors: Ambient conditions affecting the process.

12.4.2 Performance Quality Indices

Following the scheme of the statistical process control philosophy, a method using performance quality indices is proposed [4]:

1. Determine the performance objectives for each process unit in the global context of process operation.
2. Translate these objectives into a set of performance indices and related constraints.

Section 12.4 Performance Monitoring

Typically, these indices might quantify energy and material usage, emissions and effluent produced, quality (tolerance) costs, and constraints. More precisely, two types of indices are common:

A Cost Index, I_c

$$I_c = \frac{Total\ cost\ (monetary)\ of\ process\ inputs\ and\ outputs}{Process\ product\ yield}$$

A Performance Index, I_p

$$I_p = \frac{Actual\ performance\ achieved\ by\ process}{Design\ performance\ specified\ for\ process}$$

Sometimes multiple performance indices and constraints will arise for each individual process unit. While some standard mathematical quantities like quadratic error or loss functions are used, performance indices are usually process specific and reflect some special performance objectives. This is a little different from the methods of statistical process control that are usually applied to the mean value of selected process variables, for example, temperature, pressure, or a physical property like viscosity. These are primarily the steady-state variables of the process and will be highly dependent on the reference set-points applied to the process. Optimization of performance quality usually requires more than the attainment of selected setpoint values since on-line optimization exploits the degrees of freedom remaining in the process operation. To appreciate the potential of the method, it is useful to consider an example [4].

Example: *Extractive Distillation Process*

From the process industries, this example concerned the optimized performance of the process towers to extract high-purity 1,3-butadiene product from a mixture of C_4 hydrocarbons. The three towers involved each had a different process operational function as seen in Figure 12.5 and consequently different quantifiable performance objectives.

Tower 1: In the extractive distillation tower, separation is the operational objective; thus the cost index is defined as:

$$J_E = S_E/Q_E$$

where S_E is the separation factor achieved and Q_E is the total energy consumed to attain the separation level.

Tower 2: The second tower is a recycle tower used to separate the butadiene from the solvent used in the extractive process. For this tower, a performance index is devised to compare the actual energy consumption, Q_A, to the design energy consumption, Q_D, viz., $J_S = Q_A/Q_D$. Optimal operation of the solvent stripping occurs when $J_S = 1$.

Tower 3: The third tower in the process is a finishing tower with the objective of achieving the high-purity product required. A cost index was chosen for this tower as $J_F = S_F/Q_F$ where S_F is the finishing tower separation factor and Q_F is the energy consumption required to achieve the separation.

304   Chapter 12 Industrial Process Control

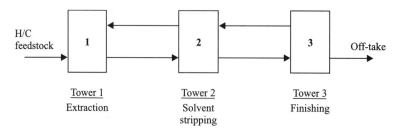

Figure 12.5 Extractive distillation of 1,3-butadiene.

Since optimal operation of the three towers does not coincide with each tower achieving an individual optimal performance value, a global cost performance index was constructed:

$$J_{Total} = \frac{Q_T F_c}{P_f}$$

where Q_T is the total energy consumption of the three towers, P_f is the product flow rate fraction, and F_c is the monetary cost per energy unit consumed. Thus index J_{Total} is the monetary equivalent of the energy cost per unit of finished product.

The three cost indices, J_E, J_F, and J_T, and the performance index, J_S, were evaluated and plotted on-line as though they were just additional measurements. On-line optimization to minimize the production costs and operate the extraction unit at its design performance was achieved by a trial-and-error process. This used accumulated plant operational experience and known empirical cause and effect relationships. However, the systematic investigation into understanding how to optimize these process performance monitoring indices brought valuable process insights, real monetary savings, and operational benefits.

As can be seen from the example, the methodology of performance quality indices follows that of statistical process control. A key difference is that the index is not the standard mathematical formula like a running mean but a quantity that is much more closely related to process economics. Formal on-line procedures to optimize such quantities remain an uncharted area. However, some general features of this study to optimize global process performance can be identified and listed:

1. Simple process models.
2. Performance quality indices for energy and material consumption costs.
3. Constraints on process conditions and operation.
4. Mixed discrete and continuous process simulation.
5. Optimized operational strategy obtained by testing different scenarios to optimize the performance quality indices.

To summarize, performance quality indices are not the usual mathematical formulas used in numerical optimization procedures but are designed to capture the objectives of good process operation. On-line or off-line, they can be used as though they were providing an additional measurement. They are usually simple and are therefore easy to display (plot) and track. These performance quality indices have three valuable uses:

1. *For heuristic on-line optimization.* In complex processes, operators and engineers can use these indices to nudge the process into a more economical condition.
2. *As early-warning indicators.* As in the control charts of SPC, performance indices might be used to detect the onset of abnormal conditions producing uneconomical process operation.
3. *As on-line and off-line management analysis tools.* Since these indices are often closely related to financial performance through operating costs, management can use them as a real-time indicator of the economic performance of process units and operational procedures.

12.4.3 Benchmarking Process Control

Benchmarking is a concept from the world of commerce and is defined as "the continuous process of measuring products, services and practices against the company's toughest competitors or those companies renowned as industry leaders" [20]. Four different types of benchmarking are often cited:

- *Internal Benchmarking*: Comparison of plant/practices within the same company.
- *Competitor Benchmarking*: Comparison of plant/practices within the same industry.
- *Parallel Industry Benchmarking*: Comparison across different industries.
- *Best Practice Company Benchmarking*: Comparison with the very best global companies

The process industries are taking note of the benchmark concept because it comes with some heavyweight recommendations, for example: "Benchmarking has been found to provide a catalyst for change with a positive influence on company morale and employee productivity" [10]. It is therefore useful to have a definition of this concept.

Definition of a Benchmark
A quantity or quality value to be used as a point of reference for performance assessment.

Benchmarking process control is a logical consequence of the ability to specify clearly and precisely the performance objectives of a process unit. Furthermore, if quantifiable performance indices can be agreed to, then the actual performance of a plant can be benchmarked and comparisons made with company performance in other plant sites. The key problem in benchmarking is that of being able to agree with production and management staff on exactly which performance index should be chosen as the benchmark against which to measure plant performance. In some cases, plant optimization is strictly related to a mathematical cost function in a very unambiguous way. For example, in power distribution network operation, optimal load flows can be closely related to quadratic programming problems. In this case the quadratic cost function provides a clear benchmark function. In other process control areas such optimal specifications are not so clear.

As international competition becomes fiercer, the need for companies to be super-efficient will increase. Tools like benchmarking will be used to ensure that businesses function in the most effective way possible. In a recent pronouncement on new research initiatives the U.K. Engineering and Physical Research Council said of performance measurement that there are now several established methods for businesses to measure themselves against the competition. While such benchmarking can show the historically most effective organization in terms of both products and flexibility, performance measures to develop and sustain competitive advantage in the future have not yet emerged. So development work remains to be done.

12.4.4 Summary Conclusions

Although the new computer network technology of process control is providing more flexibility for the control engineer, the large-scale process still has to be coordinated, synchronized and directed. For understanding, designing and implementing such large-scale system control structures, the longstanding hierarchical structures are as valuable as ever. What was found to be significant was the exciting possibility of adding practical global optimization power to this structure. Various ways in which this optimization of performance might be achieved were discussed. It was suggested that the logical outcome of these developments would eventually converge on the use of benchmarking concepts already common in the world of commercial and business practice.

Apart from these more general issues, process control technology and the industrial optimization paradigm are exploiting various advanced control methods and using them across the whole process control hierarchy. A short list of methods being used, working from the bottom of the hierarchy upward, includes:

- Automated loop controller design
- Adaptive concepts and methods
- Robustness concepts and methods
- Reliable control methods
- Prediction and predictive control methods
- Methods to inculcate intelligence
- Upper level optimization

Process control is obviously a broad subject and many different methods find a useful place in the process control engineer's toolbox. The remainder of this chapter gives a closer examination of several of these technologies starting at the bottom of the hierarchy with three-term control.

12.5 INDUSTRIAL THREE TERM CONTROL

In 1942 Ziegler and Nichols [41] sought to avoid abstruse mathematics (even then!) and devise very practical engineering procedures to address two questions:

1. How can the proper PID controller adjustments be quickly determined for any control application?

2. How can the settings of a PID controller be determined before it is installed in an existing application?

The first of these questions is an on-line controller tuning problem, while the second relates to controller commissioning. It is interesting to read in the 1942 paper that the method of *sustained oscillation* was devised to solve the tuning problem (Question 1) and that the *reaction curve* method was the proposed solution procedure for the commissioning problem (Question 2). These two methods slowly came to dominate process PID controller tuning. The following sections concentrate on the method of sustained oscillation because it is this method which has led to *automated* loop tuning technology.

12.5.1 The Sustained Oscillation Procedure

The usual unity feedback process control structure can be found in Figure 12.6 where the output z is measured and used to control the process. This measured output is fed back, compared to a reference input, r, and the error, e, is then used in the PID controller.

The controller is placed in the forward path in cascade with the industrial system. Although a textbook PID controller is used, it is given here in the dependent form, which is commonly found in industrial situations:

$$u(t) = K_c \left[e(t) + \frac{1}{T_i} \int_o^t e(\tau)d\tau + T_d \frac{de}{dt} \right]$$

where #$e(.)$ is an error signal, $e(t) = r(t) - z(t)$, $u(.)$ is the process actuator signal, K_c is the proportional gain, T_i is the integral time constant, and T_d is the derivative time constant.

The sustained oscillation procedure is based on the observation that "it is common knowledge that control with infinitely high proportional response is always unstable, oscillating continuously" [42]. This motivated the idea that if the value of the proportional gain at which the stable–unstable boundary is crossed is known, then this "ultimate" proportional gain K_u can be de-tuned by a factor to give acceptable closed-loop performance.

Ziegler and Nichols developed this idea further by relating the controller reset rate (or the integral time constant, T_i) to the period of the oscillation which occurs when K_u is applied and the stability–instability boundary is reached. Not surprisingly this data point was termed the "ultimate" period, P_u. Thus the utility of the Ziegler-Nichols procedure is that a single experiment yields two data points:

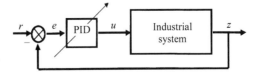

Figure 12.6 PID controller for an industrial system.

K_u = ultimate proportional gain

P_u = ultimate period

The data points K_u and P_u are then used in simple rules to yield the controller parameters K_c, T_i, and T_d. (For example, see Table 12.1).

12.5.1.1 Procedure 1: Method of Sustained Oscillation

1. The controller's existing settings must be altered so that the integral and derivative actions within the controller are switched out. This usually involves increasing T_i to a large setting, setting T_d to zero, and monitoring K_c so that the closed-loop process remains stable.
2. The proportional term in the controller is increased by advancing K_c in a series of stepped values. Each time the process output must be monitored. A purely oscillatory response (after transients have decayed) is sought.

TABLE 12.1 PID Rules

Time Response Tuning Methods

1943 Ziegler–Nichols Rules: Data K_u, P_u
PI Design	$K_c = 0.45K_u$	$T_i = 0.833P_u$	—
PID Design	$K_c = 0.60K_u$	$T_i = 0.5P_u$	$T_d = 0.125P_u$

1973 Modified Ziegler–Nichols Rules—Some overshoot: Data K_u, P_u
PID Design	$K_c = 0.33K_u$	$T_i = 0.5P_u$	$T_d = 0.33P_u$

1973 Modified Ziegler-Nichols Rules-No overshoot: Data K_u, P_u
PID Design	$K_c = 0.20K_u$	$T_i = 0.33P_u$	$T_d = 0.5P_u$

1991 Refined Ziegler–Nichols Rules: Data $K_u, P_u, K_0, K = K_0 \times K_u$
PI Design	$K_c = \dfrac{5(12+K)}{6(15+14K)}K_u$	$T_i = \dfrac{1}{5}\left(\dfrac{4K}{15}+1\right)P_u$	$1.2 < K < 15$

Time Domain Optimisation Methods

1993 Zhuang–Atherton ISTE Optimum PID Rules: Data $K_u, P_u, K_0, K = K_0 \times K_u$
PI Design	$K_c = 0.361K_u$	$T_i = 0.083(1.935K + 1)P_u$	—
PID Design	$K_c = 0.509K_u$	$T_i = 0.051(3.302K + 1)P_u$	$T_d = 0.125P_u$

1994 Pessen IAE Optimum PID Rules: Data K_u, P_u
PID Design	$K_c = 0.70K_u$	$T_i = 0.40P_u$	$T_d = 0.15P_u$

Frequency Domain Shaping

1985 Åström–Hägglund Phase/Gain Margin Rules: Data K_u, P_u
PID Design	$\phi_{PM} = 30°$	$K_c = 0.87K_u$	$T_i = 0.55P_u$	$T_d = 0.14P_u$
PID Design	$\phi_{PM} = 45°$	$K_c = 0.71K_u$	$T_i = 0.77P_u$	$T_d = 0.2P_u$
PID Design	$\phi_{PM} = 60°$	$K_c = 0.50K_u$	$T_i = 1.29P_u$	$T_d = 0.3P_u$

1995 Voda and Landau KVL Tuning: Data K_{-135}, ω_{-135} $1 < \beta \leq 2$
PID Design	$K_c = \left(\dfrac{4+\beta}{4}\right)\left(\dfrac{\beta}{2\sqrt{2}K_{-135}}\right)$	$T_i = \left(\dfrac{4+\beta}{\beta\omega_{-135}}\right)$	$T_d = \left(\dfrac{4}{(4+\beta)\omega_{-135}}\right)$

3. At the verge of instability, with a purely oscillatory output response, the ultimate proportional gain, K_u, and ultimate period, P_u, are noted.
4. The data points K_u, P_u are then used in a set of rules to give the PID controller parameters K_c, T_i, and T_d. These controller parameter rules will be related to the type of desired closed-loop system response. Ziegler and Nichols [42] gave one set of such rules.

The Ziegler and Nichols PID rules were proposed based on the time-domain experience of detuning the ultimate gain by a factor to give quarter amplitude decay in the closed-loop system response. Over the intervening years many rules have been devised with all sorts of different objectives and using different methods. A recent list is given in Table 12.1 [25]. However, it is now possible to see that two distinct aspects are inherent in the Ziegler and Nichols methodology:

1. *Nonparametric identification.* This is the design and implementation of an experiment to identify selected points on the Nyquist plot of the system.
2. *PID tuning rules.* The data relating to the selected point(s) is used in some explicit rules or formulae to give the coefficients of the PID controller.

The Ziegler–Nichols procedure remained popular for a long time, but in the early 1980s technology was changing and the opportunity to revamp the PID controller technology became feasible. It is useful to examine briefly some of the issues that have driven PID process control into the *autotune* culture.

12.5.2 Why Autotune?

Recent publications state that in process control more than 95% of loops are of the proportional-integral-derivative type, mostly PI controllers. Bialkowski [5] has described a typical Canadian paper mill as having over 2000 control loops, of which 97% use PI controllers. Simple arithmetic shows that there are more than 1940 PI loops to tune and optimize, and this is in one processing plant alone! The sheer staff resources needed to organize and accomplish this controller-tuning task are substantial. Similarly, in a recent survey of Scottish process and manufacturing companies in the Central Lowland region, Hersh and Johnson [22] found that engineers and managers were still citing PID controller tuning as a difficult problem. Clearly, a reliable, automated PID controller tuning procedure is needed to cope with the large numbers of PID loops as well as to minimize the need to use skilled staff to accomplish the task.

12.5.2.1 Problems with Ziegler–Nichols PID Tuning

The technique of sustained oscillation is manual. Although the procedure is simple, it is laborious and time consuming. The procedure also requires the loop being tuned to be operated at the verge of instability. Consequently, for some processes a safety risk may be involved. For this reason alone, skilled staff will be needed, and this means that more expensive personnel is being used on what is essentially a routine procedure. These considerations led engineers to find ways to automate the task of PID controller tuning. Among the candidate solutions, Åström and Hägglund [1] proposed a very simple, yet theoretically interesting, way of automating the method of sustained oscillation. The simplicity of their *relay method* made it particularly attractive from a practical and commercial viewpoint.

12.5.2.2 A Technology Changeover in the 1980s

In the early 1980s, a number of companies announced microprocessor-based process control products that incorporated more advanced features: self-tuning, simple adaptive controls, and automated PID controller procedures. The last named proved to be extremely attractive commercially and this feature is often given the generic name *autotuning*. In actual fact the different ideas or ways to automate PID tuning have been relatively few in number, pattern recognition (Foxboro) and relay methods (SATT Controls) being the two most well known

Apart from the market demand for a PID process controller with an autotune (or automated controller tuning) facility, a key enabling factor has been a change from controllers based on analog technology to microprocessor technology with vastly superior operator interfaces. This technology changeover occurred in the early 1980s, and controller manufacturers were keen to exploit the new computational and display potential of the devices. In this way companies like Leeds and Northrop, Turnbull Controls, Foxboro, and others were able to create distinctive devices and gain market advantages.

12.5.2.3 Process Controller Technology Today

The pushbutton "AUTOTUNE" culture is now over a decade old, and small process controllers provide some fairly advanced features. Three aspects are significant:

1. *Advanced controller structures*: Some devices use some of the more flexible controller structures. Toshiba, for instance, uses a two-degree-of-freedom control law to achieve better performance. Omron (Japan) has introduced fuzzy logic control design rules into its PID controller range.
2. *Controller schedules*: Autotune PID makes the construction of gain schedules to optimize the performance of different operational sequences fairly straightforward. However, this may require the intervention of skilled personnel to exploit this capability.
3. *Multi-loop control*: Many of the new devices have the ability to manage and tune many loops. The danger here is that true multi-loop and multivariable process control problems may be solved by inappropriate ad hoc single-loop control tuning procedures. The resulting loss in performance could be considerable.

12.5.3 The Relay Experiment

Hägglund and Åström [19] patented the idea of using a relay experiment to approximately identify the critical point, namely the ultimate gain, K_u, and ultimate period, P_u, for use in a Ziegler–Nichols PID design rule set. The basic configuration forming the complete relay experimental PID control-tuning device is shown in Figure 12.7. It is important to establish the functional decomposition blocks, for which two aspects are involved as noted:

Section 12.5 Industrial Three Term Control

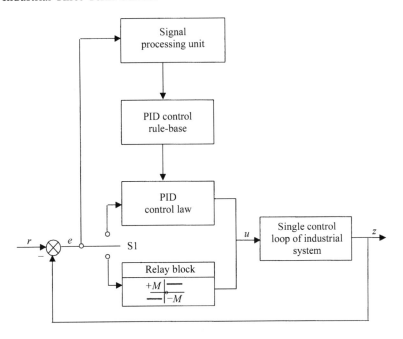

Figure 12.7 Relay experiment and system set-up.

12.5.3.1 Nonparametric Identification by Relay Experiment

This comprises the blocks:

1. On-off relay of levels, $\pm M$.
2. Signal processing unit, which uses either zero crossing or a peaks-and-troughs analysis of the oscillating signal to determine the ultimate period, P_u. The ultimate gain, K_u, will also be determined by this block using the formula $K_u = 4M/\pi a$ where a is the signal input amplitude to the delay. This formula derives from a describing function analysis [3].

12.5.3.2 PID Control

This comprises two parts:

1. *PID controller block*. This block implements the control law.
2. *Controller coefficient selection block*. This takes the output from the relay experiment, namely, the ultimate period P_u and the ultimate gain K_u, and uses a PID coefficient design rule set to produce K_c, T_i and T_d. These are passed for use in the control law of the PID controller block.

The procedure of the relay experiment is thus decoupled from the PID controller coefficient selection step.

12.5.3.3 Procedure 2: The Relay Experiment

Objective: To determine the system ultimate gain, K_u, and ultimate period, P_u.

1. The switch S1 is activated so that the PID control law block is removed from the loop and the relay block is switched into the forward path.
2. A small perturbation is applied to initiate the self-oscillation or limit cycle. This step presupposes that an oscillation will occur. The presence of system time delays or nonlinearity usually ensures that this happens. The height of the relay, M, can be selected to ensure that the size of the control signal does not unduly upset the process operation.
3. After the system transients have died away, a steady oscillation should be established on the signals z and e. The control signal u will act at a set level, with a square wave of amplitude $\pm M$ superimposed on it. The signal-processing unit can use signal e to determine the ultimate period, P_u. This can be done by analyzing peaks and troughs or by counting zero crossings. Since it is possible that relay height adjustment has taken place during the experiment, the signal processing block will also compute the ultimate gain using the formula $K_u = \frac{4M}{\pi a}$ where a is the oscillation amplitude of e or z.
4. Once consistent results for P_u and K_u have been obtained, control of the experiment moves to the routines to calculate new PID controller coefficients and switch in the new controller while switching out the relay path and block.

Åström and Hägglund [2] patented this relay experiment, and it was successfully commercialized. Many current process controllers use the relay method, and the auto-tune culture is now widespread. But, the story does not quite end there: Companies and researchers are continuing to develop the PID technology and improve its functionality and capabilities.

12.5.4 Recent Directions for Industrial PID

Though very successful, these simple relay experimental procedures initiated another decade of development and research on the topic of PID controller tuning. Quite surprisingly, the field is still very active (in Japan, Taiwan, Singapore, Australia, Canada, the United States, and the United Kingdom for example), and new methods and devices are being promoted, patented, and packaged as the PID solutions for the next 10 years.

United States: Researchers at National Instruments recently announced a modification of the relay method to the setpoint relay experiment [30]. Of course, it is the flexibility of the Virtual Instrument hardware and software which enables the method to work in a setpoint architecture. Thus with some extra analysis the PID controller does not have to be switched out of the loop, and the method provides a little more convenience.

At Honeywell, Lu [29] has tackled the problem of robust PID tuning. The new robust PID algorithm called R-PID uses the idea of choosing the PID coefficients to optimize a cost function across a family of process models. Interestingly, it is the PID tuning aspect that has not received as much attention as the nonparametric identification problem, so Lu's contribution is different and useful.

Australia: A *quantizer* is the new method arising from research at the University of Newcastle. This method extends the relay experiment idea. Instead of a relay, a quantizer is used. Furthermore, the quantizer is placed after the controller, allowing control to be maintained at all times [15,39]. As with the relay experiment device a patent application has been lodged for the new method.

Scotland: A contribution has been made to the PID tuning problem with a new methodology that finds a tradeoff between a full optimal control solution and a suboptimal fixed structure controller. The controller can easily be configured to be of PID form thereby linking an ideal solution with a PID tradeoff [17].

The relay experiment can be both inaccurate and inflexible [11]. To solve both problems, a new method is needed, and the quite different solution proposed uses a digital phase locked loop. The idea is of automated sinusoidal testing but inside a closed-loop system. The result does not depend on the describing function analysis [3] for its accuracy, and it can be easily tailored to seek other system points for use in PID tuning routines. As is common practice with control devices of commercial potential, patent protection has been sought for the method [12].

12.6 ADAPTATION AND ROBUSTNESS

12.6.1 Adaptation

Many developments have taken place in the theory of adaptive control systems, particularly regarding the control of nonlinear systems. However, few of these results can be utilized immediately in industrial applications. The most promising line of research is in *identification for control*, which involves the selection of an identification algorithm that is, in some sense, compatible with the control law calculation. To illustrate the main idea, consider a self-tuning controller in which the control law is chosen to minimize the variance of error and control signal terms. Clearly, the more effective the controller becomes in minimizing the variance of the control signal, the less excitation is present which is needed for good plant identification. In the limiting case, when the control signal is a constant, there will of course be no information on which to base the plant model estimates. It follows that the best choice of control action cannot be determined independently of the choice of identification algorithm. This subject was stimulated through the seminal text by Bitmead et al. [6].

A second area that may have practical utility lies in the subject of *limited authority adaptive control*. This involves the use of a self-tuning controller, where the full range of variation is limited, since the full optimal controller is not used. Brasca et al. [7] described a successful example of this type of algorithm that was used for voltage control in the electrical power generation industry.

12.6.2 Robustness

Numerous developments have also taken place in robust control which should have a significant impact on process control systems design. The best-known development was initiated by Zames [40] and became known as H_∞ robust control theory. For a scalar system, the H_∞ norm simply represents the largest gain of a particular transfer function that can be found by plotting the frequency response and measuring the largest value of the magnitude. If, for example, the output changes are to be robust to the disturbance inputs the maximum value of the sensitivity function can be limited. Different frequency response terms should be limited depending on the types of plant uncertainty if stability robustness is to be achieved.

Unfortunately, the type of uncertainty model employed may not characterize the physical uncertainties in process control systems very well. Moreover, high-order controllers are obtained which are often unsuitable for process control systems. Thus robust H_∞ controllers are only valuable in particular problems, where the plant may be multivariable and a high-performance, low-interaction solution is required [16].

A technique developed mainly for the aerospace industry by [23] may be more appropriate for process control design. The approach is termed Quantitative Feedback Theory (QFT) and it enables low-order controllers to be designed with plant models that may be parameterized to represent the system changes or uncertainties. For example, the plant model might be represented as:

$$G(s) = \frac{k}{s(\tau s + 1)}$$

where the gain and time constant are expected to lie in given ranges $\tau \in [100, 500]$ and $k \in [50, 100]$. The QFT design method then enables a low-order controller, such as a PID controller, to be designed which is robust to all such variations. The benefits of going to a slightly higher-order controller are also obvious from the process. The technique may be used directly with experimental results rather than with a parameterized plant model. That is, the plant can be tested at, say, 10 operating points, and the controller can then be found which will stabilize the system and satisfy performance objectives for the full range of operation.

The main disadvantage of the method is that it requires a design engineering approach rather than an on-line *test it and see* procedure. However, only very simple frequency domain ideas are required to understand and use the method, and it should offer real benefits for difficult design problems, where robust solutions are essential.

12.7 ASPECTS OF GLOBAL SYSTEM OPTIMIZATION

The optimized control and operation of large-scale industrial processes can yield significant economic benefits and savings. The industrial techniques used to pursue these benefits are still evolving, since past design and operational experience is usually the main guide for current industrial practice. There is considerable industrial interest in discrete event and continuous simulation tools like *SIMPLE++*, *WITNESS*, or *SIMFACTORY* as an inexpensive means of testing alternative operational or production strategies. Yet, the canon of optimization tools, methods, and algorithms for these problems has plenty of opportunity for growth. The command and control structure of supervisory systems is perhaps a better established area of research and development. Yet, although industrial experience and practice have been extensive, there has not been so much exchange between industry and the academic research community on these topics.

Section 12.7 Aspects of Global System Optimization   **315**

12.7.1 The Supervisory System Command Structure

Motivated by a need to define precisely the supervisory command structure for large-scale combined cycle and combined heat and power generation plants, a study produced the structure shown in Figure 12.8. The findings of the study [27] were generic for large-scale processes and may be discussed from the bottom layer of the supervisory structure upward.

12.7.1.1 Low Level Control Strategies

The technology at the unit level tends to be installed with all the local control loops supplied. In large-scale industrial plants, fine-tuning the local loops is not such a high priority; it is the global integration and optimization that is of more importance. The actual low-level control design strategies are mainly classical, being primarily multi-loop or cascade loop structures using three term controllers. Sometimes adaptation to changing process conditions is achieved by simple controller scheduling. In designing algorithms for integrating and coordinating the global set of process units, an important practical implication is that the low-level units can be regarded as a set of interacting stable (closed-loop) systems over which supervisory control is required.

12.7.1.2 Dynamic Setpoint Maneuvers

To respond to new external process load conditions or to effect feedstock changeover, the process usually has to be transferred to new operating conditions. However, it is often found that the changeover or setpoint maneuver has to occur in the presence of process operational constraints, for example:

1. *Actuator limits*: these are hard physical limits that characterize certain types of actuators. Typically, valves can only open to a full aperture, thereby limiting at a maximum flow.

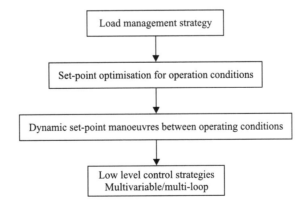

Figure 12.8 Process plant supervisory concepts.

2. *Actuator rate limits*: the actuator that might be driven by an electric drive can only react or move at a maximum speed. This gives rise to a maximum physical rate of change limit.
3. *Output constraints*: while one output is set to following a reference change, a subsidiary output might have to be constrained to lie within tight limits of its nominal value. Typically, a boiler might be absorbing a new steam-flow demand, but the temperature of the steam must not stray outside of 1% of the nominal value. Otherwise turbine blade damage might occur.

Standard industrial practice is to use past operational experience and to ramp setpoints over a preselected time period so that constraints are not violated. In the last decade, model-based predictive control (MPC) has successfully provided an optimization framework to automate this constrained changeover of operating point. If process disturbances are measurable at this level, then it is easy to incorporate a disturbance term into the optimization. The model-based approach is thereby effective for feedforward disturbance rejection.

12.7.1.3 Setpoint Optimization and Load Management Strategy

At this point in the hierarchy, the new application of constrained model-based predictive control algorithms meets the more established procedures of setpoint selection and optimization, unit allocation, and plant scheduling. Many of these procedures and strategies will have their source in the original design studies and flowsheet exercises that were conducted prior to plant construction. For this reason, often a close link exists between the supervisory control structure, the equipment sizing and the original design objectives for the plant. Thus post-plant construction process optimization is usually a constrained optimization problem. As has already been mentioned, the increasingly slow system response times that occur in the higher reaches of the hierarchy lead to a predominance of static or steady-state optimization problems at these levels.

Use of the techniques of constrained model-based predictive control in the supervisory layers of the large-scale system hierarchy has been extremely successful in some industrial sectors. In fact this has often been regarded as the advanced control success story of the last 20 years or so. For this reason, the technique is given further exposure in the next section.

12.7.2 Model-Based Predictive Control

The importance of model-based predictive control for the process industries stretches back to the late 1970s when Richalet et al. [38] reported on the first applications of the IDCOM (*id*entification and *com*mand) algorithm to industrial problems. Improved versions of the model-based predictive control philosophy followed quite quickly, as did academic analysis to determine the theoretical properties of this new type of control algorithm. The industrial evidence emerging was that these numerical optimization approaches were both versatile and economically very effective. A reputation for reducing costs and saving money is an excellent motivating force behind any new technology, and the MPC methodology is now well established in many process industries. A useful introduction to all aspects of the subject can be found in [8].

12.7 Aspects of Global System Optimization

12.7.2.1 The Basics of Model-Based Predictive Control

The fundamental difference between feedback control and model-based *predictive* control is the ability of the latter to *anticipate* the future control actions required to achieve new output setpoints. To do this, the MPC problem formulation must be fed *future* setpoint data. If a measured disturbance is also present, then this information can also be fed into the formulation, and feedforward action is obtained quite naturally in the same optimization framework. Thus in a MPC system, outputs will move in advance of setpoint changes, whereas in a feedback system the presence of a reference *error* is used to activate control action and this only comes *after* the setpoint change has occurred. The most obvious analogy is that of driving an automobile around a corner. Most drivers anticipate approaching the corner and drive *into* it. Only a very few drivers would wait until they had reached the corner and then drive *around* it. To take the corner in this way, the automobile speed has to be reduced significantly and performance is lost. The various formulations of model-based predictive control have the same basic components, as described next.

12.7.2.2 A Process Model

In MPC, it is assumed that a model of the process is available for on-line execution. This may be based on prior physical system modeling with parameter fits for the particular process being controlled, or the model may result from black-box identification tests. The model is usually linear in system parameters and may be of state space or transfer function form. The selection of model type is one factor that leads to the many different varieties of industrial MPC algorithms.

12.7.2.3 A Predictive Model Equation

It is the ability to run forward the (linear) process model for a fixed number of time steps and to predict the possible system output that gives MPC its anticipatory capability. This prediction of the future system trajectory is substituted into the cost function and controls selected to optimize the future system behavior. Any future disturbances and setpoint changes can also be included in the prediction equation.

12.7.2.4 A Process Cost Function

MPC is an optimization-based technique. The optimization cost function should have a sound physical system basis. To produce the controls automatically, a robust calculation algorithm is necessary, with well-understood numerical properties. Quadratic cost functions have a long and reliable tradition of use for the minimization of errors and are also used in MPC applications. A typical scalar cost function might be given as:

$$J(U) = E\left\{ \left[\sum_{j=N_1}^{N_2} [\hat{y}(t+j|t) - r(t+j)]^2 + \sum_{j=1}^{N_u} [\lambda_j \Delta u(t+j-1)]^2 \right] \bigg| t \right\}$$

where

$$\text{Output error term} = \sum_{j=N_1}^{N_2}[\hat{y}(t+j|t) - r(t+j)]^2$$

$$\text{Control energy term} = \sum_{j=1}^{N_u}[\lambda_j \Delta u(t+j-1)]^2$$

and $\hat{y}(t+j|t)$ is the predicted output j steps into the future, $r(t+j)$ is the reference signal j steps into the future, λ_j is the jth time step weighting, $\Delta u(t) = u(t) - u(t-1)$, the control at time t is given by $u(t)$, and N_1, N_2, N_u are tuning parameters specifying the output and control horizons over which the optimization is performed. As can be seen, this formulation puts a *cost* on the error caused when the system output does not track the desired reference signal and also penalizes the amount of control energy used to achieve the desired control objective. Thus the quadratic cost index has a physical process interpretation. There is also the added advantage that this type of cost function leads to well-known quadratic programming (QP) numerical optimization routines.

12.7.2.5 A Receding Horizon Control Philosophy

The predictive control problem is solved for a control time sequence $U = u(t+1)$, $u(t+2), u(t+3), \ldots u(t+N_u)$, and only the first of these control signals, $u(t+1)$, is actually implemented. The MPC formulation is then moved forward by one time step, and the optimization is repeated once more to find the control $u(t+2)$. However, on each optimization the moving *receding horizon* of some N_u controls is found.

12.7.2.6 Some MPC Tuning Parameters

To shape the response, there are various parameters in each of the different MPC methods. In the typical cost function given earlier can be found the output cost horizon parameters, N_1, N_2, and the control cost horizon, N_u. To adjust the balance between output and control cost, the parameters $\{\lambda_j\}$ are available. Large values of λ_j will ensure that control changes are expensive and a slow system response will result. Reduce the value of λ_j to make control action cheap, and then a lively control action will cause the output to be more responsive. These are simple examples of the way the responses of the MPC system output can be shaped, although it should be noted that each of the different methods has a particular set of design guidelines.

12.7.2.7 The Two Key Advantages of MPC

The first important advantage of the MPC paradigm is the ability to *anticipate* the future control actions required to achieve new output setpoints. The MPC optimization problem is fed *future* setpoint data to achieve this property. Furthermore, if a measured disturbance is also present then this information can also be fed into the formulation and feedforward action is obtained quite naturally in the same framework.

The second important advantage of the MPC paradigm is that it uses quadratic cost functions. These have a useful physical process interpretation but there is the equally important advantage that this type of cost function leads to well-known quadratic programming (QP) numerical optimization routines. Furthermore, it is but a small step to incorporate constraint handling to the QP problem; this makes the MPC methods especially powerful, giving them a capability that classical control methods do not have. Typical mathematical constraint descriptions might be as follows:

Operational Constraints
Control limits: $\quad u_{\min} \leq u(j) \leq u_{\max}$
Rate limits: $\quad \Delta u_{\min} \leq \Delta u(j) \leq \Delta u_{\max}$
Output limits: $\quad y_{\min} \leq y(j) \leq y_{\max}$

12.7.2.8 MPC Architectures

The last aspect of the MPC paradigm to be reported on in this presentation concerns the potential architectures that can be used with typical MPC algorithms. MPC is a self-contained controller algorithm and can be used in place of any forward path controller device. For example, as shown in Figure 12.9(a), the design could be used to replace the standard classical PID controller. However, this is not pursued so often: instead the PID controller as might be supplied by the process unit vendor is left intact, and *the MPC algorithm is placed in the supervisory level* to operate on the reference input in a cascade architecture as shown in Figure 12.9(b). In this way, the MPC acts as a supervisor looking after setpoint changeovers and disturbance rejection, and there is no need to interfere with, or retune, the PID designs that may have been provided by an external equipment supplier. The formulation of the MPC module requires only the model of the closed-loop low-level process unit as the system of interest. This series (and a similar parallel) architecture so often described in the industrial applications literature deserves more theoretical investigation by the research community.

12.7.2.9 Finally, the Industrial Varieties of MPC

The IDCOM routine of Richalet et al. [38] was followed by the Dynamic Matrix Control (DMC) algorithm of Cutler and Ramaker [13]. A plethora of routines with acronyms like EHAC, EPSAC and APCS followed [35]. Among the algorithms that have been developed over the years, the Generalized Predictive Control (GPC) algorithm of Clarke et al. [9] has received extensive research attention and wide industrial application. A notable recent development is the Robust Multivariable Predictive Control Technology (RMPCT) algorithm from Honeywell Hi-Spec Solutions. All these algorithms have exploited a predictor equation to unravel the future behavior of the process and a quadratic performance index which is physically justifiable and computationally convenient, leading to the all-important facility of control system constraint handling.

12.8 CONCLUSIONS

Future process control systems might either be based on major DCS/SCADA system suppliers' equipment or use very flexible, commonly available software and portable computers/workstations. In either event, the process control engineer will be able to use much more elegant and sophisticated control algorithms. In the past, the overriding

(a)
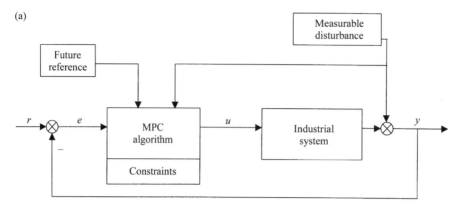

(b)
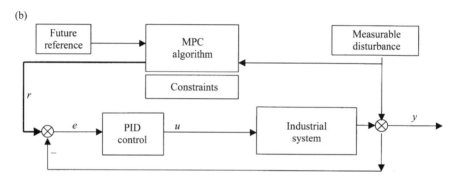

Figure 12.9 (a) Classical MPC architecture; (b) supervisory MPC architecture.

requirement was simplicity, but in the future the main imperative may be to gain a competitive advantage, which is the main driving force in most other industries. Without due diligence to the objectives of economics and effective plant operation, amusing stories of how badly tuned most of the controllers are on a given plant may coincide with less amusing stories of takeover or factory closure. There is therefore a great opportunity to exploit robust, possibly adaptive, regulators and more effective supervisory control systems, as long as they meet the essential practical requirements of being reliable, easy to commission and tune, and relatively simple to understand. Note that the underlying control theory need not be simple; it is only the ease of use that is important.

Performance monitoring techniques like statistical process control, the performance quality indices, and the new control monitoring procedures are important and valuable modern supervisory tools. These techniques have the following uses:

- Identification of good process conditions
- Detection of poor process performance
- Some fault detection capability
- On-line (nudge-nudge) optimization
- Controller tuning diagnostic applications
- Management analysis applications

For the regulating loop designs, the reliability of control is important, and the following topics should receive much more attention:

- Fault-tolerant control
- Reconfigurable control
- Reliable control
- Combined fault monitoring and control
- Limited authority adaptive control

ACKNOWLEDGMENTS

The authors would like to acknowledge invaluable discussions with industrial colleagues, James Crowe, Andrew Ogden-Swift, Andrew Riley, Mark Brewer, Claudio Brasca, and Sandro Corsi, on various aspects of process control as currently applied in industry. Academic colleagues who are thanked for their insights, thoughts, and assistance include Reza Katebi, Jacqueline Wilkie, and Andrzej Ordys.

The financial support of the European Union for the IN-CONTROL project is gratefully acknowledged.

Related Chapters

- Some control issues related to a specific process industry sector—power generation—are noted in Chapter 13.
- Chapter 16 provides an analogous overview of control systems for building automation.
- See Chapter 5 for more discussion and examples of adaptive control.
- Supervisory control in a hybrid system context is discussed in Chapter 7.

REFERENCES

[1] K. J. Åström, *Ziegler-Nichols Auto-Tuners*. Report LUTFD2/(TRFT-3167) 1-025, Department of Automatic Control, Lund Institute of Technology, S-22007, Lund 7, Sweden, 1982.

[2] K. J. Åström and T. Hägglund, *PID Controllers: Theory, Design and Tuning*. Research Triangle Park, NC. Instrument Society of America, 1995.

[3] D. P. Atherton, *Nonlinear Control Engineering: Describing Function Analysis and Design*. London, U.K.: Van Nostrand Reinhold, 1975.

[4] V. A. Bhandari, R. Paradis, and A. C. Saxena, *Using Performance Indices for Better Control*. Source unknown, ca. 1990.

[5] W. L. Bialkowski, "Dreams versus reality: A view from both sides of the gap." *Pulp and Paper Canada*, Vol. 94, no. 11, pp. 19–27, 1993.

[6] R. R. Bitmead, M. Gevers, and V. Wertz, *Adaptive Optimal Control: The Thinking Man's GPC*. Sydney, Australia: Prentice Hall, 1990.

[7] C. Brasca, V. Arcidiacono, and S. Corsi, "An adaptive excitation controller for synchronous generators: Studies and experimental results at a power station" *IEEE Conference On Control Applications*, Vancouver, Canada, 1993.

[8] E. F. Camacho and C. Bordons. *Model Predictive Control*. London: Springer-Verlag, 1999.

[9] D. W. Clarke, C. Mohtadi, and P. S. Tuffs, "Generalised Predictive Control—Part 1: The Basic Algorithm." *Automatica*, Vol. 23, no. 2, pp. 137–148, 1987.

[10] Cooper and Lybrand/CBI, *Survey of Benchmarking in the U. K.* 1993.

[11] J. Crowe and M. A. Johnson, "On a new process identification method and its application to industrial control." Submitted for publication *Proc. IEE*, 1998.

[12] J. Crowe and M. A. Johnson, *GB Patent Application No. 9802358.3: Phase Locked Loop Identification*, 1998.

[13] C. R. Cutler and B. L. Ramaker, "Dynamic matrix control—A computer control algorithm." *Proc. JACC*, San Francisco, 1980.

[14] S. B. Dolins and J. D. Reese, "A curve interpretation and diagnostic technique for industrial processes." *IEEE Trans. Ind. Applics.* Vol. 28, no. 1, pp. 261–267, February 1992.

[15] G. Goodwin and S. Crisafulli, *Method and Apparatus for Tuning of PID Controllers.* CICS Automation Report No. 96009 and Provisional Patent Application, University of Newcastle, Australia, 1996.

[16] M. J. Grimble, *Robust Industrial Control.* Hemel Hempstead: Prentice Hall, 1994.

[17] M. J. Grimble, *Restricted Structure LQG Optimal Control for Continuous-Time Systems.* Report ICC/150, Industrial Control Centre, Glasgow, U.K., 1998.

[18] M. J. Grimble, and M. A. Johnson, *Optimal Control and Stochastic Estimation Theory. Vol. 1: Deterministic Systems. Vol. 2: Stochastic Systems.* Chichester, U.K.: John Wiley & Sons, 1988.

[19] T. Hägglund and K. J. Åström, *U.S. Patent 4,549,123: Method and an Apparatus in Tuning a PID Regulator.* October 22, 1985.

[20] H. J. Harrington, *Business Process Improvement.* New York: McGraw-Hill, 1991.

[21] T. J. Harris. "Assessment of control loop performance." *Can. J. Chem. Engrg.*, Vol. 67, pp. 856–861, October 1989.

[22] M. A. Hersh and M. A. Johnson, "A study of advanced control systems in the work place." *Control Engineering Practice*, Vol. 5, no. 6, pp. 771–778, June 1997.

[23] I. M. Horowitz, *Synthesis of Feedback Systems.* New York: Academic Press, 1963.

[24] *Honeywell Hi-Spec Solutions Newsletter*, 3rd Quarter. Minneapolis, MN: Honeywell Inc., 1998.

[25] M. A. Johnson and A. S. Abdelali, "Process control PID rules: Methods for a comparative study of recent developments." *Procs. Advances in Process Control 5*, pp. 121–130, Swansea, U.K., 1998.

[26] R. E. Kalman, "When is a linear control system optimal?" *Trans. ASME, Journal of Basic Engineering*, Vol. 86, pp. 51–60, 1964.

[27] M. R. Katebi and M. A. Johnson, "Predictive control design for large-scale systems." *Automatica*, Vol. 33, no. 3, pp. 421–425, 1996.

[28] *LabVIEW Student Edition.* Version 3.1. Englewood Cliffs, NJ: Prentice Hall, 1996.

[29] Z. J. Lu, *Controllers That Determine Optimal Tuning Parameters for Use in Process Control Systems and Methods of Operating the Same.* Patent Applic., No. 120-17371, Honeywell Inc., Minneapolis, MN, 1998.

[30] R. S. Luo, J. Qin, and D. Chen. "A new approach to closed loop autotuning for PID controllers" *Proc. American Control Conf.*, Philadelphia, PA, 1998.

[31] C. L. Mamzic and T. W. Tucker, "Incorporating statistical process control within a distributed control systems." 43rd Annual Symposium: *Instrumentation for the Process Industries*, Tech. Paper 3912, Texas, 1988.

[32] D. C. Montgomery, *Statistical Quality Control.* New York: John Wiley and Sons, 1991.

[33] A. Ogden-Swift, "Maximising process profit in real time." *Int. J. Hydrocarbon Engineering*, November 1997.

[34] R. Ortega, G. Escobar, and F. Garcia, "To tune or not to tune? A monitoring procedure to decide." *Automatica*, Vol. 28, no. 1, pp. 179–184, 1982.

References

[35] A. Pike, M. J. Grimble, M. A. Johnson, A. W. Ordys, and S. Shakoor, "Predictive Control." In W. S. Levine (Ed.), *The Controls Handbook*. Boca Raton, FL: CRC Press, 1995.

[36] A. Pike and M. A. Johnson, "Simulation tools for the 90's." *Measurement and Control*, Vol. 27, pp. 185–195, July–August 1994.

[37] D. Popovic, and V. P. Bhatkar, *Computer Control for Industrial Automation*. New York: Marcel Dekker, 1990.

[38] J. Richalet, A. Rault, L. Testaud, and J. Papon, "Model predictive heuristic control: Applications to industrial processes." *Automatica*, Vol. 14, pp. 413–428, 1978.

[39] J. Welsh and G. Goodwin. *A Novel Mechanism for Autotuning Based on Quantisation, Proc. IFAC World Congress*, Beijing, 1999.

[40] G. Zames, *Feedback, Optimal Sensitivity and Plant Uncertainty via Multiplicative Semi-Norms. Proc. IFAC World Congress*, Kyoto, Japan, pp. 74–78, 1981.

[41] J. G. Ziegler and N. B. Nichols, "Optimum settings for automatic controllers." *Trans. ASME*, Vol. 64, pp. 759–768, November 1942.

Chapter 13
POWER SYSTEM CONTROL AND ESTIMATION IN A COMPETITIVE ENVIRONMENT

Christopher L. DeMarco

Editor's Summary

Electric power grids are machines that span continents, with dynamics that are correspondingly complex. Further complications to the operation of power networks have now arisen in many parts of the world as a consequence of the deregulation and competition underway in the industry. Vertical integration, centralized control, and conservative operation in the past helped ensure stability and reliability in power supply. In a competitive, deregulated environment, in which none of these strategies is feasible, assuring reliability is considerably more difficult.

The synchronous ac power grid is a unique entity in that the rotating generators connected to the grid are all dynamically coupled, regardless of geographical or topological separation. Imbalance of generation and consumption has instaneous ripple effects on all generators and loads. In the past, with conservative approaches to power plant operation, power grid loading, and the overall network control, power system behavior was sufficiently well behaved that linear control design proved reasonably effective. Today, nonlinear, decentralized control and optimization techniques are needed.

Several new technologies are being explored in this context. High-sampling rate sensors and time synchronization via the global positioning system (GPS) may permit state estimations of greater accuracy. Flexible ac transmission devices (FACTS), especially within the so-called universal power flow controllers, promise substantially greater control over the transmission systems, permitting network operators to route power independently of network impedance characteristics. New algorithmic technologies are also under active investigation, including agent-based modeling and optimization and randomized algorithms.

Christopher DeMarco is a professor in the Department of Electrical and Computer Engineering at the University of Wisconsin, Madison. He is a past chair of the IEEE-CSS Technical Committee on the Control of Power Systems.

13.1 INTRODUCTION: ELECTRIC POWER SYSTEM STRUCTURE AND FORCES FOR CHANGE

Electric power networks are among the largest human engineered systems. They display significant dynamic coupling on a continental scale, and they depend on feedback control to achieve reliable provision of the critical commodity of electric power.

Section 13.1 Introduction: Electric Power System Structure and Forces for Change

Given the importance of electric power to modern industrial societies, and the range of practically important control challenges they present, it is not surprising that such systems have been a key application area for control analysis and design for the past century.

This long history and naturally conservative engineering and control philosophies in a regulated industry serving a critical social need have contributed to a perception that the key control problems of electric power systems have been thoroughly solved. To some degree, this perception *was* correct. As more detailed discussion below will illustrate, when well-engineered and operated conservatively (or, as some economists would argue, when historically overbuilt to the point of "gold plating"), the electromechanical dynamics of power networks can behave fairly linearly. In this context, long-standing linear control design philosophies prove reasonably effective at maintaining reliable service and desirable dynamic performance characteristics. With this (usually) linear behavior, there exists a separation of time scales among various dynamic phenomena, which facilitated the evolution of a naturally structured control hierarchy. In particular, power networks have traditionally been designed with fast time scale, closed-loop controllers associated with individual pieces of equipment, using only local measurements and control inputs. More detailed discussion later in this chapter will elaborate on this point, but it is important to note here that synchronous generators constitute the most significant class of equipment on which this type of continuously acting, local feedback appears. On a slower time scale, local control signals are augmented by periodic updates of setpoint inputs. Setpoint updates are typically generated open loop, through the action of predictive optimization calculations, or by human operators. Such setpoint updates are issued as commands from a regional control center, acting to meet the operating objectives of the overall network.

Although the physical characteristics of the network and generation equipment lend themselves well to this type of control structure, it is also important to recognize that this control structure was closely tied to the institutional structure of the organizations that operated the systems. A convincing elaboration of this argument may be found in [1]; elements of the argument will be briefly highlighted here. In the United States, electric utilities were regional regulated monopolies, so that a single business entity (or a close alliance of companies, in the case of regional control centers) was responsible for local equipment and its controllers, the centralized command of setpoints, and the network reconfiguration (e.g., switched capacitor banks, transmission line switching). In other parts of the world, where electric power networks were often operated as state-owned entities, the degree of institutional integration of responsibility between local equipment control design and centralized network control and optimization was even tighter.

The description above is quite intentionally phrased in a mixture of past and present tense, because the organizational structures of power networks around the world, including that of the United States, are in transition. In the United States, the electric power system is moving to an era in which its operation will be governed by a very different regulatory and institutional structure. The regional monopolies held by electric utility companies, with vertical integration that encompasses ownership and control of local generation equipment up through the regional system operation centers, are being restructured to allow for competitive provision of electric generation. The electric power industry has not been subject to the type of judicially mandated divestiture that occurred in the U.S. telecommunications industry in the 1980s. Rather, restructuring of the power industry is coming about largely through legislative and associated regulatory actions, elements of which may be traced back to the Public Utility Regulatory Policies Act of 1978 (PURPA). Further steps toward a competitive structure came with the Energy Policy Act of 1992. This act laid the groundwork for the most direct impetus for change, in two landmark orders from the Federal Energy Regulatory Commission (FERC), numbers 888 and 889 [2, 3]. These 1996 rulings impose a number of key requirements, with the goal of opening the U.S. electric power system to competitive provision of generation. In the context of grid control, perhaps the most critical element of these rulings is the requirement for functional separation of generation activities from the central control of the transmission grid. Given the historically tight integration of these activities in the old vertically integrated utility model, this requirement indirectly mandates significant changes in the structure of control in the U.S. power grid.

The policy and technological motivations for the changes imposed in orders 888 and 889 are many-faceted, and still provide debate. In brief, the philosophical goal is to create a competitive, (near) real-time market for the provision of electric power. Although generated power is to be a competitively provided commodity, the operation of the transmission and distribution grid is to remain a regulated monopoly activity. These changes in institutional structure are proceeding rapidly and likely irreversibly. Such changes have radically altered the assumptions under which the traditional, hierarchical control structure evolved, breaking the tight integration between systemwide, centralized elements of grid optimization and control, and the design and implementation of local feedback controls on generators. Equally important, with production no longer controlled by a regional regulated monopoly, the patterns of power transmission across North America show signs of becoming more volatile, and portions of the transmission grid are periodically becoming more heavily utilized. Significant to the issue of control design is the fact that the physical dynamics to be controlled become progressively more nonlinear as the transmission system is more heavily loaded.

The dramatic shift in institutional structure and (potentially) in operating characteristics for the North American power grid, along with a range of new technologies that can significantly alter the system dynamic characteristics, force a reexamination of control analysis and design for electric power networks. In this context, the goals of this chapter are fourfold. First, to place the control issues in context, we provide a brief tutorial on the dynamics of synchronous electric power grids, the typical control objectives and available control actions, and the historic structure of hierarchical control in North America. Second, we elaborate on the above description of institutional changes in the organizational structure of the system, the proposals for restructuring responsibility for control, and for competitive provision of "ancillary services." (Several control objectives are closely tied to provision of these so-called ancillary services.) Third, we examine technological advances that have implications for electric grid control. These three elements then serve as the foundation for the fourth aspect of this chapter, which provides a perspective on the new challenges in control analysis and design that are created by these many technological and institutional changes in the power industry.

13.2 POWER SYSTEM DYNAMICS AND THE HISTORICAL STRUCTURE OF GRID CONTROL

Key to understanding the control issues in a synchronous ac power grid are two cornerstones. First, one must understand the control objectives that have traditionally been assumed to represent customer and societal needs. Second, one must grasp the dynamics of synchronous rotating generators interconnected in an ac grid, the control actions available at each machine, and the control actions at other devices that constitute elements of the grid.

13.2.1 Control Objectives in Power Systems

At a basic level, the control objectives in a power grid follow from desirable operating characteristics that customers often take for granted. Consider the customer connection point—for example, for a residential customer, the service box connection, or for a large industrial customer, a substation. Whether or not this objective is consciously formulated, a residential customer desires that, at the connection point, the electric power network behave like an ideal sinusoidal voltage source at 60 Hz, 120 volts rms magnitude.[1] Clearly, the physical reality is much more complex; but in normal operation, U.S. utilities often come extraordinarily close to meeting this ideal. Assuming transformers are moderately loaded to avoid saturation effects, transmission and distribution networks are largely linear in their voltage/current behavior.[2] Though somewhat less so, the dominant elements of loads can also be assumed to possess fairly linear voltage/current characteristics. (The exceptions are loads with power electronic controls, such as variable speed drives for induction motors, and electronic ballasts for fluorescent lighting.) As long as the sources in the network (generators) produce purely sinusoidal voltages of like frequency, the linearity of all connected components ensures that all voltages and currents in the network are sinusoidal. Hence the simplified control objective becomes one of maintaining all generators close to the target frequency of 60 Hz and maintaining voltage magnitudes in the grid so that customers receive 120 volts rms, or the rated voltage magnitude appropriate to their consumption level. These represent control objectives on a relatively fast time scale, from tens of milliseconds to minutes.

On a longer time scale, the other key customer desire is that electric power be economically delivered. To meet this objective, the traditional regulated utility operated on a cost-minimizing philosophy. Construction and operation of generating plants historically represented the dominant costs in electric power provision. Once plants are

---

[1] Historically, this assumption was crucial to allow cost-effective standardization of end-use consumer appliances and other electric energy consuming devices. The increasing penetration of electronics and computers as significant load components suggests the need to reexamine this control objective. Modern electronic power supplies can be economically constructed to be much less sensitive to the steady-state frequency and voltage magnitude of the ac source, relative to, for example, motor-driven loads. However, many electronic devices are much more sensitive than motors to fast time scale voltage dips or spikes in the (otherwise) sinusoidal supply voltage.

[2] As will be discussed further, the dynamic behavior of a power network is strongly influenced by the relation between the mechanical shaft position of generators and the electrical power flow between nodes of the network. This relation can be highly nonlinear, even when the underlying voltage versus current behavior in the network is perfectly linear.

built and paying off investment becomes a fixed cost, fuel costs become the dominant variable cost. Clearly, the rate of fuel consumption, and hence the cost of operation per hour, is a function of a generating plant's electrical power output level. To be able to meet peak load reliably, the total available capacity of generation must exceed the level of consumption at peak time periods. This implies that there will be flexibility in the choice of electrical power output levels among various generating plants in a large network. Operating cost minimization then becomes a two-stage optimization problem. Suppose that the set of generators available at an instant of time is known. Then one has the problem of determining the exact power output level for each generator that is locally feasible for the equipment, such that the sum of the power outputs meets total customer load plus transmission and distribution losses, while minimizing variable costs associated with each machine's power output. At a higher level, one has the problem of determining which set of generators should be "on-line," ready to deliver power to the grid. The latter problem is a challenging optimization problem. (Indeed, it is formally "NP-hard" when treated with realistic modeling of intertemporal constraints and non-convex cost functions.)

The lower level problem, that of tracking load with available generators while minimizing variable operating costs, can be solved to reasonable accuracy with simple computations. The details of these optimization methods are not of concern here, but their result has relevance for our control problem. In particular, the result of this optimization problem determines the generator power setpoints alluded to earlier. In a vertically integrated monopoly, the cost minimization problem (or some heuristic approximation thereof) would be solved by the grid operators. This calculation would be performed periodically at a regional control center, where computation and monitoring functions are concentrated. The result provides the setpoints for power output that each generator in the regional control area will be asked to hold over some time interval, until the next update. However, these setpoints are calculated using a prediction of instantaneous customer load and network losses. If these setpoints were rigidly maintained, they would not ensure exact balance of total generation power to total consumed power. Understanding how instantaneous power balance *is* maintained, as well as understanding the generators' role in voltage magnitude control, require examination of coupled synchronous machine dynamics. To add further challenge to the control problem, the reader can appreciate that the net flow of power between regional control areas is also a quantity of interest, which control centers would also like to regulate. Hence our starting problem of controlling generators to economically maintain sinusoidal voltage frequency and magnitude takes on several additional facets. The historic solution to the problem in the United States couples local governor feedback, with a slow time scale regional feedback control, with the open-loop update of economically attractive target output levels. The regional feedback control is generally known by the acronym AGC (automatic generation control). An overview is provided here; for more detail see [4, 5].

13.2.2 Synchronous Generator Dynamics: A Brief Tutorial

As noted earlier, appreciation of the control problems in present and future power networks requires basic understanding of coupled synchronous generator dynamics. This section provides a brief tutorial on the subject. An excellent tutorial that also characterizes the typical oscillatory electromechanical modes observed in a power network is found in [6]. The reader should note that among the future technological trends with control impact is the increasing penetration of energy sources that are not synchronous, such as photovoltaic arrays, fuel cells, and induction generator-based wind turbines. However, rotating three-phase synchronous generators constitute the *overwhelming* majority of electric energy production around the world, because of their excellent efficiency, their reliability, *and* their attractive control characteristics. The huge capital investment in these machines strongly suggests that they will continue to operate as a significant source of power for the foreseeable future, even if their ownership, institutional role, and control schemes change radically.

The physical structure of a synchronous generator begins from a single electrical coil on a shaft, wound to produce magnetic field perpendicular to the shaft's axis of rotation. This coil is known as the field winding. This field winding spins relative to three stationary coils ("stator coils" in the terminology of machine analysis), each wound to maximally link the magnetic field when the position of the rotating coil aligns with the stator coil. The rotating coil has a dc excitation applied. For three-phase sinusoidal operation, the three stator coils are wound and positioned so that as the shaft rotates, the field from the rotating coil produces a sinusoidally varying flux linkage. Hence for typical terminal connections, it produces a sinusoidally varying voltage at the machine's terminals. While a detailed analysis of this structure and its variations is beyond this presentation (and is the subject of countless texts), a few key observations are useful to appreciate the control problem. The excitation applied to the rotating coil controls the strength of the magnetic field produced. This excitation need not be purely constant dc but rather can be varied for control purposes. The variation can typically be assumed slow relative to the 60-Hz variation of the stator coils' quantities, providing one example of the many ways in which separation of time scales is exploited. This field winding consumes a relatively small amount of power compared to the electrical output of the machine, so achieving controllable excitation is feasible. The primary impact of varying this excitation, and hence the magnetic field strength, is to vary the magnitude of the sinusoidal voltages produced in the stationary coils. A secondary impact is to vary the power output of the machine, and this secondary effect can be exploited to improve the dynamic response of the machine under transient conditions. Hence, the voltage applied to the field winding is one key input to our synchronous machine.

Based on the structure we have described, consideration of magnetic field coupling should suggest to the reader that the frequency of the sinusoidal voltages at the terminals of the generator will be proportional to the speed of the rotation shaft. Design variations that "repeat" interconnected stator windings around the perimeter can be used to obtain different constants of proportionality between mechanical speed and electrical frequency and thereby allow for lower rotating speeds in hydro-driven turbines. With suitable normalization, mechanical speed and electrical frequency can be treated interchangeably.[3] This is the one subtle but critical observation that the reader *must* appreciate to understand interconnected synchronous machine dynamics.

---

[3] As one examines detailed, practical implementation of controllers, some care must be exercised in treating mechanical angle and electrical phase angle interchangeably if a broad bandwidth feedback loop is considered. As will be noted in later discussions of power system stabilizer design, a broad-bandwidth controller can excite torsional flexing and relative angular motion between different portions of the turbine/generator set shaft; a single mechanical angle no longer fully describes behavior.

To illustrate this association of mechanical behavior and electrical behavior, imagine making two types of measurements on an operating generator. Consider a machine whose design produces 60-Hz terminal voltages for a shaft speed of 1800 rpm. Further suppose the shaft has a reference marker painted on it, with the shaft illuminated by a stroboscopic lamp flashing 1800 times per minute. A human observer would see the reference line in a stationary position when the machine was rotating at constant rated speed. If the machine briefly accelerated and then returned to rated speed, the reference line would appear to advance on the shaft and then settle to a new fixed position. As an analogous electrical observation, consider measuring sinusoidal terminal voltages with an oscilloscope (though the author advises against clipping the scope probes to a 30-kV busbar). Further assume that the scope is fed an external, 60-Hz triggering signal. When operating at rated speed and frequency, the measured voltage would appear as a fixed sinusoid on the scope display. If, again, the machine briefly accelerated and then returned to rated speed, the sinusoidal wave shape would appear to have shifted its phase on the horizontal display axis. Our first critical observation is as follows: The mechanical angle relative to the "synchronous reference," as indicated by the position of the strobe illuminated mark on the shaft, is directly proportional to the phase shift of the observed sinusoidal voltage. The significance of this point will become apparent momentarily.

To describe the interaction of generators interconnected by a transmission network, one can assume (to good approximation) that the voltage versus current behavior of the network is linear. In normal operation, there exist 60-Hz sinusoidal voltages of nearly equal magnitude at each end of the transmission line. (This condition is imposed by the requirement of holding customer voltage magnitude close to rated values.) The transmission line will appear as a series impedance between these voltages. The current flow that results will be a sinusoid of like frequency, and its magnitude is determined *by the phase difference between the sinusoidal voltages at each end*. Next, consider the flow of average power on this line. A straightforward calculation reveals that the flow of average *power* from one side of the line to the other is proportional to the sine of the voltage phase angle difference. This sinusoidal dependence of power exchange on phase angle differences is the fundamental nonlinearity in the electromechanical dynamics of power systems.

The last step in this analysis relates to the mechanical behavior of rotating generators, which are subject to Newton's law in rotational form. In particular, the rotational acceleration of a machine is proportional to the net torque acting on its shaft. The mechanical torque from a prime mover (e.g., turbine) tends to accelerate the machine. The fact that we are removing electrical power from the machine implies that the magnetic fields interacting between the field winding and stator coils must produce a torque that opposes motion, the so-called electrical torque. Ignoring internal losses in the generator, we find that these torques must balance to zero when the applied mechanical shaft power equals the electrical power delivered from the machine to the network. When applied mechanical power exceeds delivered electrical power, the machine accelerates; conversely, when electrical power delivered is larger, the machine decelerates.

In the strobe/oscilloscope experiment above, we noted that when the machine accelerates, its electrical phase angle advances. This creates a larger phase angle difference relative to the other end of the transmission line(s) connected to that machine, causing more electrical power to be drawn from the machine. This excess electrical power drawn creates a "restoring" torque that tends to decelerate the machine, bringing it back toward equilibrium speed. In a system of many interconnected generators, the question of whether the system recovers to a stable (and acceptable) equilibrium following a disturbance is a critical one; such studies are termed "transient stability analysis." Given the sinusoidal nature of the power flow nonlinearity, it should not be surprising that there exist many plausible disturbance scenarios in which the power system does not return to an acceptable equilibrium. Moreover, individual pieces of equipment are protected by circuit breakers that disconnect from the grid when voltages, currents, or mechanical speed/acceleration exceed safe limits. Hence the typical blackout scenario is one of a large transient in system state variables following a disturbance (e.g., a lightning strike to a transmission line), followed by cascading tripping of protective breakers until a significant portion of a grid is deenergized. The dynamic performance of control systems is critical in this process, for these control systems influence the magnitude of state variable excursions and thereby influence whether or not protective relays act to disconnect a particular piece of equipment in a given disturbance scenario. In power system controller design, dynamic performance under transient conditions must be carefully considered along with a controller's contribution to steady-state regulation about the normal operating point.

13.2.3 Grid Frequency Regulation

With this dynamics description, we return to the issue of the use of generators for frequency control. The first observation is that an interconnected grid is truly in equilibrium only if all generators are at the same frequency and that at such an equilibrium there must be a systemwide balance between generated power and load consumption plus losses. Any mismatch in power production relative to power consumption drives a change in speed, and hence instantaneous frequency, at one or more generators. Hence, for a system at equilibrium, or varying quasistatically, frequency serves as a systemwide, "shared" signal that indicates the relative balance between total generation power and total consumption. In a synchronously connected grid, frequency decreases when total power consumption exceeds total production, and it increases when production exceeds consumption. It is this inherent feature of the dynamics that allows electric power grids to maintain systemwide balance without instantaneous measurement of power consumption and production at all points.

To complete this overview of electromechanical dynamics, it is important to recognize that frequency dependence of power consumption in some loads, as well as small rotational losses at generators, creates a positive, roughly linear correlation of power consumption to frequency deviation. This is a natural restorative effect that provides damping, and it can allow the system to "find" a new equilibrium when there are minor variations in frequency-independent components of load consumption. For example, if a frequency-independent component of load consumption (which may be viewed as an exogenous input) were to increase slightly, and no controller acted to change the mechanical power feeding generators, the system frequency would gradually decrease until a new equilibrium was reached at which the decrease in frequency-dependent load (and losses) balanced the original increase in the frequency-independent load.

The natural damping effect of the frequency-dependent load is small, and long-term system frequency variations would be unacceptably large if this were the only corrective mechanism. Therefore, as a first step toward frequency correction and power balance, consider a local control loop that dictates incremental changes in mechanical power from the prime mover (e.g., turbine), based on local measurements of that machine's mechanical speed (proportional to its electrical frequency). This is typically termed the speed governor loop. It is important to recognize that the mechanical power command signal produced by the governor loop may not be the only signal contributing the prime mover power command. The governor loop operates with relatively broad bandwidth; other signals contributing to mechanical power command typically arise from slower control loops or from periodically updated open-loop setpoint commands. One simple rule related to governor control design was learned early in the history of interconnected grid operation: Local governor control for generators should not strive for zero steady-state error in frequency on multiple machines. The smallest difference between setpoint signals, which will be widely separated geographically, creates a dynamic system whose steady state does not display a uniform frequency. For example, with multiple integral control loops acting, often a steady-state oscillation in system frequency, a "hunting mode," results. To avoid this, governor feedback is typically dominated by a simple proportional term. The gain constant of the proportional feedback is inversely specified as normalized constant, the percentage "droop." Droop describes the percentage change in frequency that, acting through this proportional feedback, would yield a commanded change in mechanical power equal to the rated power of the generator. The dynamics of the loop are complicated by the fact that there is a nontrivial dynamic transfer characteristic for the prime mover, relating commanded change in mechanical power to actual mechanical power achieved at the shaft. Moreover, given the natural load damping effect described here, governor loops often have a small, intentional deadband. The design philosophy here is to allow the system to find a new equilibrium without any change in prime mover power outputs if the resulting deviation in steady-state frequency is sufficiently small. For a control engineer examining the power system, it is important to remember that costs of mechanical wear are associated with continuous governor action that are hard to quantify, yet not insignificant. One may willingly sacrifice some control performance in exchange for fewer control actions.

For the next higher level in the generation control hierarchy, it is useful first to consider a simple approach to systemwide frequency correction, recognizing that the proportional control of the local governor loops alone allows steady-state frequency error, away from the desired 60-Hz setpoint. In this simple scheme, a single "master machine" has a small gain, integral control term added to its governor control loop, so that this machine controls to zero steady-state frequency error. The nature of the interconnected dynamics then ensures that this equilibrium frequency is imposed on the whole interconnected area. However, while conceptually useful for illustration, this simple scheme is not practical for the large synchronous interconnections that exist in North America.

As the size of synchronous interconnected regions grew in the United States, it became clear that assigning frequency control to a single master machine was infeasible. For a long period, there was no closed-loop system to ensure systemwide balance between power consumption and production, and with it, zero steady-state frequency error. Instead, manual control of power setpoints was exercised in such a way as to keep average frequency error near zero [5]. However, in the 1960s there developed in the United States an approach toward automating systemwide frequency correction and power balance. Although the term does not have a unique definition, the family of control techniques developed generally comes under the title of Automatic Generation Control, or AGC. Extending from the simple master machine concept, it is important to recognize that only a subset of generators in the system needs to participate in AGC; that is, only a subset of machines have supplementary signals added to the prime mover power command. These supplementary signals are not local but often are computed centrally, for a portion of the grid and a corresponding set of generators that lies within a defined "control area." Currently, there exist 136 control areas within the North American grid. These control areas are administratively defined by the North American Electric Reliability Council (NERC).[4] Physically, they represent disjoint subsets of the North American transmission grid, whose union covers (essentially) the entire grid.

Transmission lines that connect between control areas, known as tie-lines, are typically required to have measurement devices that allow monitoring of the flow of power on these lines. These tie lines are operated with agreed-upon schedules that dictate the desired net power flow between any connected pair of control areas; this is known as an "interchange schedule." The interchange schedules provide a setpoint for the measured output quantity of net flow of power on the tie lines between two areas. Regulating operation at or near these setpoint values becomes an added control objective for the AGC level of generator control. The approach employed allows a tradeoff between the objective of maintaining systemwide power balance and frequency regulation versus this second objective of maintaining tie line flow. Although variations exist, the general approach is to construct an indicator output variable within each control area, known as the Area Control Error, or ACE. The ACE is simply a weighted sum of instantaneous frequency deviation within the area (typically measured at an individual generator or another preselected reference point), and the net tie line flow errors for each outside control area to which the control area computing ACE is connected. Then, for those generators within the control area that are participating in AGC, a supplementary control signal is added to their power command input. This supplementary AGC signal is a weighted integral of the ACE signal for that area. Since this is a pure integral feedback (albeit with small gain), one would expect the equilibrium value of the ACE signal to be driven to zero. However, since there are exogenous, time-varying inputs (customer power demand) that also drive this system, it does not display true equilibrium behavior on the time scale for which the AGC has significant effect (a time scale of minutes). Instead, it tends to display a stochastic variation about a zero value. The control objective for AGC, as administratively monitored by NERC, becomes one of keeping the maximum excursion of the ACE signal within specified bounds and ensuring that it crosses through a zero value periodically. (For a more detailed discussion, see [7].) Being a weighted sum of the two quantities, a zero value of the ACE signal does not precisely guarantee zero tie line error, or zero frequency error. In practice, the quality of frequency regulation in North America is extraordinarily good; NERC publishes average frequency deviations on a monthly basis, with typical values less than ±0.003 Hz. Hence a zero ACE signal does indicate that tie line flow deviations are close to zero. Historically, the integral of tie line inter-

---

[4] For a wealth of information relating to the administration, operation, and historic performance of the North American power grid, as well as standards for its control, NERC maintains an extensive set of resources on the Internet at www.nerc.com.

change errors (which indicate net energy deviation from scheduled exchange) was monitored, and an after-the-fact accounting was done to settle the financial impact of this "inadvertent interchange" of energy between control areas. Not surprisingly, this financial use of the ACE construct is undergoing scrutiny and modification in the transition to a competitive environment.

This review of generation control is necessarily somewhat superficial; the interested reader is strongly advised to seek more detailed accounts. However, this review and historical perspective are intended to emphasize the significant institutional structure and history that underlie current control practice. The actual control algorithms are relatively simple, but the dynamics of the physical system they act upon, and the institutional arrangements that determine the control objectives and possibilities, are exceedingly complex. To understand the challenges and opportunities created for new control approaches, it is important to understand the nature of this historical, and largely still current, control structure.

13.2.4 Stability-Enhancing Controls in Power Systems

Before leaving this tutorial on control in power systems, we must touch on several topics beyond that of generator governor control and AGC. Stepping back and taking a somewhat abstract view of the power system as a dynamical system, one must recognize that the primary objective of the control loops described here is to allow states of the power system to quasistatically "track" desirable operating points in response to the (mostly) slowly varying exogenous input of customer load demand.[5] Hence these loops are typically designed by treating the slowly varying inputs as a sequence of frozen "snapshots," at which the input is viewed as constant and the system model becomes autonomous. The control design goal becomes one of ensuring that the equilibrium points corresponding to these snapshots have desirable properties in terms of the values that state variables and outputs take at these equilibria. For example, one wants operating points that yield minimum fuel cost at generators, that keep frequency at or near 60 Hz, that maintain transmission line flows within specified limits or near setpoint values, and so on. But none of these considers the dynamic response of the system in any significant detail. An uninitiated reader may rightly ask whether any part of the control design within a power system is concerned with dynamic response characteristics and stability properties. The answer is most certainly yes but perhaps to a lesser extent than one might expect. In a sense, the control actions that are available to respond on a fast time scale, effective in improving dynamic response characteristics, are somewhat limited. This lessens (but certainly does not eliminate) the interest in control *design* for improving dynamic response characteristics. Conversely, the power system is sufficiently nonlinear that its dynamic response characteristics can change radically with the operating point.

---

[5] A more realistic model of aggregate load variation measured at a distribution substation starts from an underlying, slowly varying component that is highly correlated to weather; given weather data, this underlying component can be quite accurately predicted on an hourly basis. Superimposed on this underlying component are instantaneous fluctuations that arise from thousands of individual pieces of equipment switched on and off under customer control. This yields an additional randomly varying, zero mean stochastic component that might typically have a variance on the order of a few percent of the total load at the substation.

As noted previously, one may roughly say that the system response is worsened, and stability margins are compromised, as the network is progressively more heavily loaded. Hence there has historically been considerable interest in analysis techniques that characterize the relative stability of the system based on changes in operating point. Often, operators improve the dynamic response and stability characteristics of the power system by moving to a new operating point, either by quasistatically "steering" to such a point or by discrete switching of components connected to the grid. This class of corrective controls to steer the operating point is only loosely closed loop and often involves significant intervention on the part of human operators at regional control centers. Faster time scale feedback control algorithms[6] were typically kept relatively simple and designed for robustness in their dynamic response characteristics over a wide range of operating points rather than choosing a higher performance controller that might adapt to system operating conditions. This choice will bear re-examination in a more competitive business environment, where limits on utilization of the grid to maintain conservative stability margins will be weighed against the denial of profitable opportunities to competitors wishing to ship power.

Despite the preceding comments, some control actions are available that can act to improve dynamic response characteristics and stability margins,[7] and design of feedback controllers to achieve these goals is an important topic. Moreover, as will be described, a number of promising new technologies offer improved opportunities for stability enhancement. In reviewing existing control techniques, among the most notable feedback design problems for stability enhancement is that of power system stabilizers (PSS). This seemingly generic term has a very specific definition in the power field. It refers to a local feedback that measures the output variable of generator speed or frequency deviation, and from this signal produces an additive input to vary the excitation of the generator field winding. Being a purely electrical control input, variation of field-winding excitation does not carry with it the penalties of mechanical wear and tear that come with varying prime mover power. Also, provided the source of excitation voltage is sufficiently responsive, the speed of response of such a system can be quite good. Many modern exciters use a controllable solid-state rectifier, taking energy from the generator's ac terminals, and hence *are* responsive on fast time scales.

---

[6] The process of steering to a new operating point could itself be implemented via feedback, but the analysis of a model that simultaneously captures the fast time scale dynamics being stabilized and the slow time scale quasistatic adjustment of operating point is daunting. Hence the steering of operating point for improved stability is generally not treated as a feedback design problem.

[7] The comment above begs the question of precisely defining what is meant by "improve dynamic response characteristics and stability margins." While this classification greatly oversimplifies, the problem of improving dynamic response is approached by linearizing the power system model about a nominal operating point, or family of operating points, and applying any of a range of linear system analysis tools to quantify the quality of performance. Such linearized, local analysis about an operating point(s) is typically termed "small disturbance" stability analysis in the power system. Characterizing stability margins is typically approached using a full nonlinear model. The basin of attraction for a power system operating point is never the entire state space, so determining whether a disturbance or change in configuration yields an initial condition outside the domain of attraction of the desired equilibrium is an important problem. The typical approach starts from a preselected list of plausible large disturbances; this is termed a contingency list. The stability margin of the system is judged adequate if analysis (usually time domain simulation of a detailed dynamic model) shows that the system returns to an acceptable, stable operating point in all of the selected disturbance scenarios.

While the earlier overview of power system dynamics suggested that mechanical speed and electrical frequency could be treated interchangeably, in the context of PSS analysis more care must be exercised. Large generator/turbine shafts can have significant torsional dynamics, and broad bandwidth measurement of rotational speed will reflect torsional oscillations of the shaft at the point of measurement. For an in-depth discussion of PSS and the current state of the art in design methods, the text [8] is recommended; the (slightly dated) bibliography of [9] provides a sampling of research in improved PSS design methods.

Among other technologies that offer the possibility for fast time scale control are those based on power electronics. Here a "fast" time scale would refer to control actions with significant impact on the order of a few cycles of the 60-Hz sinusoidal frequency. The dominant traditional applications in this category are direct current transmission lines and controllable capacitive devices known as static Var compensators. In the case of dc transmission, controllable rectifiers and inverters constructed with high-power thyristors offer the opportunity to control steady-state current or power flow on the line, relatively independent of ac voltage magnitudes and phase angles on sending or receiving ends. Moreover, the ability to modulate the flow of power allows opportunities to influence the dynamic response characteristics of the network. At present, the number of dc transmission installations in the United States is relatively small, and this opportunity is exploited in only a few locations. However, there has been recent interest in developing much lower cost implementations of dc transmission in order to allow greater control of steady-state flow and to exploit market opportunities in a competitive system. If this so-called dc transmission lite technology were to see widespread adoption in the U.S. grid, it would bring with it much greater opportunity for (and perhaps necessity for) modulation control to improve dynamic response characteristics.

With regard to static Var compensators, the opportunities for dynamic stability enhancement are conceptually similar to those in dc transmission, but the control action available (variation of effective capacitance connected from a bus to ground) tends to have less impact on dynamic performance. The motivation to use these devices primarily for enhancement of the steady-state operating condition has been strong. To understand this from a control perspective, consider the dynamics of a power network, as represented in a linearization about a typical operating point. One commonly has a subset of the eigenvalues of the system which are lightly damped complex conjugate pairs. Physically, these modes are associated with behavior in which groups of generators show lightly damped oscillations of frequency and phase following small disturbances. As a rough rule of thumb, one may say that such electromechanical modes show a much higher measure of controllability from control inputs such as prime mover mechanical power, or from power modulation control of a dc line, and a lower degree of controllability from a control input such as a static Var compensator. However, it is dangerous to generalize too broadly because, as noted previously, the dynamic characteristics of a power network are highly dependent on its loading pattern and operating point. Static Var compensators have been studied for their potential contribution as dynamic control elements. Moreover, while improved damping of electromechanical oscillations has been a traditional goal in enhancing dynamic performance, the last decade and a half has seen growing concern for loss of stability modes in which the divergence from an acceptable equilibrium appears primarily as a collapse in sinusoidal voltage magnitudes. This is the so-called voltage stability or voltage collapse problem.

The literature on this topic is huge; see [10, 11] and the references therein. In the context of voltage stability, the dynamic control impact of static Var compensators and other devices that influence the injection of reactive power into the network can be much more significant [12, 13].

13.3 INSTITUTIONAL CHANGES IMPACTING CONTROL TECHNIQUES

The previous review of system dynamics and existing control techniques in power systems has already alluded to a number of the new control challenges being created by changes in the regulatory structure and organizational structure overseeing the North American power grid. Incremental shifts toward a competitive market for electric power provision in the United States have been underway for many years, and significant restructuring in Great Britain predated that of the United States. As previously noted, the landmark event that set into motion much of the current activity in the United States was the issuance by the Federal Energy Regulatory Commission (FERC) of its orders number 888 and 889. The observation that a competitive generation market would bring significant control challenges was certainly anticipated before FERC's orders, particularly in light of earlier moves toward a competitive market in Britain. (For a sampling of related discussions, see [14].) A true competitive market for generation was judged to be predicated on the ability of all suppliers to access customers through the transmission network. The fact that U.S. utilities typically owned and controlled both the regional transmission grid and the generators was seen as an impediment. Therefore, FERC's action began the process to allow "nondiscriminatory" access to the U.S. transmission network. The FERC orders place particular emphasis on functionally separating generation and transmission, and on the means of communicating the transmission network state in an open fashion. This is intended to allow generation companies to assess the potential for long-distance power transactions on a continually updated basis. The technology is given the acronym OASIS, for *O*pen *A*ccess *S*ame-time *I*nformation *S*ystem, a term used in order 889's title. FERC's orders also discuss some of the support functions necessary to maintain quality of service and desirable dynamic performance (e.g., frequency regulation, voltage support) and define the necessary engineering functions as "ancillary services." However, the FERC orders do not give commensurate attention to the role of feedback control systems intended to enhance the stability of the network. The absence of clear regulatory guidance on the issue of dynamic controls has left many open questions; [15] provides an excellent discussion of these questions posed in the context of the summer 1996 outages in the western United States.

In the context of this chapter, the goal of this section is to provide further background on these institutional changes so that the reader will have an appreciation for their potentially dramatic impact on this critical infrastructure and its control. The topic is immense, and interpretation of the many technological and policy elements that have contributed to these changes is a daunting task. Equally challenging, and potentially more controversial, is any attempt to answer questions of cause and effect in changes that have both policy and technological components. That is: is a technological change the driving force behind a policy shift, or does a philosophically motivated policy shift open the door to technological changes to follow? Clearly in a technological and institutional infrastructure as large as the North American power grid, cause-and-effect relations between policy change and technological change are mixed in a complex fashion, and historic decisions made with specific goals often lead to unforeseen consequences. The following brief overview represents only the author's limited perspective on these topics and is undoubtedly open to challenge. Independent of the driving forces discussed, an obvious caution offered to engineering readers is to avoid the intellectual trap of technological determinism. Technological change is a strong social force, but there is no unique public policy consequence of a given technological development.

13.3.1 Power Grid Control Structures: If They're Not Broken, Why Fix Them?

Given the mature, reliable system of electric power production and delivery in the United States, it is a natural question for a pragmatic engineer to ask, "If it is not broken, why fix it?" Such a question greatly oversimplifies the many facets of electric power production in North America. In some aspects, the historic utility and regulatory structure in the United States has been extraordinarily successful. One can convincingly argue that given the physical hardware in place, the operation and control of that equipment has been performed extremely well over the past several decades. Although further opportunities for improving control technology and practice certainly exist in the traditional utility structure, these appear largely in the context of improving the system's ability to recover from relatively rare events and operating conditions rather than making significant improvements in the day-to-day operation. However, viewed nationwide, the procedure of the traditional industry and regulatory structure for making major capital investments has proven cumbersome and at times quite flawed. If capital investment decisions are flawed, the equipment built (or not built) inevitably yields a less than optimal system. In the context of this chapter, a reader may rightly ask why this criticism should be of concern to a control engineer. If one accepts that the historic flaws were largely in choices of capital investment, while control of equipment was carried out quite effectively, were the flaws and their repair not exclusively in the domain of the economists? Elements that argue for an answer in the negative have already been outlined, but let us reiterate.

The fix being implemented in the United States chooses to let competitive market forces largely replace regulated central planning in the decisions for capital investment, particularly with regard to electrical generation. This fix is certainly in keeping with the general philosophic direction in policy for many major U.S. industries over the past decade or more. In the power grid, this implies that generators will be placed in the hands of entities independent of the central grid operators and that these entities controlling generation will compete to provide power to customers. However, recall our earlier discussion of the nature of electric grid dynamics and its traditional control mechanisms. Generators are the dominant vehicles for exercising control over the grid, in order to achieve systemwide objectives such as frequency regulation and stable dynamic response. In many ways, a synchronous electric power grid presents dynamic features unlike those of any other market. Rival sets of generating units do not interact through the market alone. Rather, their electromechanical dynamics are tightly coupled through the transmission grid. Therefore, the dynamic governor and excitation control exercised at one machine can have a large impact on the dynamic response of other generating units and on the network as a whole. A competitive solution puts control of generators into the hands of independent, profit-maximizing organizations. This could pose the risk of severely compromising systemwide control objectives, *unless new control techniques and technologies are brought in to maintain acceptable dynamic performance and other systemwide objectives without the historic structure of central control coordination.*[8]

13.4 NEW TECHNOLOGIES IMPACTING RESTRUCTURING AND CONTROL IN A COMPETITIVE ENVIRONMENT

13.4.1 The Impact of Efficient Gas Turbines

Among the technological developments that are contributing factors to the end of regulated, regional monopolies in electric generation, the most obvious is the advance in combined cycle gas turbines as mechanical power sources for generators. One element of the traditional argument for monopoly ownership of generators was that of significant economies of scale in generation technology. For much of the twentieth century, the hardware deployed to generate electricity appeared to verify this assumption: Power could be produced with greater efficiency in ever larger generating stations. Representative of the culmination of this trend are large coal-fired generating stations of mid-1970s vintage; 38% net efficiency from chemical energy content of the fuel to electric energy output from the generator is typical in plants of approximately 1000 MW in size. The trend toward greater efficiency in larger sized plants was generally perceived to hold true in nuclear plants as well, though the overall thermal efficiencies achieved are somewhat lower, because of lower achievable steam temperatures. However, offsetting the reduced operating (fuel) cost of large plants were the huge capital costs and long lead times in their construction.

The availability of cost-effective combined cycle gas turbine plants in the last decade and a half, coupled with the availability of low-cost natural gas, has significantly altered this picture. Combined cycle gas turbine technology can yield thermal efficiencies well in excess of 50%, in plants of modest size and capital cost. As a result, such plants can pay for themselves on a much shorter time horizon, making them more attractive prospects for investment by an unregulated, for-profit entity. Perhaps not surprisingly, the overwhelming majority of new generation installed in the United States over the past several years has employed this technology [17, 32]. From a control standpoint, a key impact of the greater penetration of gas turbines as a primary power source for generation is the *potential* that these units offer for more responsive control of their prime mover mechanical power. However, institutional arrangements can

---

[8] New control technologies are not the only mechanism to ensure that competitive generators contribute to systemwide goals of frequency regulation, stable dynamic response, grid integrity, and reliability. There is much to be done to develop market-based and regulatory incentives to induce competitive generator owners to operate in ways that contribute to systemwide goals. However, in many cases, new control, measurement, and communication technologies will be necessary to facilitate the implementation of these incentive systems.

negate technological opportunity; gas supplies for such units are sometimes contracted on a take-or-pay basis, discouraging the use of the machine as a controllable, variable power source. A secondary dynamic impact of gas-turbine-based generators will be the relatively smaller rotational inertia of individual machines.

The significant coupling of control and policy questions for new generators is this: Given that many of these sources are being installed by nonutility entities, will there exist financial incentives to install and operate these units with the engineering necessary to exploit their control opportunities? The actual control hardware is a relatively small portion of capital cost, so many would argue that convincing plant owners to include such additions will be easy. Indeed, there exist ongoing activities within NERC, the IEEE Power Engineering Society, and elsewhere to consider standards for such supplemental controllers of traditional form (such as power system stabilizers). However, as yet, there appears to be no universal incentive system (or enforceable regulation) to encourage installation of such stability-enhancing generator controls. The fate of such generator controls in a highly competitive market for commodity power provision remains to be seen.

13.4.2 The Role of New Information and Measurement Technologies

One of the questions in a restructured, competitive power network is the role of grid information. The FERC orders require that information regarding the power transfer capability of the grid be made widely available and auditable in order to allow evaluation of the potential for power transfers. In contrast to this, in a competitive market, individual generator owners will want to guard data regarding their production resources as proprietary information.[9] Studies of dynamic performance characteristics of the power system and associated control systems design require both types of data: the transmission system parameters and configuration, and detailed dynamic characteristics of generating units. Who will possess both types of data? In the evolving institutional structure, it seems generally agreed that there must remain a central body overseeing the grid, usually termed the Independent System Operator, or ISO. The California market has an ISO, and the formation of ISOs appears essentially complete (as of early 1999) for the Pennsylvania–New Jersey–Maryland interconnection, for New England, and for New York. Notably, these are regions that had strong regional control centers coordinating the resources of multiple utilities before the advent of orders 888 and 889; such multicompany regional control centers are not common to all regions of the United States. Other regions of the United States await agreements to form ISOs.

Based on the examples in place so far, the ISO is typically given strong administrative powers, and will likely be in a position to *collect* both types of data described above. This body is also likely to be closely involved in engineering analyses to ensure desirable dynamic performance. But even if the ISO has administrative power to collect

---

[9] In one of the ironies inherent in a transition from a regulated monopoly to competitive markets, until recently the U.S. Department of Energy Information Administration (EIA) cataloged and published data on significant generating units in the U.S., including their fuel use, cost, and production efficiency. Until there is more turnover of the capital stock of generation, evaluation of the nature and efficiency of a competitor's production facilities may require little more than a trip to the local library [16]. The EIA continues to publish an annual catalog of major generating facilities and their capabilities.

proprietary generator dynamic data from individual owners, will it have the resources to validate this huge data set? As reported in [15], the dynamic study models in existence in the western United States (arguably among the most advanced in the world) had significant inaccuracies prior to the blackouts experienced in the summer of 1996. In particular, had the exact initiating events been studied in advance, simulation tools using the (then) best available data would have failed to correctly predict the occurrence of major blackouts. A major postmortem engineering effort later corrected model parameters to a degree that the simulation tools did match actual occurrences with some fidelity. We may naturally ask how much more severe this situation could become in the future, if we rely on traditional methods to gather data and assemble dynamic models.

One may speculate that if ISOs must carry out time domain simulations of network dynamics to ensure system security and perform control design based on these studies, supplemental means of collecting dynamic data will become increasingly critical. In particular, there will be a need for improved tools for estimating both steady-state operating point and dynamic parameters from physical measurements of the system. The former task, that of estimating current system operating point and grid configuration from on-line measurements, has a long history in power systems. In power systems terminology, this is termed the state estimation problem. However, a control engineer encountering this term must understand that its typical usage refers to estimation of the values of system variables at a steady-state operating point rather than values along a transient trajectory (and indeed, not all the quantities estimated are truly state variables of a dynamic model). Estimation of equilibrium values of system variables from noisy measurements that are non-linear functions of those variables is often formulated iteratively as a least-squares estimation problem. This steady-state problem becomes more challenging if one also recognizes that various components in the grid may be switched in or out of service and that direct knowledge of the switch status may not be guaranteed. Therefore, the steady-state estimation problem may be augmented to attempt estimation of network configuration and switch status from other indirect measurements. With growth in the number of entities whose competitive position is impacted by system operating condition and grid configuration, it is easy to predict that interest in this traditional form of power system state estimation may also grow.

Among the new technologies impacting these developments, most notable is the growing availability of low-cost, high sampling rate "phasor measurement" units. The reader should recall our earlier discussion regarding the importance of voltage phase angle differences between nodes in the grid. This is a key quantity determining the flow of power. Yet accurate measurement of phase angle differences between sinusoidal voltages that are hundreds or even thousands of kilometers apart is a challenging technical problem. Judging relative phase is critically dependent on a precise time reference. In recent years, the global positioning satellite system (GPS) has provided a low-cost means of acquiring precisely synchronized time references at remote measurement points. In the power system, this has created the opportunity to precisely and directly measure relative phase angles of geographically dispersed sinusoidal voltages in the grid. These are often termed "wide area" phasor measurements. This adds a very valuable measurement to the set available for (steady-state) state estimation. It also creates opportunities for improved system protection and dynamic control. (For a recent sampling of these ideas, see [18,19].)

13.4.3 Control Opportunities for Flexible AC Transmission Systems

New applications of high-power electronics in the transmission grid are often grouped under the heading of Flexible AC Transmission, or FACTS [20]. More specifically, the term FACTS refers to a family of circuit configurations that use high-power semiconductor switching elements, such as thyristors or gate turnoff devices (GTOs), to vary the duty cycle of passive elements such as capacitors and inductors. Such configurations can functionally approximate a variable, controllable impedance at the 60-Hz fundamental, or in more advanced configurations, they can transfer power between series-connected elements in the transmission grid and shunt-connected elements in the grid. As a result, FACTS devices make the transmission grid itself much more dynamically and continuously controllable rather than leaving it to operate only as a passive circuit.

In the eyes of many observers, FACTS technologies have represented a potential revolution that has continued to wait in the wings for a number of years. While a number of interesting demonstration projects have been completed or are on-going, significant penetration of this technology into the high-voltage transmission grid has yet to occur. This delay is perhaps not surprising given the institutional restructuring of the U.S. grid. FACTS devices are relatively high-cost elements that will not contribute to economic power generation directly, but rather, indirectly, through more efficient control and utilization of the transmission system. In the aftermath of FERC's 1996 orders, open questions remain regarding the means for recovery of investment in the transmission grid, as well as organizational questions about the form of Independent System Operators (or other entities) for some portions of the United States. Hence the failure to see large investment in significant new transmission control technologies is not completely surprising. However, as issues relating to transmission investment are resolved, it is likely that the FACTS revolution will come, and with it, a range of interesting new control opportunities and challenges.

Power electronic controllers can present challenging nonlinear problems because, fundamentally, these devices are composed of circuits in which controlled switches are the primary regulating element. In transmission applications, one is typically attempting to control the 60-Hz fundamental component of a current or voltage waveform, or of an impedance, by switching within an appropriate circuit topology. When the switching frequency is significantly above that of the fundamental, as is the case in low-power applications, averaging techniques provide a fairly tractable, usually linear, model for control design. However, present solid-state technologies for high power are limited in their switching frequency by loss effects. The limitations on switching frequency create much more complex dynamic behavior and challenges to control design. Moreover, combinations of new circuit topologies and devices in the so-called Universal Power Flow Controller [21] create an opportunity for significantly enhancing steady-state power flow in a manner that could make the economics of such FACTS technology much more attractive. Once this technology is deployed in the grid, it will also open the door to many interesting opportunities in control design for dynamic performance enhancement.

13.5 A PERSPECTIVE ON FUTURE DIRECTIONS FOR POWER SYSTEM CONTROL DEVELOPMENT AND RESEARCH

Perhaps the first key control challenge in the immediate future of power systems is one alluded to several times in our earlier review: that of rethinking the existing hierarchical system of systemwide frequency control, the AGC system. This is as much a problem of administration as it is one of control design. Effective economic incentives must be found to encourage the participation of competitive units in "global" frequency regulation. The NERC Web site (see footnote 4) provides up-to-date documentation of the U.S. perspective on the next generation of AGC. However, beyond the administrative and economic aspects, significant opportunities exist for conceptual innovation in the controller designs. In the overview of AGC provided in [7], the authors and various discussants allude disparagingly to attempts in the 1970s to apply optimal control design concepts to the frequency regulation. These optimal control design techniques were critiqued as grossly unrealistic, neglecting the many practical constraints on equipment response rates and bounds, and issues of wear and tear on steam valve systems. However, recent work such as [22] has begun to reexamine the use of optimal control in frequency regulation for steam-driven electric generators, bringing in much more realistic representations of the steam-flow system and its constraints.

More broadly, the AGC problem encapsulates the general nature of challenges that will likely be recurring themes in control design as power systems move toward a competitive generation market. In particular, how does one migrate from a control structure predicated on centralized ownership and unified administration of generation and transmission control equipment that existed in the past? As this chapter's review attempted to indicate, generators are among the most effective elements for achieving systemwide control objectives of frequency regulation and stable dynamic response, and to a lesser degree, voltage control. Yet these "control resources" (generators) will be owned and administered by independent, profit-maximizing entities, divorced from the Independent System Operators that have responsibility for the transmission. What new structures of control and what economic incentives will serve to align the individual profit-maximizing objectives of generation owners with systemwide control objectives?

Looking to the future, we find that considerable interest is being generated by agent-based concepts [23]. The agent approach appears to provide a reasonable model for many examples in nature in which fairly simple local control laws and logic apparently succeed in generating quite complex global behavior in the aggregate (e.g., flocking behavior in birds). Clearly, this is an appealing concept when one is faced with the challenge of obtaining desirable global dynamic performance from relatively simple local control actions in a competitive power system. Recognizing that NP-hard optimization problems abound in power systems control and optimization problems, it is also notable that agent-based approaches are proving quite powerful in improving solution algorithms for several classic NP-complete problems. At present, agent-based approaches are being adopted for power systems applications not directly in control design, but in closely related problems of design of bidding/offering strategies for competitive generating units.

Beyond the agent-based approach, a range of more established control theoretic concepts remains promising if the system sees a strong shift toward fully decentralized controllers. One of the clear possibilities in a competitive generation market is that of much more volatile patterns of generator commitment. That is, individual generators may be connected and disconnected from the grid in less predictable patterns and with increasing frequency. In such a scenario, one has a particular structure of robust control problem in which the structure of the system may vary over a huge range of configurations, with each configuration having a different set of controllers (on individual generators) active or not. Robust decentralized control has a long history in power system research; the control needs in a competitive environment have further strengthened the desirability of this effort. In this context, controller designs based on passivity and dissipativity ideas hold considerable promise. Indeed, it may be argued that several traditional design methods for power system stabilizers produce controllers with these properties. Closely related to passivity-based designs are controllers based on identifying a system Lyapunov function, or family of such functions, and designing for improved dissipation relative to the Lyapunov function(s). This approach has been proposed for use with multiple FACTS devices closely interacting in a network.

To the extent that a central body, such as the Independent System Operator, will continue to tackle systemwide optimization, dynamic control design, and performance validation, the issue of NP-hard computational problems in power systems remains significant. Designing control systems that provide acceptable dynamic performance over wide-ranging operating conditions and grid configurations is a recurring challenge, and a number of works have sought to transfer concepts from the robust stability and controller design literature to power systems applications. (See, for example, [24].) However, power systems have long been recognized as suffering from the "curse of dimensionality"; the developments of computational complexity allow one to formally classify many of the robust stability problems found in power systems to be NP-hard [25]. In the robust control literature and in a range of control design problems, there has been a recent recognition of the power of probabilistic methods in treating NP-hard analysis and design problems [26]. Transfer of these concepts to power systems control design is an extremely appealing avenue for future work.

On the same theme of computationally challenging problems, another aspect of control design relevant to power applications is the potential for strong interaction between continuously acting feedback controllers and discontinuous, discrete switching events, such as protective relays. Given the huge computational challenge these problems present, traditional approaches in power systems have been rather ad hoc, with initial control design efforts largely ignoring protective relay action, and, at best, followup simulation efforts to test whether relay thresholds are encountered in foreseeable fault and system disturbance scenarios.[10] Clearly, this approach is severely limited by the fact that only those disturbance events and grid configurations anticipated in the "contingency list" are studied for interaction. It is almost a folk theorem that major

---

[10] Autonomously acting discrete switching events (in contrast to those commanded by human intervention from a control center operator) are largely associated with protective devices. These act when system states deviate widely from their acceptable operating range. Therefore, it is reasonable to suppose that interaction between these effects and normal feedback controls should occur mostly under fault or disturbance conditions. However, among the common contributing disturbances may be an inappropriate triggering of one of the protective devices themselves.

failures in complex engineering systems, such as power grids, result from the simultaneous occurrence of several rare events, or unusual operating conditions, the combination of which would not have been identified as a plausible subject for study *a priori*. Ideally, one would like a probabilistic, dynamic simulation in which random actions occur periodically, so that the simulation may "unearth" unexpected interactions of discrete events and continuously acting controllers. However, in a system of large dimension, in which the events to be identified are extremely rare, direct computational implementation of this approach is completely intractable. With suitable modeling, the occurrence of the rare failure mode appears as a "large deviation" in the state of the system. Similar issues appear in control and coordination of communication networks, in which one seeks to identify possible failure modes that have extremely low probability [27]. To improve computational tractability, the techniques of importance sampling have proven promising in the study of communication networks, and the control community is playing an active role in the continuing development of related methods. Such methods are beginning to see application in the study of power system protective relays [28]. Such methods could be critical to ensure reliability in the development of control and protection technology for the future U.S. power grid.

Closely related to the issue treating the interaction of discontinuous switching events and network reconfiguration, continuously acting feedback controls, and stochastically varying inputs is the application of discrete event and hybrid systems concepts in the power systems context. Design methodologies to fully coordinate the consideration of the various types of phenomena in accurate models will be extremely challenging to develop, but this mix of features is hardly unique to the power system application, and progress on general methods is being made [29]. Research into these topics is growing as competitive pressures demand less conservative operating margins in power networks.

Many of the topics for future development and research in power systems control represent new perspectives on long-standing control problems, being motivated by restructuring and the emergence of competitive markets in the power industry. However, competitive markets themselves, and certainly the interaction of physical dynamics with market-driven events, are important new topics for study within the power systems domain. It is widely recognized that Wall Street has seen a significant influx of advanced technical talent to study stochastic market behavior over the last decade, often drawing on individuals with a control systems orientation. A power exchange proves a most interesting market for study, given the many time scales that are spanned by this market's activity, with strategic decisions to be made all the way from long-term futures markets, down to second-by-second balancing of instantaneous generation and load. Work in [30] provides an overview of how this mix of market and control structures is achieved in the structure of the California Independent System Operator. More broadly, there are a range of interesting questions with control aspects raised when market decisions by individual grid participants contribute as feedback elements to the overall dynamic behavior of the grid.[11] Extending the perspective slightly beyond pure control problems, there are also a host of interesting questions relating to the prediction of market participants' behavior that may be well-formulated as game-theoretic problems. Predicting the dynamic

---

[11] These issues were anticipated in the pioneering work of Schweppe, on the so-called Frequency Adaptive Power Energy Rescheduler, or FAPER, in the late 1970s. See [31] for a description of this concept and its interaction with longer time scale market phenomena.

performance and reliability levels of the future electric power grid will depend critically on the ability to successfully integrate these many elements into system control analysis and design tools.

The discussion here has, at best, scratched the surface of the huge range of control-oriented issues and challenges that are motivated by the restructuring of the electric power industry in the United States and around the globe. The author recognizes that the references provided are a very small, and inevitably inadequate, sampling of the field, and offers apologies to the many fine engineers and researchers contributing to power system control development who are not acknowledged here. However, this chapter will have fulfilled its objective if it has sparked interest in the new problems of electric power system control and estimation among individuals currently focused on other application areas. Indeed, the process and ultimate outcome of current restructuring in the power industry should interest any individual concerned with the direction of our technological society. Control engineers have important skills and perspectives to offer to this task.

Related Chapters

- An overview of process control systems can be found in Ch. 12.
- See Ch. 10 for another agent-based application concept.
- Issues related to the interaction of continuous dynamics and discrete switching are discussed in Ch. 7.

REFERENCES

[1] L. H. Fink, "New control paradigms for deregulation." In M. Ilic, F. Galiana, and L. Fink (eds.), *Power System Restructuring: Engineering and Economics*. Boston: Kluwer Academic Publishers, 1998.

[2] FERC Order no. 888, Final Rule. "Promoting wholesale competition through open access non-discriminatory transmission services by public utilities; Recovery of stranded costs by public utilities and transmitting utilities." Docket #RM95-8-000, issued April 24, 1996; available via http://www.ferc.fed.us/news1/rules/pages/order888.htm

[3] FERC Order No. 889, Final Rule. "Open access same-time information system and standards of conduct." Docket # RM95-9-000, issued April 24, 1996; available via http://www.ferc.fed.us/news1/rules/pages/order889.htm

[4] A. J. Wood, and B. F. Wollenberg, *Power Generation, Operation, and Control*, 2nd ed. New York: John Wiley & Sons, 1996.

[5] N. Jaleeli et al, "Understanding automatic generation control." *IEEE Trans. on Power Systems*, Vol. 7, no. 3, pp. 1106–1122, August 1992.

[6] G. Rogers, "Demystifying power system oscillations." *IEEE Computer Applications in Power*, Vol. 9, no. 3, pp. 30–35, July 1996.

[7] N. Jaleeli and L.S. VanSlyck, "Tie-line bias prioritized energy control." *IEEE Trans. on Power Systems*, Vol. 10, no. 1, pp. 51–59, February 1995.

[8] P. Kundur, *Power System Stability and Control*. New York: McGraw-Hill, 1994.

[9] J. R. Smith, G. Andersson, and C. W. Taylor, "Annotated bibliography on power system stability controls: 1986–1994." *IEEE Trans. on Power Systems*, Vol. 11, no. 2, pp. 794–800, May 1996.

[10] Y. Mansour (ed.), "Suggested techniques for voltage stability analysis." IEEE Power Engineering Society, Publication #93TH0620-5PWR, 1993.

References

[11] C. W. Taylor, *Power System Voltage Stability*. New York: McGraw-Hill, 1994.

[12] T. J. E. Miller, *Reactive Power Control in Electric Systems*. New York: John Wiley & Sons, 1982.

[13] "Application of static Var systems for system dynamic performance." IEEE Power Engineering Society, Publication #87TH0187-5-PWR, 1987.

[14] Proceedings, *The Impact of a Less Regulated Utility Environment on Power System Control and Security*. C. L. DeMarco (ed.), workshop sponsored by National Science Foundation, Engineering Systems/Power Systems Program, Madison, WI, April 19–20, 1991.

[15] J. F. Hauer and C. W. Taylor, "Information, reliability, and control in the new power system." *Proc. American Control Conference*, pp. 2986–2991, Philadelphia, PA, June 24–26, 1998.

[16] *Electric Plant Cost and Power Production Expenses 1990*. DOE/EIA-0455(90), Energy Information Administration, U.S. Department of Energy, Washington, DC, June 1992.

[17] *Annual Energy Review 1997*. DOE/EIA-0384(97), Energy Information Administration, U.S. Department of Energy, Washington, DC, July 1998.

[18] J. S. Thorpe and A. G. Phadke, "Protecting power systems in the post-restructuring era." *IEEE Computer Applications in Power*, Vol. 12, no. 1, pp. 33–37, January 1999.

[19] I. Kamwa, L. Gerin-Lajoie, and G. Trudel, "Multi-loop power system stabilizers using wide area synchronous phasor measurements." *Proc. American Control Conference*, pp. 2963–2967, Philadelphia, PA, June 24–26, 1998.

[20] N. G. Hingorani and K. E. Stahlkopf, "High-power electronics." *Scientific American*, Vol. 269, no 5, pp.78–85, November 1993.

[21] L. Gyugyi, "Unified power flow control concept for flexible transmission systems." *IEE Proceedings-C*, Vol. 139, no. 4, pp. 323–331, July 1992.

[22] C-K. Weng and A. Ray, "Robust wide-range control of steam-electric power plants." *IEEE Trans. on Control Systems Technology*, Vol. 5, no. 1, pp. 74–88, January 1997.

[23] A. M. Wildberger, "Complex adaptive systems: concepts and power industry applications." *IEEE Control Systems Magazine*, pp. 77-88, December 1997.

[24] M. H. Khammash, V. Vittal, and C. D. Pawloski, "Analysis of control performance for stability robustness of power systems." *IEEE Trans. on Power Systems*, Vol. 9, no. 4, pp. 1861–1867, November 1994.

[25] C. L. DeMarco, "Computational complexity results in parameteric robust stability analysis with power systems applications." In J. H. Chow, P. V. Kokotovic, and R. J. Thomas (eds.), *Systems and Control Theory for Power Systems*. New York: Springer-Verlag, 1995.

[26] M. Vidyasagar, "Statisical learning theory and randomized algorithms for control." *IEEE Control Systems Magazine*, Vol. 18, no. 6, pp. 69–85, December 1998.

[27] P. Glasserman, K. Sigman, and D. Yao (eds.), *Stochastic Networks: Stability and Rare Events*, Lecture Notes in Statistics Vol. 117, New York: Springer-Verlag, 1996.

[28] J. S. Thorp, A. G. Phadke, S. H. Horowitz, and S. Tamronglak, "Anatomy of power system disturbances: Importance sampling." *Electrical Power & Energy Systems*, Vol. 20, no. 2, pp. 147–152, August 1997.

[29] M. S. Branicky, V. S. Borkar, and S. K. Mitter, "A unified framework for hybrid control: Model and optimal control theory." *IEEE Transactions on Automatic Control*, Vol. 43, no. 1, pp. 31–45, January 1998.

[30] Z. Alaywan and J. Allen, "California electric restructuring: Broad description of the development of the California ISO." *IEEE Trans. on Power Systems*, Vol. 13, no. 4, pp. 1445–1452, November 1998.

[31] F. C. Scweppe, M. C. Caramanis, R. D. Tabors, and R. E. Bohn, *Spot Pricing of Electricity*. Boston: Kluwer Academic Publishers, 1988.

[32] *The Changing Structure of the Electric Power Industry: Selected Issues, 1998*. DOE/EIA-0562(98), Energy Information Administration, U.S. Department of Energy, Washington, DC, July 1998.

Chapter 14 | INTELLIGENT TRANSPORTATION SYSTEMS: ROADWAY APPLICATIONS

Ümit Özgüner

Editor's Summary

Intelligent Transportation Systems (ITS) is a new field, encompassing all modes of transportation of people and goods. The focus of this chapter is on intelligent road transportation and on the role of control technologies in this interdisciplinary area.

Two primary control-relevant topics in ITS are traffic control technologies and intelligent vehicles. The former can be further classified into street traffic and highways. Street traffic applications include signaling for both single intersections and networks of intersections, where the control variables are traffic light timings, and routing, in which traffic flow along different roads is manipulated to optimize some overall efficiency or congestion criterion. Highway traffic control applications range from the relatively mundane one of ramp control or metering to the exotic prospect of Automated Highway Systems (AHS) in which multiple vehicles operate as platoons. The grander visions of traffic control require a number of practical considerations to be addressed, including identification, sensing, and actuation over roadways with wireless networks.

Intelligent vehicles require the incorporation within automobiles of intelligence and autonomy capabilities. (See Chapter 5 for a general discussion of intelligent control.) Limited autonomy is exhibited even in today's automobiles with cruise control and antilock braking systems—these loops automate throttle control and brake control. More meaningful autonomy requires closing a third loop, that of steering. (This requires a drive-by-wire capability that is currently lacking in production automobiles.) Furthermore, the overall problem must be considered a multivariable one. The chapter presents mathematical formulations and solution approaches for advanced cruise control, which includes automatic braking as well as acceleration, lane keeping with a sliding-mode-like nonlinear controller (Chapter 8 discusses sliding-mode control in detail), and lane changing.

Ümit Özgüner is a professor in the Department of Electrical Engineering at the Ohio State University and the president of the IEEE Intelligent Transportation Systems Council.

14.1 INTRODUCTION

This chapter presents an overview of automatic control in the Intelligent Transportation Systems (ITS) area. ITS covers the totality of computer, communication, and control technologies, as well as techniques for transportation of goods and people in an optimal way. As such, ITS is an interdisciplinary area, and we will describe the reliance of ITS on control and other engineering topics and on different technolo-

(a)

(b)

Figure 14.1

(c)

Figure 14.1 (continued)

gies. ITS covers all modes of transportation; because of space limitations, this chapter focuses on roadway transportation.

Historically, the separation of ITS from standard transportation science may be traced to the identification of Automated Highway Systems (AHS) as a specific focus area. AHSs are dedicated highway systems in which specially equipped cars can take their occupants to their declared destinations automatically, without driver involvement. With the inclusion of vehicles other than cars and use of technologies beyond those needed for vehicle automation, the computer-based roadway transportation area (still having AHS at its core) began to be known as Intelligent Vehicle Highway Systems, or IVHS for short.

The switch in terminology from IVHS to ITS indicates a major expansion in interest. ITS covers all aspects of computerized and computer-aided transportation. It has been an exciting active research area in recent years.

The 1997 National Automated Highway System Technology Demonstration in San Diego, California, illustrated many aspects of the state-of-the-art of control research in ITS. In this one-week event, known as Demo'97, different teams from industry and academia demonstrated different concepts and technologies that could lead to an Automated Highway System. Figure 14.1 shows two fully automated cars: the two white Honda Accords in the middle lane, following a human–driven car on I-15 during the demonstration. As designed by the Ohio State University (OSU) team, the middle car in this convoy eventually *decides* to pass the manual car and accomplishes an

Section 14.2 Traffic-Related Issues

automated lane change. The following car automatically fills the gap to keep a fixed distance to the car in front.

We shall return to examples from Demo'97 in the last part of this chapter. However, as we mentioned above, AIIS is not the only area in which ITS technologies are used, and it is also not the only area that should be of interest to control engineers.

The following discusses ITS in terms of (1) traffic-related issues and (2) intelligent vehicle (IV) issues. Obviously, both topics are interrelated, although the first has received comparatively less attention from electrical engineers.

14.2 TRAFFIC-RELATED ISSUES

14.2.1 Signalization

The first problem to be considered in the area of transportation as a control application is intersection control, or signalization. The control variable is signal *split*, that is, the specification of the ratio of green to red time, given a fixed *cycle time* for an intersection. In a controlled intersection, the split can be adjusted continuously based on balancing the queue lengths or waiting times for incoming traffic. Constraints can be applied on minimum green time, and turn options can further complicate this basic problem.

For a typical intersection as shown in Figure 14.2, the control problem, in which the cycle time of the traffic light is fixed as T and the only variable that can be adjusted is the split c, a model and a criterion can be developed.

Let q denote the queue lengths in the direction indicated by their subscripts. Let u similarly denote incoming and outgoing traffic at the intersections. The *undersaturated intersection* is modeled in the following way:

$$\begin{cases} \dot{q}_x = u_x^i(t) - u_x^o(t) \\ \dot{q}_y = u_y^i(t) \end{cases} \quad \text{if } 0 < t \leq cT$$
$$\begin{cases} \dot{q}_x = u_x^i(t) \\ \dot{q}_y = u_y^i(t) - u_y^o(t). \end{cases} \quad \text{if } cT < t \leq T \tag{14.1}$$

All these equations say is that, if more traffic arrives at the intersection than departs, queues build up. Usually, there also exists a constraint on c

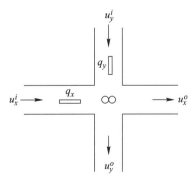

Figure 14.2

$$0 < c_{min} \leq c \leq c_{max} < 1, \tag{14.2}$$

which means that you will get a green light during every cycle, and although it may be a short one, it will not be too short.

When the intersection is *oversaturated*, the model can be simplified as

$$\begin{cases} \dot{q}_x &= u_x^i(t) - cf_x \\ \dot{q}_y &= u_y^i(t) - (1-c)f_y \end{cases} \tag{14.3}$$

during their corresponding green times, where f_x and f_y are the maximum flow capacities in the x-direction and y-direction, respectively. These equations imply that we will pump as many vehicles out as the outgoing street will take, during all of our green time.

In either case, an optimal solution can be found by defining some *cost criterion* such as

$$\min_c J = \int_0^T (q_x(t) + q_y(t))dt, \tag{14.4}$$

which minimizes total queue lengths over the cycle time. In the oversaturated case, using Pontrayagin's maximum principle, the solution can be found to be a "bang-bang" control strategy, that is,

$$c = \begin{cases} c_{max} & f_x \geq f_y \\ c_{min} & \text{otherwise} \end{cases}. \tag{14.5}$$

If the intersection is not oversaturated, other solutions can be given. Assuming the queues at the intersection could be cleared in a cycle of the traffic light with the maximum flow capacities in the x-direction and y-direction, f_x and f_y, we can use some reasonable ratio depending on the initial queue lengths and flows such as

$$c = \frac{q_x(0)f_y}{q_x(0)f_y + q_y(0)f_x}. \tag{14.6}$$

The point we wanted to make in this simple exposition is that, even in the most basic ITS application, one can find a control problem to be solved.

14.2.2 Networks of Intersections

Embedding a single intersection into a network, where many roads and many other intersections exist, complicates various issues. When dealing with networks of intersections, one also has to model both splitting roadways and turning vehicles at intersections (both possibly with the traffic ratio predetermined). But the real problem now is no longer clearing a single intersection but a whole network. Decisions at one intersection can affect queue lengths at others.

Two basic questions need to be answered:

Section 14.2 Traffic-Related Issues   353

1. What is the overall goal? Is there some relationship between the overall goal and local goals?
2. What information transmittal framework is to be imposed?

These are standard questions in the large-scale dynamic systems and decentralized control literature. Indeed, approaches ranging from the decentralized servocompensator setting to parallel dynamic programming can be attempted.

14.2.3 Routing

Extensive research has been reported for dynamic routing, route guidance, or traffic assignment, as it is sometimes called in the transportation literature.

The queues in the single-destination routing problem are assumed to be modeled as a store-and-forward network as shown in Figure 14.3. The mathematical model can be given as

$$\dot{q}_n = q_n + r_n + \sum_{l \in L_n^i} u_l(t - t_l) - \sum_{l \in L_n^o} u_l(t), \quad (14.7)$$

where q_n is the queue length at time t in node n, r_n denotes the traffic that arrives at node n from outside the network at time t, L_n^i and L_n^o denote the index set of branches entering and leaving node n, respectively, and $u_l(t)$ is the traffic along the link l.

Assuming that each branch has a capacity C_l and the traveling time along link l is t_l, we find that the following inequalities exist:

$$q_n(t) \geq 0 \quad and \quad 0 \leq u_l(t) \leq C_l. \quad (14.8)$$

A cost function that the routing strategy may attempt to minimize is the aggregate delay up to a specified time T, that is,

$$J(u(t)) = \int_0^T \left[\sum_{n=1}^N q_n(t) + \sum_{l=1}^L t_l u_l(t) \right] dt. \quad (14.9)$$

The q_n terms in this equation give the delay due to congestion of the network, and the u_l terms give the transit delays or costs. Different constant weights can also be attached to the terms, and other cost criteria can be envisaged.

Solutions to the above problem, or similar problems that may be defined, will depend on the information structure allowed. If all measurements are available at some central location, we will have a different control strategy (and different perfor-

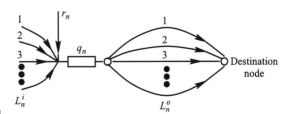

Figure 14.3

mance) than when only local queue lengths are measured. A solution in which estimates of *cost-to-go* are passed upstream was provided in [4] for the single-destination case. Extensions to multi-destination cases and the use of different cost criteria have been reported through the years.

14.2.4 Control of Traffic on Highways

14.2.4.1 Convoys, Platoons, et al.

In the 1990s, extensive research was done on highway automation as a result of the continuously increasing number of vehicles on the roads and the high cost of building new roads. One of the early steps leading to Automated Highway Systems was the development of the concept of convoys of vehicles with a leader and numerous automated followers. The problems associated with control for such a convoy can be split into longitudinal and lateral problems.

The longitudinal control problem for car following has been studied extensively by many researchers for a long time. In 1967, Levine and Athans considered the problem of controlling a string of vehicles so as to keep them moving with constant speed and separation using optimal control theory in a decentralized manner [5]. Chu compared several feedback structures to determine the effects of supplied information for the same example [6] based on the optimal decentralized regulation theory. Between 1964 and 1980, several studies for highway automation, including the implementation of longitudinal and lateral controllers for the lane-tracking and car-following problems, were conducted at the Ohio State University. A detailed overview of these studies can be found in [7]. The experience of researchers at PATH, a California alliance of universities including the University of California at Berkeley and others, was summarized in [8].

The PATH group contributed extensively to a framework in which the Automated Highway System relied heavily on a concept of tightly packed convoys of cars moving at high speed as a single unit called a *platoon*. Platoons could split, merge, change lanes together, and so on. The tight spacing and high-speed expectation in platoons implied fairly demanding constraints for the control loops (say, for speed regulation) for the individual cars. It also led to the need for reliable intervehicle communication links. In addition to the lower end control design specifications, the concept of platoons generated the need for tools to analyze (and design for) the higher level group operations. This need was met with concurrent research developments in the *hybrid systems control* area. Although the expansion of research in hybrid systems can also be attributed to needs in modeling manufacturing systems and process control, or the natural outcome of developments in discrete-event dynamic systems, it is clear that control engineers in AHS needed tools to analyze high-level decision making and scenario resolution.

14.2.4.2 Ramp Control and Merging

Ramp control is one of the first problems, together with determination of signal split, which has been addressed in terms of control. In standard operation, ramp control simply controls admission of vehicles into the freeway in a regular way. This forestalls the infusion of a large burst of traffic into the freeway artery, which may lead to an unwanted stop-and-go operation. Drivers usually prefer to wait at the entrance ramp rather than be subject to variable speed situations once on the road.

One conceivable way of doing ramp control is to admit vehicles into gaps in freeway traffic. This would require a means of measuring oncoming traffic on the freeway in real time, say, by a vision system. Obviously, such a system would work if the freeway artery was not saturated, and drivers already on the freeway could be relied on to keep the gaps constant as the admitted vehicles accelerate on the ramp. Indeed, in an Automated Highway System where the speeds of the vehicles on the freeway, and in fact the gaps between the vehicles, would be regulated, ramp control would basically be done as above.

An early approach that was used in considering speed control on the freeway was a *conveyor* analogy. The virtual conveyor would have regular segments, and each vehicle would have to adjust its speed to remain in its assigned virtual segment as it moved. Some segments would be empty, and the vehicles entering on the ramp would be assigned to those. One way of thinking of this situation is in terms of a moving sequence of ones and zeros, where a one denotes a full segment on the virtual conveyor. This image simplifies creation of the merging string from the ramp in a systematic way.

14.2.4.3 Automated Highway Systems

The fully Automated Highway System, in which all vehicles on designated highways have their speeds and headway regulated, has been advocated for a long time for providing an optimal solution for future transportation problems. Two facts strengthen this argument:

1. It is becoming very expensive to add new roads to the highway system.
2. Full control implies that vehicles at high speed can be packed closer together on the highways, resulting in higher throughput.

It is this second assertion that signifies that control theory and practice has a very important role to play. The controllers designed for the automated highway system have to be precise, practical, robust, and reliable.

Through the years, many control groups have worked as parts of teams that have developed working demonstration systems illustrating parts of AHS. The largest of these was organized in August 1997 on I-15 in San Diego by the National Automated Highway Systems Consortium (NAHSC) and was referred to as Demo'97. A number of different teams (PATH-GM, CMU-Metro Houston, Ohio State University, Honda, Toyota, and Eaton-Vorad) had different vehicles or groups of vehicles performing different automated scenarios on I-15, demonstrating both different concepts of an AHS and different technologies that can be utilized. Control played an important role in all of them.

14.2.5 Some Practical Concerns

We have already mentioned that ITS is an interdisciplinary science. We will certainly notice this again when we overview the topic of intelligent vehicles a little later in the chapter. But the control engineer with traffic-related ITS problems will also rely on other disciplines. The issue of getting real-time data for decision making for a whole network is especially important.

Individual intersection signal control is a local feedback loop. Once the intersection is part of a full network, the intersection has to be controlled regionally or centrally (see Figure 14.4). The present practice is to reset splits occasionally (based on statistics related to time of day), so that control signals are not transmitted continuously to the intersection. Local measurements are passed back to the regional center; therefore, whether loop closure is at a high or low rate, the physical communication link needs to exist and probably constitutes the majority of the cost of the control setup.

Fully automated routing, if it were to be implemented, would at least require the same communication infrastructure. Added to that is the need of each vehicle to declare its destination, if optimal regulation is expected. A suboptimal approach could be based on statistical analysis, which would provide decisions based on average demand for different destinations. Similarly, if a totally automated vehicle guidance system is not implemented, it is not clear how individual drivers can be induced to follow the optimal route calculated by the central traffic center. Presumably, optimal routing selections could be transmitted to the vehicle, either on an in-vehicle display or on a panel at the intersection (node), and a certain percentage of drivers would accept and abide by these suggestions.

The optimal routing problem setting also implies identification and tracking of individual vehicles as they make their way through the network. One possibility is cars checking in at each intersection since locations are not needed while traveling along a link. On the other hand, the technology for locating vehicles already exists and is used in Global Positioning System (GPS)-based vehicle location systems. The wireless portion in such systems, presently informing a dispatcher of the vehicle's location, can also be used for other information transmittal.

One class of technologies that needs to be mentioned at this point is that used for counting and possibly identifying vehicles as they pass along a roadway or stop at an intersection. These range from the classic detection loop embedded in the road to new radar-based or laser-based or image-processing-based technologies. Recent applications, specifically tested on trucks, would have a transponder on the vehicle providing

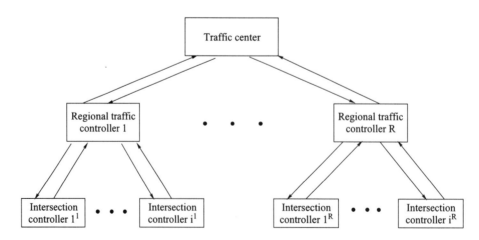

Figure 14.4

information to roadside units. All these technologies could conceivably be utilized in closing the loop in any kind of automated routing system.

The fully Automated Highway System of the future provides many more challenges. Overall routing information needs to be provided to individual vehicles, along with either roadway-based speed settings or speed directives to platoons. Ingress and egress information also needs to be provided, possibly at entry–exit locations. Although there may be simpler solutions for the latter, in general a wireless network seems to be needed. A wide-scale application of AHS or routing in general and the associated wireless communication network introduces new problems. For example, the question of *handoff* as information related to individual vehicles as well as links to them change dynamically provides interesting control theory challenges.

The vehicles in an AHS may actually need more infrastructure aid, not just for macro-level movement (traffic control), but also micro-level movement (automated lane tracking). We will discuss this topic in the next section.

14.3 INTELLIGENT VEHICLES

14.3.1 Pre-IV Autonomy: Cruise Control and ABS

Just as the topical area of intelligent control refers to designing systems with some amount of autonomy, the term "intelligent vehicle" is used for a car (or other means of transportation) that has some capability of autonomous mobility.

Vehicles where some aspect of human driving is taken over exhibit some autonomous behavior beyond simply having automatic control feedback loops. Viewed in this light, both cruise control (the human driver is no longer regulating the speed) and ABS (the driver is no longer adjusting the brakes) have characteristics of autonomy.

Cruise control is basically a throttle control feedback loop, whereas ABS is a brake control feedback loop. From the pure control viewpoint, the realization of an intelligent vehicle would require two additional capabilities: the ability to jointly control multiple loops and the ability to close the third loop, steering. As we mentioned before, more capabilities of situation analysis and decision making will have to be added at this point, and the need for additional sensing, especially with respect to the effects of the environment, becomes important.

14.3.1.1 Preliminary Needs: Drive-by-Wire Vehicles

As mentioned earlier, the control-related requirements of an intelligent vehicle would involve the ability to close the throttle, brake, and steering feedback loops through a decision mechanism, presumably implemented on a computer. This implies that drive-by-wire capability is needed. Electronic throttle control is not a new issue, and simple changes in cruise control ECUs (electronic control units) can provide full capability. ABS units are somewhat more self-contained, dedicated units that cannot readily be turned into full electronic brake controllers. However, the technology is not particularly remote, and most vehicle companies have gone through the development of electronic throttle and brake control units. Steer-by-wire, on the other hand, is not standard. Few cars on the road have full electrical steering, although the technology does exist in the market.

During Demo'97, various companies (GM, Honda, Toyota) demonstrated cars that had full drive-by-wire capability. Figure 14.5 shows one of the Honda vehicles which the Ohio State University team used in Demo'97. The figure also shows various sensing systems (radar, laser, camera) installed by the OSU researchers in developing an intelligent vehicle. The same technologies have been reported on, and similar capabilities demonstrated, by many car companies and research groups in Europe and Japan.

14.3.2 Car Following and Advanced Cruise Control

Although both ABS and cruise control have attributes of autonomy, one point of transition from *standard* vehicles to so-called *intelligent* vehicles relates to the ability to slow down in cruise control mode, when a vehicle detects a slower vehicle ahead. This feature is referred to as Advanced Cruise Control (ACC) and Cruise Assist Systems or, less popularly, as Adaptive Cruise Control and Intelligent Cruise Control.

ACC implies that there exists a means by which a vehicle ahead can be detected, its distance measured and its relative velocity measured or calculated. Various technologies exist to do this at this time. Popular among them are laser and radar systems, although vision-based techniques can also be considered. ACC can be used either for cars or trucks, or it can be part of a full convoy operation (where lateral control is also implied).

In this section, we are concerned with the longitudinal control aspects of the convoying problem. For the time being, it is assumed that the convoy consists of only two vehicles (trucks) in a leader-follower configuration as depicted in Figure 14.6, although the generalization to more vehicles is straightforward. The follower has to be able to detect the leader, estimate the distance to it, and estimate the leader's speed. Vision-, radar-, or laser-based systems can be utilized to accomplish these tasks. In Figure 14.6 we have indicated a *patch*, providing the follower's (forward-looking) sensor system an easily detectable target.

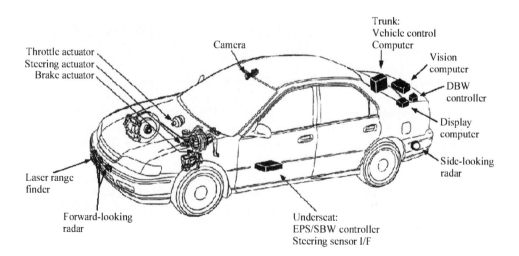

Figure 14.5

Section 14.3 Intelligent Vehicles

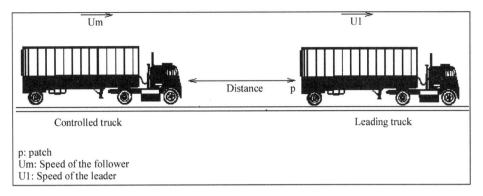

Figure 14.6

The main task of a longitudinal controller is to keep the distance between the vehicles at a desired safety level. Let the longitudinal velocities of the follower and the leader be U_M and U_L, respectively. The measured headway d and the safety distance d_s can be defined as

$$d = x_L - x_M$$
$$d_s = hU_L + d_o, \tag{14.10}$$

where x_L and x_M are the longitudinal positions of the leader and the follower, respectively, h stands for headway (the time it takes for the leader to stop), and d_o provides an additional safety margin. The velocity difference ΔU is given by:

$$\Delta U = U_M - U_L = -\dot{d}. \tag{14.11}$$

Consider Figure 14.7. The strategy for regulation is as follows: The recommended velocity of the follower should be chosen in such a way that the velocity vector of the solution trajectory in the $(d, \Delta U)$ plane is directed to the $(d_s, 0)$ point at any time. This choice enforces state trajectories toward the goal point on a straight line whose slope is determined by the initial position of the system in the $(d, \Delta U)$ plane and guarantees that the velocities of the vehicles become equal when the desired safety distance is achieved. The slope of the line on which the trajectory slides to the goal point determines the convergence rate. We divide the $(d, \Delta U)$ plane into six regions. The collision region and the constant velocity region are also included in Figure 14.7. In the constant velocity region, the follower keeps its current velocity until the distance between the vehicles becomes less than a user-defined critical distance d_c. Figure 14.7 also includes a relative acceleration curve that gives the least possible value of d at which the follower should begin to decelerate at its maximum rate to be able to reach the goal point for a given ΔU, assuming a constant velocity for the leader. The figure also includes a minimum convergence rate line (MCRL) whose slope is chosen by considering the minimum admissible convergence rate.

In Region 2 and Region 5, it is physically impossible to enforce the trajectories toward the goal point on a straight line. So, the controller should decelerate (accel-

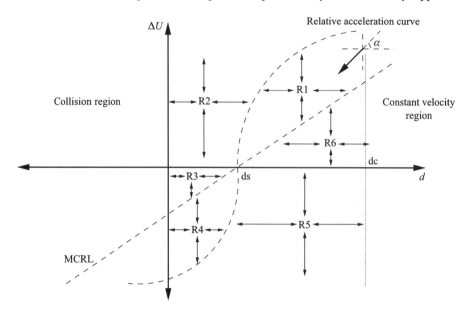

Figure 14.7

erate) the follower at the maximum deceleration (acceleration) rate in Region 2 (Region 5). In Region 3 and Region 6, it is possible to steer the trajectories to the $(d_s, 0)$ point through a straight line between the initial point and the goal point. However, the convergence rate would be smaller than the minimum admissible convergence rate because the slope of the line is less than the slope of the MCRL. So, in Region 6 (Region 3) we prefer first accelerating (decelerating) the follower toward the MCRL at its maximum acceleration (deceleration) rate and then sliding the trajectories to the goal point through this line.

In Region 1 and Region 4, the desired velocity can be calculated as follows:

$$\begin{aligned} m &= \tan(\alpha), \\ m &= \frac{\Delta \dot{U}}{\dot{d}} = -\frac{a_M - a_L}{\Delta U}, \\ m_{des} &= \frac{\Delta U}{d - d_s}, \\ m &= m_{des} \Rightarrow a_M = -\frac{(\Delta U)^2}{d - d_s} + a_L, \end{aligned} \qquad (14.12)$$

where m is the slope of the trajectory velocity vector, m_{des} is the desired slope, and a_M, a_L are the accelerations of the follower and the leader, respectively.

Equation (14.12) gives the necessary acceleration for the follower that ensures the exact convergence of the solution trajectory to the goal point on a straight line. However, it may not always be possible to obtain this acceleration due to the acceleration and jerk limits of the vehicle. The bounds on the acceleration are determined by the physical capacity of the vehicle, whereas jerk limits are mainly determined by riding comfort.

In the other regions, the above argument also holds except that a_M is taken as a_{max} (a_{min}) in Region 6 and Region 5 (Region 3 and Region 4) instead of using Eq. (14.12).

14.3.3 Lane Tracking

One of the key goals of an automated vehicle is the ability to perform automatic steering control. Steering control is a nontrivial design problem. Two of the major control subproblems associated with lateral control are "lane keeping" and "lane changing." We discuss lane keeping in this section and lane changing in the following one.

There are various external disturbances (wind gusts, bumps on the road, and sensor noise) and unmodeled dynamics (due to model simplifications, uncertain parameters, and actuator nonlinearities) that affect the closed-loop performance of the feedback system. Robustness with respect to modeling uncertainties and disturbances, rider comfort, and safety are crucial design concerns.

Steering control is a fundamental design challenge, and the approaches taken to obtain a stabilizing robust controller design vary significantly based on the available set of sensors and the performance of the actuators involved. A measure of the vehicle's orientation and position with respect to the road must be available to the controller. Among the most commonly used techniques are vision-based lane marker detection (preferred by many because of its simplicity in terms of the required machinery and implementation convenience), radar-based offset signal measurement (developed and used by OSU researchers exclusively), and the magnetic nail-based local position sensing (used by PATH researchers). Vision- and radar-based systems provide an offset signal at a preview distance ahead of the vehicle that contains relative orientation information. The vision system directly processes the image of the road and detects lane markers. Therefore, it does not require any modifications to current highway infrastructures. The radar system requires that an inexpensive passive frequency selective stripe (FSS) be installed in the middle of the lane, in which case the radar is capable of providing preview information similar to a vision system. Most other sensor technologies provide only local orientation and position information. It has been pointed out that control of vehicles without preview distance measurements poses a difficult control problem at high speeds. Indeed, the experience of researchers using look-down-only sensors is that road curvature information must be provided to the lateral controller, usually by encoding it in the sensor components installed on the road. Thus we see that sensors are an integral part of the design and that the performance of the sensor system directly impacts the closed-loop system stability and performance.

The lane change problem is even more challenging than lane keeping. Unless the sensor system can provide reliable data to guide the vehicle completely from one lane to another, a portion of the lane change maneuver needs to be performed "open-loop." Major problems arise due to the following: (1) sensors are noisy, (2) road curvature and super-elevation (the road as a surface in 3-D space) are unknown, (3) a sensor blind transition period requires open-loop control, and (4) robustness with respect to wind disturbances, unmodeled dynamics, and uncertainties in the system becomes more important and requires extra consideration.

14.3.3.1 Vehicle Model

It is assumed that the vehicle is operating on a flat surface and that a linearized bicycle model is capable of describing the motion of a vehicle effectively. The standard linearizing small-angle assumptions are made for the tire slip angles and the front tire steering angle. A wind disturbance is modeled that affects the lateral and yaw motions of the vehicle. The corresponding model is depicted in Figure 14.8. The variables represent the following physical quantities: $u(t)$, $v(t)$, and $r(t)$ are the longitudinal velocity, lateral velocity, and yaw rate, respectively, $\delta(t)$ is the actual steering angle of the front tires, $\psi(t)$ is the yaw angle with respect to the road, $y_{cg}(t)$ is the deviation of the vehicle's center of gravity from the lane center, $o(t)$ is the offset signal at the look-ahead point, f_f and f_r are the lateral tire forces on the front and rear tires, respectively, a and b are the distances from the center of gravity of the vehicle to the front and rear axles, respectively, l_w is the position at which a wind disturbance force of f_w laterally affects the vehicle motion, d is the sensor preview distance, and $\rho(t)$ is the road curvature at the look-ahead point. All distance measurements are in meters, and all angles are in radians.

The vehicle dynamics are represented by the following set of linear system equations

$$\dot{v}(t) = a_{11}v(t) + a_{12}r(t) + b_1\delta(t) + d_1 f_w, \tag{14.13}$$

$$\dot{r}(t) = a_{21}v(t) + a_{22}r(t) + b_2\delta(t) + d_2 f_w, \tag{14.14}$$

$$\dot{y}_{cg}(t) = v(t) + u\psi(t), \tag{14.15}$$

$$\dot{\psi}(t) = u\rho(t - t_o) - r(t), \tag{14.16}$$

$$\ddot{z}(t) = u^2[\rho(t) - \rho(t - t_o)] - du\dot{\rho}(t - t_o), \tag{14.17}$$

$$o(t) = y_{cg}(t) + d\psi(t) + z(t), \tag{14.18}$$

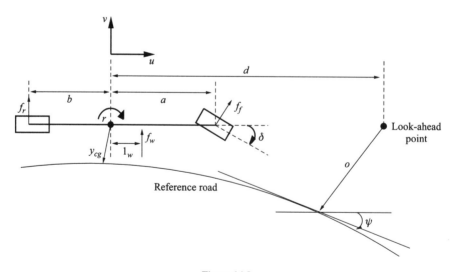

Figure 14.8

Section 14.3 Intelligent Vehicles

where $z(t)$ is a dummy variable that is necessary to characterize the transient response of the offset signal precisely. In this context, $o(t)$ is the measured offset from lane center at the look-ahead point (positive to the left of the lane center), and the vehicle center of gravity $y_{cg}(t)$ (also positive to the left of the lane center) is to be regulated to zero for all possible road curvature reference inputs $\rho(t)$ (positive for clockwise turns) defining the desired path to be followed using the front-wheel steering angle $\delta(t)$ (positive for clockwise turns). The linearized model is valid at the operating longitudinal velocity u (positive for forward motion), which is assumed to be kept constant by means of a decoupled longitudinal controller, and for small values of $\rho(t)$. The sensor delay t_o depends on the operating velocity and the preview distance d and is given by $t_o = d/u$. The other parameters of the vehicle model are determined from

$$a_{11} = -\frac{k_f + k_r}{mu}, \quad a_{12} = u + \frac{ak_f - bk_r}{mu}, \quad b_1 = -\frac{k_f}{m}, \quad d_1 = \frac{1}{m},$$

$$a_{21} = \frac{ak_f - bk_r}{uI_z}, \quad a_{22} = -\frac{a^2 k_f + b^2 k_r}{uI_z}, \quad b_2 = \frac{ak_f}{I_z}, \quad d_2 = -\frac{l_w}{I_z},$$

where $k_f > 0$ and $k_r > 0$ are the lateral tire stiffness coefficients of the front and rear tires, respectively, m is the (virtual) mass of the vehicle, and I_z is the (virtual) moment of inertia around the center of mass perpendicular to the plane in which the vehicle is located. The remaining variables are as previously defined.

Typical parameter values approximating those of the OSU vehicles are given in Table 14.1.

14.3.3.2 A Nonlinear Lane-Keeping Controller

The vehicle model provided in the previous section is fairly generic. Here, on the other hand, we provide a specific controller (that used by the OSU team during Demo'97) simply to illustrate the type of nonlinear controller design that may be needed in an IV application.

The lateral control law that is employed to steer the vehicle, which consists of multiple terms that are functions of the measured signals, is

TABLE 14.1 Typical Model Parameters for OSU Vehicles

a	1.35 m	cg to front axle distance
b	1.37 m	cg to rear axle distance
m	1569 kg	Total mass of the vehicle
k_f	5.96×10^4 N/rad	Front tire cornering stiffness
k_r	8.66×10^4 N/rad	Rear tire cornering stiffness
I_z	272.4 Ns/rad	Moment of inertia along z-axis
u	$[1, 40]$ m/s	Range of longitudinal velocity
d	8.1 m	Preview distance
G	1/19160	Actuator gain
ω_n	22.94 rad/s	Actuator natural frequency
ζ	0.517	Damping coefficient
t_1	0.03 s	Actuator delay

$$\delta_{buf}(t) = K_d \cdot \dot{\hat{o}}(t) + K_s \cdot \hat{o}(t)|\hat{o}(t)| + K_\psi \cdot \Psi_{reset}(t)$$

$$+ K_r(r(t) - r_{ref}(t)) + K_i \cdot \text{sat}\left(\int_0^t \hat{o}(\tau)d\tau\right)$$

$$+ K_m \cdot |p| \cdot \text{sign}(\text{deadzone}(o(t))), \tag{14.19}$$

$$\delta_{com}(t) = \text{sat}(\delta_{buf}(t)), \tag{14.20}$$

where $\hat{o}$ and $\dot{\hat{o}}$ are the Kalman observer estimates of the offset signal and its derivative and K_d, K_s, K_ψ, K_r, K_i and K_m are gains of appropriate dimensions and signs. $\Psi_{reset}(t)$ is defined as

$$\Psi_{reset}(t) = \int_0^t r(\tau)d\tau, \tag{14.21}$$

such that $\Psi_{reset}(t) = 0, \forall t = 0.5k, k \in \mathbb{N} \cup \{0\}$.

Each component of the nonlinear steering signal given in Equation (14.19) has a particular significance. The derivative of the offset signal helps suppress the otherwise noticeable limit cycles. The quadratic term generates a large penalty for large deviations from the lane center. The resetting yaw angle periodically corrects the orientation of the vehicle and aligns it with the road. The integral term is used to minimize the tracking error at steady state, and the saturation helps reduce oscillatory behavior during transients. The last term accounts for a sliding-mode-like switching assist toward the lane center upon necessity. A crossover detection algorithm along with a resettable timer runs continuously. If the vehicle deviates from the lane center for more than a specified time period and if its peak deviation, p, exceeds a threshold value, then an additive steering term nudges the vehicle toward the lane center until a crossover occurs. Under normal operating conditions where the vehicle is tracking the road center closely, this term has no contribution. During normal driving, the dominant component in the steering command is the term based on yaw error (the difference between a reference and the actual yaw rates). The overall steering command to the steering motor is saturated in order to satisfy safety requirements.

The parameters used for normal highway driving are shown in Table 14.2. A different set of parameter values are required for high-performance (high-speed, large-curvature, winding, or slalom course) driving.

TABLE 14.2 Lateral Controller Parameter Values

K_d	12.00	K_s	46.00	K_ψ	−10.00
K_r	−1200.00	K_i	12.00	K_m	75.00
K_{ref}	0.03	K_δ	−25.00	κ	2.00
T	0.01	B_1	1.00	B_2	100.00
P	0.00437	μ	1.00	γ	0.10
M_u	1.50	Δ_1	1.00	Δ_2	1.00

14.3.4 A Lateral Lane Change Controller

In any application-oriented controller design, the reasoning behind the design path pursued lies in the plant to be controlled and the available sensors and actuators. Modularity and flexibility are always desirable, but the controller must work on the system at hand. In this case, the choice of a design procedure was mandated by the fact that preview sensor information (which is used in our lane-keeping algorithm) cannot be measured continuously during the transition from one lane to another using either the vision or the radar reflective sensors. There is a dead-zone period when the preview sensing systems do not provide useful data. This creates a transition period that must be handled "open-loop" with respect to lateral position information. Attempts to generate a true open-loop time series steering angle command profile failed because of wind and super-elevation disturbances, nonsmooth actuator nonlinearities, unmodeled vehicle–road interactions, and uncertainties in the (possibly time-varying) plant parameters. Most of these dynamics and disturbances can be bypassed through yaw rate measurement. Thus for the lane change a vehicle yaw rate controller was designed and used to implement a desired time series yaw rate profile, which would bring the vehicle to the center of the next lane and preserve the vehicle's angular alignment with the road.

The lane change problem can be summarized as follows: While maintaining lane orientation at a longitudinal speed u, the vehicle travels a specified distance (a full lane width) along the lateral axis with respect to its body orientation within a finite time period and aligns itself with the adjacent lane at the end of the maneuver such that the lane-keeping task can be resumed safely and smoothly. The autonomous lane change problem deals with the generation of the appropriate steering signal to cause the vehicle to accomplish the above described task without driver assistance. The major design assumptions are: (1) only the yaw rate r and the steering angle δ are measured, (2) vehicle parameters are known within a bounded neighborhood of some nominal values, and (3) the road curvature does not change significantly during the lane change maneuver.

Studies have been performed to estimate the ideal lateral jerk, acceleration, velocity, and displacement signals that the vehicle's center of gravity should follow to perform a lane change maneuver while preserving passenger comfort. However, in practice the only input to the vehicle is commanded steering angle. Therefore, these results must ultimately be used to generate steering angle commands. This can be accomplished by generating a reference yaw rate signal and applying a yaw rate controller to generate steering angle commands.

14.3.5 Hybrid Systems and Scenario Resolution

We have considered some of the basic, control-related issues that are relevant to an intelligent vehicle. Depending on what the IV is to do and what intelligence it is to show, a level of real-time situation-analysis and decision making will have to be implemented next.

There are many different, technical ways of addressing situation-analysis and decision making. As far as producing an implementable controller is concerned, we will usually end up with a hybrid system. One approach to modeling the discrete decision making portion of the hybrid system is to use finite state machines. For illustration purposes, we will provide a few examples of finite state machine representations of situations in IV problems. (The reader is warned, however, that these are contrived examples; the true situation is much more complex.)

We first consider the standard cruise control situation. Assume that there are only two *states*, manual and cruise. Transitions can occur, depending on a number of external events, as shown in Figure 14.9(a). The states and transitions are associated with a set of external measurements and imply certain lower level control actions. In the standard cruise control situation, the external measurements causing state transitions are somewhat limited, in fact, and hence so are the number of states.

Figure 14.9(b) illustrates an Advanced Cruise Control (ACC) situation. The number of states has increased, the situations covered have expanded, and a larger sensor suite is implied. A speed-up state (which could have existed in the standard situation also) has now been distinctly identified.

The ACC outlined above and in Figure 14.9(b) is a rather simple one. We have not dwelt on issues like possible delays before speeding up when a car in front of us disappears, or whether brakes will be used in slowing down, or distinctions in required deceleration, and so on.

14.4 CONCLUSIONS

14.4.1 Related Problems

In this chapter we have concentrated on a few specific control problems in the ITS area. The selection was somewhat subjective. Indeed, many other problems would be of interest to the control engineer. We will briefly mention some of them, although no claim of completeness is made for this list either.

14.4.1.1 Precision Movement

The evaluation of the concept of fully automated driving has increased interest in other areas where an automated road vehicle may exercise its capabilities. One area is in precise automated movement for busses in approaching bus stops, otherwise labeled as *docking*. The bus may or may not be driving autonomously on the roadway. However, control is switched to automatic as the bus stop is approached, and a precise stop is achieved. Docking is envisaged for direct wheelchair access to a bus, but the issues are similar to any such movement by trucks for loading purposes.

14.4.1.2 Coupled Systems

The movement of a convoy of vehicles, especially trucks with multiple trailers, provides interesting new stability problems. The full 3-D movement of such systems, with due attention being paid to delayed actuation, generally distributed braking problems, and so on, still provide fertile ground for control research.

14.4.1.3 Autonomy versus Full Information Exchange

The problem of autonomy versus full information exchange has control-design-related implications for both general traffic problems and individual vehicle motion.

Section 14.4 Conclusions

(a)
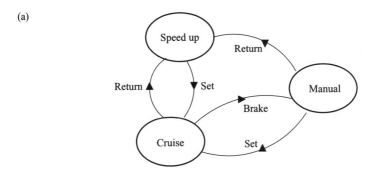

(b)
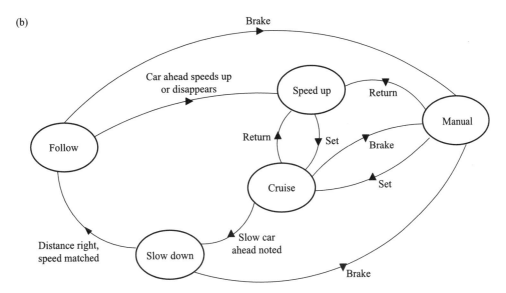

(c)
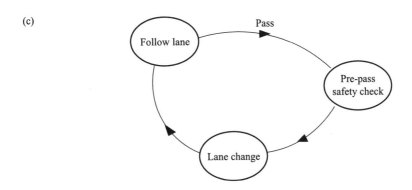

Figure 14.9

There are two aspects to this problem. The first is related to the imposition of an information exchange structure. In the United States, a National ITS Architecture has been adopted, which defines the paths of information exchange and the location of various pieces of real-time traffic and transportation data. Use of such data for control purposes imposes certain constraints on the system design.

Second, the availability and cost of sensor information also impose constraints on the system designer.

14.4.1.4 Fault Tolerance/Safety

One of the key issues in present-day ITS research is safety enhancement. As the intelligent vehicles of the future are being developed, more demands are being placed on the complex systems controlling both the flow of the traffic and the regulation of each moving platform.

A number of technologies are presently being pursued for warning devices. Lane change safety warning, rear collision avoidance, lane departure warning, and weaving vehicle (sleeping driver) warning devices are all initial steps toward what could become control systems in the future. The error-free operational requirements for these warning devices today imply very high safety expectations once the control loops are closed tomorrow. These concerns will lead to extensive investment in fault-tolerant control in the ITS area.

14.4.2 And Technology Keeps Marching On . . .

Undoubtedly, ITS applications and ITS research in control infrastructure rely on the choice of technologies available. For example, the automated steering control design would be different for an IV that uses magnets inserted in the roadway and utilized as the vehicle passes over them than for an IV that would look ahead to identify its heading from a roadway-placed, radar reflective stripe.

Again, if a car has access to information that a stopped vehicle is in the lane ahead, it would need a decision strategy and a controller that would execute a possible lane change. (A communication system was demonstrated in Demo'97 where a leading car provided just such information to following cars.)

Yet another example would be the utilization of GPS and precise road map databases in aiding the movement of a car. A lane departure control system that would nudge a truck back into the lane would be very different if it were based on a GPS/map system or a vision system.

Finally, the supply of information to a car, not just for its micro-level movements, but also for its general decision making, can affect overall traffic patterns. For example, if information about a full parking lot is supplied to a moviegoer, she may choose to drive to a theater in a different part of town.

So, as technology changes, develops and becomes cheaper, different options will become available to the control engineer practicing in the ITS area. In this chapter we have tried to provide a glimpse of some of the interesting and exciting possibilities.

> **Related Chapters:**
>
> - New developments in air transportation, specifically air traffic management, are discussed in Chapter 10.
> - See Chapter 8 for a tutorial on sliding-mode control.
> - Several popular intelligent control techniques are described in Chapter 5.
> - More details on hybrid systems are available in Chapter 7.

REFERENCES

[1] *Concise Encyclopedia of Traffic & Transportation System.* New York: Pergamon Press, pp. 478–483, 1991.

[2] E. J. Davison and Ü. Özgüner, "Decentralized control of traffic network." *IEEE Trans. System, Man, and Cybernetics*, Vol. SMC-13, pp. 476–487, 1983.

[3] D. C. Gazis, "Network modeling and control: Store-and-forward approach." *Concise Encyclopedia of Traffic & Transportation System.* New York: Pergamon Press, pp. 278–284, 1991.

[4] P. E. Sarachik and Ü. Özgüner, "On decentralized dynamic routing for congested traffic networks." *IEEE Trans. Auto. Control*, Vol. AC-27, pp. 1233–1238, 1982.

[5] W. S. Levine and M. Athans, "On the optimal error regulation of a string of moving vehicles." *IEEE Transactions on Automatic Control*, Vol. AC-11, pp. 355–361, 1966.

[6] K. C. Chu, "Decentralized control of high-speed vehicular strings," *Trans. Sci.*, no. 8, pp. 361–383, 1974.

[7] R. E. Fenton and R. J. Mayhan, "Automated highway studies at the Ohio State University." *IEEE Trans. on Vehicular Technology*, pp. 100–113, 1991.

[8] S. E. Shladover, C. A. Desoer, J. K. Hedrick, M. Tomizuka, J. Walrand, W. Zang, D. H. McMahon, H. Peng, S. Sheikholeslam, and N. McKeown, "Automatic vehicle control developments in the PATH program. " *IEEE Trans. on Vehicular Tech.*, pp. 114–130, 1991.

[9] Ü. Özgüner, K. A. Ünyelioğlu, C. Hatipoğlu, "An analytical study of vehicle steering control." *Proc. IEEE Conference on Control Applications*, Albany, NY, pp. 125–130, 1995.

Chapter 15

AUTOMOTIVE POWERTRAIN CONTROLLER DEVELOPMENT USING CACSD

K. Butts, J. Cook, C. Davey, J. Friedman, P. Menter,
S. Raman, N. Sivashankar, P. Smith, and S. Toeppe

Editor's Summary

Reducing the cost and time associated with developing and deploying new control systems is a challenge facing all industries that rely extensively on control technology. The automotive industry furnishes a prime example, especially for powertrain control development. Computer-aided control system design (CACSD), a topic discussed in some generality in Chapter 3, provides the enabling tools for meeting this challenge. (For another example of an important application domain for CACSD, see Chapter 11 on flight control.)

This chapter discusses in detail the process of developing automotive powertrain controller software. Powertrain control systems can be highly complex, integrating a number of different subsystems: fuel injection, throttle control, idle speed control, vehicle speed control, engine torque management, emissions control, knock detection, electronic transmission control, on-board diagnostics, and numerous others. The overall design must satisfy constraints and criteria imposed by government (such as fuel economy standards and emission thresholds), the corporate environment (which may demand support for legacy systems in addition to cost minimization and rapid time to market), and consumer preferences.

To address the complexity of powertrain controller development, Ford Motor Company employs a systems engineering process. The emphasis is on validation or verification of products of all intermediate development stages, on the widespread implementation of feedback mechanisms, and on the continuing availability of support from engineering analysis and design teams. A requirements-based and CACSD-enabled development process has been implemented which incorporates requirements capture; architecture design; control feature design, implementation, and verification; and software validation, verification, and integration.

The chapter also highlights the role of effective project management in large-scale control system development. Project management metrics can be integrated within the development process, and in the case of Ford these helped justify the CACSD project investment. Statistical data demonstrating the improvement in control software quality due to automation supported by CACSD tools is included.

The authors are all with powertrain control systems groups in Ford Motor Company.

15.1 INTRODUCTION

There are two stages of automotive powertrain controller development. First, the fundamental control laws that provide robust powertrain system performance in the presence of parameter and environmental variation must be developed. This work is typically the responsibility of powertrain research and advanced engineering activities and is well documented in the literature. A very incomplete reference list includes [1, 3, 20, 32, 33]. Second, prior to product release, more design detail must be added to the fundamental control laws. This design detail ensures seamless integration with the legacy powertrain control systems, reuse across powertrain product families, and well-defined start-up, shutdown, and diagnostics behavior. The detailed design augmentation and its corresponding software realization are the domain of the production powertrain controller development organization. In this chapter we describe how computer-aided control system design (CACSD) tools can be used to support the large-scale automotive powertrain controller software development organization.

15.1.1 The Role of the Powertrain Control System

The challenge for any business that wants to be competitive in a global marketplace is to rapidly and efficiently develop innovative products that meet the needs of worldwide customers. In the automotive industry, customers demand high-value, reliable personal mobility encompassing a wide spectrum of use and lifestyle. These customer requirements must be achieved without neglecting society's imperatives manifested by government regulations on emissions, fuel economy, and safety. To the designer of automotive powertrain control systems, these needs translate into system constraints and performance requirements that must be met quickly and at minimum cost for a multitude of powertrain options and vehicles.

It is an important requirement that safe and reliable performance be maintained over a wide range of environments and operating conditions for the life of the vehicle, often many more than 100,000 miles. Certainly, the vehicle must provide good fuel economy, not only to satisfy the customer's desire for low operating costs but, in the United States, to meet legislated Corporate Average Fuel Economy (CAFE) standards. By far the most conspicuous system requirements are the regulations governing exhaust emissions. Beginning in 1999, U.S. federal standards are attempting to reduce ozone formation by regulating nonmethane organic gas emissions. (NMOGs are total hydrocarbons, less methane, plus aldehydes, keytones, alcohols, and ethers.) These standards define three categories of passenger cars and light duty trucks. Those with the lowest emissions are called Ultra Low Emission Vehicles, or ULEVs. These vehicles have requirements of 0.04 grams/mile NMOG, 1.7 grams/mile carbon monoxide, and 0.2 grams/mile oxides of nitrogen, all after 50,000 miles of use. In California, regulations have been proposed that will sharply reduce even ULEV standards. Worldwide, new regulations are anticipated that will enforce emission reductions in every market. In addition, legislated on-board diagnostics (OBD) require the embedded powertrain controller to monitor all emission-related subsystems, inform the driver of a subsystem malfunction by illuminating an indicator on the instrument panel, and store a descriptive code in the on-board computer memory to be cleared when the vehicle is repaired.

Developing and implementing a management system to achieve these objectives is a complex and multifaceted task. The modern automobile powertrain consists of many interacting subsystems that must be coordinated to achieve the required system performance. Naturally, technology improvements that reduce emissions often increase system complexity by adding new subsystems with additional sensors and actuators. Invariably, transient system performance is crucial, and robustness with respect to parameter variations caused by age and environment is mandatory. Finally, controllers and estimators must be designed to minimize *in-situ* calibration to speed up product development, and the embedded control software must be reusable across different vehicle powertrains.

Powertrain control systems vary widely in scope. The low-end systems consist of simple port injected four-cylinder engine controllers with integrated idle speed control, spark timing control, and manifold pressure-based air estimation. The high-end systems integrate port or direct fuel injection systems for up to 10 cylinders, with exhaust gas oxygen sensors before and after catalytic converters. Electronic throttle control subsystems may be employed with integrated idle speed control, vehicle speed control, and engine torque management. Exhaust gas recirculation, evaporative emissions controls, electronic ignition systems, knock detection and control systems, electronically controlled returnless fuel delivery systems, and engine auxiliary controls (e.g., air conditioning, alternator) are also common. Electronic transmission controls, including pressure control, torque converter slip controls, and shift controls for up to six speeds, are fully integrated with the aforementioned engine control subsystems. In addition, the on-board diagnostics requirements can increase the complexity by as much as 100% because every input and output to the control system must be continuously verified for proper operation.

15.1.2 The Powertrain Controller Development Organization

The organizational structure to develop powertrain control systems can take on many different implementations. However, the following tasks are common to each.

- System specification
- Hardware specification
- Hardware design
- Software specification
- Software design
- System integration and calibration

System specification refers to the task of converting the many governmental regulations and customer requirements into a first-level decomposition of the overall control system. This task may include choosing general control system sensor and actuator combinations that meet cost, weight, functional, and other targets. The production development group typically limits itself to choices that have been proven ready for implementation by an advanced engineering activity. From the system specification, detailed hardware and software specifications can be developed through the cascading of system or subsystem requirements. On the software side, CACSD-based processes and tools can be employed to develop and realize these specifications. Once the realized software has been verified, it is calibrated, or tuned, for the specific vehicle in which it will be used.

Organizational structures that support this range of activities for a major automotive manufacturer or supplier of control systems are typically complex and distributed geographically. The organizations can focus on projects or control features (e.g., idle speed control, fuel control, ignition timing control). Project-focused organizations tend to provide good customer support at the expense of customized solutions to generic problems. On the other hand, organizations that focus on control features try to optimize resource requirements and provide generic solutions for all supported vehicle applications.

Given a control feature-focused organization, care must be taken to partition the overall control system into chunks that can be maintained effectively. Chunks chosen too small result in excessive interfaces between control features. Chunks chosen too large or coarsely can result in low productivity due to individual workload constraints and unstructured control/software solutions within the chunk. We have chosen to partition a high-end control system into about 100 control features. Some engineers will maintain only one large feature (e.g., ignition timing), whereas others may maintain 10 or 15 smaller features (e.g., input processing features).

Our organizational structure mirrors the chosen control system architecture. We combine individual control feature engineers into subgroups for administrative as well as functional reasons. These subgroups are organized, for example, to collect all the fueling-related features together. It is imperative that those individuals responsible for design and maintenance of the core control features also be responsible for the related on-board diagnostic features. In addition, design experience has shown that isolating the specific input/output devices, microcontrollers, and operating systems from the core control features is beneficial. For example, the fuel control features should not have specific knowledge of or be dependent on the air measurement technology. Instead, the fuel features are designed with a standard interface to the air measurement system so that the fuel system features can be developed in parallel with the air measurement feature. This feature architecture management facilitates reuse across powertrain applications, even when implementation technologies differ.

The production groups responsible for control system implementation focus primarily on maintenance of the existing system. That is, new work must conform to legacy software application architectures and implementations. This seemingly rigid constraint is driven by vehicle development program demands for quality, productivity, and timing. In particular, today's automakers strive to complete vehicle programs in 24 months or less. This goal leaves very little time for redesign of system architectures or new feature development. In fact, in this organizational model, a strong advanced engineering activity is required for new and complex control structures. The advanced engineering organization employs rapid prototyping techniques such as automatic code generation in conjunction with high-powered development platforms to develop and bookshelf new technologies [14]. These groups must be able to transfer their technology, with minimal or no rework, into the production development machinery.

15.2 THE SYSTEMS ENGINEERING PROCESS

We have found the *IEEE Trial-Use Standard for Application and Management of the Systems Engineering Process* [18] to be instructive and helpful. In the standard's terminology, our subject matter, automotive powertrain controller software, would most likely be classified as a subassembly of the powertrain-control-module component. Thus powertrain controller software development is far down the systems engineering process hierarchy. Even so, the engineering effort and software behavioral complexities warrant a carefully considered development process that emphasizes requirements validation and synthesis verification. Consider the abstract systems engineering process presented in Figure 15.1. We emphasize three critical ideas that we have adopted for our powertrain controller software development process:

1. Each work product is validated (to ensure compliance with system requirements) or verified (to ensure faithful implementation of system functional requirements) prior to delivery to the next development stage.
2. There is always a feedback mechanism to previously executed development stages.
3. Engineering analysis and design support development at every stage.

15.2.1 The Powertrain Controller Development Process

Poor understanding of requirements has been cited as a problem area in software engineering since the 1960s [4, 11, 12, 25, 28]. Requirements defects cost 2 to 13 times more to correct during the testing phase and 5 to 50 times more to correct during the deployment phase than during the requirements phase [12]. Internal measurements of our old development process show that as much as 60% of our engineering effort was spent reworking powertrain controller software requirements and design defects. The new approach is to emphasize powertrain control system design, analysis, and design validation early in the development process. Thus the design engineer is responsible for

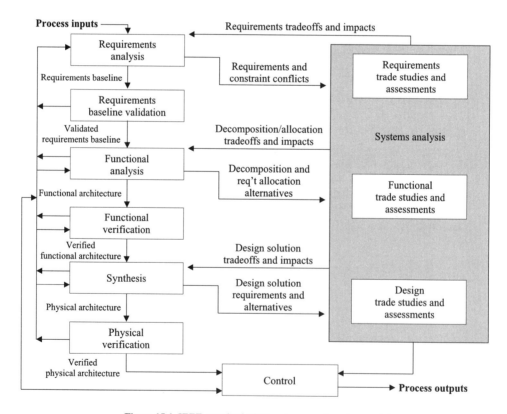

Figure 15.1 IEEE standard 1220 systems engineering model.

ensuring that the control software design meets the system's needs. We compensate for this additional engineering effort with efficient software synthesis, automated software verification, and reduced rework. The new requirements-based CACSD development process is described below.

Software Requirements Capture: The process of capturing and documenting the software requirements. These must be testable, and they form the foundation of the Functional Validation test scenarios. They should be linked to higher-level system/subsystem requirements and lower-level Software Application Architecture and Control Feature Design components.

Software Application Architecture Design: The process of allocating software requirements to control feature design components. The architect strives to reuse legacy control feature components whenever possible. New components and their interfaces to the legacy architecture are defined when necessary. Most aspects of algorithm scheduling are accomplished during this stage.

Control Feature Design and Validation: The act of transforming allocated requirements into an executable Control Feature specification model that details each and every expected behavior of the production software. The output of this step is a detailed, executable, and structured model. Once the algorithm model is prepared, exhaustive simulations should be used to validate the model against the requirements. These test scenarios must be structured in a way that allows for reuse during Control Feature Functional Verification. When this stage is executed by an advanced engineering activity, rapid-prototyping is used to validate the fundamental control law for robustness to plant uncertainty and environmental variation.

Software Application Validation: The compilation and analysis of the complete system model to validate system behavior. Executed after detailed control feature design and prior to committing software development resources, this analysis ensures that the integrated control application meets system-level requirements. The test scenarios developed in this stage are reused in the Software Application Structural Verification and Software Application Functional Verification stages.

Control Feature Software Design: The elaboration of the control feature specification model with software engineering design detail. Additional information such as signal data-typing for target microcontroller implementation, function partitioning, and file allocation is included. Software engineering design metrics, such as fan-in, fan-out, complexity, and cohesion, are generally used to assess the design.

Control Feature Software Implementation: The conversion of the elaborated control feature specification model to production code. Local coding practices guide the software engineer in this conversion. There is an opportunity to automate this step, which should reduce development time and dramatically improve initial software quality.

Control Feature Structural Verification: The comparison of control feature specification model components and software code component behaviors. These non-real-time, host-based simulation tests verify compliance between code and algorithm model at the lowest levels. We strive for 100% statement and branch coverage of the algorithm model. Automation plays a large role in test suite preparation and execution.

Control Feature Functional Verification: The behavioral evaluation of a control feature software realization or a collection of control feature software realizations. The specification model behavior is captured via non-real-time, host-based simulation. The models are then replaced by the equivalent production code, subjected to the same test scenarios, and simulated. Simulation results from the two systems are compared to ensure all monitored signals in the code and the model agree. The test scenarios are reused from the earlier Control Feature Design and Validation stage. This testing, by association, verifies that the code meets the stated requirements.

Software Application Structural Verification: The comparison of application assemblies of the control feature specification models and the corresponding software realizations according to the software application architecture definitions. Data-flow and control-flow analysis techniques are employed to ensure that the application code behavior is consistent with the application model behavior. Particular emphasis is placed on inter-control-feature connectivity, signal latencies, and execution order.

Software Application Functional Verification: The evaluation of application software realizations against software system functional requirements via realistic drive scenario testing. Test scenarios defined in the Software Application Validation stage are reused in this non-real-time, host-based simulation. Once again, the model and code behaviors are compared for consistency.

Software/Module Integration Verification: The assembly of the application software with a target microcontroller module in a hardware-in-the-loop system. In this system, the powertrain and vehicle dynamics are simulated in real-time, and the sensor/actuator electrical signatures are emulated. Thus the target microcontroller module is stimulated as if installed in the target vehicle application. The realistic drive scenarios are rerun in this environment to ensure there are no detrimental results due to real-time software performance, low-level driver interactions, or target microcontroller compiler issues.

15.3 COMPUTER-AIDED CONTROL SYSTEM DESIGN FOR POWERTRAIN CONTROLLER DEVELOPMENT

We rely on CACSD tools to meet the quality and efficiency objectives of the process outlined in the previous section. Advanced development groups within the automotive industry have leveraged these tools for several years [13, 16, 21, 23, 32, 35]. Our more recent challenge was to show that the tools could be used for detailed software specification in a large-scale production environment. Toward this end, we have formulated tool requirements for the automotive powertrain controller development community [7, 8, 34]. Briefly, these requirements include the following:

- A commercially supported and seamless environment for information transfer between the various stages of powertrain controller development.
- An environment for performing dynamic systems analysis and controller synthesis.

- An environment for modeling and simulating hybrid (mixed continuous and discrete-state) dynamic systems [9].
- An environment for modeling detailed-software-behavioral specifications within a real-time-structured-analysis methodology [17].
- An environment for performing controller validation and verification via rapid-controller-prototyping and hardware-in-the-loop experimentation.
- An environment for automated document and report generation.
- An open environment with documented interfaces to model data and tool features.

We expand on these requirements and their application to the powertrain controller development process in the remainder of this section.

15.3.1 Software Requirements Capture

Functional and nonfunctional requirements need to be determined for each of the identified features; this is the first step of the CACSD process. Individual requirements should exhibit the qualities outlined succinctly in [22]. Requirements thus are feature capabilities that are necessary, concise, implementation free, consistent, unambiguous, and verifiable.

The software functional requirements include those that are necessary for normal and safe operation of the vehicle, as well as requirements pertaining to mandated diagnostic information storage and visibility. The functional requirements address performance attributes such as steady-state and dynamic response trajectories, rise and settling times, and allowable error budgets based on the subsystem and sensor/actuator budgets. In addition, design constraints (e.g., integration with legacy software architecture) that the system must satisfy are also captured. Requirements are the essential building blocks in the development of a model-based software validation methodology.

Over time the intent is to generate a complete set of requirements. Many requirements will be "reverse-engineered" from the current control systems implementations. A requirements capture template, which includes the rationale and the validation procedure, is used by the control feature development teams to document feature requirements. The requirements specify the necessary behavior at start-up, shutdown, and during normal, aged, or degraded operating conditions of the plant, sensors, and actuators. The requirements document is reviewed by a team of experienced engineers for content and rationale.

The generation, refinement, and tracking of requirements can be accomplished using commercial requirements management tools. In the future, these tools will be used to link software requirements to the higher-level system requirements and to provide links to the relevant control feature specification model objects (e.g., data elements, model components, and test procedures). Thus it will be possible to trace (in both directions) the requirements cascade to the design level. The same capability will be available for managing validation and verification procedures.

15.3.2 Software Application Architecture Design

Software design determines the proper software architecture, module organization, multitask scheduling, software packaging, and detailed design specification. It is important to select or establish a methodology that supports the unique organizational and business issues and then to consistently apply the methodology. Two of the most popular design methodologies include structured methods and object-oriented methods. Structured methods [17, 36] are commonly used for embedded control systems. Newer object-oriented methods [5] have been successfully applied in some application domains but are less common in embedded control systems. We discuss architecture, module organization, and multitask scheduling in the remainder of this section. Software packaging and detailed design are discussed subsequently.

Software architecture provides the overall structure and organization for a software design. For embedded control systems, the architecture needs to permit numerous control algorithms to be combined and scheduled. The architecture also addresses how the algorithms, utilities, communication interfaces, input and output processing, and other necessary support software are integrated using uniform approaches. Finally, the architecture needs to be structured to support expansion, maintenance, and microprocessor retargeting. Because of its pervasive nature, the architecture should be consistent with all phases of powertrain-controller software development.

Module organization defines how a system is decomposed into major functional partitions. Well-selected functional partitions permit the system to be efficiently implemented, maintained, and extended. There are several approaches to determining the proper functional partitioning. The best approach is to be aware of each of the potential modularization criteria, establish tradeoffs to be made, and then apply the techniques that best fit the project's overall needs. Some projects may be resource constrained and cannot afford the overhead of extensive data hiding. Other projects may require significant integration of numerous modules that need well-defined and rigorously enforced interfaces.

Multitask scheduling is crucial to establishing the proper algorithm concurrency, execution timing, and interactions. Reactive control systems are particularly sensitive to problems associated with scheduling. The first step is to determine the concurrent threads of execution within an application. A concurrent thread is any calculation sequence with an established output that needs to occur in response to a controller stimulus. All of the processing steps between the start point and the final output are part of the thread. Once all the threads of execution have been established, it is necessary to assign them to specific tasks. Generally, an efficient approach is to combine several threads of similar temporal characteristics within the same task. It is necessary to determine the optimal relative ordering of threads within the task. This analysis is based primarily on data-flow dependencies and data latency requirements. Synchronization between tasks needs to be developed when algorithms running in different tasks require data sharing or coordination. Resource-sharing approaches need to be carefully designed to avoid deadlocks.

More traditional approaches to software engineering would require that the algorithm requirements be specified independently of the implementation. Once the algorithms are specified, a separate software design step would organize the algorithm specification into a design specification. However, our organizational and business requirements admit a more optimized approach. Our organization's primary work task is maintenance of the control algorithms and software for numerous powertrain applications. Completely new feature development is a secondary activity. Given this work structure, it is not necessary to redesign the architecture, module partitioning, or task scheduling each time a revision is needed. It is therefore advantageous to use a fixed architecture, module partitioning, and task schedule. Only when a significant number of changes are made is it necessary to consider changes to these designs. The CACSD process described in this chapter combines the algorithm requirements definition (Control Feature Design) and the software design (Software Application Architecture Design) into a single specification. The combined approach permits a single CACSD model to be used throughout the development process.

Module partitioning is strongly enforced by the concept of the control feature. The control feature is a stand-alone module that delivers some algorithmic service to the overall powertrain controller application. Many features can be combined and mixed together based on the standard architecture. CACSD diagrams are used to capture the entire feature specification. Each feature is broken down into execution context hierarchies. The execution context hierarchies are concurrent threads of execution, which serve as the fundamental components of task scheduling. The resulting control feature design model has a strong behavioral correlation with the software realization because the algorithm requirements and software design are combined into a single specification model. This is useful for debugging, verification, and validation efforts. However, it does require that the specification engineer take into account some basic software engineering concepts; otherwise subsequent steps will suffer from poor software design.

15.3.3 Control Feature Design and Validation

Based on the software system-level requirements and architecture analysis, requirements for the individual control features are identified. The goal of the control feature design process is to meet each of these requirements. This goal can be achieved by designing a new algorithm or by updating and modifying an existing design to meet the new requirements. This step of the design process constitutes the core element of the entire process since the control algorithms that get into the final production software are designed, analyzed, and validated in this step.

The control algorithm is modeled in CACSD tools such as the MathWorks'[1] tool suite. These design tools help not only in visualizing a design concept but also in validating the design requirements through simulation and analysis. Several structured design approaches are discussed in the literature. It is important to adopt one of these approaches to have a systematic, efficient, and unambiguous design process. Our approach is based loosely on the Hatley–Pirbhai [17] design method. Data-flow is modeled in Simulink®, while the control-flow is modeled in Stateflow™. The intent of these models is to specify the production software behavior. The feature interface is clearly defined with unique signal names that are managed through an organization-wide data dictionary. Signal labels used in the production software will match those defined in the algorithm models.

The control system designer can use a top-down or a bottom-up approach (or a combination of the two) to model the control algorithm. As described previously, the fundamental algorithm has already been designed at an advanced engineering organization. At this process step, this fundamental algorithm is refined to make it more robust; to ensure integration with other control features within the application; and to satisfy all regulatory requirements. It is a good design practice to validate each subcomponent of the control feature before integration. After the model for the algorithm

---

[1] Matlab and Simulink are registered trademarks, and Stateflow is a trademark of The MathWorks, Inc.

is developed, the scheduling constraints are derived from the control system requirements. These constraints, together with real-time software architecture constraints, will ultimately dictate scheduling for the different pieces of the control feature.

After the controller is designed (and implemented as a model), it is validated in a non-real-time simulation environment. The validation tests are designed to check each requirement for the control feature. Some of the validation tests are done using plant models with feedback from the control feature under test. Since numerous subsystems need to be validated using non-real-time simulation, plant models that are appropriate for each control feature need to be developed, implemented, and maintained. Most of the plant model components are common to all of the subsystems, while only a few models must be customized. Hence it is very inefficient to develop and maintain individual custom plant models to meet the validation requirements for each control feature. Instead, if a large-scale flexible modeling environment is developed, then each of the control designers can "plug-in" appropriate component models into this common environment for their individual validation studies. A successful implementation of such a flexible environment is very resource intensive and has a steep learning curve. However, creating this flexible environment can have productivity returns through improved model reuse and reduced maintenance requirements.

A control-oriented powertrain plant model has been implemented in such a flexible, multi-user modeling framework in Matlab®, Simulink®, and Stateflow™ to support the control feature validation process. This plant model captures the essential dynamics required to test control algorithms at the feature design step. This multilevel powertrain model consists of interconnected component models that are linked to elements in component libraries. The use of model libraries not only facilitates sharing of models between two designers but also gives the user a number of model choices (if they exist) for a single component. In the flexible modeling environment, a control designer may choose to "plug-in" a simple model of a component during the early stages of the design process and later "plug-in" a detailed model of the component (in place of the simple model) as the design matures. The development and management of plant model libraries and plant model architecture is undertaken by a select group of plant modelers. These plant modelers help control designers configure the appropriate instantiation of the plant model to meet their validation and analysis needs. Much of the flexible modeling environment, system, and component models are reused in other downstream process steps such as Control Feature Functional Verification and Software Application Functional Verification.

The functional validation of a specific control feature requires not only the plant models but also models of other control features with which it interacts. When necessary, it is possible to use the full detailed models of these other features for validation. However, the use of simplified idealized representations is often warranted to facilitate analysis. The control designers use large-scale plant models with feedback from the primary control feature (under test) and supporting control features to carry out the validation tests. The simulation configuration for such a test is shown in Figure 15.2. To partially automate this process, validation blocks are also designed in some instances to depict the expected behaviors of system variables. The specific instantiation of the model architecture (i.e., the library links), the system initial conditions, and the system inputs describe the different validation tests. These tests are captured in an executable script so that they can be used in a regression-testing environment for subsequent

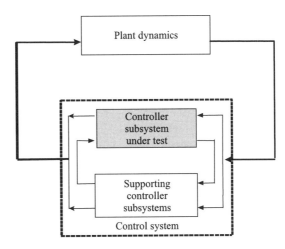

Figure 15.2 Control feature validation test harness.

design releases. The test scripts and results are also reused in downstream software verification procedures.

Debugging of the control algorithm when observing unreasonable performance is more of an art than a science. This process is where engineers add a lot of value based on their knowledge of the intrinsic behavior of the feature under test. There are no standard procedures to quickly identify the source of the problem. The control feature may have a number of tunable parameters that need to be set to reasonable values for the validation tests. One of the common reasons for test failures is that the parameters chosen for the controller may not match the characteristics of the plant model. Hence it is important to pay close attention to plant and controller parametrization during the validation phase.

Non-real-time simulation tests cannot be exhaustive. The tests are designed primarily to validate each of the functional requirements. The discrete-event nature of embedded control systems makes it possible to use formal verification methods to verify that the design (model) implementation meets the (conceptual) design intent. Verification methods that use an intelligent exhaustive mechanism may help identify some design flaws that may not be found using simple non-real-time simulation tests. However, some of the known formal verification methods are very hard to use in a production environment. This is an area of active research [30] and has good potential for the future.

Rapid prototyping can also be used to validate functional requirements for a control feature on a prototype vehicle or in a dynamometer facility. This method could be used when it is difficult to model the system behavior with reasonable accuracy. The difficulty in modeling could be due to complex physical processes and uncertainty in the initial state of the system that is being modeled. Modern-day CACSD tools provide a user-friendly means to seamlessly transfer the control algorithm to a rapid prototyping target processor. The results of the test can also be instantly analyzed using the CACSD tools.

The design and validation process steps and methodology described above are repeated until the designer and his customers are satisfied with the results. This process step takes more time when a control feature is designed from scratch than when modifications are made to an existing design. The designer has a validated model of the control feature that meets all the functional requirements at the end of this process step. This control feature model can now serve as the control feature specification and is provided to the software design engineers and application system engineers.

15.3.4 Software Application Validation

Two significant functions are performed in this validation stage. First, all of the control feature specification models are integrated into an application-level executable model. This model includes an abstract model of the scheduling and tasking definitions established in the Software Application Architecture Design stage. The validation engineer must integrate any new control feature execution contexts (threads) into the tasking definitions. In addition, all control feature and software architecture interface connections are checked for consistency.

Functional validation, from the software system perspective, commences when this consistent and complete application model is available. The validation engineer stimulates the application model with standard operational scenarios to ensure that the control features and software application architecture are functionally well integrated. The engineer analyzes the simulation results and establishes acceptance criteria (objective and subjective) for downstream software structural and functional verification.

One standard operating scenario is the Key On Engine Off (KOEO) mode, during which the controller initializes itself and checks for functional capability. Another test scenario injects sensor and actuator faults to validate the application model's on-board diagnostic and failure-mode effects behaviors. The validation tests are executed in non-real-time in the large-scale model environment.

Special application management tools automatically combine (place and interconnect) control feature models into a complete application model. The tools are fully integrated into the CACSD tool suite and incorporate extensive structural analysis and other design aids. A scheduling analysis tool that aids in preparation of the tasking definitions is one example. In the future, computing resource estimation and automated measurement technologies will be used to further enhance the software architecture validation effort.

15.3.5 Control Feature Software Design

Detailed software design specification provides all of the extra information required for implementation. Special cases associated with a particular microprocessor target implementation need to be identified and addressed. Precise definition of all algorithmic details must be refined to address all possible operating conditions and default cases.

Software packaging is a lower-level extension of modularization and is concerned with how specific functions are defined within a broader defined module. Software packaging involves establishing functions, function prototypes, file organization, data organization, and data scoping.

Recall that the Simulink® and Stateflow™ diagrams serve as the detailed design specification in our CACSD development process. Details of variable names, data definitions, data scoping, state-machine behavior, execution rates, and order of execution have been fully specified in the Software Application Architecture Design and Control Feature Design and Validation steps. The structure and detail of the diagrams remain unchanged for all subsequent software development stages. However, it is possible to add target specific implementation information at a hidden model layer. Thus the base specification can be reused for multiple implementations.

Future tool development efforts will permit structure chart representations of the "as wired" design to be generated. Metrics associated with the structure chart organization will permit the user to determine if the "as wired" design is adequate for long-term maintenance [36].

15.3.6 Control Feature Software Implementation

Software implementation is the translation of the detailed design specification into the target language. A straightforward translation will simplify testing and debugging; however, optimization may be necessary for resource constrained applications. Traditional approaches suggest that optimization should be deferred until translation is completed and tested. However, if the application domain has known resource constraints and certain design constructs occur frequently, it may be more practical to directly apply the optimized implementation approach. The key is to use standard optimization approaches that are documented and understood by all software engineers associated with the development effort.

The software implementation should be true to the detailed design, and packaging specifications such as function partitions and data encapsulation should be followed. There should not be any modifications to the algorithm's functional intent during the software implementation phase. A software implementation should also follow a coding standard that defines a style and set of rules for implementation. A standard style is important for reliability, understandability, readability, and maintenance. Specific rules are necessary to satisfy safety issues or other mandated coding requirements. These rules may also address optimization issues.

The CACSD process for software implementation focuses on a coding standard that ensures a safe implementation that is maintainable and efficient. The software coder currently determines function partitioning, function prototypes, and file organization. The core software engineering design activities have occurred primarily during the specification and detailed design stages, so variable names, data types, data scoping, and module partitioning are all predetermined and not subject to change. However, documented methods to address code efficiency are available to the software coder. Code quality is augmented by the use of automated coding standard checking utilities and code inspections.

In the future, software will be provided by automatic code generators. The Simulink® and Stateflow™ diagrams will be developed using the current practices. However, packaging and detailed design information will be added to the hidden model layers prior to code generation. Once all of the information is present, the code generator will produce efficient, function-level code. The automatically generated functions will be integrated per the architecture defined during the Software Application Architecture Design and Control Feature Software Design steps.

15.3.7 Control Feature Structural Verification

Control feature structural verification involves unit and integration testing of control feature components. Unit testing is conducted on small components that typically correspond to a single function. Integration testing is conducted on several small components that have been integrated. Integration testing takes place after successful completion of unit testing.

Unit testing determines if all aspects of the code implementation produce correct results. The most common form of unit testing ensures that all statements and possibly all branches have been tested (covered) [2]. However, this form of testing does not always ensure that all operating conditions have been fully tested. Therefore, additional rigor is needed to ensure the domain coverage of predicate equations and the operating regions of the algorithm equations. It is impossible to achieve complete unit testing coverage because there are many different aspects to address. A practical solution is to select a level of coverage that is adequate for organizational needs and the budget.

Integration testing is designed to test whether the integration of several components provides the desired functionality. Since the components have already passed unit testing, it is not necessary to reestablish all paths of operations. However, it is necessary to test all of the interfaces between the components. It is necessary to verify that state-machine transition sequences, task schedules, and execution thread sequences are correctly implemented.

The CACSD process for unit testing uses automatically generated test harnesses within the Simulink® environment. Random numbers are generated and passed to both the model and the software implementation. The results are compared and differences noted as failures. Each failure is investigated and resolved. The random numbers are constrained within the operating ranges of the input variables. Discrete and enumerated variables use discrete random numbers. This random number approach results in approximately 75% statement and branch coverage. Manual methods are applied when more rigorous coverage levels are needed. Currently, the CACSD process for integration testing is based on tests conducted at the bench level, dynamometer cell, or in the vehicle. A series of experiments conducted by several engineers provides the necessary level of integration coverage.

Future unit testing will be fully automated wherein the test vectors will be determined by the structural characteristics of the model. Modified condition/decision coverage (MC/DC) will be achieved [10]. MC/DC coverage ensures that all predicate equation variables can independently alter the outcome of the decision. Most decision-related errors can be detected with MC/DC coverage. Tests to stress the predicate and algorithmic equations will also be generated. Additional tests will be generated to verify lookup table, fixed-point math, loop, state-machine, and feedback loop implementations.

Future integration testing will utilize the structural characteristics of the model to generate the test vectors. The vectors will focus on data-flow relationships associated with equations and interfaces. Test vectors will also be generated to test state-machine traversal and to establish initial operating states for simulation and analysis. Additional test vectors will be generated to verify the relative order of execution is as expected.

15.3.8 Control Feature Functional Verification

The goal of control feature functional verification is to verify that the software implementation of a control feature meets its design intent. Recall that the control feature models were functionally validated using closed-loop non-real-time simulation in the Feature Design and Validation step (see Figure 15.2). This Control Feature

Functional Verification step is very similar, with the exception that production code replaces the model as the controller in the feedback.

The various test scenarios and results from the design and validation step are reused in this software verification step. The executable test script file that was generated during the design step is executed, and the results are compared against the design validation results. Any discrepancy is attributed to software implementation of the control algorithm. If specific functional requirements are associated with the software implementation, additional tests are designed and added to the validation test script. The software engineer verifies the results of these tests.

Coding standards have to be adopted so that engineers can "Plug and Test" the controller code in the design validation simulation model. These standards include retaining the sequence of input and output variables from the control feature model, as well as the declaration and usage of global variables and shared calibration parameters. No major structural issues are anticipated at this step because the code has already passed the structural verification step. The CACSD tool needs to be flexible enough to support easy integration of code into the modeling and simulation environment. Automatic procedures using Matlab® script files have been used to embed the production code into a C-MEX S-function block that is placed in the control feature model library. The substitution of the controller model with the code is a simple task of changing a link in the large-scale flexible modeling environment.

Once again, the debugging of any undesirable control system performance is challenging. Although problems can be attributed to the software implementation, the resolution requires interaction between algorithm designer, software designer, software implementer, and the tester. At the end of this step, a functionally verified code realization of the control feature is available. The system application integrator can now use this code. The verification results and the test scenarios are again saved for downstream process usage.

15.3.9 Software Application Structural Verification

Software application structural verification involves integration testing of a complete application or system. A complete application or system is made up of all algorithms, tasks, input/output drivers, support utilities, and the real-time operating system. Structural testing focuses on verifying that the model behavior and software behavior match under a variety of test cases derived from the modeled system structure. The test vectors focus on verifying end-to-end data-flow, scheduling order, and system states and modes traversal. Structural testing complements functional testing because functional verification mainly tests the primary operational scenarios of the system. However, functional testing may not cover all data-flow paths or scheduling sequences. It is therefore necessary to fill in the gaps with additional structural tests.

End-to-end data-flow testing is intended to verify that all data-flow sources are properly implemented. In some cases, it is possible for a particular output to be calculated by a number of different equations depending on the system modes and states. End-to-end data-flow testing ensures that each possible equation (or data flow path) is exercised. Care must be taken to ensure that correct system modes are established to permit the equation to be activated. This approach permits a very rigorous checkout of each of the many possible equations. Various coverage metrics are possible. As with control-flow coverage, it is necessary to determine the level of coverage that is adequate for organizational needs and the budget.

Schedule order testing is intended to verify that the relative order of execution within each task and the temporal characteristics of all of the tasks are correctly implemented. External measurement equipment is necessary to perform these timing measurements. Depending on the microprocessor technology, it may be necessary to instrument key software in order to obtain visibility. Task timing can be determined by measuring the time between variable updates or discrete output updates. Relative ordering within a task can be directly measured or accomplished with unique tag outputs. Technology is commercially available that will aid in automating these measurements.

System states and modes testing is intended to verify that the primary system states and modes are properly implemented. These states and modes are identified, and test vectors are established to exercise all of the transitions.

Currently, the CACSD process does not directly address software system structural verification. However, as with the Control Feature Structural Verification, the process accomplishes each desired test scenario via tests conducted by several engineers. Future structural system testing will be highly automated. Analysis of the model structure will permit test vectors to be established for end-to-end testing, scheduling order testing, and state-machine traversal. Technology under evaluation will permit automatic measurement of task schedule performance and relative order of execution on the target processor.

15.3.10 Software Application Functional Verification

This section is concerned with verifying the functional behavior of the integrated control features. Although the previous section concentrated on "plugging" the subsystems together, this section discusses verifying that the subsystems can "play" together.

Functional verification is accomplished through regression testing in which the code is substituted for the model in the simulation environment, and the functional tests run on the algorithm model are repeated. The results are compared to the results from the model validation tests. This stage of testing can be largely automated and run in the non-real-time host environment. It is important to note that the goal of system verification is not to repeat subsystem testing on the entire system but to test functional dependences and system level requirements.

15.3.11 Software/Module Integration Verification

The controller hardware and algorithms are developed in parallel and are integrated at this final stage. Often, systems-level testing is accomplished through the use of open-loop testers and prototype vehicles. Open-loop testers have limited capabilities since feedback loops are not simulated. Prototype vehicles are expensive and difficult environments to achieve repeatable test scenarios. A closed-loop tester (i.e., "virtual vehicle") can be created through use of hardware-in-the-loop (HIL) technology; this provides a cost-effective laboratory environment for integration verification and subsystem validation.

Different HIL configurations have been described in the literature [26], based on the "hardware" that is in the loop. In this classification, the microcontroller module and software are the components in the loop, while the plant dynamics and sensor/actuator behaviors are simulated on the computational platform. Both phenomenological and system identification formulations of the models are used. The lumped formulations of conservation of mass, momentum, and energy typically are sufficient, resulting in the phenomena being expressed either as ordinary differential equations or algebraic expressions. In the system identification type of models, the input and output signals are measured, and the model is adapted to process behavior by minimization of the errors between the model and process [19].

The other components of the system are the standard and custom interface cards that allow the controller to exchange information with the "virtual vehicle" and the interface through which the user interacts with the HIL system. The details will be described in [29]. "Test scenarios" generated earlier in the CACSD process are reused in an automated fashion to exercise the controller software. This step identifies any "regression" of the software behavior from the algorithm behavior after it has been embedded in the microcontroller.

Questions that can be answered in this stage include: Does the integrated software behave in an acceptable manner under the control of the real-time operating system or scheduler? Are software units executed within their allocated times? Does the software run for an extended period of time without generating overflow conditions? Can the software start the simulated engine? Is the software robust to electrical noise on the sensor and actuator signals? Have errors been introduced due to the microprocessor target compiler?

An HIL system is currently being developed to validate the software for a six-cylinder application. Processes are being developed so that model parameters can be readily generated from corporate databases with component, powertrain, and vehicle characterizations. Eventually, we hope to provide a flexible HIL system that can be easily and quickly configured for the entire family of powertrain applications.

15.3.12 User Documentation

User documentation is often an after-thought in a systems development process. This can lead to documentation that is inaccurate and difficult to maintain. Automating the documentation process can mitigate this tendency. In our process, a few background documents are manually produced, while the detailed design documents are automatically generated from the executable specification model. The resulting documentation set is made available to the engineering organization on our intranet. Thus the generally static supporting information and the highly dynamic design information can be accurately maintained at an affordable cost.

The control feature documentation set is comprised of several components. First, we present those components that are manually prepared. The feature functional description is a background-oriented document that describes the functional operation of the feature in a concise manner. The calibration crib sheet is an abstracted graphical representation that emphasizes the relationship between the tunable parameters and the nominal control algorithm. The crib sheet is intended to provide quick reference for the calibration engineer who is familiar with the detailed operation of the control feature. The feature release notes section documents the change history of the control feature.

The final document section is automatically derived from the feature control model. This document includes a hierarchical map of the feature, a list of parameters, and the full definition of the parameters. Bookmarks, notes, and hotlinks allow the reader to navigate the document in several ways.

These documents are combined to produce the feature user documentation package. This package gives a comprehensive overview of the functionality and operation of the control system. All sections, whether manually or automatically generated, are automatically translated into a document format that is readily printed or posted on our intranet system.

15.3.13 Configuration Management

Configuration management systems are massive storehouses of the various documents needed to produce an automotive control system. These systems act as libraries from which multiple developers can share and modify documents in a controlled manner. Many configuration management systems support high levels of process and process measurement automation. They enable large organizations to manage a high-frequency software release schedule. Our organization achieves the rate of one or more development or production releases per day.

In any production automotive software development organization, control of the various documents, models, data, and source code is important for many reasons. The production group must be able to reproduce the production release software for some amount of time into the future. This need may arise when a postproduction service fix is required or a recall is mandated and control system software must be modified. In addition, production organizations typically need to have reproducibility for any number of internal development software releases. Most internal releases tend to be small to moderate modifications of an existing control system. Configuration management systems enable the developers to reuse large portions of proven, validated, and verified portions of the control system.

Requirements, specification models, reusable component libraries, parameterizations, test scenarios, and test results can all be captured in file format. Thus we have found that the work products of our model-based development process can be easily integrated into our configuration management system, at least at a rudimentary level.

15.3.14 Software Engineering Project Management

The primary purpose of the CACSD-based process is to support the design, development, and delivery of a control system project that fully meets or exceeds the customer's functional and quality requirements at minimal cost. This capability alone, however, is insufficient. The process must also have the integrated capability to (1) accurately predict and track estimates of project effort, cost, and quality, and (2) continually assess and improve the process's capability to deliver its required work products. Toward this end, fundamental product and process measurements (known as metrics in the software engineering discipline) have been integrated into the CACSD development process. This section considers CACSD tools in the context of a typical software engineering project management process [6, 15, 27], as seen in Figure 15.3.

Section 15.3 Computer-Aided Control System Design for Powertrain Controller Development   **389**

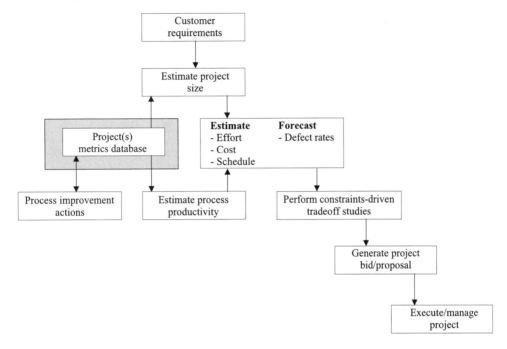

Figure 15.3 Typical software project management model.

The value of the process and project metrics database cannot be overstated. It is key in determining process productivity: a factor used to estimate project resource, cost, timing and quality. During the development and maintenance phases of a project, the metrics database helps the project manager ensure that the project is within organizational targets for quality, cost, and productivity. Finally, the metrics provide management with data to calculate the return on investment of process improvement initiatives. In fact, these data were used to justify the CACSD process project expenditures prior to deployment.

Figure 15.4 indicates the initial benefits gained from implementing a formal algorithm-style review process step. This process step reviews the algorithm design to ensure that it meets the appropriate style-guide standards. Early removal of noncompliant algorithm-design style enhances the downstream process by ensuring consistent designs with consistent interpretations. The style inspection process step can be largely automated and provide the immediate population of the metrics database with design defect identification data.

In Figure 15.4 we can also see the benefits experienced when moving from a formal but manual code-inspection process to an automated unit-test process. The software implementation defects found during the manual inspection process were significant with respect to algorithm design defects found, yet after the implementation of an automated unit-test tool, the software defects found increased by a factor of three. The tool can also be used to gather key metrics such as the percentage of code path test coverage and the number of deviations found. These measurements of the implementation process can also automatically populate the metrics database.

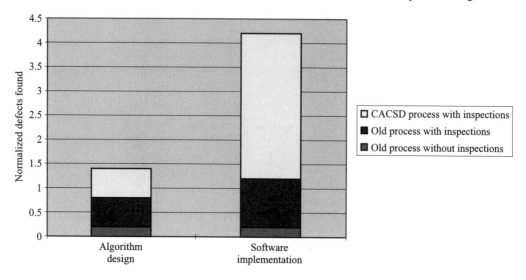

Figure 15.4 Selected CACSD process metrics.

15.4 CONCLUSION

Automotive powertrain controller development organizations are experiencing demands for improved performance, higher quality, reduced cost, and reduced development time (better, cheaper, faster). Software industry measurements indicate that requirements and model-based engineering processes are critical to the fulfillment of these demands. We have found that modern CACSD tools are capable of supporting such a requirements- and model-based process within a large-scale, multi-user, powertrain controller development organization. When coupled with the appropriate engineering expertise, the inherent analysis and design capabilities of the CACSD tools ensure that the investment in requirements and modeling yields adequate return. Furthermore, the open nature of the CACSD tools allows for organizational customization and a high degree of process automation. We expect to realize even greater return on our investment in CACSD-based processes as these automation technologies mature.

Related Chapters

- Chapter 3 provides a general discussion of control design automation with software tools.
- The application of CACSD tools for flight control is discussed in Chapter 11.
- Chapter 7, on hybrid supervisory control, reviews state machines and verification methods for hybrid systems.

REFERENCES

[1] M. Ashhab, A. Stefanopoulou, J. Cook, and M. Levin, "Camless engine control for robust unthrottled operation." *Special Publication SP-1346*, Paper No. 981031, Society of Automotive Engineers, 1998.

[2] B. Beizer, *Software Testing Techniques*, 2nd ed. London: International Thomson Computer Press, 1990.

[3] A. Beydoun, L. Wang, J. Sun, and N. Sivashankar, "Hybrid control of automotive powertrain systems: A case study in hybrid systems." *Computation and Control*, Berlin, Heidelberg, New York: Springer-Verlag, April 1998.

[4] B. Boehm, "Software engineering." *IEEE Transactions on Computers*, Vol. C-25, no. 12, December 1976.

[5] G. Booch, *Object Oriented Design with Applications*. Reading, MA: Addison-Wesley, 1994.

[6] F. Brooks, *The Mythical Man-Month: Essays in Software Engineering*. Reading, MA: Addison-Wesley, 1975.

[7] K. Butts, D. Stetson, and J. Cook, "Computer aided engineering for automotive powertrain controller development." *Advanced Automotive Technologies*, DSC-Vol. 56/DE-Vol. 86, Proceedings of ASME International Engineering Congress and Exposition, San Francisco, November 1995.

[8] K. Butts, "An application of integrated CASE/CACSD to automotive powertrain systems." *Proceedings of the 1996 IEEE International Symposium on Computer-Aided Control System Design*, pp. 339–345, September 1996.

[9] K. Butts, L. Kolmanovsky, N. Sivashankar, and J. Sun, "Hybrid systems in automotive control applications." In S. Morse (ed.), *Proceedings of the Block Island Workshop on Control Using Logic Based Switching*. Berlin, Heidelberg, New York: Springer-Verlag, 1996.

[10] J. Chilenski and S. Miller, "Applicability of modified condition/decision coverage to software testing." *Software Engineering Journal*, September 1994.

[11] M. Cusumano, *Japan's Software Factories: A Challenge to U.S. Management*. New York: Oxford University Press, 1991.

[12] *Parametric Cost Estimating Handbook: Joint Government/Industry Initiative*, Chapter 5—Software Parametric Cost Estimating, sponsored by the United States Department of Defense, http://www.jsc.nasa.gov/bu2/pcehg.html, Fall 1995.

[13] M. DePoyster, J. Hoying, and K. Majeed, "Rapid prototyping of chassis control systems." Computer-Aided Control System Design, *Proceedings of the 1996 IEEE International Symposium on Computer-Aided Control System Design*, pp. 141–145, September 1996.

[14] R. Dorey and D. Maclay, "Rapid prototyping for the development of powertrain control systems." *Proceedings of the 1996 IEEE International Symposium on Computer-Aided Control System Design*, pp. 135–140, September 1996.

[15] P. Drucker, *Management: Tasks, Responsibilities, Practices*. New York: Harper & Row, 1973.

[16] D. Godbole and S. Karahan, "Automotive powertrain modeling, simulation and control using Integrated System's CASE tools." Society of Automotive Engineers, Paper No. 940180, International Congress & Exposition, 1994.

[17] D. Hatley and I. Pirbhai, *Strategies for Real-Time System Specification*. New York: Dorset House, 1987.

[18] *IEEE Trial-Use Standard for Application and Management of the Systems Engineering Process*, IEEE Std 1220–1994, Institute of Electrical and Electronics Engineers, February 1995.

[19] R. Isermann, S. Sinsel, and J. Schaffnit, "Modeling and real-time simulation of diesel engines for control design." Society of Automotive Engineers, Paper No. 980796, International Congress and Exposition, 1998.

[20] M. Jankovic and I. Kolmanovsky, "Robust nonlinear controller for turbocharged diesel engines." *Proceedings of the 1998 American Control Conference*, Philadelphia, PA, June 1998.

[21] K. Jung-Ho, J. Byeong, D. Dong-Il, and K. Hoyoun, "AUTOTOOL, a PC-based object-oriented automotive powertrain simulation tool." *Proceedings of the IEEE Intelligent Transportation Systems Conference*, pp. 753–758, 1997.

[22] P. Kar, Documented prepared by the Requirements Working Group of the International Council of Systems Engineering, Presented at the 1996 INCOSE Symposium.

[23] H. Krohm and V. Gheorghiu, "Hardware-in-the-Loop simulation for an electronic clutch management system." *Special Publication SP-1080*, Society of Automotive Engineers, Paper No. 950420, International Congress & Exposition, 1995.

[24] J. Moskwa, J. Anthony, G. Babbitt, and Z. Rubin, "Synthesis of software and hardware in PCRL for powertrain design and development." *IEEE Control Systems Magazine, Special Issue on Powertrain Control*, 1998.

[25] P. Naur and B. Randell (eds.), "Software engineering: A report on a conference sponsored by the NATO Science Committee." Brussels: Scientific Affairs Division, NATO, January 1969.

[26] B. Powell, N. Sureshbabu, K. Bailey, and M. Dunn, "Hardware in the loop vehicle and powertrain analysis and control design issues." *Proceedings of the 1998 American Control Conference*, Philadelphia, PA, June 1998.

[27] L. Putnam and W. Myers, *Controlling Software Development: An Executive Briefing*, New York: IEEE Press, 1996.

[28] C. Ramamoorthy, A. Prakash, W. Tsai, and Y. Usuda, "Software engineering: Problems and perspectives." *Computer*, Institute of Electrical and Electronics Engineers, p. 205, October 1984.

[29] S. Raman, N. Sivashankar, and W. Stuart, "HIL simulators for powertrain control system software development." *Proceedings of the 1999 American Control Conference*. San Diego, CA, June 1999.

[30] M. Rausch and B. Krogh, "Symbolic verification of Stateflow logic." *Proceedings of the International Workshop on Discrete Event Systems* (WODES '98), pp. 489–494, August, 1998.

[31] A. Stefanopoulou, J. Cook, J. Freudenberg, J. Grizzle, M. Haghgooie, and P. Szpak, "Modeling and control of a spark ignition engine with variable cam timing." *Proceedings of the 1995 American Control Conference*, Seattle, WA, June 1995.

[32] A. Stefanopoulou, I. Kolmanovsky, and J. Freudenberg, "Control of variable geometry turbocharged diesel engines for reduced emissions." *Proceedings of the 1998 American Control Conference*, Philadelphia, PA, June 1998.

[33] J. Sun and N. Sivashankar, "Issues in cold start emission control for automotive IC engines." *Proceedings of the 1998 American Control Conference*, Philadelphia, PA, June 1998.

[34] S. Toeppe, S. Ranville, and K. Butts, "Specification and testing of automotive powertrain control system software using CACSD tools." *Proceedings of the 17th Digital Avionics Systems Conference*, Seattle, WA, November 1998.

[35] R. Weeks and J. Moskwa, "Automotive engine modeling for real-time control using MATLAB/SIMULINK." Society of Automotive Engineers, Paper No. 950417, International Congress & Exposition, 1995.

[36] E. Yourdon and L. Constantine, *Structured Design: Fundamentals of a Discipline of Computer Program and Systems Design*. Englewood Cliffs, NJ: Prentice-Hall, 1979.

Chapter 16 | BUILDING CONTROL AND AUTOMATION SYSTEMS

Albert T.P. So

Editor's Summary

In industrialized societies, the comfort and productivity of people are influenced substantially by the quality of automation and control provided in the buildings in which they live and work. Accordingly, the history of building automation and control shows steady progress in complexity and capability. The first generation of systems used localized, stand-alone pneumatic controls. A major development in the 1950s, driven by the development of pneumatic sensor transmitters and receiver controllers, was pneumatic centralization. Electromechanical multiplexing systems, introduced in the 1960s, substantially reduced installation and maintenance costs and enabled automatic control of air-handling units for the first time. Minicomputers and programmable logic controllers (PLCs) became popular after the oil crisis in 1973 and helped spur the development of energy management systems. Today, the personal computer has revolutionized building control systems. Heating, ventilation, and air-conditioning (HVAC), lighting, elevators and escalators, and fire and security can now be integrated within one building automation system.

In terms of basic control technologies, the PID controller remains dominant, especially for HVAC systems. PLCs are widely used for event-driven and sequencing operations such as start-up of chillers. With the maturing of local area network technologies and their widespread deployment for building automation, digital control loops can now be implemented throughout buildings. The infrastructure is available for implementing advanced algorithms.

Much of this advanced technology is inspired by developments in artificial intelligence. Expert systems, neural networks, and fuzzy logic have all been used for some building control applications. (Tutorials on these methods can be found in Chapter 5.) Finally, the availability of inexpensive cameras and high-speed image processing electronics is permitting the use of vision-based sensing in building automation (and also in other fields—see Chapter 18 for a discussion of visual servoing in robotics). This could be used, for example, to estimate the number of residents within an air-conditioned space and regulate accordingly in response to ad hoc changes in the air-conditioned environment and to conserve energy.

Albert So is an associate professor in the Department of Building and Construction in the City University of Hong Kong and chair of the IEEE-CSS Technical Committee on Control Electronics.

16.1 INTRODUCTION

It is generally accepted that people spend almost 80% of their lives in buildings. Except for holidays to the countryside, the destination of most people traveling outside a building is another building. We live, work, and entertain ourselves inside buildings.

Therefore, a comfortable, healthy, and work-effective environment within buildings is critical for ensuring the efficiency and quality of our daily activities. That is why interior environmental control and building automation are so important.

The first environmental control systems for large buildings were pneumatic. They were capable of maintaining acceptable environmental conditions in a building and of performing some relatively complex control sequences. However, since they were hardware-intensive, the initial installation costs and maintenance requirements could be substantial. There were also problems of limited accuracy, mechanical wear, and inflexibility. In recent decades, the integration of building systems such as heating, ventilation, air-conditioning, lighting, fire safety, and security has proven economically advantageous, while simplifying system interaction. First, let us have a look at the history of building control development [1].

There was a general expansion in the construction industry after World War II. A desire to improve comfort within new, larger buildings resulted in more complex mechanical systems, and so better heating and cooling control systems were developed. Pneumatic controls and electrical switches were mounted everywhere, while large numbers of panels were installed near equipment-controlled areas. The involvement of human operators to monitor the status of systems and to record readings became necessary.

In the 1950s, the introduction of the pneumatic sensor-transmitter permitting local indication and remote signaling plus the receiver-controller with optional remote adjustment led to pneumatic centralization. The number of local control panels was thus reduced to a more or less single center that was located in a control room. Another trend, miniaturization, resulted in the reduction of the physical size of instruments. The use of electronic sensors and analog control loops by the end of that decade resulted in a hardwired centralized control center.

In the 1960s, the growth of control companies for commercial buildings promoted the development of new technologies. Electromechanical multiplexing systems were introduced, resulting in reduced installation costs and maintenance. Wires were reduced from hundreds to a few dozen per multiplexer. The control center panel was transformed into a control center console. Commercial digital indication and logging systems were available on the control center console to permit the automatic recording of selected parameters during unusual conditions and to provide information regarding these selected parameters. Automatic control of systems, like air-handling units (AHUs), became possible. Temperature, flow, pressure, and other equipment parameters were monitored on the console. Intercom systems and phones were also part of the console. The first computerized building automation control center was marketed late in this decade, and data communication was done by means of coaxial cables or twisted pairs. Up to this stage, building automation technology was based on the concept of a centralized control and monitoring system (CCMS).

The use of minicomputers or central processing units (CPUs) and programmable logic controllers (PLCs) in building automation systems increased dramatically as a result of the oil crisis of 1973. People began to appreciate the importance of energy conservation. A new term, energy management system (EMS), was derived and became a standard in control manufacturers' sales brochures. New application software packages were incorporated into basic automation systems. Some packages such as duty cycle, demand control, optimum start/stop, optimum temperature, day/night control, and enthalpy control were introduced. In addition, fire and security systems were emerging from the fundamental infrastructure of building automation. The building owner could directly oversee the systems by keeping track of energy usage and cost. These new tools helped management make better predictions and compare relative costs of products. By the mid-1970s, the cost of hardware began to decrease, systems became "user-friendly," and it was possible to program and generate new databases on the same system. Printers with keyboards (KBs) and cathode ray tubes (CRTs) with KBs were the primary man–machine interface with the CPUs. "Dumb" multiplexers were becoming "smart." The small microprocessor embedded inside some multiplexers could "stand-alone," providing analog alarm detection that reduced communication transactions. Field interface devices (FIDs) appeared and were the remote processing units compatible with the CPUs.

In the 1980s, the introduction of personal computers (PCs) revolutionized the control industry. The comparatively low cost of chips was the principal cause of the development of new technology in building automation and energy management. The resultant rapid change motivated manufacturers to engage in research and development rather than investing in their existing hardware and software. Users accepted the production of individual microprocessor-based distributed direct digital control (DDDC) because of the popularity of PCs. The DDDC systems were replacing conventional pneumatic control systems. The building operator console (BOC) became the major man–machine interface, and all programming was done through high-level languages such as Pascal or C. The BOC was directly linked to remote local microprocessor control panels (LMCPs) using proprietary local area network (LAN) protocols.

16.2 EXISTING BUILDING CONTROL TECHNOLOGIES

Buildings can roughly be categorized into four major types: commercial, residential, institutional, and industrial. Commercial buildings include office buildings, restaurants and shopping centers. Institutional buildings can include school or university campuses, libraries, hospitals, and public transportation terminals. Industrial buildings refer to all the industrial plants and factories. The building control and automation technologies discussed in this chapter apply to all four types of buildings, although some special-purpose industrial facilities (e.g., nuclear power stations and military research centers) are outside the scope of the discussion.

Major building systems include heating, ventilation, and air-conditioning (HVAC), illumination, vertical transportation, electrical distribution, life safety such as fire detection and fighting and security, office automation, and plumbing and drainage. Normally, HVAC systems consume about 40% to 50% of the total energy load of buildings, while lighting systems consume 15% to 20%. Vertical transportation systems consume about 3% to 5%. Hence most applications of control technologies have been on these three types of systems. Office automation is often provided by the tenants themselves and so is normally not intrinsic to the original design of the building. However, there is a trend that the communication part of office automation will soon be integrated as one standard facility in modern buildings, in particular those calling themselves "intelligent buildings."

16.2.1 Applications of PID Loops

Control using proportional, integral, and derivative (PID) loops has been well proven. PID control is relatively simple, robust, and straightforward. It is used mostly in HVAC systems, for speed control of fans, air volume control with air-handling units (AHUs), variable air volume (VAV) boxes, and so on. It is also employed in the speed control of elevators and voltage control of generators.

The air-conditioning process is highly nonlinear; the interaction between the temperature and humidity control loops is significant, and the constraints imposed by the nonideal behavior of actuators are considerable. Conventionally, a cascaded, multiloop PID control structure has been used. Brandt and Shavit [2] simulated the response of a PID-controlled discharge air temperature control system to a step change input. PID control has been considered a successful implementation in HVAC control since most practical systems available today employ this conventional technique.

An example of a typical AHU is given here, which will be used in other sections of this chapter. The schematic model of the AHU is shown in Figure 16.1. AHUs are widely used in a centralized HVAC system. Normally, one AHU serves one story, and sometimes up to three stories with the AHU installed at the midstory of the building. It receives a cooling or heating medium from a remote plant and produces either cool (in summer) or warm (in winter) air to the air-conditioned space via an air duct distribution network. Without loss of generality, only cooling is considered here. Chilled water is supplied from a centralized chiller system via water pipes to the cooling coil inside the AHU. Air from the conditioned space returns to the AHU as return air through the air ducts. Most of the return air (Q_p) (the remaining being exhausted through the exhaust air duct) mixes with fresh outdoor air in a proportion controlled by adjusting the exhaust and fresh air dampers. The damper of the mixing air chamber is interlocked with both the exhaust (Q_{ep}) and fresh air (Q_e) dampers. The mixed air then passes through the filter where dust is removed and arrives at the chilled water-cooling coil where both the air temperature (T_c) and humidity ratio (X_c) are reduced. The humidity ratio is also known as absolute humidity or moisture content. It represents the absolute amount of water vapor within a unit mass of dry air. Accurate control of humidity is

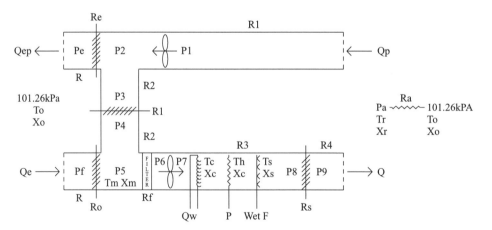

Figure 16.1 Air handling unit model.

accomplished by the selective operation of either the reheater or the humidifier. Conditioned air is then supplied to the air-conditioned space via air ducts, as supply air. The pressure of the supply air (Q) is controlled by the variable speed supply air fan.

The aim of the control action on the whole is to achieve the desired room temperature (T_r) and humidity levels (X_r) inside the air-conditioned space with minimum delay time and energy consumption. Pressure at any point inside the AHU is denoted by P, with the suffix indicating the location, and the air resistance at any point is denoted by R, with the suffix indicating the location. The speed of the supply air fan is controlled by a variable-speed variable-frequency (VVVF) motor drive to fix the pressure behind the supply air damper (P_8) to a constant setpoint (P_{8s}). Modern motor drives usually employ the VVVF technique through which speed can be controlled precisely and energy can be saved. The idea behind VVVF is to maintain a more or less constant magnetic flux inside the motor so that the general shape of the torque-speed curve of the motor remains unchanged during the variation of speed. The heater and the humidifier are controlled to maintain the room's relative humidity at a desired setpoint (RH_s). The chilled water-flow rate is controlled by the regulating valve to maintain the supply air temperature at a constant setpoint (T_{ss}). Dampers and the speed of the return air fan are coordinated to maintain the fresh air supply rate at a desired value. Normally, an AHU is not equipped with a supply air damper, the R_s, as shown in Figure 16.1, and it is included here to model the VAV action. For energy conservation, the VAV system is commonly adopted in most commercial buildings. Supply air from the AHU will reach the VAV boxes through air ducts. The VAV box automatically adjusts the volume flow rate of cool air to the conditioned space based on the deviation of the instantaneous room temperature as sensed by a thermostat and the desirable room temperature as adjusted by the occupants. For the sake of completeness, the VAV action is also built inside the AHU model in our study, serving the same function of controlling the temperature of the air-conditioned space, that is, the room.

In PID control, the deviation, $e(k)$, of the control parameter, say, the supply air temperature, T_s, from the setpoint, T_{ss}, that is, the desired supply air temperature, at the kth time step, is used to control the actuating command at the $(k+1)$th time step, $ac(k+1)$. The actuating command can be the percentage of opening of the valve of the chilled water coil. Then, the chilled water-flow rate, Q_w, at the $(k+1)$th time step can be controlled according to

$$\Delta ac(k) = K_p e(k) + K_d[e(k) - e(k-1)] + K_i \sum_{i=1}^{k} e(i)$$

$$ac(k+1) = ac(k) + \Delta ac(k) = Q_w(k+1)$$

$$e(k) = T_s(k) - T_{ss}$$

(16.1)

Here, K_p, K_d, and K_i are the gains of the proportional, derivative, and integral control loops, respectively. Equation (16.1) formulates a discrete controller; the continuous control operations are similar. Although PID control is simple and straightforward to use, these three gains need to be manually adjusted for optimal operation of the controllers. Very often, their settings greatly depend on the experience of the designer, or they are just fine tuned by trial and error. Even though a PID controller may be well tuned, its effectiveness is based on the assumption that the system model parameters do not change much. When a well-tuned PID controller is applied to another system with different model parameters, the response is likely to become poor and energy consumption is likely to increase. This is because the gain settings of most PID controllers are based on the normal operating ranges of the plants themselves, or, in other words, the settings are not adaptable to a changing environment.

16.2.2 Programmable Logic Control

When control systems in buildings were implemented half a century ago, electromechanical devices, that is, relays and contactors, were commonly used. Although we are now in the age of high-speed electronics and microprocessors, technicians of building systems are very familiar with this conventional equipment. The introduction of programmable logic controllers (PLCs) helped to bridge the gap between the old and new technologies. PLCs are generally used for event-driven and time-sequence operations. Examples include the sequential re-starting of motors after a major electric power failure and the starting sequence of all components within a chiller system.

A PLC is a microprocessor-based device designed to perform logic functions previously accomplished by relays and mechanical timers. It is smaller, faster, more reliable, and easier to modify compared with the conventional circuits of relays, which are complicated and expensive. The facilities provided by a PLC are basically AND/OR/NOT logic, timers, and counters. Programming by relay ladder diagrams is offered and is often performed through a separate portable computer. A PLC has two main operating modes, stop and run. In stop mode, the PLC is powered up but is not performing any control function; in run mode, it executes all the instructions contained in the memory. Each logic statement is called a step. Every step is numbered so that the instructions can be treated in a definite order and recalled when necessary. The processor scans each instruction in quick succession and logically assembles a list of outputs to be turned on. The whole process of scanning the inputs/outputs/program and finally updating the outputs is known as a cycle. A ladder logic diagram is shown in Figure 16.2, which corresponds to the following program:

LDI   001; load inverse contact 001, i.e., the normally closed button, "stop"
LD   002; load contact 002, i.e., the normally open button, "start"
OR   201; perform a logical "or" between 002 and the self-holding contact A
ANB   ; perform a further logical "and" with 001
OUT   201; output the result to the output terminal of 201

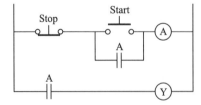

Figure 16.2 A typical ladder logic diagram.

16.2.3 Direct Digital Controls

Signals from building equipment, such as thermostats, light sensors, tachometers, and pressure transducers, have historically been analog in nature. Owing to the popularity of analog control techniques involving pneumatics up to the 1970s, electronics dealing with continuously changing signals were developed and implemented. As the application of computers became widespread in the late 1970s, there was a problem with the conversion between the analog signals from building equipment to the computer interface, which was absolutely digital. In addition, since computers carried out tasks in sequence, it was impossible for the computer to convert the measured value into a digital form, carry out calculations, and convert the digital results back into an appropriate analog output value simultaneously and continuously. Therefore, the digital sampling principle was called in. The input values are not measured continuously, but a measurement is carried out repeatedly with a given interval, or sampling period, and the results are converted into digital signals. During the interval, the computer can process the signal and produce an output before the next sample is available. Throughout the interval, the output is more or less kept constant until the output signal is available toward the end of that interval. This is the basis of direct digital control (DDC) [1]. Recently, in line with the concept of DDC, sensors with digital outputs and actuators receiving digital signals are seeing common use. Thousands of control devices have been installed in modern commercial buildings, and, hence, the concept of distributed DDC (DDDC) is very popular where each device is equipped with a direct digital controller.

A DDDC system has many benefits. However, in order to ensure high performance, the distributed architecture must behave as a single integrated entity and not as a series of separate controllers. The communication network then becomes the integrator, providing features to make every system point or variable accessible for programs, trends, or displays in controller and operator terminals throughout the system. A powerful and flexible operator control language (OCL) must be available that offers control for each point through a single, comprehensive program, powerful and effective program editing features, and a full range of mathematics and logic functions, and that permits flexible operator override for all system points and variables.

16.3 INFORMATION TECHNOLOGY FOR BUILDING SYSTEMS CONTROL

Decades ago, controls in buildings adopted the "centralized" concept: Everything was supervised by a central controller. Today the trend is toward distribution of functions by employing modern networking techniques. Integration between control devices is only feasible with a comprehensive capability for exchanging real-time information between them. This relies on the provision of a high-speed "bus" or "network." A bus or network is the link between computer components in a data communication environment. The term "bus" is used mainly when referring to links between components or parts of a single computer system, while the term "network" is most often used in referring to links between several separate computer systems. The application of local area networks (LANs) in building system control is now quite mature.

16.3.1 Control Networks

One of the oldest networking product lines is ARCnet [3], which was developed by the Datapoint Corporation in the 1970s for use with its own proprietary line of computers and released commercially in 1977. It is still being used for the management/automation levels of building automation systems (BASs) in some popular systems such as Metasys from Johnson Controls. ARCnet technology is inexpensive and extremely simple to install and use. The basic configuration of an ARCnet LAN is a token-passing bus. This unique format is fully documented in IEEE standard 802.4. Two or more PCs are connected to a hub (a hub is a device responsible for broadcasting data packages to and from all ports connected to the LAN), and the hubs themselves are connected along the single bus. This might be more accurately described as a bus interconnection of stars. Each hub is the center of a star, and each hub is attached to the bus. Most hubs are active, meaning that they contain repeaters. However, passive hubs can be used to connect three to four nodes over very short distances. The cable used in ARCnet LANs is RG-62A/U, which has a characteristic impedance of 93Ω, offers very low loss, and is small, flexible, and inexpensive. BNC (bayonet-locking) connectors are used. The cable run between two active hubs cannot exceed 610 m, and the total length of the bus must not exceed 6 km. ARCnet systems are limited to 256 nodes, while the longest permissible cable run between a passive hub and one of the PC nodes is 30 m. Until recently, the speed of data transmission on the ARCnet system was 2.5 Mbps. However, the newest version of ARCnet transmits at 20 Mbps.

Another commonly used LAN type in buildings is Ethernet. Ethernet, which was developed by the Xerox Corporation at its Palo Alto Research Center in the 1970s, was based on the Aloha wide-area satellite network implemented at the University of Hawaii in the late 1960s. In 1980, Xerox joined with DEC and Intel to sponsor a joint standard for Ethernet. The collaboration resulted in a definition that became the basis for the IEEE 802.3 standard. Ethernet uses the bus topology. Network nodes simply tap into the bus cable, which may be a large coaxial cable, like RG-8/U, or a small coaxial cable, like RG-58/U, or a twisted pair. Information to be transmitted from one user to the other can move in either direction on the bus, but only one node can transmit at any given time. The bus coaxial cable has a special terminating connector at each end containing a resistor whose value is equal to its characteristic impedance. This prevents signal reflections that cause signal loss and significant data errors. For the RG-8 and RG-58 cables used in Ethernet, this value is 53Ω. An RG-8/U cable is approximately 10 mm in diameter, and so it is referred to as a thick cable, and it is usually bright yellow in color.

Ethernet systems using thick coaxial cables are generally referred to as 10Base-5 systems, where 10 means a 10-Mbps speed; Base means baseband operation (i.e., baseband is the normal frequency of signal transmission), and the 5 designates a 500-m maximum distance between nodes, transceivers, or repeaters. Ethernet LANs using thick cable are also referred to as Thicknet. Ethernet systems implemented with thinner coaxial cables are known as 10Base-2, or Thinnet systems; here the 2 indicates the maximum 200-m (actually 185-m) run between nodes or repeaters. The most widely used thin cable, more flexible and easier to use, is RG-58/U, which is around 6 mm in diameter. Recent versions of Ethernet use twisted-pair cables. The twisted-pair version of Ethernet is referred to as a 10Base-T network, where T stands for twisted pair. The twisted pair used in 10Base-T systems is standard 22-, 24- or 26-gauge solid copper wire with RJ-45 modular connectors. Physically, a 10Base-T LAN looks like a star, but the bus is implemented inside the hub itself. It is usually easier and cheaper to install 10Base-T LANs than it is to install coaxial Ethernet systems, but the transmission distances are generally more limited. With twisted pairs, the maximum distance between nodes is 100 m and the maximum total permissible length with repeaters is 2500 m. The most recent development in Ethernet technology is a modified version that permits a data rate of 100 Mbps, 10 times the normal Ethernet rate. Two versions are available, namely, 100Base-T and 100VG-AnyLAN. Both use twisted pair and have the same access method and packet size. The 100VG-AnyLAN version has the IEEE standard number 802.12. The trend for all new designs is to adopt 100-Mbps Ethernet, and it is anticipated that the high-speed LANs used for building system control will be upgraded to 100Base-T or 100VG-AnyLAN in the near future.

Fiber Distributed Data Interface (FDDI) is a high-speed fiber-optic cable network offering a data transmission rate of 100 Mbps. Though not yet a formal standard, FDDI is well defined, and its use is growing. The wiring consists of two fiber-optic cables that are bundled together. The basic topology is a ring, and the access method is token passing. Network interfacing cards (NICs) installed in each computer contain fiber-optic transmitters and receivers that repeat the data transmission. Only one of the fiber-optic cables is used at any given time, and data circulate from node to node around the ring. The other cable is primarily a reserve path that is used if the main ring fails. The second ring has its own set of transmitters and receivers, but the data travel in the opposite direction from the primary ring. The first advantage of FDDI is speed, allowing more users to access and transmit high volumes of data with little loss of network performance. The second advantage is security, for it is impossible to tap into or monitor information on an FDDI ring. Fiber-optic cables are also completely immune to electrical noise. When a network must be implemented over long distances in noisy environments, FDDI is an excellent choice. However, the downside is its high cost. Therefore, for building automation and control, 100Base-T will probably be widely adopted in the future, while FDDI will likely be used for communications outside buildings or for office automation only. Another application for FDDI is its use as a backbone LAN for supporting or interconnecting two or more other LANs belonging to BASs of different brands within a mega-building being developed under different phases.

With the installation of proper LANs within a building, the concept of networked DDC can be realized. Networked DDC systems have three general levels of hierarchy or "architecture," namely, distributed control, building-wide or island host-level control, and information management. Each level serves an important purpose. The added value of networked DDC is that all three levels are interconnected.

At the distributed control level, microprocessor-based controllers monitor sensors and regulate devices to meet the needs of a specific application. A distributed control device is a complete control system inside a box with all the necessary inputs, outputs, and control processing logic. These controllers are usually used for traditional HVAC functions such as discharge air temperature, space temperature, humidity, and fan control. Distributed control in an intelligent integrated network has three major benefits. The first one is repeatable performance where multiple control strategies for a specific comfort zone can provide consistent and repeatable results such as switching from occupied to unoccupied, then reverting back to the exact same comfort conditions for the next occupied mode. The second benefit is individual control where we can control a specific zone whatever the appropriate comfort level may be. The system allows individual employees to set their own comfort levels, and all these devices can be tied to the BAS network, allowing data to be collected so that a certain comfort index can be measured. The final benefit is employee productivity where productivity of personnel can be enhanced by improving the comfort control of the indoor environment. An increase of 5% to 15% productivity with improved environmental control may be achieved [4].

The building-wide or host-level control coordinates building-wide control strategies through which a higher level of integrated control is possible by tying together all distributed controllers via a communication network. This creates a reporting path that allows information from one controller to be passed to another. Through coordinated control sequences, the entire BAS can be monitored and its various functions optimized.

At the information management level, data collected from variable points on the DDC system are transformed into usable information. Data from thousands of I/O points throughout a BAS can be accessed quickly to assist in management decision making. With the proper communication architecture, access to system information can take place at any one personal computer workstation. The same information can be available at the building-wide control level and at the stand-alone controllers. Information management is important for several reasons:

- Responsiveness to occupant problems
- Regulatory compliance and risk management
- Financial decision making
- Public perception
- Quality assurance

16.3.2 Protocols

Even with the appropriate hardware, information exchange on LANs is only possible with strong software support, that is, the provision of effective protocols. Today's state-of-the-art buildings contain computer-based control equipment that helps them to be energy-efficient and user-friendly, but very often pieces of equipment operate independently from one another. This is especially true when the building control systems have been completed in phases so that they are of different brands. One of the greatest impediments to the acceptance and growth of computer-based control has been the lack of a protocol that enables different manufacturers to "interoperate" together. In 1991, committees were formed in Europe to create a standard for a common communications protocol, ideally at one level of system complexity or, failing that, at a limited number of levels. As always in the standards-making process, politics played a prominent role. FND was a German communications protocol which had already been used. Batibus was French inspired, and it was backed by Merlin Gerin. In response to industry demand, ASHRAE developed Standard 135-1995—also known as BACnet—a data communication protocol for building automation and control networks. These protocols prescribe a detailed set of rules and procedures that govern all aspects of communicating information from one cooperating machine to another. All these systems work at different levels of complexity.

The European committee was tasked to select one standard at each level, and finally, FND was chosen for level 3, BACnet for levels 2 and 3, and EIB and Batibus for level 1. While all this was going on, the American company Echelon poured millions of dollars into creating a market for its own open communications protocol—LonWorks. Whereas BACnet relies on a freely available software protocol, LonWorks is hardware-dependent. We shall look at BACnet and LonWorks a little bit more closely.

BACnet is a communications and data protocol suite defined by the ASHRAE 135P committee for use in connecting building automation components from various manufacturers. The effort began in 1987, and BACnet was adopted as an ASHRAE standard in June, 1995. In December, 1995, BACnet was also adopted by ANSI and is now an American National Standard (ANSI/ASHRAE 135-1995). Nearly every major vendor of BASs in North America has demonstrated support for BACnet. With BACnet, two or more compliant computers may share the same networks and ask each other to perform various functions on a peer-to-peer basis. There are two key concepts that are critical. First, BACnet is applicable to all types of building services systems. The same mechanism that gives BACnet this flexibility has two important benefits: vendor independence and forward compatibility with future generations of systems. These are accomplished using an object-oriented approach for representing all information within each controller. The second key idea is that BACnet can be used on different communication hardware media and low-level protocols, namely, ARCnet, Ethernet, Echelon LonTalk, RS485 MS/TP and RS232 point-to-point. However, if some BACnet-compliant devices use Ethernet and others use RS232, for example, the integration or communication will not be direct. In any event, BACnet provides a model for integrated building control, consisting of five components:

- Objects to represent system information and databases, along with a uniform method for accessing both standardized and proprietary information.
- Services that allow BACnet devices to ask each other to perform various functions in standardized ways.
- LANs that provide transport mechanisms for exchanging messages across various types of networks and communication media.
- Internetworking rules that permit the construction of large networks composed of different LAN types.
- Conformance rules that define standardized ways of describing systems in BACnet terms.

The second communication protocol, LonTalk, originally a proprietary one, is now an open protocol. The term "proprietary" means that the technology was initially owned by a single proprietor, that is, Echelon. The LonTalk protocol uses some advanced ideas that are unique, and thus, a special type of communication "chip" was developed to suit the application. Using this chip and the appropriate software, users can completely absorb much of the burden of implementing LonTalk, freeing the rest of the system for application tasks. This chip is called Neuron and is manufactured by Motorola and Toshiba. LonTalk is like a very simple mailing system that provides system designers with some basic mechanisms for transporting messages between systems. In and of itself, LonTalk does not define what these messages contain. For the message system to be useful in a given application, the sender and receiver need to agree on the content of these messages. Since Echelon's designers had some idea of the types of applications that LonTalk might be used for, they developed a second protocol that could be used to define the content of application messages. This "one size fits all" protocol represents the session, presentation, and application layers of LonTalk and is often referred to as LonWorks [5].

Controllers making use of LonWorks can communicate with each other through what LonWorks calls standard network variable types (SNVTs). The SNVT method is a different approach to defining data objects and requires detailed knowledge on the part of the sender and receiver of the structure of each SNVT. SNVTs are identified by a code number that the receiving controller can use to determine how to interpret the information presented in each SNVT. The open-ended nature of SNVTs is both a strength and a liability. Different vendors can use the same SNVT code to mean different things. At best, this causes confusion when these systems are coupled together, and, at worst, it can cause inappropriate actions to be mistakenly taken. To help solve this problem, a consortium of Echelon vendors, known as the LonMark Consortium, was formed to try to agree on rules for LonWorks applications.

16.4 BUILDING AUTOMATION SYSTEMS (BASs)

So far, we have discussed the basic control technologies adopted in modern buildings and the information technology employed to accomplish the concept of distributed direct digital control. The BAS can be considered the heart of every "intelligent building." The major functions of a BAS are twofold—monitoring/control and energy management—although most BASs also include several ancillary features such as risk management and asset management. There are two aspects of a BAS to discuss, the hardware structure and the software features. In this section, Metasys of Johnson Controls is used as an example to illustrate a modern BAS. Actually, BASs of different manufacturers, such as Honeywell, Landis & Staefa, and Alerton, have more or less similar configurations and features.

16.4.1 Hardware Structure

Consider Figure 16.3. The basic architecture consists of multiple programmable control panels, called network control units (NCUs) and operator workstations (OWSs), which communicate with each other over a high-speed communication network called the N1 LAN. The OWS is normally a standard personal computer, such as a Pentium-based computer. Each NCU manages a physical area within the building, such as a mechanical equipment room. The capacity of an NCU can be enhanced with remote panels called network expansion units (NEUs). The NCUs and NEUs directly control central plant equipment, while the management of smaller air handlers, heat pumps, lighting circuits, and other building services systems is delegated to a family of application-specific controllers (ASCs). The ASCs and NEUs communicate with the NCUs over a secondary communication network, called the N2 bus. The architecture is unique in the way that the control functions are mainly distributed but yet remain tightly coupled. ASCs and NCUs provide stand-alone control capability for HVAC, fire management, access control, and lighting control wherever they are needed, giving maximum fault tolerance and reliability. When these controllers are interconnected on the N1 LAN and N2 bus, all parts of the facility's operation are coordinated with each

Section 16.4 Building Automation Systems (BASs)

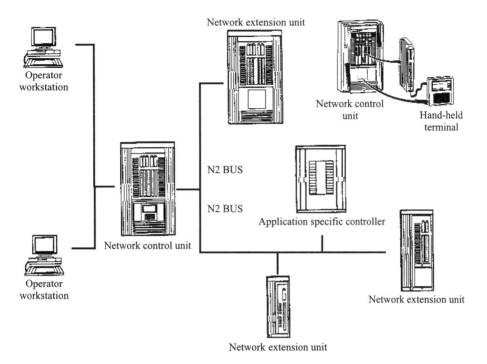

Figure 16.3 Structure of a standard BAS.

other so that the building operators can have complete and consistent information about the facility from all input/output devices. The N1 LAN is the communication backbone for the whole BAS, and it allows the functions of multiple NCUs and OWSs to be integrated into a facility-wide control and information network. The original design using ARCnet had a communication rate of 2.5 mega-baud over a combination of media, including coaxial cable, twisted pair, or fiber optics. The N2 bus connects point interfaces and remote ASCs within an equipment room, or within occupied spaces of the facility to the relevant NCUs. It was initially designed to use the Opto-22 Optomux bus operating at a maximum communication rate of 9600 baud, the de facto industry standard for many industrial automation and control applications.

The NCU is the heart of the network, and it has access to all information from every control device anywhere on the network, allowing it to perform control tasks with efficiency and intelligence. NCUs can work alone, and thus the OWS becomes the man–machine interface between the system and the human operator. There are various operationally stand-alone ASCs, such as the air-handling unit (AHU) controller, the intelligent lighting controller (ILC), the variable air volume box (VAV) controller, the unitary equipment (UNT) controller, the intelligent access controller (IAC), the intelligent fire controller (IFC), and so on. The HVAC system can consume up to half of the total energy of a building, and, therefore, the AHU, VAV and UNT controllers are very important. The controller software allows them to maximize the control of heating, cooling, economizer, preheating, humidification, dehumidification, static pressure, electric or hot-water reheating, fan assistance, and the like. Lighting is the second largest consumer of electrical energy in a facility—up to 20% of the total building energy consumption in most countries. ILCs provide flexible zone control and after-hours override capability. The network terminals (NTs) provide convenient interfaces to NCUs because of their portability.

16.4.2 Software Features

DDC control loops for damper operation are available to provide ventilation or to utilize outdoor air for cooling. For ventilation control, there are mainly three schemes: fixed quantity of outdoor air, mixed air control, and economizer control of mixer air. For heating control, five schemes are available: constant temperature hot-water control, hot-water reset control, discharge air control, discharge air reset control, and space temperature control. For cooling/heating/humidification/dehumidification control, five schemes are available: chilled water control, dehumidification control, humidification control, heating-cooling sequencing, and humidification-dehumidification sequencing. Static pressure control and VAV system terminal box control are also quite popular.

The original design concept of BAS is for building energy management. Hence, lots of relevant features are available inside a modern BAS. The duty cycle program reduces electrical energy consumed by the fan by cycling it on and off. The power demand limiting program monitors electrical consumption during each and every demand interval and sheds assigned loads as required to reduce demand. The unoccupied period program, or night cycle program, is primarily a heating season function, but it can also maintain a high space temperature limit during the cooling season, if desired. The optimum start-stop program of chillers is an adaptive energy-saving program that uses intelligence and the flywheel effect (energy retention capacity) of a building to save a considerable amount of energy with the program clock. The unoccupied night purge program can be applied to most HVAC systems that are capable of using 100% outdoor air when the temperature of outdoor air drops considerably at night. The enthalpy program monitors the temperature and relative humidity or dewpoint of the outdoor and return air and then positions the outdoor air and return air dampers to use the air source with the lowest total heat or least enthalpy. The load reset program controls heating and/or cooling to maintain comfort conditions in the building while consuming a minimum amount of energy. The zero-energy band program saves energy by avoiding simultaneous heating and cooling of air delivered to spaces.

Lighting, as already mentioned, consumes quite a large amount of energy. The occupied-unoccupied lighting control is a time-based program that schedules the on/off time of luminaires for a building or zone to coincide with the occupancy schedules. Another way to reduce the costs associated with lighting is to control the level of lighting in a building or building zone in terms of time schedule. Lighting level control is accomplished by two different methods: multilevel lighting and modulated lighting. To achieve the goal of control, control ballasts are separately designed for different methods.

For fire protection, the present concept is partial integration where there is an existing fire alarm system due to the legal requirements in most countries. When hardwired to a BAS controller, the fire alarm system behaves as a few input points, either digital or analog, to the NCUs. However, the trend is toward the total integration concept. This implies that a single central host computer serves the fire alarm system as well as the other building systems. The fire alarm control panels and the ASCs communicate with the OWSs over separate communication buses. An advantage here is that, for example, if a fire occurs on one floor of a multistory building, the HVAC units can be used to prevent the smoke from spreading by opening exhaust dampers and closing the outdoor air intake dampers of the fire floor. The integration of security and access control and other building services systems into a BAS can provide both economic and operational benefits. First, initial installation work, such as electric wiring, can be consolidated, resulting in cost savings. Substantial paybacks can be generated through HVAC energy management and lighting programs, thereby offsetting some of the costs involved in the integration process. Second, the cost of on-site human guard services can be greatly reduced.

The BAS can also be a tool to assist the facility management and operating personnel of a building. The computerized maintenance management programs provide facility management personnel with the tools needed to protect equipment, control costs, schedule workloads, review historical trends, manage materials, and plan budgets. Maintenance scheduling includes work order printout, maintenance history, material inventory, financial analysis, and management information. With a BAS, the concept of condition-based maintenance (CBM) can be realized versus the breakdown maintenance and planned preventive maintenance schemes conventionally adopted. With CBM, machines will be maintained only when demanded. The utility's metering program provides the means to dynamically monitor and record a facility's energy consumption on a real-time basis while a tenant energy monitoring program is also available. The heating/cooling plant efficiency program can continuously monitor the efficiency of the central HVAC plants because a small reduction in the operating efficiency of these large central systems can result in a significant increase in energy consumption and its associated costs.

16.5 ADVANCED BUILDING CONTROLS TECHNOLOGIES

Throughout the 1990s, techniques in artificial intelligence have widely been applied to control systems in buildings. Some of them are briefly discussed here because they are still under research and are not yet available in the market. Readers interested in these technologies are encouraged to consult the works cited in relevant sections.

16.5.1 Applications of Expert Systems

Vertical transportation is one very important service in modern high-rise buildings. Elevators are provided in groups, and the assignment of a particular elevator car to serve a particular landing call is carried out by the supervisory control system. The philosophy of supervisory control based on traffic sensing and rule-based expert systems was developed in 1992 [6]. The system was implemented using standard packages, built on a spreadsheet in the first instance. Simulated input traffic was generated and dynamically linked to the simulator, showing car movements. An expert system linked to the traffic sensing system continuously calculated optimal car movements. It needs expert knowledge to develop the expert system, and thus, the objectives of optimization must be clearly defined. Another approach consisted first of a better definition of objectives—that is, factors related to passengers were quantified; second, an evaluation module was designed; and finally, the evaluation module was integrated into a target system [7]. One newer approach [8] addressed the problem of finding optimum routes for a multi-lift system, with the objective of reducing the overall trip time for passengers by executing an exhaustive search for all possible moves. Alternatively, the blackboard architecture is a powerful expert system architecture and a model of problem solving that can be used to deal with large amounts of diverse and incomplete knowledge. The designer is not committed to either forward or backward chaining modes of reasoning by treating the blackboard as a central data store within the system. The blackboard architecture was implemented [9] for the control and supervision of group automatic operation of elevators using Prolog. A "channeling" approach [10] was adopted that took maximum advantage of "coincident destinations" by directing passengers with similar destinations into the same car. This was done by restricting the number of floors served on any trip to a small subset of the total number of upper floors.

A similar system [11] that is becoming more and more popular is one using the "hall call allocation" approach in which no car call panel is available inside the car. Each passenger needs to register his or her destination floor at the landing hall, and a car will be selected to serve this call. This system makes use of the additional information of destination hall calls to furnish data for car allocation. Then, assignment indicators are used to direct passengers' boarding. However, this system does not allow the passenger any mistake. If the passenger keys in a wrong destination call or enters the wrong elevator, it will take her quite a long period of time to arrive at the destination. This concept of channeling was further developed to give higher flexibility in up-peak, down-peak, and interfloor situations by introducing the idea of full dynamic zoning [12].

16.5.2 Neural Network-Based Control

The use of artificial neural networks (ANNs) has been proliferating with remarkable speed. The application of these networks is an attempt to simulate simple biological networks by joining together "cells" (or nodes) in a cascaded fashion, all interconnected. Mathematically, ANNs provide a parameterized structure for nonlinear function approximation and nonlinear classification applications. An ANN-based approximator or classifier is developed by "training" on a data set representative of the problem. Several applications to building control systems have been proposed. Anstett and Kreider [13] applied an ANN to predict the energy use in a complex institutional building without the need for a data acquisition system. Curtiss et al. [14] used ANNs for predictive control of a hot-water coil to warm an airstream. This work provided a very good basis for further research in this area of application. In our air-handling unit, shown in Figure 16.1, there are large numbers of plant status parameters and control actuators, such as the supply air damper position, the water valve position, the fan speed, and the power output of the reheater. These are heavily interconnected with one another so that the control action becomes very complicated. Such a system calls for a multiple-input multiple-output approach. The ANN, shown in Figure 16.4, serves as the identifier of the AHU and the controller as well [15]. R_s stands for the resistance of the supply air damper; P for the power of the reheater; $WetF$ for the humidification process; N_s for the speed of the supply air fan; Q_w for the rate of the chilled water supply; T_r for the room air temperature; RH for the room relative humidity; $P8$ for the pressure in the supply air duct; T_s for the supply air temperature; Q for the supply airflow rate; Q_p for the return airflow rate; and X_s for the moisture content of the supply air.

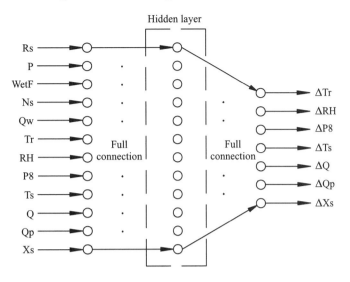

Figure 16.4 Air handling system identifier-controller.

There are a total of twelve input nodes consisting of the seven plant status parameters and the five actuating actions at time t. The normalized values, rather than the absolute values, are fed into the ANN by appropriate normalization functions. It is better to normalize all values to within a range such as from 0 to 1 before they are fed into the input nodes. There are twelve hidden nodes at the second layer of the ANN identifier/controller. For the output layer, there are seven nodes comprising the seven plant status parameters at one time step later, that is, at $t + \Delta t$. The number of hidden layers and the number of hidden nodes can be varied during neural network development, depending on the speed of the computational machine and the expertise of the designer. In our case, Δt is always set to a value of one second. Originally, the normalized absolute values of the seven plant status parameters were used. However, it has been found that the system becomes very insensitive to changes since within a time step of one second, all changes are actually small changes. Thus, a modification has been made so that the output nodes contain the normalized values of the deviations of the seven plant status parameters. In this way, any small changes can be appropriately amplified so as to have a wide span to cover the whole range from 0 to 1. In our case, a value of 0.5 of a certain plant status parameter implies that the parameter has not changed at all during that time step. A value of 1 implies a peak change in the positive direction, while a value of 0 indicates a peak change in the negative direction.

At the same time, this identifier functions as a controller. At time t, the identifier updates itself by using the normalized absolute plant status parameters and the normalized absolute actuating actions at time $t - \Delta t$ together with the normalized deviation of the plant status parameters at time t as its first training example. Of course, if memory space is not a problem, it is preferable that past training data be incorporated as well since every time a new set of data is available, the ANN will slightly forget the past data. In order to keep the ANN updated around the current operating point of the plant or system, training by using previous data constitutes good practice. Fortunately, the plant characteristics change slowly compared with the training rate of the ANN due to the large inertia of the physical system. In addition, our target is always on steady-state control and the whole system stays at a particular operating point for quite a long period of time. Under these circumstances, significant adjustments of the ANN's internal weighting functions are deemed unnecessary, and there is no genuine control problem even if the ANN forgets the plant characteristics of a few hours ago. Once the ANN is trained around a particular operating point, the computational effort to slightly adjust the individual weighting function inside the ANN subject to minor changes of the plant characteristics is quite minimal. After the updating process, the ANN becomes a controller that suitably adjusts the five actuating actions at time t so as to arrive at a desirable control at time $t + \Delta t$.

The control algorithm developed for this ANN is based on the minimization of a performance index (PI) designed for two aspects: setpoint error minimization and total energy consumption minimization. It is possible to put the concentration on either aspect to accomplish the desirable control result. The performance index is defined as:

$$PI = \frac{K_1}{K_1 + K_2}\left[A_1\left(\frac{T_r - T_{rs}}{T_{rs}}\right)^2 + A_2\left(\frac{RH - RH_s}{RH_s}\right)^2 + A_3\left(\frac{P_8 - P_{8s}}{P_{8s}}\right) + A_4\left(\frac{T_s - T_{ss}}{T_{ss}}\right)^2\right]$$
$$+ \frac{K_2}{K_2 + K_2}[(T_{wo} - T_{wi})D_w C_{pw} Q_w + P + (P_7 - P_6)Q]$$

The A_is are used to distinguish the differences in significance between the four control variables, T_r, RH, P_8, and T_s. D_w is density, and C_{pw} is specific heat capacity of chilled water. The adjustment of two Ks allows emphasis to be placed on either setpoint error minimization or energy optimization. T_r is the dry-bulb temperature of the conditioned space; RH is the relative humidity; P_8 is the air pressure at the supply air duct; and T_s is the temperature of the supply air. Another additional suffix "s" refers to the setpoint of the variable. T_{wo} and T_{wi} are, respectively, output temperature and input temperature at the chilled water coil; P is the power of the reheater; Q_w is chilled water-flow rate; C_{pw} is heat capacity. P_7 and P_6 are air pressure on both sides of the supply air fan; and Q is the supply airflow rate.

16.5.3 Fuzzy Logic-Based Control

The AHU must be designed to cope with a wide range of operating conditions since the weather and occupants' activities are subject to significant, periodic changes from day to night and from season to season and the air-conditioning process is highly nonlinear. Earlier we discussed the real-time modeling of the whole system by an ANN. Because air-conditioning is a complicated process and the air-conditioned space is often subject to external disturbance, the system model is usually not well known beforehand, and it will be quite time consuming to fine-tune the model. To execute control actions immediately after commissioning, past experience on similar machines needs to be relied on, and, therefore, fuzzy logic-based control [16] may perhaps be the best alternative solution. Fuzzy control is based on the valuable operational experience of human experts, and thus the system is robust with respect to ad hoc changes in the environment. In the example discussed here, the same AHU as shown in Figure 16.1 is used.

The first step of fuzzy control is fuzzification. Fuzzification is the process of converting a real-world parameter into a corresponding set of membership functions of the associated fuzzy sets. In this case, all input variables are subtracted by their reference set values to form error signals and then converted into membership functions of seven linguistic fuzzy subsets, namely, VN (very negative), MN (medium negative), SN (small negative), ZR (zero), SP (small positive), MP (medium positive), and VP (very positive). The fuzzifying functions for the error signal and the rate of change of error signal of the return air temperature (i.e., actual temperature inside the conditioned space) are shown in Figures 16.5(a) and 16.5(b) as examples.

The second step is the provision of a rule base. Basically, two types of control actions are associated with the AHU: positive action and negative action. For positive action, the control action has to be positive when both the error and the rate of change of error are positive. Controllers for the cooling coil and reheater fall into this class. For example, if the supply air temperature is high and there is a tendency for further increase, that is, error and rate of change of error are positive, the chilled water-flow rate has to be increased, that is, is positive, as well. The rule base is listed in Table 16.1, and is applicable to all five fuzzy controllers in the system, namely, that of the supply air fan, supply air damper, chilled water valve, reheater, and humidifier.

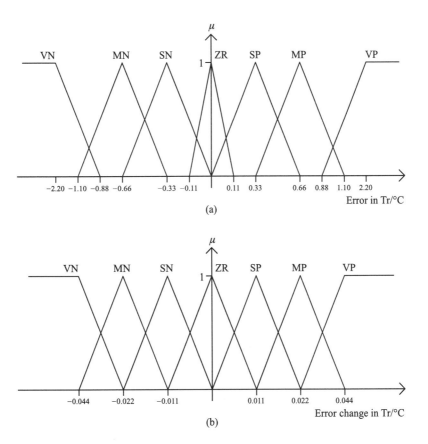

Figure 16.5 (a) Fuzzifying functions of error of return air temperature. (b) Fuzzifying functions of change of error of return air temperature

TABLE 16.1 Positive Control Action

Error	Error Change						
	VN	MN	SN	ZR	SP	MP	VP
VN	VN	VN	VN	VN	MN	SN	ZR
MN	VN	VN	VN	MN	SN	ZR	SP
SN	VN	VN	MN	SN	ZR	SP	MP
ZR	VN	MN	SN	ZR	SP	MP	VP
SP	MN	SN	ZR	SP	MP	VP	VP
MP	SN	ZR	SP	MP	VP	VP	VP
VP	ZR	SP	MP	VP	VP	VP	VP

The rule base is normally built inside the inference engine of the fuzzy controller. The chilled water coil controller is described as an illustrative example. The chilled water-flow rate is governed by the error in the supply air temperature, T_s, and the aim is to maintain a constant supply air temperature, T_{ss}. The rule "If $(T_s - T_{ss})$ is MP and rate of change of $(T_s - T_{ss})$ is SP, then Action is VP" gives a membership function for VP (μ_{VP}) as:

$$\mu_{VP} = \min\left\{\mu_{MP}[T_s - T_{ss}], \quad \mu_{SP}\left[\frac{\partial(T_s - T_{ss})}{\partial t}\right]\right\}$$

where μ_{MP} is the degree of belonging of $(T_s - T_{ss})$ in the fuzzy subset entitled MP and μ_{SP} is the degree of belonging of the rate of change of $(T_s - T_{ss})$ in the fuzzy subset entitled SP. After the whole evaluation process, we obtain 49 membership functions from all the rules of each controller. A certain number of rules output identical fuzzy subsets; for example, 10 rules return membership functions for VP, five rules return membership functions for MP, and so on. The contribution from each rule is summarized in the defuzzification procedure, which converts a value from the fuzzy environment back to the real world.

Each of the 49 rules returns a membership function μ_i, for $i = 1, \ldots, 49$ from the inference engine. A crisp, nonfuzzy value can be assigned to the output of each fuzzy rule, C_i. The actual output, OP, of the fuzzy controller can then be computed by the center-of-gravity method as shown by the following equation:

$$OP = \frac{\sum_{i=1}^{49} C_i \mu_i}{\sum_{i=1}^{49} \mu_i}.$$

This output is then fed to the actuator for the proper control action.

16.5.4 Computer Vision-Based Control

Conventional control for HVAC relies on measuring devices such as thermostats and humidistats to monitor the temperature and humidity of the supply and return air of the air-conditioned space. Various control algorithms, such as PID, adaptive, self-tuning, and fuzzy logic-based, have been incorporated in the control of AHUs for thermal comfort. However, it is well recognized that slow response rate is the major drawback of most commonly used measuring devices. When an ad hoc change of considerable magnitude in the load demand occurs, there is usually a rather long delay before the controller can take any subsequent action. For example, the VAV box only opens after the thermostat confirms an increase in the return air temperature, which is a slow action because of the high room inertia as well as the intrinsic delay in the thermostat's response. Although it may not be too serious bearing in mind the usual noncritical nature of control applications in most commercial and domestic buildings, the long transient period may sometimes imply extra energy consumption in bringing the system back to steady state. By adopting computer vision [17] that can estimate the number of residents within an air-conditioned space, it is possible to identify any abrupt changes within seconds. Conventional transducers are, in no way, comparable to this new approach. Furthermore, better zone control can be accomplished if through computer vision we can identify the location of each occupant. Significant improvements are possible with respect to both response rate and energy savings.

An experimental system developed at City University of Hong Kong involves a stereoscopic camera system in which two standard charge coupled device (CCD) cameras are placed side by side. Calibration of each camera needs to be carried out so that 11 parameters associated with the geometry of the camera, such as the focal length, center of the focus, and lens distortion, can be precisely estimated. After the process of calibration, any point (x_w, y_w, z_w) in the world coordinates can be appropriately matched to a point (X_0, Y_0) in the memory of the frame grabber. The frame grabber is a device that retrieves signals from the camera to produce a corresponding image file. Then, the process of depth from motion using an optical flow technique is carried out so that the absolute position in world coordinates of any point seen on the image file can be estimated. Optical flow estimates the velocity vector of any point on the image file based on successive images taken within a very short period of time, say fractions of a second. The velocity vector helps in doing pixel correspondence so that the same point in space can be accurately identified on the multiple image files from the two cameras. Based on simple geometry, the absolute position of that point in space is calculated. If such a point is on the surface of any occupant, the number of occupants and their positions in space can be found. Appropriate HVAC control actions can then be executed.

16.6 DIFFICULTIES WITH BUILDING SYSTEMS CONTROL

The state-of-art technologies related to building control and automation have seen significant advances for different applications. Two major problems that require urgent solution are the implementation of full integration and financial viability.

In a previous section on protocols, we studied the effort made by various organizations on the development of open systems and the emphasis on interoperability. The fundamental idea is to manufacture control devices that talk in an identical and universal language regardless of the brand and model of each device. This dream did come true in certain restricted levels within a BAS. More should be done in this aspect, bearing in mind the commercial concerns of most manufacturers in the industry. Actually, a number of open protocols are available, and it will take time for designers to agree on one common set of open protocols. A common language for all control devices may imply free replacement of components of different manufacturers by the users, thus affecting the profit that is almost guaranteed to certain suppliers who have been dominating the market for a long period of time. Even though true interoperability at all networking levels within BASs can be achieved, the integration is still confined to the machine environment. However, the main goal of providing BASs in modern intelligent buildings is to serve the occupants in terms of human comfort, human safety, and human efficiency, instead of merely demonstrating high technologies. Comprehensive and convenient information flow between the BAS and the occupants should be the key feature in the near future. Existing media for presenting and inputting information, such as numeric codes, long strings, and computer languages, are not user-friendly enough, making occupants, who are usually laypersons to advanced technology, try to escape close contact with the building control system. Therefore, more research and development needs to be done in the area of man–machine interfaces, not solely for system operators but for the normal occupants as well. This is what is meant by full integration at both the machine and human levels.

The building industry can, in general, be regarded as a more conventional industry. The rate of advancement of technology in the field of building and construction is relatively low compared with that of the electronics and control industries or information technology. In this regard, developers of buildings are usually more reluctant to make huge investments in building systems control, whereas they very often intend to put more resources into improving the architectural layout such as interior design and furnishing. In this way, the development strategy for building control systems must be significantly different from that for, say, aerospace or military systems, where quality and effectiveness are the major concerns. The installation of a BAS has to be financially viable, and the performance should emphasize reduction of operational cost and enhancement of productivity of the building as a whole.

16.7 CONCLUSION

In this chapter, we have first considered existing, well-proven control technologies in modern buildings, especially intelligent buildings. The most important development, the information highway and its integration into building control, has been discussed in some detail. Local area networks, in terms of hardware and software, have also been described. The structure and features of a typical building automation system can be considered the heart of every intelligent building. As a practical example, one comprehensive intelligent building system can be found in Germany, the Munich International Airport [18]. This system, developed by Honeywell, has the capacity to control 200,000 points, integrating 13 major subsystems from nine different vendors. It provides control of systems including the power plant, HVAC, people-moving systems, interior lighting control, runway lights, baggage handling, elevators, and aerobridges, for a total of 112,000 physical points. The airport, opened in 1992, was named the "Intelligent Building of the Year" in 1993 by the Intelligent Buildings Institute Foundation. The project demonstrates that a set of functions, no matter how widespread, diverse, or complex, can be smoothly and effectively integrated together for maximum control and productivity.

Finally, various advanced control techniques and their applications in building services systems have been briefly introduced, including control based on expert systems, artificial neural networks, fuzzy logic, and computer vision. The difficulties that the building automation industry will face are also discussed to provide designers with some vision on the road that lies ahead. It is hoped that this chapter will build up control engineers' and scholars' interests in the research and development of building control systems.

Related Chapters

- An analogous overview of control systems in the process industries appears in Chapter 12.
- Neural networks for function approximation are discussed in detail in Chapter 6.
- Tutorial material on intelligent control, with examples on expert systems, neural networks, and fuzzy logic, can be found in Chapter 5.
- Chapter 12 also reviews a number of PID tuning methods.
- The structural control of buildings—as distinct from the control of their interior environments—is the topic of Chapter 17.

REFERENCES

[1] R. A. Carlson and R. A. Di Giandomenico (eds.), *Understanding Building Automation Systems*. Kingston, MA: R. S. Means Co., 1991.

[2] S. G. Brandt and G. Shavit, "Simulations of the PID algorithm for direct digital control application." *Proc. Workshop on HVAC Controls, Modelling and Simulation*, G.I.T., Atlanta, 1984.

[3] A. T. P. So, A. C. W. Wong, and W. L. Chan, "The role of high speed communication in building automation." *Proceedings of Mainland Hong Kong HVAC Seminar '98*, ASC, HKIE, CIBSE, ASHRAE, Beijing, March 1998, pp. 6–11.

[4] D. P. Wyon, "Healthy buildings and their impact on productivity." *Proc. Indoor Air '93, 6th Int. Conf. on Indoor Air and Climate*, Helsinki, Finland, July 1993.

[5] M. Lockareff, "A control networking solution for the utility industry." *LonWorks Technology and the LonMark Standard*, 1996.

[6] R. W. Prowse, T. Thomson, and D. Howells, "Design and control of life systems using expert systems and traffic sensing." In G. C. Barney (ed.), *Elevator Technology* 4, IAEE, pp. 219–226, 1992.

[7] P. Chenais and K. Weinberger, "New approach in the development of elevator group control algorithms." In G. C. Barney (ed.), *Elevator Technology* 4, IAEE, pp. 48–57, 1992.

[8] A. F. Alani, P. Mehta, J. Stonham, and R,. Prowse, "Performance optimisation of knowledge-based elevator group supervisory control system." In G. C. Barney (ed.), *Elevator Technology* 6, IAEE, pp. 114–121, 1995.

[9] G. K. H. Pang, "Elevator scheduling system using blackboard architecture." *IEEE Proceedings D*, Vol. 138, no. 4, pp. 337–346, 1991.

[10] B. A. Powell, "Important issues in up-peak traffic handling." In G. C. Barney (ed.), *Elevator Technology* 4, IAEE, pp. 207–218, 1992.

[11] J. Schröder, "Advanced dispatching, destination hall calls and instant car-to-call assignments." *Elevator World Educational Package and Reference Library*, Vol. 4, pp. IV8–IV14, 1994.

[12] A. T. P. So and W. L. Chan, "Comprehensive dynamic zoning algorithms." In G. C. Barney (ed.), *Elevator Technology* 8, pp. 98–107, April 1997, reprinted in *Elevator World*, pp. 99–103, September 1997.

[13] M. Anstett and J. F. Kreider, "Application of neural networking models to predict energy use." *ASHRAE Transactions: Research, Pt. 1*, pp. 505–517, 1993.

[14] P. S. Curtiss, J. F. Kreider, and M. J. Brandemuehl, "Adaptive control of HVAC processes using predictive neural networks." *ASHRAE Transactions: Research, Pt. 1*, pp. 496–504, 1993.

[15] A. T. P. So, T. T. Chow, W. L. Chan, and W. L. Tse, "A neural network based identifier/controller for modern HVAC control." *ASHRAE Transactions*, Vol. 102, Pt. 1, pp. 14–31, 1996.

[16] A. T. P. So, T. T. Chow, W. L. Chan, and W. L. Tse, "Fuzzy air handling system controller." *Building Services Engineering Research and Technology*, CIBSE, Vol. 15, no. 2, 1994, pp. 95–105.

[17] A. T. P. So, W. L. Chan, and T. T. Chow, "A computer vision based HVAC control system." *ASHRAE Transactions*, Vol. 102, Pt. 1, 1996, pp. 661–678.

[18] M. Ancevic, "Intelligent building system for airport," *ASHRAE Journal*, pp. 31–35, November 1997.

Chapter 17

CONTROLLING CIVIL INFRASTRUCTURES

B. F. Spencer Jr. and Michael K. Sain

Editor's Summary

Controls is well-established in most of the major engineering disciplines—electrical, chemical, mechanical, aerospace. Historically, an important exception has been civil engineering, and, as this chapter illustrates, recent developments are bridging the gap. The importance of understanding the dynamics of civil structures has been recognized since the 1940 Tacoma Narrows bridge collapse, but feedback control of buildings, bridges, towers, and other structures is a relatively recent development. The concept of active control for such systems was first introduced in 1972. Since then, a vast literature has been generated on the topic, and, more impressively, a number of successful implementations have been completed (the first full-scale one in 1989). Many of the largest applications have been to buildings in Japan, driven by the desire to achieve protection against earthquakes.

The first implementation of structural control was based on active mass dampers (AMDs). An AMD system couples an auxiliary mass to the structure through an actuator. Sensor measurements of building movement and stresses are used in a control algorithm to move the auxiliary mass relative to the building. Such systems are versatile and capable, but issues of reliability and power consumption have driven the search for improvements. The next significant development was controllers that employed a combination of active and passive devices. These hybrid active/passive control systems (no relationship to the hybrid discrete/continuous systems discussed in Chapter 7) rely on one of two approaches: hybrid mass damping and hybrid base isolation. The former is especially popular. The largest building in Japan, the Yokohama Landmark Tower, incorporates two hybrid mass dampers (HMDs), each weighing 170 tons.

The most recent innovation is the semiactive control device. These devices cannot inject mechanical energy into the structure but have properties that can be manipulated to achieve structural disturbance rejection. In many cases, they can operate on battery power; this is a significant advantage since seismic events can interrupt main power supplies. Examples of semiactive devices include variable-orifice fluid dampers, variable-stiffness devices, variable-friction devices, controllable and tuned liquid dampers, and magnetorheological dampers. The last topic is discussed at some length in this chapter, and experimental results are shown.

B. F. Spencer Jr. is a professor in the Department of Civil Engineering and Geological Sciences at the University of Notre Dame. Michael K. Sain is the Frank M. Freimann Professor of Electrical Engineering at the same institution.

17.1 INTRODUCTION

The protection of civil structures, including their material contents and human occupants, is a worldwide priority of the most serious importance. Such protection may range from reliable operation and comfort, on the one hand, to survivability on the other. Examples of such structures that readily leap to mind include buildings, offshore rigs, towers, roads, bridges, and pipelines. In like manner, the events that require such protective measures are earthquakes, winds, waves, traffic, lightning, and—today, regrettably—deliberate acts. Control methods will make a genuine contribution to this problem area, which has such great economic and social implications. In this chapter, we review recent developments that have been rapidly occurring in the area of controlled civil structures, including full-scale implementations and actuator types and characteristics, as well as trends toward the incorporation of more modern algorithms and technologies.

Buildings and other physical structures, including highway infrastructures, have traditionally relied on their strength and ability to dissipate energy to survive under severe dynamic loading. In recent years, worldwide attention has been directed toward the use of control and automation to mitigate the effects of these dynamic loads on these structures [1–3]. In fact, several buildings in Japan, including a 70-story hotel and a 52-story office complex, are currently employing active control strategies for motion control. Active systems are also used temporarily in the construction of bridges or large span structures (e.g., lifelines, roofs) where no other means can provide adequate protection.

Figure 17.1 provides a schematic diagram of the structural control problem. The basic task is to determine a control strategy that uses the measured structural responses to calculate an appropriate control signal to send to the actuator that will enhance structural safety and serviceability. To better understand the problem, consider control of the tall building depicted in Figure 17.2 using an active mass damper (AMD) system. For this control system, a small auxiliary mass, which is usually less than 1% of the total mass of the structure, is installed on one of the upper floors of the building and an actuator is connected between the auxiliary mass and the structure. Responses and loads at key locations on the building are measured and sent to the control computer. The computer processes the responses according to the control algorithm and sends an appropriate signal to the AMD actuator. The actuator then reacts against the auxiliary mass, applying inertial control forces to the structure to reduce the structural responses in the desired manner. A wealth of structural control studies have been conducted since Yao (1972)[1] first introduced

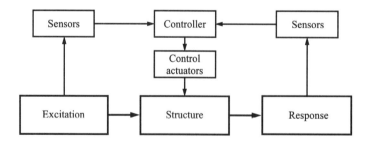

Figure 17.1 Schematic diagram of the structural control problem.

---

[1] Space does not permit a complete listing of references. The reader may obtain the specific citations from the earlier version of this chapter titled "Controlling Buildings: A New Frontier in Feedback," published in IEEE *Control Systems Magazine*, December 1997; or they may be accessed at www.nd.edu/~quake.

Section 17.1 Introduction

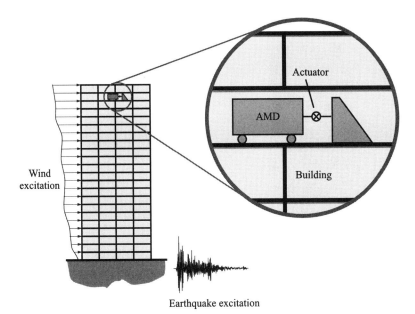

Figure 17.2 Concept of the AMD control system.

the concept of active control of civil engineering structures. These include, for example, H_2/H_∞ control (Suhardjo et al., 1992; [4]; Jabbari, Schmitendorf, and Yang, 1995; Kose et al., 1996); sliding-mode control (Nonami et al., 1994; Yang et al., 1995; Adhikari et al. 1996; Singh et al., 1996); saturation control (Chase and Smith, 1996; Agrawal, Yang, et al., 1997); reliability-based control (Spencer, Sain, et al., 1992; [5]; Field et al., 1995, 1996a, b, c, 1997); fuzzy control (Nagarajaiah, 1994; Subramaniam, Reinhorn, et al., 1996; Faravelli and Yao, 1996; Casciati et al., 1996, 1997); neural control (Venini and Wen, 1994; Ghaboussi et al., 1995); modeling and identification ([6]; Skelton and Lu, 1996; Nishitani et al., 1996; Smith et al., 1996); nonlinear control ([7]; Spencer et al., 1996; Agrawal and Yang, 1996, 1997; Yang et al., 1996); implementation issues (Chung et al., 1989; Reinhorn et al., 1993; Quast et al., 1995; Dyke et al., 1996; [8]; Yang et al., 1996); and benchmark studies (Spencer et al., 1997, to appear).

The first full-scale application of active control to a building was accomplished by the Kajima Corporation in 1989 ([9]; Sakamoto, Kobori et al., 1994). The Kyobashi Seiwa building shown in Figure 17.3 is an 11-story (33.1 m) building in Tokyo, Japan, having a total floor area of 423 m^2. A control system was installed, consisting of two AMDs—the primary AMD is used for transverse motion and has a mass of 4 tons, while the secondary AMD has a mass of 1 ton and is employed to reduce torsional motion. The role of the active system is to reduce building vibration under strong winds and moderate earthquake excitations and consequently to increase the comfort of the building's occupants.

Although nearly a decade has passed since the construction of the Kyobashi Seiwa building, a number of serious challenges remain to be resolved before feedback control technology can gain general acceptance by the civil engineering and construction professions at large. These challenges include: (1) reducing capital cost and maintenance,

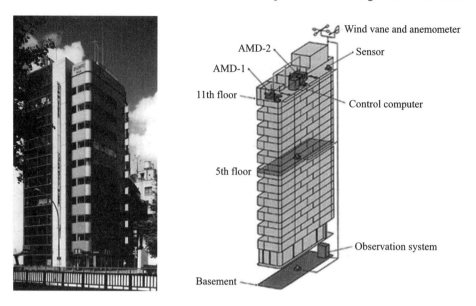

Figure 17.3 Kyobashi Seiwa Building with AMD installation.

(2) eliminating reliance on external power, (3) increasing system reliability and robustness, and (4) gaining acceptance of nontraditional technology. Hybrid and semiactive control strategies are particularly promising in addressing a number of the challenges to this technology.

17.2 HYBRID CONTROL SYSTEMS

Hybrid control strategies have been investigated by many researchers to exploit their potential for increasing the overall reliability and efficiency of the controlled structure [10]. A hybrid control system is typically defined as one that employs a combination of passive and active devices. Because multiple control devices are operating, hybrid control systems can alleviate some of the restrictions and limitations that exist when each system is acting alone. Thus, higher levels of performance may be achievable. In addition, the resulting hybrid control system can be more reliable than a fully active system, although often it is also somewhat more complicated. To date, over 30 buildings and 10 bridges (during erection) have employed feedback control strategies in full-scale implementations (see Tables 17.1 and 17.2). The vast majority of these have been hybrid control systems. Research in the area of hybrid control systems has focused primarily on two classifications of systems: hybrid mass damper systems, and hybrid base isolation.

17.2.1 Hybrid Mass Damper

The hybrid mass damper (HMD) is the most common control device employed in full-scale civil engineering applications. It is a combination of a tuned mass damper (TMD) and an active control actuator. The ability of this device to reduce structural responses relies mainly on the natural motion of the TMD. The forces from the control

TABLE 17.1 Summary of Actively Controlled Buildings/Towers

Full-scale structure	Location	Year completed	Scale of building	Control system employed	AMD/HMD No.	AMD/HMD Mass (tons)	Actuation mechanism
Kyobashi Seiwa	Tokyo, Japan	1989	33 m, 400 ton, 11 stories	AMD	2	5	hydraulic
Kajima Research Institute KaTRI No. 21 Building	Tokyo, Japan	1990	12 m, 400 ton, 3 stories	Active Variable Stiffness System (6 devices)	—	—	hydraulic
Sendagaya INTES	Tokyo, Japan	1992	58 m, 3280 ton, 11 stories	AMD	2	72	hydraulic
Applause Tower	Osaka, Japan	1992	161 m, 13943 ton, 34 stories	HMD	1	480	hydraulic
Kansai Int. Airport Control Tower	Osaka, Japan	1992	86 m, 2570 ton, 7 stories	HMD	2	10	servomotor
Osaka Resort City 2000	Osaka, Japan	1992	200 m, 56980 ton, 50 stories	HMD	2	200	servomotor
Yokohama Land Mark Tower	Yokohama, Kanagawa, Japan	1993	296 m, 260610 ton, 70 stories	HMD	2	340	servomotor
Long Term Credit Bank	Tokyo, Japan	1993	129 m, 40000 ton, 21 stories	HMD	1	195	hydraulic
Ando Nishikicho	Tokyo, Japan	1993	54 m, 2600 ton, 14 stories	HMD (DUOX)	1	22	servomotor
Hotel Nikko Kanazawa	Kanazawa, Ishikawa, Japan	1994	131 m, 27000 ton, 29 stories	HMD	2	100	hydraulic
Hiroshima Riehga Royal Hotel	Hiroshima, Japan	1994	150 m, 83000 ton, 35 stories	HMD	1	80	servomotor
Shinjuku Park Tower	Tokyo, Japan	1994	227 m, 130000 ton, 52 stories	HMD	3	330	servomotor
MHI Yokohama Bldg.	Yokohama, Kanagawa, Japan	1994	152 m, 61800 ton, 34 stories	HMD	1	60	servomotor
Hamamatsu ACT Tower	Hamamatsu, Shizuoka, Japan	1994	212 m, 107500 ton, 46 stories	HMD	2	180	servomotor

(continued)

TABLE 17.1 (continued)

Full-scale structure	Location	Year completed	Scale of building	Control system employed	AMD/HMD No.	AMD/HMD Mass (tons)	Actuation mechanism
Riverside Sumida	Tokyo, Japan	1994	134 m, 52000 ton, 33 stories	AMD	2	30	servomotor
Hikarigaoka J-City	Tokyo, Japan	1994	110 m, 29300 ton, 26 stories	HMD	2	44	servomotor
Miyazaki Phoenix Hotel Ocean 45	Miyazaki, Japan	1994	154 m, 83650 ton, 43 stories	HMD	2	240	servomotor
Osaka WTC Bldg.	Osaka, Japan	1994	252 m, 80000 ton, 52 stories	HMD	2	100	servomotor
Dowa Kasai Phoenix Tower	Osaka, Japan	1995	145 m, 26000 ton, 28 stories	HMD (DUOX)	2	84	servomotor
Rinku Gate Tower North Bldg.	Osaka, Japan	1995	255 m, 75000 ton, 56 stories	HMD	2	160	servomotor
Hirobe Miyake Bldg.	Tokyo, Japan	1995	31 m, 273 ton, 9 stories	HMD	1	2.1	servomotor
Plaza Ichihara	Chiba, Japan	1995	61 m, 5760 ton, 12 stories	HMD	2	14	servomotor
TC Tower	Kao Hsung, Taiwan	1996	85 stories	HMD	2	350	servomotor
Rinku Gate Tower North Bldg.	Osaka, Japan	1996	255.3 m, 56 stories	HMD	2	80	servomotor
Herbis Osaka Bldg.	Osaka, Japan	1997	190 m, 40 stories	HMD	1	158	servomotor
Itoyama Tower	Tokyo, Japan	1997		HMD			
Nanjing Tower	Nanjing, China	1997/98	310 m	AMD	1	60	hydraulic
Japan OTIS Elevator Test Tower	Chiba, Japan	1998		HMD			
JR-Odakyu Communication Bldg.	Tokyo, Japan	1998		AMD			
Bunka Fashion College	Tokyo, Japan	1998	93 m, 24000 ton	HMD + TMD	2	3	servomotor
Oita Oasis Hibora Bldg.	Oita, Japan	1998	101 m, 20000 ton, 21 stories	HMD	2	3	cylindrical linear-induction-servomotor
Sotetsu Takashimaya Bldg.	Japan		115 m	HMD	2	61	

Section 17.2 Hybrid Control Systems

TABLE 17.2 Summary of Bridge Towers Employing Active Control During Erection

Name of bridge	Years employed	Height, weight	Frequency range (Hz)	Moving mass, mass ratio (%[1])	Control algorithm	No. of controlled modes
Rainbow Bridge						
Pylon 1	1991–1992	119 m 4800 tonf	0.26–0.95	6 ton × 2 0.6	Feedback control	3
Pylon 2	1991–1992	117 m 4800 tonf	0.26–0.55	2 ton 0.14	DVFB[2]	1
Tsurumi-Tsubasa Bridge[3]	1992–1993	183 m 3560 tonf	0.27–0.99	10 ton × 2 0.16	Optimal regulator DVFB	1
Hakucho Bridge						
Pylon 1	1992–1994	127.9 m 2400 tonf	0.13–0.68	9 tonf 0.4	sub-optimal feedback control	1
Pylon 2	1992–1994	131 m 2500 tonf	0.13–0.68	4 ton × 2 0.36	DVFB	1
Akashi Kaikyo Bridge Pylons 1 & 2	1993–1995	293 m 24,650 tonf	–0.127–	28 ton × 2 0.8	Optimal regulator DVFB	1
Meiko-Central Bridge[3]						
Pylon 1	1994–1995	190 m 6200 tonf	0.18–0.42	8 ton × 2 0.98–1.15	H_∞ Feedback control	1
Pylon 2	1994–1995	190 m 6200 tonf	0.16–0.25	0.17–0.38		1
1st Kurushima Bridge						
Pylon 1	1995–1997	112 m 1600 tonf	0.23–1.67	6 ton × 2 0.15–2.05	Sub-optimal regulator control	3
Pylon 2	1995–1997	145 m 2400 tonf	0.17–1.70	10 ton × 2 0.3–2.6	H_∞ Feedback control	3
2nd Kurushima Bridge						
Pylon 1	1994–1997	166 m 4407 tonf	0.17–1.06	10 ton × 2 0.41	DVFB/H_∞	2
Pylon 2	1995–1997	143 m 4000 tonf	0.20–1.45	10 ton × 2 0.54–1.01	Fuzzy control	more than 3

(*continued*)

TABLE 17.2 (continued)

Name of bridge	Years employed	Height, weight	Frequency range (Hz)	Moving mass, mass ratio (%[1])	Control algorithm	No. of controlled modes
3rd Kurushima Bridge						
Pylon 1	1995–1996	179 m 4500 tonf	0.13–0.76	11 ton × 2 0.3–2.4	Variable gain DVFB	1
Pylon 2	1994–1996	179 m 4600 tonf	0.13–0.76	11 ton × 2 0.3–2.4	H_∞ output feedback control	1
Nakajima Bridge[3]	1995–1996	71 m 580 tonf	0.21–1.87	3.5 ton × 2 1.0–10.6	Fuzzy control	3

[1] Percent of first modal mass.
[2] Direct velocity feedback.
[3] Cable-stayed bridge. Others are suspension bridges.

actuator are employed to increase the efficiency of the HMD and to increase its robustness to changes in the dynamic characteristics of the structure. The energy and forces required to operate a typical HMD are far less than those associated with a fully active mass damper system of comparable performance.

Many researchers have made significant contributions to the development of HMDs that are compact, efficient, and practically implementable. A number of innovative, long-period devices have been reported. For example, Tanida et al., 1991, developed an arch-shaped HMD that has been employed in a variety of applications, including bridge tower construction, building response reduction, and ship roll stabilization. An arch-shaped hybrid mass damper (see Figure 17.4) was used during erection of the bridge tower (height = 119 m) of the Rainbow suspension bridge in Tokyo to reduce large-amplitude vortex-induced vibration expected to occur at a wind speed of 7 m/s (Tanida et al., 1991, 1995). The mass ratio for the hybrid damper used for the Rainbow bridge tower was 0.14% of the first modal mass of the structure, whereas a comparable passive TMD would require a 1% mass ratio to achieve a similar level of performance. Figure 17.5(b) shows an extension of the arch-shaped HMD, the V-shaped HMD (Koike et al., 1994), which has the advantage of having an easily adjustable fundamental period. Three of these devices were installed in the Shinjuku Park Tower, the largest building in Japan in terms of square footage (see Figure 17.5(a)).

Two multistep pendulum HMDs, each with a mass of 170 tons (Yamazaki et al., 1992), have been developed and installed in the Yokohama Landmark Tower, (Figure 17.6), the tallest building in Japan. The process of constructing the Landmark Tower provides yet another interesting and attractive application of active control, which is associated with the way construction cranes were used during its erection. Active control of the position of the crane was carried out by two fans (see Figure 17.7). These fans prevented excessive displacement and rotation of the building panels while hoisting and installing them, even under strong winds. Moreover, the overall efficiency of the crane work was significantly improved and resulted in reduced construction time for the Tower.

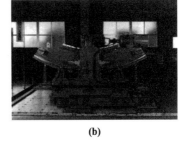

Figure 17.4 (a) Rainbow Bridge Tower while under construction, (b) HMD Employed during Tower erection.

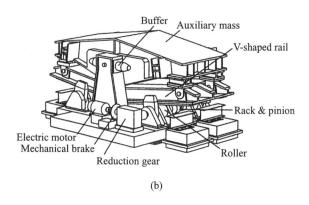

Figure 17.5 (a) Shinjuku Park Tower, (b) V-shaped hybrid mass damper employed in the Shinjuku Park Tower.

The DUOX HMD ([9]; Ohrui et al., 1994), which attains high control efficiency with a small actuator force, has also been proposed and employed in two buildings (see Figure 17.8). Devices similar to the DUOX HMD were also studied by Iemura and Izuno, 1994. Otsuka et al., 1994, conducted experiments in which a roller-pendulum-based HMD was applied to control a tower experiencing seismic excitation. Similar full-scale structural control implementations employing HMDs have been well documented (e.g., see Sakamoto et al., 1994; Koike et al., 1994; Higashino et al., 1993; Hirai et al., 1993; Ohyama et al., 1994; Suzuki et al., 1994; [11]; Fujino et al., 1994; Fujita et al., 1994a, b; Nakamura et al., 1994; Shiba et al., 1994; Yamamoto et al., 1994; Sakamoto and Kobori 1996; Iemura et al., 1996).

The hybrid mass damper is also effective for retrofit applications. Figure 17.9 depicts the Nanjing Tower, a 340-meter high television transmission and observation tower recently constructed in Nanjing, China. The tower has two observation decks, the uppermost being at 240 meters. During storms, excessive vibration occurs, and accelerations at this upper deck can exceed the human comfort limit of 0.15 m/sec^2. Cheng et al., 1994, proposed using an HMD system, combining a control actuator with a passive tuned liquid damper to control wind-induced vibration of the tower. Because the structure already existed, numerous physical constraints had to be accommodated in the control system design process. Cao et al., 1997, and Riley et al., 1997, have designed an innovative active mass damper system as a means of bringing the response of the tower to within acceptable limits. Their design, employing a 60-ton ring-shaped mass on

Section 17.2 Hybrid Control Systems

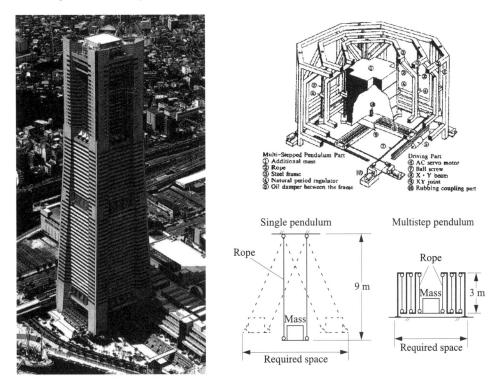

Figure 17.6 Multistep pendulum damper used in the Yokohama Landmark Tower.

Figure 17.7 Actively controlled crane used during construction of the Yokohama Landmark Tower.

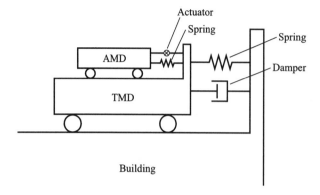

Figure 17.8 Concept of the DUOX system.

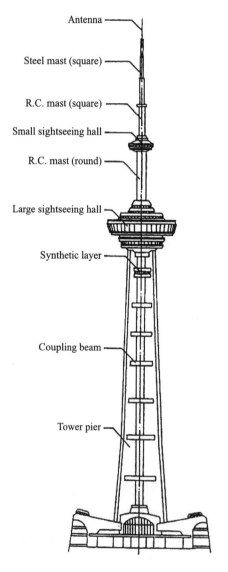

Figure 17.9 Nanjing Tower elevation.

sliding friction bearings, was shown to adequately reduce the structural response via a nonlinear control policy, while not violating the constraints. Wu and Yang, 1997, considered continuous sliding-mode control of the Nanjing Tower. This research was conducted as part of the U.S.–People's Republic of China cooperative program through the National Science Foundation.

A number of other interesting ideas employing the mass damper concept have been proposed. Seto et al., 1994, 1996, investigated the possibility of using active or passive forces acting between two adjacent structures to reduce the seismic response of both structures. As viewed from actual construction, many modern buildings might be divided into two or more adjacent substructures with connecting elements. Mita et al., 1994a, b, and Chai and Feng, 1996, presented studies of mega-subcontrol systems for tall buildings. The control system takes advantage of the megastructure configuration by designing the substructures contained in the megastructure to act as multi-degree-of-freedom tuned mass dampers. This approach implies that the subsystems act as vibration absorbers, and hence no additional mass is required as would be the case with a more conventional design. Craig et al., 1993, showed that hybrid control schemes, combining a simple active mass damper with the passive damping provided by cladding-structure interaction (Pinelli et al., 1995), doubled the reduction in peak response due to passive damping alone.

Researchers have investigated various control methods for HMDs. For example, Shing et al., 1994, Kawatani et al., 1994, Petti et al., 1994, Suhardjo et al., 1992, and Spencer et al. [4], have considered optimal control methods for HMD controller design. Tamura et al., 1994, proposed a gain-scheduling technique in which the control gains vary with the excitation level to account for stroke and control force limitations. Similarly, Niiya et al., 1994, proposed an ad hoc control algorithm for HMDs to account for the limitations on the stroke. Adhikari and Yamaguchi, 1996, and Nonami et al., 1994, applied sliding-mode theory to control structures with HMD systems.

17.2.2 Hybrid Base Isolation

Another class of hybrid control systems that has been investigated by a number of researchers is found in the active base isolation system, consisting of a passive base isolation system combined with a control actuator to supplement the effects of the base isolation system. Base isolation systems have been implemented on civil engineering structures worldwide for a number of years because of their simplicity, reliability, and effectiveness. Excellent review articles of base isolation systems are presented by Kelly, 1981, 1986; Buckle and Mayes, 1990; and Soong and Constantinou, 1994. However, base isolation systems are passive systems and have only a limited ability to adapt to changing demands for structural response reduction. With the addition of an active control device to a base isolated structure, a higher level of performance can potentially be achieved without a substantial increase in the cost (Reinhorn et al., 1987), which is very appealing from a practical viewpoint. Since base isolation by itself can reduce the interstory drift and the absolute acceleration of the structure at the expense of large absolute base displacement, the combination with active control is able both to achieve low interstory drift and to limit the maximum base displacement with a single set of control forces. A robust controller for uncertain linear base-isolated structures was proposed by Kelly et al., 1987 and more recently by Yoshida et al., 1994, Schmitendorf et al., 1994, and Yang et al., 1996.

Several small-scale experiments have been performed to verify the effectiveness of this class of systems in reducing the structural responses. Reinhorn and Riley, 1994, performed analytical and experimental studies of a small-scale bridge with a sliding hybrid isolation system in which a control actuator was employed between the sliding surface and the ground to supplement the base isolation system.

Also mentioned in this context is another type of *hybrid* base isolation system that employs a semiactive, friction-controllable fluid bearing in the isolation system. Feng et al., 1993, employed such bearings in a hybrid base isolation system in which the pressure in the fluid could be varied to control the amount of friction at the isolation surface. Yang et al., 1995a, b, investigated the use of a continuous sliding-mode control and variable structure system for a base isolated structure with friction-controllable bearings.

Because base isolation systems exhibit nonlinear behavior, researchers have developed various nonlinear control strategies including fuzzy control (Nagarajaiah, 1994), neural network-based control (Venini and Wen, 1994; Ghaboussi et al., 1995), and robust nonlinear control (Luo et al., 1996). In addition, Inaudi et al., 1993, studied the use of frequency domain shaping techniques in designing controllers.

17.3 SEMIACTIVE CONTROL SYSTEMS

Control strategies based on semiactive devices appear to combine the best features of passive and active control systems and to offer the greatest likelihood for near-term acceptance of control technology as a viable means of protecting civil engineering structural systems against earthquake and wind loading. The attention received in recent years can be attributed to the fact that semiactive control devices offer the adaptability of active control devices without requiring the associated large power sources. In fact, many devices can operate on battery power, which is critical during seismic events when the main power source to the structure may fail.

According to presently accepted definitions, a semiactive control device is one that cannot inject mechanical energy into the controlled structural system (i.e., including the structure and the control device) but has properties that can be controlled to optimally reduce the responses of the system. Therefore, in contrast to active control devices, semiactive control devices do not have the potential to destabilize (in the bounded input/bounded output sense) the structural system. Preliminary studies indicate that appropriately implemented semiactive systems perform significantly better than passive devices and have the potential to achieve the majority of the performance of fully active systems, thus allowing for the possibility of effective response reduction during a wide array of dynamic loading conditions ([12]; Dyke et al., 1996a, b). Examples of such devices will be discussed in this section, including variable-orifice fluid dampers, variable-stiffness devices, variable-friction devices, controllable tuned liquid dampers, controllable fluid dampers, and controllable impact dampers.

17.3.1 Variable-Orifice Dampers

One means of achieving a variable-damping device is to use a controllable, electromechanical, variable-orifice valve to alter the resistance to flow of a conventional hydraulic fluid damper. A schematic of such a device is given in Figure 17.10. The concept of applying this type of variable-damping device to control the motion of

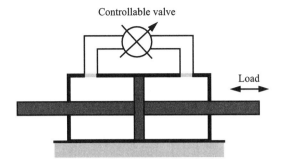

Figure 17.10 Schematic of a variable-orifice damper.

bridges experiencing seismic motion was first discussed by Feng and Shinozuka, 1990; Kawashima and Unjoh, 1993; and Kawashima et al., 1992. Subsequently, variable-orifice dampers have been studied by Symans et al., 1994, and Symans and Constantinou, 1996, at the National Center for Earthquake Engineering Research in Buffalo, New York.

Sack and Patten, 1994, conducted experiments in which a hydraulic actuator with a controllable orifice was implemented in a single-lane model bridge to dissipate the energy induced by vehicle traffic (see also Patten, et al., 1994). Figure 17.11 shows a full-scale experiment being conducted by Sack and Patten on a bridge on interstate highway I-35 in Oklahoma to demonstrate this technology. This experiment constitutes the first full-scale implementation of structural control in the United States.

The effectiveness of variable-orifice dampers in controlling seismically excited buildings has been demonstrated through both simulation and small-scale experimental studies (Hrovat et al., 1983; Mizuno et al., 1992; Sack et al., 1994; Patten et al., 1994; Kurata et al., 1994; Liang et al., 1995; Yang et al., 1996; Iwan et al., 1996; Inaudi et al., 1997). Kobori et al., 1993, and Kamagata and Kobori, 1994, implemented a full-scale variable-orifice damper in an active variable-stiffness system to investigate adaptive control methods for an active variable-stiffness system at the Kobori Research Complex. The results of these analytical and experimental studies indicate that this device is effective in reducing structural responses.

17.3.2 Variable-Friction Dampers

Various semiactive devices have been proposed which utilize forces generated by surface friction to dissipate vibratory energy in a structural system. Akbay and Aktan, 1990, 1991 and Kannan et al., 1995, proposed a variable-friction device that consists of a friction shaft which is rigidly connected to the structural bracing. The force at the frictional interface was adjusted by allowing slippage in controlled amounts. A similar device was considered at the University of British Columbia (Cherry, 1994; Dowdell and Cherry, 1994a, b). Through analytical studies, the ability of these semiactive devices to reduce the interstory drifts of a seismically excited structure was investigated. In addition, a semiactive friction-controllable fluid bearing has been employed in parallel with a seismic isolation system by Feng et al., 1993 and Yang et al., 1995.

17.3.3 Controllable Tuned Liquid Dampers

Another type of semiactive control device utilizes the motion of a sloshing fluid or a column of fluid to reduce the responses of a structure. These liquid dampers are based

Figure 17.11 Full-scale experiment on Interstate 35 in Oklahoma.

on passive tuned sloshing dampers (TSD) and tuned liquid column dampers (TLCD). As in a tuned mass damper (TMD), the TSD uses the liquid in a sloshing tank to add damping to the structural system. Similarly, in a TLCD, the moving mass is a column of liquid that is driven by the vibrations of the structure. Because these passive systems have a fixed design, they are not very effective for a wide variety of loading conditions, and researchers are looking toward semiactive alternatives for these devices to improve their effectiveness in reducing structural responses (Kareem, 1994). Lou et al., 1994, proposed a semiactive device based on the passive TSD, in which the length of the sloshing tank could be altered to change the properties of the device. Haroun et al., 1994 and Abe et al., 1996, presented a semiactive device based on a TLCD with a variable orifice.

17.3.4 Controllable Fluid Dampers

All of the semiactive control devices discussed so far in this section have employed some electrically controlled valves or mechanisms. Such mechanical components can be problematic in terms of reliability and maintenance. Another class of semiactive devices uses controllable fluids. The advantage of controllable fluid dampers is simplicity; they contain no moving parts other than the piston.

Two fluids that are viable contenders for development of controllable dampers are: (1) electrorheological (ER) fluids and (2) magnetorheological (MR) fluids. The essential characteristic of these fluids is their ability to reversibly change from a free-flowing, linear viscous fluid to a semisolid with a controllable yield strength in milliseconds when exposed to an electric (for ER fluids) or magnetic (for MR fluids) field. Although the discovery of both ER and MR fluids dates back to the late 1940s (Winslow, 1947, 1949; Rabinow, 1948), research programs have to date concentrated primarily on ER fluids. A number of ER fluid dampers (see Figure 17.12) have recently been developed, modeled, and tested for civil engineering applications (Ehrgott and Masri, 1994; Gavin et al., 1996a, b; Gordaninejad et al., 1995; Burton et al., 1996; McClamroch et al., 1995).

Recently developed MR fluids appear to be an attractive alternative to ER fluids for use in controllable fluid dampers [13–15] (see also www.rheonetic.com/mrfluid/ and www.nd.edu/~quake/). MR fluids have an inherent ability to provide a simple and robust interface between electronic controls and mechanical components. Much of the current interest in MR fluids can be traced directly to the need for reliable, fast-acting valves necessary to enable semiactive vibration control systems (Crosby et al., 1973; Karnopp et al., 1974; Ivers et al., 1994). MR fluid technology provides the means to enable such a valve.

A typical magnetorheological fluid consists of 20 to 40% by volume of relatively pure, soft iron particles, for example, carbonyl iron, suspended in an appropriate carrier liquid such as mineral oil, synthetic oil, water, or a glycol. MR fluids made from iron particles exhibit a yield strength of 50 to 100 kPa for an applied magnetic field of 150 to 250 kA/m ($\sim$ 2–3 kOe). MR fluids are not highly sensitive to contaminants or impurities such as are commonly encountered during manufacture and usage. Furthermore, because the magnetic polarization mechanism is not affected by the surface chemistry of surfactants and additives, it is relatively straightforward to stabilize MR fluids against particle-liquid separation in spite of the large density mismatch. Antiwear and lubricity additives can also be included in the formulation without affecting strength and power requirements (Weiss et al., 1993; Lord Corporation, 1995).

As a controllable fluid, the primary advantage of an MR fluid stems from the large, controlled yield stress it is able to achieve. Typically, the maximum yield stress of an MR fluid is an order of magnitude greater than that of the best ER fluid, while their viscosity is comparable. This has a profound impact on ultimate device size and dynamic range because the minimum amount of active fluid in a controllable fluid

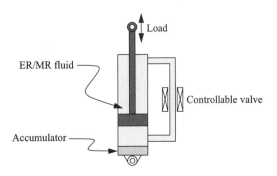

Figure 17.12 Schematic of controllable fluid damper.

device is proportional to the plastic viscosity and inversely proportional to the square of the maximum field-induced yield stress [13, 15]. This means that for comparable mechanical performance the amount of active fluid needed in an MR fluid device will be about two orders of magnitude smaller than that of an ER device.

From a practical application perspective, an advantage of MR fluids is the ancillary power supply needed to control the fluid. Although the total energy and power requirements for comparably performing MR and ER devices are approximately equal [13, 15], only MR devices can be powered directly from common, low-voltage sources. Furthermore, standard electrical connectors, wires, and feedthroughs can be reliably used, even in mechanically aggressive and dirty environments, without fear of dielectric breakdown. This aspect is particularly important in cost-sensitive applications.

Another advantage of MR fluids is their relative insensitivity to temperature extremes and contaminants. Carlson and Weiss [14] indicated that the achievable yield stress of an MR fluid is an order of magnitude greater than its ER counterpart and that MR fluids can operate at temperatures from −40 to 150°C, with only slight variations in the yield stress. This arises from the fact that the magnetic polarization of the particles, and therefore the yield stress of the MR fluid, is not strongly influenced by temperature variations. Similarly, contaminants (e.g., moisture) have little effect on the fluid's magnetic properties. A summary of the properties of both MR and ER fluids is given in Table 17.3.

The future of MR devices for civil engineering applications appears to be quite bright. More details regarding the application of MR technology to control of civil engineering structures are given in section 17.4.

17.3.5 Semiactive Impact Dampers

Passive impact dampers have been around for many years and have been used very successfully to reduce vibration and noise in turbines and gear cases. Studies of multiparticle dampers under random excitation (Papalou and Masri, 1994) have shown that significant vibration reduction can be achieved in lightly damped systems with a relatively small multiparticle impact damper. Single particle dampers of the same total mass give greater vibration reduction in certain frequency bands but may have little or no effect in other frequency bands. To remedy this defect, semiactive control has been

TABLE 17.3 Summary of the Properties of Today's MR and ER Fluids

Property	MR Fluids	ER Fluids
Maximum yield stress, $\tau_{y(field)}$	50–100 kPa	2–5 kPa
Maximum field	~ 250 kA/m	~ 4 kV/mm
Plastic viscosity, η_p	0.1–1.0 Pa-s	0.1–1.0 Pa-s
Operable temperature range	−40 to 150°C	+10 to 90°C
Stability	Unaffected by most impurities	Cannot tolerate impurities
Response time	milliseconds	milliseconds
Density	3 to 4 g/cm^3	1 to 2 g/cm^3
$\eta_p/\tau_{y(field)}^2$	10^{-10}–10^{-11} s/Pa	10^{-7}–10^{-8} s/Pa
Maximum energy density	0.1 Joules/cm^3	0.001 Joules/cm^3
Power supply (typical)	2–25 V 1–2 A	2000–5000 V 1–10 mA

applied to impact dampers, such that only favorable impacts are permitted (Caughey et al., 1989; Masri et al., 1989, 1994).

17.4 SEMIACTIVE CONTROL OF CIVIL ENGINEERING STRUCTURES

Magnetorheological dampers are one of the most promising realizations of semiactive dampers for application to full-scale civil structures. Spencer et al., 1996, 1997; [16], Dyke et al., 1996a, b [12], and Carlson and Spencer, 1996, have recently conducted pilot studies to demonstrate the efficacy of MR dampers for semiactive seismic response control. Simulations and laboratory model experiments show that an MR damper, used in conjunction with recently proposed acceleration feedback strategies, significantly outperforms comparable passive damping configurations, while requiring only a fraction of the input power needed by the active controller. Moreover, the technology has been demonstrated to be scalable to devices sufficiently large for implementation in civil engineering structures. This section summarizes these efforts.

17.4.1 Scale-Model Studies

Figure 17.13 is a diagram of the three-story model building that was employed in the pilot MR damper studies conducted at the Structural Dynamics and Control/Earthquake Engineering Laboratory at the University of Notre Dame (see www.nd.edu/~quake/). The test structure used in this experiment is designed to be a scale model of the prototype building discussed in Chung et al., 1989, and is subject to one-dimensional ground motion. A single magnetorheological (MR) damper is installed between the ground and the first floor, as shown in Figure 17.13. The MR damper employed here, the Lord SD-1000 linear MR fluid damper, is a small, monotube damper designed for use in a semiactive suspension system in large on- and off-highway vehicle seats. The SD-1000 damper is capable of providing a wide dynamic range of force control for very modest input power levels. The damper is 3.8 cm in diameter, 21.5 cm long in the fully extended position, and has a ±2.5-cm stroke. An input power of 4 watts is required to operate the damper at its nominal maximum design current of 1 amp.

Because of the intrinsically nonlinear nature of all semiactive control devices, development of control strategies that are practically implementable and can fully utilize the capabilities of these unique devices is a challenging task. Various nonlinear control strategies have been developed to take advantage of the particular characteristics of the semiactive devices, including bang-bang control (McClamroch, et al., 1995), clipped optimal control (Dyke et al., 1996a, b [12]; Patten et al., 1994a, b), bi-state control (Patten et al., 1994a, b), fuzzy control methods (Sun et al., 1994), modulated homogeneous friction (Inaude, 1997), and adaptive nonlinear control (Kamagata and Kobori, 1994). Caughey, 1993, proposed a variable-stiffness algorithm that employed a semiactive implementation of the Reid spring (Reid, 1956) as a structural element that could provide large amounts of damping for a very small expenditure of control energy.

To evaluate the effectiveness of the semiactive control system employing the MR damper, acceleration feedback control strategies (Dyke et al., 1996a, b, [12]) based on H_2 performance measures were implemented on the laboratory structure.

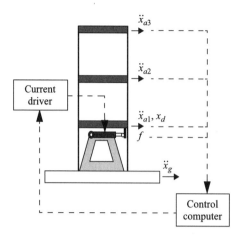

Figure 17.13 Diagram of MR damper implementation.

The three-story model structure was subjected to a scaled version of the N-S component of the 1940 El Centro earthquake, and the measured responses were recorded. Figure 17.14 shows the uncontrolled (i.e., without the MR damper attached) and semiactively controlled responses for the tested structure. The effectiveness of the proposed control strategy is clearly seen, with peak third-floor displacement being reduced by 74.5% and the peak third-floor acceleration being reduced by 47.6%.

The semiactive control systems performed significantly better than two passive configurations that were simultaneously considered. A 24.3% reduction in the peak third-floor displacement and a 29.1% reduction in the maximum interstory displacement were achieved as compared to the best passive case. Moreover, these results were obtained while also achieving a modest reduction in the maximum acceleration over the comparable passive case. These results demonstrate the significant potential for the use of MR technology in dynamic hazard mitigation.

17.4.2 Full-Scale Seismic MR Damper

To prove the scalability of MR fluid technology to devices of appropriate size for civil engineering applications, a full-scale, MR fluid damper has been designed and built (Spencer et al., 1997; Carlson and Spencer, 1996). For the nominal design, a maximum damping force of 200,000 N (20-ton) and a dynamic range equal to ten were chosen. A schematic of the large-scale MR fluid damper is shown in Figure 17.15. The damper uses a particularly simple geometry in which the outer cylindrical housing is part of the magnetic circuit. The effective fluid orifice is the entire annular space between the piston outside diameter and the inside of the damper cylinder housing. Movement of the piston causes fluid to flow through this entire annular region. The damper is double-ended; that is, the piston is supported by a shaft on both ends. This arrangement has the advantage that a rod-volume compensator does not need to be incorporated into the damper, although a small pressurized accumulator is provided to accommodate thermal expansion of the fluid. The damper has an inside diameter of 20.3 cm and a stroke of ±8 cm. The electromagnetic coil is wound in three sections on the piston. This results in four effective valve regions as the fluid

Section 17.4 Semiactive Control of Civil Engineering Structures 437

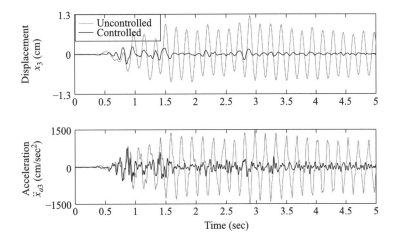

Figure 17.14 Controlled and uncontrolled structural responses due to El Centro earthquake.

flows past the piston. The coils contain a total of about 1.5-km magnetic wire. The completed damper is approximately 1 m long and has a mass of 250 kg. The damper contains approximately 5 liters of MR fluid. The amount of fluid energized by the magnetic field at any given instant is approximately 90 cm^3. A summary of the parameters for the 20-ton damper is given in Table 17.4.

Figure 17.16 shows the experimental setup at the University of Notre Dame for the 20-ton MR fluid damper. The damper was attached to a 7.5-cm thick plate that was grouted to a 2-m thick strong floor. The damper is driven by a 560-kN actuator configured with a 305-lpm servo-valve with a bandwidth of 80 Hz. A Schenck-Pegasus 5910

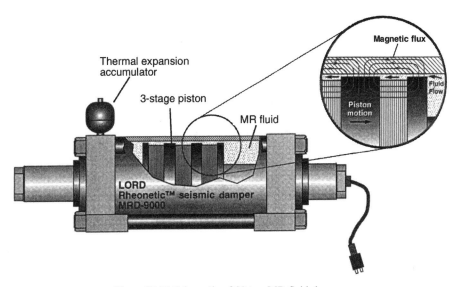

Figure 17.15 Schematic of 20-ton MR fluid damper.

TABLE 17.4 Design Parameters for 20-ton Seismic Damper

Stroke	±8 cm
F_{max}/F_{min}	10.1 @ 10 cm/s
Cylinder bore (ID)	20.32 cm
Maximum input power	< 50 watts
Maximum force (nominal)	200,000 N
Effective axial pole length	8.4 cm
Coils	3 × 1050 turns
Fluid $\eta_\rho/\tau^2_{y(field)}$	2×10^{-10} s/Pa
Fluid η_ρ	1 Pa-s
Fluid $\tau_{y(field)}$ max	70 kPa
Gap	2 mm
Active fluid volume	~ 90 cm³
Wire	16 gauge
Inductance (L)	6.6 henries
Coil resistance (R)	3 × 7.3 ohms

Figure 17.16 Experimental setup for 20-ton MR fluid damper.

servo-hydraulic controller is employed in conjunction with a 200 MPa, 340-lpm hydraulic pump.

Figure 17.17 shows the measured performance for the damper at 5 cm/sec (triangular displacement). The maximum force measured at full magnetic field strength is 201 kN at a piston velocity of 5 cm/sec, which is within 0.5% of the analytically predicted result (Spencer et al., 1997). Moreover, the dynamic range of the damper is well over the design specification of 10.

Because of their mechanical simplicity, low-power requirements, and high-force capacity, magnetorheological (MR) dampers constitute a class of semiactive control devices that meshes well with the demands and constraints of civil infrastructure applications and will likely see increasing interest from the engineering community as a viable means of mitigating the devastating effects of severe dynamic loads on civil structures.

17.5 CONCLUSIONS

Protecting civil structures from natural and other types of unwanted dynamic influences is continuing to move steadily up the list of high-priority needs of the world community. The structures alone represent a huge investment of resources. Moreover, they are platforms that carry within them very expensive equipment, irreplaceable records, and priceless human cargo.

As our readers have seen over and over again, the traditional methods of dealing with these exigencies are being reconsidered and are beginning to give way to the influence of more recent technologies. Of course, along with these technologies comes the possibility of more advanced design goals, more modern algorithms, and more state-of-the-art implementations.

Full-scale buildings are being controlled successfully, and attention is turning toward the features of a whole new family of actuators, especially those of semiactive

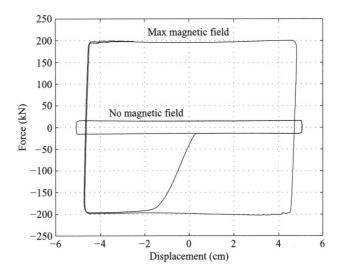

Figure 17.17 Measured performance for 20-ton MR fluid damper at 5 cm/sec.

type. Controllable fluid dampers provide a fascinating class of instances, with the magnetorheological fluids offering attractive properties.

It turns out that models for such devices lead one into issues of hybrid control and hysteresis, both of which are topics of considerable current interest in the controls community.

In summary, the modern thrust toward control of civil structures is providing a new opportunity for control engineers to make their work more understandable to the public, while at the same time making a genuine technical, economic, and social contribution.

And there are hundreds of interesting ideas to ponder!

ACKNOWLEDGMENTS

The research efforts of the authors are supported in part by National Science Foundation Grant Nos. CMS 95–00301 and CMS 95–28083. The work of the second author is supported in part by the Frank M. Freimann Chair in Electrical Engineering. The authors are grateful for the contributions of Professors T. T. Soong and A. M. Reinhorn of the State University of New York at Buffalo, Professor Y. Fujino of the University of Tokyo, Japan, Professor K. Yoshida of Keio University, Japan, Professor A. Nishitani of Waseda University, Japan, and Professor K. Seto of Nihon University, Japan. The authors would also like to thank the reviewers for their careful reading of the manuscript and their helpful comments.

Related Chapters

- The automation and control of building environments is discussed in Chapter 16.

REFERENCES

[1] T. T. Soong, *Active Structural Control: Theory and Practice,* London: Longman Scientific and Technical, 1990.

[2] Y. Fujino, T. T. Soong, and B. F. Spencer Jr., "Structural control: Basic concepts and applications." *Proc. ASCE Structures Congress XIV,* Chicago, Illinois, pp. 1277–1287, April 1996.

[3] G. W. Housner et al., "Structural control: Past, present and future." *J. Engrg. Mech., ASCE,* September 1997.

[4] B. F. Spencer Jr., J. Suhardjo, and M. K. Sain, "Frequency domain optimal control strategies for aseismic protection." J. Engrg. Mech., *ASCE,* Vol. 120, no. 1, pp. 135–159, 1994.

[5] B. F. Spencer Jr., M. K. Sain, C.-H. Won, D. C. Kaspari Jr., and P. M. Sain, "Reliability-based measures of structural control robustness." *Struct. Safety,* Vol. 15, pp. 111–129, 1994.

[6] S. J. Dyke, B. F. Spencer Jr., P. Quast, and M. K. Sain, "The role of control-structure interaction in protective system design." *J. Engrg. Mech., ASCE,* Vol. 121, no. 2, pp. 322–338, 1995.

[7] D. P. Tomasula, B. F. Spencer Jr., and M. K. Sain, "Nonlinear structural control for limiting extreme dynamic responses." *J. Engrg. Mech., ASCE*, Vol. 122, no. 3, pp. 218–229, 1996.

References

[8] S. J. Dyke, B. F. Spencer Jr., P. Quast, M. K. Sain, D. C. Kaspari Jr., and T.T. Soong, "Acceleration feedback control of MDOF structures." *J. Engrg. Mech., ASCE*, Vol. 122, no. 9, pp. 907–918, 1996.

[9] T. Kobori, "Future direction on research and development of seismic-response-controlled structure." *Proc. 1st World Conf. on Struct. Control*, Los Angeles, California, Panel:19–31, August 1994.

[10] T. T. Soong and A. M. Reinhorn, "An overview of active and hybrid structural control research in the U.S." *The Struct. Dyn. Design of Tall Buildings*, Vol. 2, pp. 192–209, 1993.

[11] Y. Fujino, "Recent research and developments on control of bridges under wind and traffic excitations in Japan." *Proc. Int. Workshop on Struct. Control*, pp. 144–150, 1994.

[12] S. J. Dyke, B. F. Spencer Jr., M. K. Sain, and J. D. Carlson, "Seismic response reduction using magnetorheological dampers." *Proc. IFAC World Congress*, San Francisco, CA, June 30–July 5, 1996.

[13] J. D. Carlson "The promise of controllable fluids." *Proc. of Actuator 94* (H. Borgmann and K. Lenz, eds.), AXON Technologie Consult GmbH, pp. 266–270, 1994.

[14] J. D. Carlson and K. D. Weiss, "A growing attraction to magnetic fluids." *Machine Design*, pp. 61–65, August 1994.

[15] J. D. Carlson, D. M. Catanzarite, and K. A. St. Clair, "Commercial magneto-rheological fluid devices." *Proc. 5th Int. Conf. on ER Fluids, MR Fluids and Associated Technology*, University of Sheffield, UK, 1995.

[16] B. F. Spencer Jr., S. J. Dyke, M. K. Sain, and J. D. Carlson, "Phenomenological model of a magnetorheological damper." *J. Engrg. Mech., ASCE*, Vol. 123, no. 3, pp. 230–238, 1997.

Chapter 18 ROBOT CONTROL

Bruno Siciliano

Editor's Summary

Where robotics is concerned, controls has been ahead of its time. Robotic controls has been a popular research area for some time, yet the practical impact of this research has been limited. The robotics industry remains small, and most production robots incorporate only basic control schemes such as PIDs. But a number of encouraging signs herald progress. For the first time, robot shipments in North America exceeded $1 billion recently. European manufacturers now sell tens of thousands of robots per year, and new control technologies are starting to be exploited by industry.

This chapter discusses a number of techniques for robot control. In kinematic control, the inverse kinematics of the robot are approximated so that the joint variables (e.g., joint angles) required for desired robot effector position can be obtained. These joint angles can then be used as setpoint inputs to a feedback loop that issues torque commands to the robot. Substantial performance improvements can be gained by dynamic control in which a dynamic model for the robot is used. Desired trajectories for velocities and accelerations of joint variables can then be specified and tracked. The next major step is force control. This requires several innovations: on-line computation of inverse kinematics within the feedback loop, coordinate transforms for control in the operational space rather than the robot joint space, and a model for the environment. The benefit is that the force employed by the robot end effectors can now be regulated, allowing, for example, precise tasks in elastic or compliant environments to be undertaken.

If commercial robots are to perform the same sorts of activities that humans perform on a routine basis, robot control laws must be integrated with vision feedback. The chapter discusses new developments in vision-based control, specifically visual servoing in which a visual feedback control loop continuously steers the robot-end effector toward the target. Hardware and software limitations that have kept such techniques from being practical have now been overcome.

Bruno Siciliano is a professor in the Department of Computer and Systems Engineering at the University of Naples. From 1996 to 1999 he chaired the IEEE CSS Technical Committee on Manufacturing Automation and Robotic Control.

18.1 A HISTORICAL PERSPECTIVE

Robotics is concerned with the study of those machines that can replace human beings in the execution of a task, as regards both physical activity and decision making. As such, robotics has attracted an ever increasing number of control researchers in the last 20 years, producing a visible cross-fertilization between the two fields. This is rather evident from the number of publications and annual conferences devoting much space to control problems in robotics.

As automation becomes more prevalent in industry and as typical bulky robots are replaced with new systems that are smaller, faster, lighter, and smarter, traditional PID control will no longer be a satisfactory means of control in many situations. Optimum performance of industrial automation systems, especially if they include robots, will demand the use of such technologies as robust control, adaptive control, and intelligent control.

Despite many years of robotics research involving a large number of scientists and engineers, there has been some disappointment about the fairly slow progress in robotics compared to human performance. Nowadays, industrial robots in nearly all applications are purely position-controlled devices, still far removed from the human arm's performance with its amazingly low own weight against load ratio, force/torque-controlled muscular actuation, and on-line sensory feedback through vision and touch.

For many years, robot manufacturers have not integrated available research results into their robot controllers. Nevertheless, a number of factors indicate that the field is becoming mature for a transfer of technology from the robotics control research community to the robotics industry.

First of all, industrial robots are much cheaper now—approximately a factor of four—than 10 years ago, while the peripheral costs, for example, for precise part feeding devices, have remained substantially the same. Interestingly enough, there are no big North American manufacturers left—with the noticeable exception of Adept—while the big European manufacturers ABB and Kuka have been gaining excellent positions selling as many as 10,000 robots per year. Just recently, European manufacturers have started to focus on control improvements [1]. The specific areas of improvement include:

- Intuitive programming with six–degree-of-freedom manual devices.
- Model-based dynamic control.
- Open control architectures for on-line sensory feedback.

In what follows, some key issues in robot control technology are discussed. These offer the potential for the design of a new generation of enhanced industrial robot controllers. For the purpose of discussion, the focus will be on conventional robot manipulators consisting of a sequence of rigid links connected by joints with an end effector performing the task required of the robot.

18.2 KINEMATIC CONTROL

The problem of controlling a robot is to determine the time profile of the generalized forces (forces or torques) to be developed by the joint actuators so as to guarantee execution of the commanded task while satisfying given transient and steady-state requirements. The task may involve either the execution of specified motions for a robot operating in free space, or the execution of specified motions and contact forces for a robot whose end effector is constrained by the environment.

Several techniques can be employed for controlling a robot. The technique followed, as well as the way it is implemented, may have a significant influence on the robot performance and on the possible range of applications. For instance, the need for trajectory tracking control in the Cartesian space may lead to hardware/software implementations that differ from those allowing point-to-point control where only reaching of the final position is of concern. On the other hand, the robot mechanical design influences the control scheme utilized. For instance, the control problem of a Cartesian robot is substantially different from that of an anthropomorphic robot, that is, a robot whose mechanical structure resembles that of the human arm.

Regardless of the specific type of mechanical robot, task specification (end-effector motion and forces) is usually carried out in the so-called operational space, whereas control actions (joint actuator generalized forces) are performed in the joint space. Therefore, an inverse kinematics problem arises which consists of computing the joint motion corresponding to the given end-effector motion.

The kinematic model of a robot can be written in the form

$$x = k(q) \qquad (18.1)$$

where q denotes the vector of joint variables, x denotes the vector of task variables—typically three position coordinates and three Euler angles, hereafter called pose—and $k(\cdot)$ is the direct kinematics function that can be derived for any robot structure according to well-established procedures, for example, based on the Denavit–Hartenberg parameters [2].

The inverse kinematics problem—computing q given x in (18.1)—cannot be solved in closed form for robots that do not possess a simple geometry, for example, a six-joint robot with a spherical wrist. In such cases, it is worth considering the differential kinematics in the form

$$\dot{x} = J(q)\dot{q} \qquad (18.2)$$

where $J = \partial dk/\partial dq$ is the robot Jacobian. By exploiting the linearity of (18.2) in $\dot{q}$, the inverse differential kinematics problem can be solved for any robot geometry; joint velocities can be obtained at the current joint configuration (resolved-rate motion [3]), and then joint positions can be computed by integrating the velocity solution over time with known initial conditions.

Assuming that J is square and full-rank, the implementation of the *inverse kinematics algorithm* leads to the scheme in Figure 18.1, where a *closed loop* has been introduced to avoid typical solution drift owing to numerical integration. A feedback correction term is used, based on the task space error between the desired and actual end-effector poses $(x_d - x)$, with K being a positive definite matrix gain [4].

This approach allows a natural treatment of singularities (of the matrix J) and redundancy, that is, when more joint variables than task variables are available. The most general solution can be cast in the form [5]

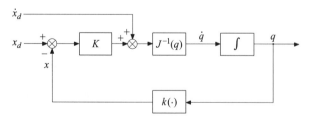

Figure 18.1 Closed-loop inverse kinematics algorithm.

Section 18.2 Kinematic Control

$$\dot{q} = J^{\dagger}(q)\dot{x} + (I - J^{\dagger}(q)J(q))\dot{q}_0 \tag{18.3}$$

where $J^{\dagger}$ denotes a right pseudo-inverse of the Jacobian—for instance, $J^{\dagger} = J^{\mathrm{T}}(JJ^{\mathrm{T}})^{-1}$, but a damped least-squares inverse is to be used in the neighborhood of singularities [6]. The homogeneous term $(I - J^{\dagger}J)\dot{q}_0$ is available to meet additional task requirements—through the choice of $\dot{q}_0$—in the case of redundancy [7]. Typically, a local optimization of a performance index can be carried out by choosing $\dot{q}_0$ as the gradient of such index.

Despite its simplicity and applicability to any geometry, it should be stressed that the inverse kinematics problem for most industrial robots is not solved at the differential kinematics level. For this reason, considerable errors typically occur for robots with complex geometry, for example, with a nonspherical wrist as in several current designs. Similarly, singularities and redundancy are not handled in the robot controller. Only recently, a successful implementation of a solution of the type shown in Figure 18.1 can be found in the so-called Space Mouse developed at DLR, which Stäubli and Kuka have now integrated in the programming panels of their robots.

No matter what technique is used, a joint space control scheme can be designed that allows tracking of the reference inputs as obtained by the inverse kinematics solution. However, this two-stage solution has the drawback that a joint space control scheme does not influence the task space variables that are controlled in an open-loop fashion through the robot mechanical structure. It is then clear that any uncertainty of the structure (construction tolerance, lack of calibration, gear backlash, elasticity) or any imprecision in the knowledge of the end-effector pose relative to an object to manipulate causes a loss of accuracy in the task space variables. Therefore, a joint space controller performs well for robot motion in free space, whereas an operational space controller should be adopted for constrained robot motion, as is discussed later in this chapter.

The driving system of the joints also has an effect on the type of control strategy used. Most industrial robots are actuated by electric motors with reduction gears of high ratios. The presence of gears tends to linearize system dynamics and thus to decouple the joints in view of the reduction of nonlinearity effects. The price to pay, however, is the occurrence of joint friction, elasticity, and backlash. These factors can limit system performance to a greater extent than inertial and Coriolis forces, that is, those forces generated by configuration-dependent inertias. Therefore, a typical industrial robot controller adopts a so-called kinematic control strategy according to which each joint of the robot is regarded as an independent (linear) system and coupling effects between joints are treated as disturbances [8], that is, only the kinematics is taken into account in the first stage, and no attempt is made to account for the dynamic interaction due to varying configurations during motion. A block scheme of kinematic control is illustrated in Figure 18.2 where standard PID controllers are used to generate the input driving torques τ at the joints. The PID gains are usually pretuned by the manufacturer for optimum dynamic and static performance. In this respect, ABB was among the first to perform gain-scheduling in their controllers to adjust the gains for various inertial configurations of the robots. With this strategy, the joint servos are voltage- (or velocity-) controlled and are typically not accessible to the user.

On the other hand, a robot actuated with direct drives eliminates the drawbacks due to friction, elasticity, and backlash, but the effects of nonlinearities and couplings

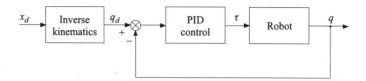

Figure 18.2 Kinematic control.

between the joints become relevant. Very few industrial robots with anthropomorphic structure adopt this kind of actuation, while direct drives can be found in SCARA (Selective Compliance Articulated Robot Arm) robots. A breakthrough in tracking performance, even for gear-driven robots operating at high speeds, can be gained by resorting to a so-called dynamic control strategy where the robot dynamic model is explicitly used [13].

18.3 DYNAMIC CONTROL

The dynamic model of a robot can be written in the form [2]

$$B(q)\ddot{q} + C(q, \dot{q})\dot{q} + g(q) = \tau \tag{18.4}$$

where B is the symmetric and positive definite inertia matrix, $C\dot{q}$ is the vector of Coriolis and centrifugal torques (with C a suitable factorizing matrix), and g is the vector of gravitational torques.

A notable physical property of the dynamic model is the linearity in the parameters, that is, the equations of motion (18.4) can be rewritten in the form [9]

$$Y(q, \dot{q}, \ddot{q})\pi = \tau \tag{18.5}$$

where π is a vector of dynamic parameters—in general, the mass, the three components of the first moment of inertia, and the six components of the inertia tensor for each link—and Y is a factorizing matrix (called regressor), which is a function of joint positions, velocities, and accelerations. Notice that not all the dynamic parameters for each link explicitly appear in the dynamic model, and a suitable minimization can be sought by recognizing linear combinations of more parameters [10].

Since the dynamic parameters of an industrial robot are not known, property (18.5) can be exploited to pursue a linear *dynamic parameter identification,* which is a natural prelude to designing a dynamic control scheme. On the assumption that the kinematic parameters in the matrix Y are known with good accuracy, for example, as a result of a kinematic calibration [11], measurements of joint positions q, velocities $\dot{q}$, and accelerations $\ddot{q}$ are required. Joint positions and velocities can be measured while numerical reconstruction of accelerations is needed. This operation can be performed on the basis of the position and velocity values recorded during the execution of suitable motion trajectories imposed on the robot. These should preferably be of polynomial type but should not excite unmodeled dynamic effects such as joint elasticity or link flexibility. As regards joint torques, since no torque sensors are available at the joints, they can be evaluated from current measurements in the typical case of electric

Section 18.3 Dynamic Control

actuators. Eventually, they can be computed from wrist force measurements, if such sensors are available.

If measurements of joint torques, positions, velocities, and accelerations have been obtained at given time instants $t_1, \ldots, t_N$ along a given trajectory, then

$$\bar{\tau} = \begin{bmatrix} \tau(t_1) \\ \vdots \\ \tau(t_N) \end{bmatrix} = \begin{bmatrix} Y(t_1) \\ \vdots \\ Y(t_N) \end{bmatrix} \pi = \bar{Y}\pi. \tag{18.6}$$

The number of time instants sets the number of measurements to perform and should be large enough to avoid ill-conditioning of matrix $\bar{Y}$. Solving (18.6) by a least-squares technique leads to the solution in the form

$$\pi = (\bar{Y}^T \bar{Y})^{-1} \bar{Y}^T \bar{\tau} \tag{18.7}$$

where $(\bar{Y}^T \bar{Y})^{-1} \bar{Y}^T$ is a left pseudo-inverse matrix of $\bar{Y}$.

The technique presented here can be applied even for a reduced number of parameters—say that some of them are accurately known—and eventually to identify the dynamic parameters of an unknown, rigidly grasped payload at the robot's end effector. In that case, it is sufficient to regard the payload as a structural modification of the last link.

An alternative way to write the dynamic model of a robot is in terms of energy conservation (Hamiltonian form), that is,

$$\frac{1}{2}\frac{d}{dt}(\dot{q}^T B(q)\dot{q}) = \dot{q}^T(\tau - g) \tag{18.8}$$

where the lefthand side is the time derivative of the kinetic energy and the righthand side represents the power generated by the torques acting on the joints (driving torques minus gravitational torques). Expression (18.8) does not mean that Coriolis and centrifugal terms in (18.4) have disappeared, but merely that they are now accounted for via the time derivative of the inertia matrix. In fact, it can be shown that a proper factorization of the matrix C can be found so that the matrix $\dot{B} - 2C$ is skew-symmetric, which constitutes another notable property of the dynamic model of a robot.

This alternative form leads to deriving a simple control scheme to regulate the joint variables q to a desired constant set point q_d. The key point is to choose a controller so as to mimic the effects of equipping each joint with a passive mechanical device of the spring and damper type. In that case, the associated mechanical energy can be written as

$$V = \frac{1}{2}(\dot{q}^T B(q)\dot{q} + \tilde{q}^T K_P \tilde{q}) \tag{18.9}$$

where $\tilde{q} = q_d - q$ and K_P is a positive definite (usually diagonal) matrix. Using such virtual mechanical energy as a candidate Lyapunov function, it is worth taking the time derivative of (18.9), which in view of (18.8) can be written as

$$\dot{V} = \dot{q}^{\mathrm{T}}(\tau - g(q) - K_P\tilde{q}). \tag{18.10}$$

At this point, choosing a *PD control with gravity compensation* as [12]

$$\tau = K_P\tilde{q} - K_D\dot{q} + g(q) \tag{18.11}$$

with K_D a positive definite (diagonal) matrix, gives

$$\dot{V} = -\dot{q}^{\mathrm{T}}K_D\dot{q} \le 0 \tag{18.12}$$

which is negative semi-definite. Observing that $\dot{V}$ is identically zero only if $\tilde{q} = 0$ (via LaSalle's theorem) implies that $\tilde{q} \to 0$ asymptotically.

Apart from the sole model-based requirement of gravity compensation, the controller (18.11) matches the physical intuition that a joint PD controller can stabilize the mechanical system in spite of all the nonlinearities and coupling effects! This outstanding result is a major one in the robot control literature. It serves to demonstrate why a linear independent joint control, as is to be found in most industrial robot controllers, guarantees good positioning accuracy. If the desired trajectory is time-varying, the simple adjustment of adding a velocity feedforward action $K_D\dot{q}_d$—and even an acceleration term $\ddot{q}_d$—to (18.11) guarantees reasonably good performance. Furthermore, since gravity is usually not known, the replacement of the gravity compensation with an integral action on the joint error $K_I\int^t \tilde{q}d\varsigma$—leading to a complete *independent joint PID control*—is the typical recipe adopted in industrial robot controllers for all practical purposes. Indeed, it is possible to prove that with a suitable choice of the PID gains, asymptotic stability can still be guaranteed [12]. A scheme for PID control with *velocity and acceleration feedforward* is illustrated in Figure 18.3. Clearly, if joint velocity transducers are not available (as in most industrial robots), a truly derivative action on the joint position error is present in lieu of the proportional action on the joint velocity error.

Whenever the required operational speeds become too high, or better tracking accuracy is desired, a simple PID controller no longer suffices, and a full dynamic model-based control is to be designed. Dynamic model compensation can be carried out in two ways: in a feedforward fashion and in a feedback fashion. The former leads

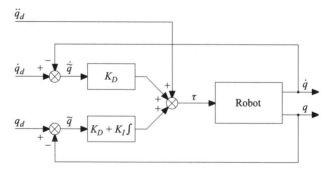

Figure 18.3 PID control with velocity and acceleration feedforward.

Section 18.3 Dynamic Control

to the so-called computed torque strategy [14], while the latter can be cast in the framework of the inverse dynamics control strategy for nonlinear mechanical systems.

The essence of *computed torque control* is illustrated by the block scheme in Figure 18.4 where a nonlinear model-based feedforward action is added to the linear feedback control action and the linear feedforward action. This further action is computed on the basis of the desired joint trajectory (position, velocity, and acceleration) and compensates the nonlinear coupling terms due to inertial, Coriolis, centrifugal, and gravitational torques that vary during robot motion. Even in the case of imperfect dynamic modeling, the computed torque technique has the advantage of alleviating the disturbance rejection task for the feedback control structure. In other words, the gains of the linear action need not be so large for good rejection since the actual disturbance is much smaller than without the nonlinear model-based action. Nonetheless, only a partial feedforward action may be performed so as to compensate those terms of the dynamic model that give the most relevant contributions during robot motion. Since inertial and gravitational terms dominate velocity-dependent terms (at operational joint speeds not greater than a few radians per second), a partial compensation can be achieved by computing only the gravitational torques and the inertial torques due to the diagonal elements of the inertia matrix. In this way, only the terms depending on the global robot configuration are compensated, while those deriving from motion interaction with the other joints are not.

For high tracking performance, the "best" controller can be found through a nonlinear model-based feedback action (*inverse dynamics control*) as [15]

$$\tau = B(q)y + n(q, \dot{q}) \tag{18.13}$$

with $n(q, \dot{q}) = C(q, \dot{q})\dot{q} + g(q)$. Thanks to the positive definiteness of the inertia matrix, the control law (18.13) provides a global feedback linearization of the system described by (18.4) into the equivalent system

$$\ddot{q} = y \tag{18.14}$$

which is thus linear and decoupled. The design of the new control input y can be carried out according to well-established techniques for linear systems, for example, pole placement, and takes on the general form

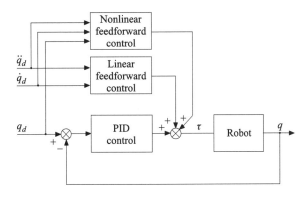

Figure 18.4 Computed torque control.

$$y = \ddot{q}_d + K_D \dot{\tilde{q}} + K_P \tilde{q} \qquad (18.15)$$

with K_P and K_D positive definite (diagonal) matrices. The resulting block scheme is illustrated in Figure 18.5, in which two feedback loops are represented: an inner loop based on the robot dynamic model and an outer loop operating on the tracking error. The function of the inner loop is to obtain a linear and decoupled input/output relationship, whereas the outer loop is required to stabilize the overall system. The controller design for the outer loop is simplified since it operates on a linear and time-invariant system. Notice that the implementation of this control scheme requires computation of the inertia matrix and the vector of Coriolis, centrifugal, and gravitational terms. These terms must be computed on-line since control is now based on nonlinear feedback of the current system state. Thus it is not possible to precompute the terms off-line as can be done for computed torque control.

This technique of nonlinear compensation and decoupling is very attractive from a control viewpoint since the nonlinear and coupled robot dynamics is replaced with a set of linear and decoupled second-order subsystems. Nonetheless, this technique is based on the assumption of perfect cancellation of dynamic terms, and then it is quite natural to raise questions about sensitivity and robustness problems due to unavoidably imperfect compensation.

Implementation of a dynamic control strategy is based on achieving torque control at the joint servos, which is virtually impossible to realize since industrial robot controllers are typically not accessible to the user. Even then, the parameters of the dynamic model should be accurately known and the equations of motion should be computed in real time. These conditions are difficult to verify in practice. On one hand, the model is usually known with a certain degree of uncertainty due to imperfect knowledge of robot mechanical parameters, existence of unmodeled dynamics, and model dependence on end-effector payloads not exactly known and thus not perfectly compensated. On the other hand, inverse dynamics computation is to be performed at sampling times of the order of a millisecond so as to ensure that the assumption of operating in the continuous time domain is realistic. This may pose severe constraints on the hardware/software architecture of the control system. In such cases, it may be advisable to lighten the computation of inverse dynamics and compute only the dominant terms.

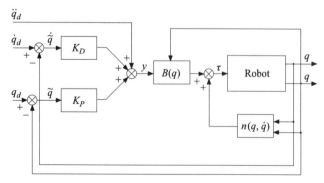

Figure 18.5 Inverse dynamics control.

In order to cope with imperfect dynamic compensation, a great body of robot control research has been devoted to the design of robust and adaptive controllers; see [17] and [18], respectively, for two valuable surveys of the two classes of controllers. These control schemes are not treated in detail here, because only very recently have industrial robot manufacturers become receptive to dynamic control. Before advanced control techniques can become widespread in industrial robots, open control architectures will likely need to be adopted. In this respect, it should be mentioned that an early attempt to realize an open controller was carried out by Tecnospazio a few years ago for the controllers of Comau industrial robots [16]. More recently, an open control architecture is being developed at DLR using the industrial standard real-time multi-tasking operating system VxWorks. Hopefully, such an auspicious trend will be followed up by several robot manufacturers.

For the purpose of the present chapter, however, it is worth emphasizing the conceptual difference between robust and adaptive control, although both attempt to accomplish basically the same goal, that is, control under uncertainty. An adaptive controller incorporates some sort of on-line parameter estimation, whereas a robust controller is usually a fixed controller designed to satisfy performance specifications over a given range of uncertainty. Therefore, in general, robust controllers provide a natural rejection to unmodeled dynamics and external disturbances, whereas adaptive controllers rely on an accurate analytical dynamic model, the only uncertainty being in the knowledge of the dynamic parameters. On the other hand, some adaptive controllers can learn from past experience and usually give a smoother time behavior of the control torques, whereas some estimate of the uncertainty needs to be found for a robust controller so as to impose control inputs that the mechanical structure can bear. Robust controllers can be of the high-gain [19], sliding-mode [20], Lyapunov-based [21], or dynamic compensation [22] type, while most adaptive controllers exploit the property of linearity in the dynamic parameters expressed by (15.5) and can be of direct [23], indirect [24], or composite [25] type.

18.4 FORCE CONTROL

One of the fundamental requirements for the success of a manipulation task is the capability to handle interactions between robot and environment. Typical examples of manipulation tasks are mechanical part mating, object contour surface tracking, and employment of tools for machining mechanical parts.

The success of an interaction task undertaken with motion control algorithms depends entirely on planning accuracy and control performance. To this end, it is crucial to have a detailed model of both robot (kinematics and dynamics) and environment (mechanical features and geometry). A model of the robot can be known with enough precision, but a detailed description of the environment is difficult to obtain. Planning errors may result in a trajectory assigned to the end effector which is no longer suitable for correct task execution. To understand the importance of this implication, it is sufficient to observe that to perform a mechanical part mating with a positional approach, the relative positioning of the parts should be guaranteed with an accuracy of an order of magnitude greater than part mechanical tolerance. Once the absolute position of one part is exactly known, the robot should guide the motion of the other with the same accuracy.

When the robot is governed by position control algorithms, any deviation of the actual trajectory from the reference one provokes a reaction of the control system. This tends to minimize such deviation independently of the generating cause. Hence, if the deviation from the planned trajectory is due to the interaction of the robot with the environment, reaction forces arise and the position control attempts to reduce the deviation as it would for any disturbance opposing the end-effector motion. In this case, however, the effect of the control action may be an increase in contact force which is not accompanied by a decrease in deviation. This situation may lead to an increase of contact force until the natural limit set by saturation of robot actuators is encountered, or mechanical crisis of one of the elements of the interaction takes place. The higher the environment stiffness and position control accuracy, the more easily an unstable contact case can occur. In fact, large constraint reaction forces result from deformation of a stiff environment under a strong position control action.

The quantity that describes the state of interaction more effectively is the contact force at the robot's end effector. High values of contact force are generally undesirable because they may stress both the robot and the manipulated object [26]. For appropriate handling of interactions, it is then necessary to consider *force control* strategies, either in an indirect way via a suitable use of position control laws or in a direct way via truly force control laws [27].

Since contact forces are naturally described in the operational space, it is convenient to refer to an operational space control strategy [28]. This is radically different from the above kinematic and dynamic control strategies which require the pre-inversion of the end-effector motion into equivalent joint motions. Since the user typically specifies the motion in terms of operational space variables, the measured joint space variables can be transformed into the corresponding operational space variables through direct kinematics relations. Comparing the desired variables with the reconstructed variables allows designing feedback control loops where trajectory inversion is replaced with a suitable coordinate transformation embedded in the feedback loop. This transformation typically requires the computation of the robot Jacobian. In particular, the static model of the robot, which is dual to the differential kinematics model in (18.2), is

$$\tau = J^{T}(q) f \qquad (18.16)$$

where f is the equivalent force at the end effector.

It is not difficult to show that the counterpart of the joint space PD control with gravity compensation (18.11) in the operational space is [29]

$$\tau = J^{T}(q) K_P \tilde{x} - K_D \dot{q} + g(q) \qquad (18.17)$$

where $\tilde{x} = x_d - x$ denotes the error between the desired and actual task space variables. Differently from (11), the virtual spring no longer acts at the joints but at the end effector. Hence the elastic force $K_P \tilde{x}$ needs to be transformed into equivalent joint torques through a relationship similar to (18.16). Asymptotic stability still holds as long as the Jacobian is full rank. Similarly, the counterpart of the joint space inverse dynamics control (18.13, 18.15) in the operational space differs only for the choice of the resolved acceleration [30] as

Section 18.4 Force Control

$$y = J^{-1}(q)(\ddot{x}_d + K_D \dot{\tilde{x}} + K_P \tilde{x} - \dot{J}(q, \dot{q})\dot{q}) \quad (18.18)$$

where the time derivative of the differential kinematics model (15.2) has been exploited. As above, singularities of J may give trouble, and the inverse needs to be replaced with a pseudo-inverse for a redundant robot.

Compared to joint space control schemes, operational control schemes suffer from considerable computational complexity in view of the necessity to handle the inverse kinematics inside the feedback loop, that is, on-line. Therefore, such schemes have not yet been implemented on industrial robot controllers. Nevertheless, they are fundamental to design force control schemes as shown next and as a premise for on-line sensory feedback control schemes such as the visual servoing strategy presented in the next section.

If the end effector is in contact with the environment, an additional term $-J^T(q)f$ arises on the righthand side of (18.1) where f is the vector of contact forces exerted by the robot's end effector on the environment. Hence, for the PD control scheme with gravity compensation (18.17), at the equilibrium the task space error is

$$\tilde{x} = K_P^{-1} f \quad (18.19)$$

under the assumption of a full-rank Jacobian.

For a better understanding of the interaction between robot and environment, it is necessary to have an analytical description of contact forces. A real contact is a naturally distributed phenomenon in which the local characteristics of both robot and environment are involved. In addition, friction effects between parts typically exist which greatly complicate the nature of the contact itself. A detailed description of the contact is demanding from a modeling viewpoint. To point out the fundamental aspects of interaction control, it is convenient to resort to a simple but significant model of contact. To this purpose, a decoupled elastically compliant environment is considered, which is described by the model

$$f = K(x - x_e) \quad (18.20)$$

where K is the positive semi-definite stiffness matrix and x_e denotes the rest pose of the undeformed environment. Folding (18.20) into (18.19) gives at the equilibrium:

$$x_\infty = \left(I + K_P^{-1} K\right)^{-1}(x_d + K_P^{-1} K x_e) \quad (18.21)$$

$$f_\infty = \left(I + K K_P^{-1}\right)^{-1} K(x_d - x_e). \quad (18.22)$$

These equations reveal that the control (18.17) behaves as a *stiffness control* [32] in that the interaction of the robot with the environment is influenced by the mutual weight of the active stiffness K_P imposed by the controller versus the passive stiffness K offered by the environment. For a given environment stiffness, according to the prescribed interaction task, one may choose large values of the elements of K_P for those directions along which the environment has to comply and small values of the elements of K_P for those directions along which the robot has to comply.

The dynamic extension of stiffness control is straightforwardly derived by considering the operational space inverse dynamics control (18.13, 18.18) and modifying τ and

y so as to cope with the presence of contact forces. On the assumption of having a wrist force sensor to measure the contact force, the following *impedance control* [31] can be introduced:

$$\tau = B(q)y + n(q, \dot{q}) + J^{T}(q)f \qquad (18.23)$$

with

$$y = J^{-1}(q)M_d^{-1}\left(M_d\ddot{x}_d + K_D\dot{\tilde{x}} + K_P\tilde{x} - M_d\dot{J}(q,\dot{q})\dot{q} - f\right). \qquad (18.24)$$

This leads to the error dynamic equation

$$M_d\ddot{\tilde{x}} + K_D\dot{\tilde{x}} + K_P\tilde{x} = f \qquad (18.25)$$

which remarkably makes the end effector behave as a linear and decoupled active mechanical impedance characterized by a desired equivalent mass M_d, damping K_D, and stiffness K_P, in spite of all the nonlinear and coupled dynamics of the robot. The resulting block scheme of a robot in contact with an elastic environment under impedance control is illustrated in Figure 18.6

In these schemes, the interaction force could be indirectly controlled by acting on the reference value of the robot motion control system. Interaction between robot and environment is anyhow directly influenced by the environment stiffness and by either the stiffness or the impedance of the robot. On the other hand, if it is desired to accurately control the contact force, it is necessary to devise control schemes that allow directly specifying the desired interaction force. The development of a direct force control scheme, in analogy to a motion control scheme, would require the adoption of a stabilizing PD control action on the force error, besides the usual nonlinear compensation actions. Force measurements may be corrupted by noise, and then a derivative action could be troublesome in practical implementation. The stabilizing action is to be provided by suitable damping of velocity terms. As a consequence, a force control system typically features a control law based not only on force measurements but also on velocity measurements and eventually position measurements as well.

The realization of a true *force control* scheme can be entrusted to the closure of an outer force regulation feedback loop [33] generating the control input for the motion

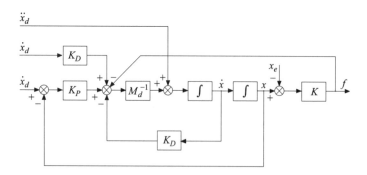

Figure 18.6 Robot in contact with elastic environment under impedance control.

Section 18.4 Force Control

control scheme with which the robot is usually endowed. Therefore, force control schemes are presented below which are based on the use of an inverse dynamics position control. Nevertheless, notice that a force control strategy is meaningful only for those directions of the operational space along which interaction forces between robot and environment may arise.

With reference to the inverse dynamics law with force measurement (18.23), the new input can be chosen as

$$y = J^{-1}(q)M_d^{-1}\left(-K_D \dot{x} + K_P(x_F - x) - M_d \dot{J}(q,\dot{q})\dot{q}\right) \quad (18.26)$$

with

$$x_F = C_F(f_d - f) \quad (18.27)$$

where f_d is a desired value of contact force and C_F is a diagonal matrix whose elements give the control actions to perform along the operational space directions of interest. The resulting block scheme of such force control, illustrated in Figure 18.7, suggests choosing a PI action on the force error, that is,

$$C_F = K_F + K_I \int^t (\cdot)\, d\varsigma \quad (18.28)$$

where the matrices K_F and K_I have to be properly tuned, together with K_P and K_D, to ensure good stability margins and bandwidth for the equivalent third-order system. On the assumption that a stable equilibrium is reached, it is

$$Kx_\infty = Kx_e + f_d \quad (18.29)$$
$$f_\infty = f_d \quad (18.30)$$

confirming that force regulation is achieved, while the actual end-effector pose depends on the amount of environment stiffness and desired force.

It may be argued that, if the position feedback loop in Figure 18.7 is opened, then x_F would represent a velocity reference so that a simple proportional force controller would suffice—with a simplified control design for the equivalent second-order system. However, the absence of an integral action in the force controller would not ensure

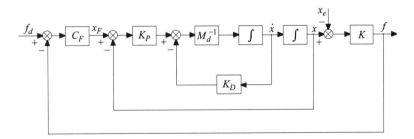

Figure 18.7 Robot in contact with elastic environment under force control.

reduction of the effects due to unmodeled dynamics. Thus such a solution is not further pursued.

If it is desired to specify a desired end-effector pose x_d as in pure motion control schemes, the scheme of Figure 18.7 can be modified by adding the reference x_d to the input where positions are summed, leading to the block scheme in Figure 18.8. This is termed *parallel force/position control* [34], in view of the presence of a position control action $K_P\tilde{x}$ in parallel to a force control action $K_P C_F (f_d - f)$. It is easy to verify that, in this case, the equilibrium is

$$x_\infty = x_d + C_F \big(K(x_e - x_\infty) + f_d \big) \qquad (18.31)$$
$$f_\infty = f_d. \qquad (18.32)$$

Assuming that the desired force is assigned consistently with the constrained task space directions, these equations reveal that the desired force f_d is reached at steady state and that the desired pose x_d is reached by x along the unconstrained directions. Along the constrained directions, the adoption of an integral action in C_F as for the scheme of Figure 18.7 results in a position error on x depending on the environment stiffness.

Whenever an accurate geometric description of the environment is available, it is possible to predetermine the directions along which to control a position and the directions along which to control a force [35]. This leads to the so-called hybrid position/force control formalism [36] in which either a force or a position is controlled along each task space direction according to a selection mechanism. However, in those situations when hybrid control has to operate under imperfect task planning, the system behavior may become quite critical. For instance, consider the extreme case when a hybrid controller governs robot motion in a situation of unplanned impact. Clearly, it is not possible to modify the behavior of the control scheme based on what actually happens in the environment, as the selection has canceled part of the force sensor measurements on the assumption that this information is not useful to the controller.

To conclude the discussion about force control, it should be noticed that stiffness control could be easily implemented in available industrial robot controllers if the user were offered the possibility of adjusting the gains of the PD action. Obviously, the PD action in the joint space differs from that required in the operational space, but through knowledge of robot kinematics (Jacobian), it is possible to find a relationship between the joint stiffness and the end-effector stiffness and then establish some good ranges of gains for ensuring a more or less compliant interaction. On the other hand, impedance

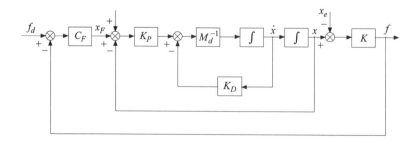

Figure 18.8 Robot in contact with elastic environment under parallel force/position control.

control and force control rely on the possibility of implementing a dynamic control, and more crucially of mounting a force/torque sensor at the robot's wrist. In reality, the force sensor is not strictly required of an impedance controller, but then the resulting impedance would be coupled and configuration dependent. Such a device is still judged by robot manufacturers to be rather expensive in the face of the cost of the robot—although prices have been decreasing recently as more products have become available on the market—and not fully reliable in terms of the available force and (especially) torque ranges. Again, it is to be stressed that in order to make a robot more intelligent and capable of autonomous interaction with the environment, an open control architecture is needed. This architecture should permit interfacing external sensors, for example, the force sensor just mentioned and other sensors such as the vision sensor discussed next.

18.5 VISUAL SERVOING

A great many tasks routinely performed by humans (for example, machine control, driving, assembly, or fruit picking) are based on visually perceived information. In order for robots to perform such tasks, without extensive instrumentation or reengineering of the environment, they must also have the ability to perceive and act upon visual information. Computer vision is therefore an important sensory modality for robotic systems because it mimics the human sense of vision and permits noncontact measurement of the environment. Limited vision capability has been available in commercial robot controllers for many years now. It is used for tasks such as inserting parts with respect to fixed marks on printed circuit boards or for grasping random parts moving on conveyor belts. Typically, these systems adopt a look-and-move strategy—a well-calibrated camera and vision system determines the desired robot end-effector pose, and the robot system is commanded to make the appropriate motion. See [37] for a valuable tutorial on visual servo control. The accuracy of the resulting motion clearly depends directly on the quality of the camera calibration [38] and the accuracy of the robot. The systems in operation today are able to achieve the necessary precision using high-quality and expensive components and good systems engineering. A visual system may include more than one camera, and the cameras can be placed either on the robot observing the target (the so-called eye-in-hand configuration [39]) or in the world observing the robot and the target (the so-called hand-eye coordination [40]).

A distinction can be made in commonly used vision-based systems regarding the presence or absence of joint level feedback. For several reasons, nearly all implemented systems adopt joint level feedback. First, the relatively low sampling rates available from vision makes direct control of a robot's end effector with complex, nonlinear dynamics an extremely challenging control problem. Using internal feedback with a high sampling rate generally presents the visual controller with idealized axis dynamics [41]. Second, many robots already have an interface for accepting Cartesian velocity or incremental position inputs to the internal position controller.

An alternative promising approach to increasing the performance of the overall system is to use a vision system to continuously guide, or steer, the robot end effector toward the target. Such a closed-loop position control structure for a robot end effector is referred to as a *visual servoing system*. A visual feedback control loop, like any feedback control system, will increase closed-loop accuracy and robustness to error in the sensor or the robot.

According to the taxonomy of visual servo systems introduced in [42], it is worth distinguishing *position-based* visual servoing from *image-based* visual servoing. With reference to the block scheme in Figure 18.9, in position-based servoing, a feature vector φ is extracted from the image vector i—typically the coordinates of the centroid of the target—and used in conjunction with a geometric model of the target and the known camera model to estimate the pose of the target with respect to the camera. Feedback is computed on the estimated target pose x, which is compared to the desired pose x_d to form an error used by an operational space controller.

On the other hand, with reference to the block scheme shown in Figure 18.10, in image-based servoing, feedback of the feature vector is accomplished and the error is computed directly in the image space. This presents a significant challenge to controller design since the overall system is nonlinear and highly coupled. In this respect, a useful tool to transform the image-space error $\tilde{\varphi} = \varphi_d - \varphi$ (typically expressed in terms of pixels in the image plane) into an equivalent operational space error $\tilde{x}$ is the so-called image Jacobian J_i relating the rate of change of the target pose to the rate of change of the image feature, that is,

$$\dot{\varphi} = J_i(x)\dot{x}. \qquad (18.33)$$

In this way, for instance, it is possible to design a PD control action with gravity compensation in the image space analogous to (15.17) in the form [43]

$$\tau = J^T(q)K_P J_i^{-1}(x)\tilde{\varphi} - K_D(q) + g(q), \qquad (18.34)$$

which gives successful regulation to a desired pose, provided that a stable behavior can be guaranteed. On the other hand, the design of a tracking controller may be cumbersome because of the limitation imposed by the analog video signal field rate (25 or 30 Hz), which in turn prevents accurate tracking at fast speeds.

To conclude the discussion, benefits from technology trends are pointed out [44]. The fundamental technologies required for visual servoing are image sensors and computing. Fortunately, the price to performance ratios of both technologies are improving as a result of continuing progress in microelectronic fabrication density and the convergence of video and computing driven by consumer demands. Cameras may become so cheap as to become ubiquitous. Rather than using expensive robots to position cameras, it may be cheaper to add large numbers of cameras and switch between them as required.

The performance of early and current visual servo systems has been constrained by broadcast TV standards, with limitations discussed earlier in this chapter. In the last

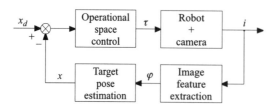

Figure 18.9 Position-based visual servoing.

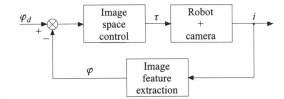

Figure 18.10 Image-based visual servoing.

few years, nonstandard cameras have come onto the market which provide progressive scan (noninterlaced) output and tradeoffs between resolution and frame rate. Digital output cameras are also becoming available, which have the advantage of providing more stable images and requiring a simpler computer interface. The field of electro-optics is also booming, with phenomenal developments in laser and sensor technology. Small-point laser range finders and scanning laser range finders are now commercially available. More recently, frame grabber boards to be installed directly in the controller (e.g., a PC) are being developed so as to minimize the total computation time of the control law.

18.6 THE FUTURE

This chapter has surveyed some technologies that offer a potential for the design of smarter robot controllers. Classical applications of industrial robotics in well-structured environments did not justify developing new designs or controllers. Recently, with the advent of the so-called advanced or service robots, some companies have apparently been reconsidering the issue and are willing to develop customized, special-purpose robotic devices with marked characteristics of autonomy for operation in hostile environments (space, underwater, nuclear, military, etc.) or to execute service missions (domestic applications, medical aids, assistance to the disabled, agriculture, etc.). The key to success in these applications is the incorporation of enhanced sensory feedback capabilities into the control unit of a robotic system. In this respect, the rapid technological progress in sensors and computers of the last decade gives concrete hopes that the robots of the new millennium will eventually be more intelligent and autonomous. The outlook for the future therefore seems bright, with two provisos. On the academic side, researchers should be able to clearly demonstrate real progress in robot technology rather than prizing their results. At the same time, on the industrial side, control engineers should overcome their typical reluctance to bring in concepts and methods from academia because of their apparent complexity. These are a challenge and an opportunity!

> **Related Chapters**
>
> - A general discussion on adaptive control appears in Chapter 5.
> - Lyapunov methods are also reviewed in Chapters 6 and 8.
> - Chapter 7 uses a robotic system to illustrate hybrid dynamical system concepts.

REFERENCES

[1] G. Hirzinger, J. Bals, B. Brunner, R. Koeppe, and M. Schedl, "Towards a new robot generation." *Proc. 5th Int. Symp. on Methods and Models in Automation and Robotics*, Miedzyzdroje, Poland, pp. 747–762, 1998.

[2] L. Sciavicco and B. Siciliano, *Modeling and Control of Robot Manipulators*. New York: McGraw-Hill, 1996.

[3] D. E. Whitney, "Resolved motion rate control of manipulators and human prostheses." *IEEE Trans. on Man-Machine Systems*, Vol. 10, pp. 47–53, 1969.

[4] B. Siciliano, "A closed-loop inverse kinematic scheme for on-line joint-based robot control." *Robotica*, Vol. 8, pp. 231–243, 1990.

[5] A. Liégeois, "Automatic supervisory control of the configuration and behavior of multibody mechanisms." *IEEE Trans. on Systems, Man, and Cybernetics*, Vol. 7, pp. 868–871, 1977.

[6] S. Chiaverini, B. Siciliano, and O. Egeland, "Review of the damped least-squares inverse kinematics with experiments on an industrial robot manipulator." *IEEE Trans. on Control Systems Technology*, Vol. 2, pp. 123–134, 1994.

[7] Y. Nakamura, *Advanced Robotics: Redundancy and Optimization*. Reading, MA: Addison-Wesley, 1991.

[8] J. Y. S. Luh, "Conventional controller design for industrial robots: A tutorial." *IEEE Trans. on Systems, Man, and Cybernetics*, Vol. 13, pp. 298–316, 1983.

[9] F. Nicolò and J. Katende, "A robust MRAC for industrial robots." *Proc. 2nd IASTED Int. Symp. Robotics and Automation*, Lugano, Switzerland, pp. 162–171, 1983.

[10] M. Gautier and W. Khalil, "Direct calculation of minimum set of inertial parameters of serial robots." *IEEE Trans. on Robotics and Automation*, Vol. 6, pp. 368–373, 1990.

[11] J. M. Hollerbach, "A survey of kinematic calibration." In O. Khatib, J. J. Craig, and T. Lozano-Pérez (eds.), *The Robotics Review 1*, pp. 207–242. Cambridge, MA: MIT Press, 1998.

[12] S. Arimoto and F. Miyazaki, "Stability and robustness of PID feedback control for robot manipulators of sensory capability." In M. Brady and R. Paul (eds.), *Robotics Research: The First International Symp.*, pp. 783–799, Cambridge, MA: MIT Press, 1984.

[13] C. H. An, C. G. Atkeson, and J. M. Hollerbach, *Model-Based Control of a Robot Manipulator*. Cambridge, MA: MIT Press, 1988.

[14] A. K. Bejczy, *Robot Arm Dynamics and Control*. Memo. 33-669. Jet Propulsion Laboratory, California Institute of Technology, 1974.

[15] K. Kreutz, "On manipulator control by exact linearization." *IEEE Trans. on Automatic Control*, Vol. 34, pp. 763–767, 1989.

[16] F. Dogliani, G. Magnani, and L. Sciavicco, "An open architecture industrial controller." *Newsl. of IEEE Robotics and Automation Soc.*, Vol. 7, no. 3, pp. 19–21, 1993.

[17] C. Abdallah, D. Dawson, P. Dorato, and M. Jamshidi, "Survey of robust control for rigid robots." *IEEE Control Systems Mag.*, Vol. 11, no. 2, pp. 24–30, 1991.

[18] R. Ortega and M. W. Spong, "Adaptive motion control of rigid robots: A tutorial." *Automatica*, Vol. 25, pp. 877–888, 1989.

[19] S. Jajasuriya and C. N. Hwang, "Tracking controllers for robot manipulators: A high-gain perspective." *ASME J. of Dynamic Systems, Measurement, and Control*, Vol. 110, pp. 39–45, 1988.

[20] J.-J. E. Slotine, "The robust control of robot manipulators." *Int. J. Robotics Research*, Vol. 4, no. 2, pp. 123–138, 1985.

[21] M. Corless, "Tracking controllers for uncertain systems: Application to a Manutec r3 robot." *ASME J. of Dynamic Systems, Measurement, and Control*, Vol. 111, pp. 609–618, 1989.

[22] M. W. Spong and M. Vidyasagar, "Robust linear compensator design for nonlinear robotic control." *IEEE J. of Robotics and Automation*, Vol. 3, pp. 345–351, 1987.

References

[23] J.-J. E. Slotine and W. Li, "On the adaptive control of robot manipulators." *Int. J. of Robotics Research*, Vol. 6, no. 3, pp. 49–59, 1987.

[24] R. H. Middleton and G. C. Goodwin, "Adaptive computed torque control for rigid link manipulators." *Systems & Control Lett.*, Vol. 10, pp. 9–16, 1988.

[25] J.-J. E. Slotine and W. Li, "Composite adaptive manipulator control." *Automatica*, Vol. 25, pp. 509–519, 1989.

[26] D. E. Whitney, "Force feedback control of manipulator fine motions." *ASME J. of Dynamic Systems, Measurement, and Control*, Vol. 99, pp. 91–97, 1977.

[27] B. Siciliano and L. Villani, *Robot Force Control*. Boston: Kluwer Academic Publishers, 1999.

[28] O. Khatib, "A unified approach for motion and force control of robot manipulators: The operational space formulation." *IEEE J. of Robotics and Automation*, Vol. 3, pp. 43–53, 1987.

[29] M. Takegaki and S. Arimoto, "A new feedback method for dynamic control of manipulators." *ASME J. of Dynamic Systems, Measurement, and Control*, Vol. 102, pp. 119–125, 1981.

[30] J. Y. S. Luh, M. W. Walker, and R. P. C. Paul, "Resolved-acceleration control of mechanical manipulators." *IEEE Trans. on Automatic Control*, Vol. 25, pp. 468–474, 1980.

[31] N. Hogan, "Impedance control: An approach to manipulation: Parts I–III." *ASME J. of Dynamic Systems, Measurement, and Control*, Vol. 107, pp. 1–7, 1985.

[32] J. K. Salisbury, "Active stiffness control of a manipulator in Cartesian coordinates." *Proc. 19th IEEE Conf. on Decision and Control*, Albuquerque, NM, pp. 95–100, 1980.

[33] J. De Schutter and H. Van Brussel, "Compliant robot motion I–II." *Int. J. of Robotics Research*, Vol. 7, no. 4, pp. 3–33, 1988.

[34] S. Chiaverini and L. Sciavicco, "The parallel approach to force/position control of robotic manipulators." *IEEE Trans. on Robotics and Automation*, Vol. 4, pp. 361–373, 1993.

[35] M. T. Mason, "Compliance and force control for computer controlled manipulators." *IEEE Trans. on Systems, Man, and Cybernetics*, Vol. 6, pp. 418–432, 1981.

[36] M. H. Raibert and J. J. Craig, "Hybrid position/force control of manipulators." *ASME J. of Dynamic Systems, Measurement, and Control*, Vol. 103, pp. 126–133, 1981.

[37] S. Hutchinson, G. Hager, and P. Corke, "A tutorial on visual servo control." *IEEE Trans. on Robotics and Automation*, Vol. 12, pp. 651–670, 1996.

[38] R. Y. Tsai and R. K. Lenz, "Techniques for calibration of the scale factor and image center for high accuracy 3D machine vision metrology." *IEEE Trans. on Pattern Analysis and Machine Intelligence*, Vol. 10, pp. 713–720, 1988.

[39] B. Nelson and P. K. Khosla, "Increasing the tracking region of an eye-in-hand system by singularity and joint limit avoidance." *Proc. 1993 IEEE Int. Conf. on Robotics and Automation*, Atlanta, GA, Vol. 3, pp. 418–423, 1993.

[40] G. D. Hager, "A modular system for robust hand-eye coordination." *IEEE Trans. on Robotics and Automation*, Vol. 13, pp. 582–595, 1997.

[41] P. I. Corke, *Visual Control of Robots*. New York: Research Studies Press and John Wiley, 1996.

[42] A. C. Sanderson and L. E. Weiss, "Image-based visual servo control using relational graph error signals." *Proc. IEEE*, pp. 1074–1077, 1980.

[43] R. Kelly, "Robust asymptotically stable visual servoing of planar robots." *IEEE Trans. on Robotics and Automation*, Vol. 12, pp. 759–766, 1996.

[44] P. I. Corke and G. D. Hager, "Vision-based robot control." In *Control Problems in Robotics and Automation*, B. Siciliano and K.P. Valavanis (eds.), London: Springer, pp. 177–192, 1998.

Chapter 19 | CONTROL OF COMMUNICATION NETWORKS

R. Srikant

Editor's Summary

Two other chapters in this volume discuss new application areas for control: intelligent transportation systems (Chapter 14) and civil structures (Chapter 17). This chapter focuses on a third, and equally exciting, one: the control of communication networks. It is a sign of the enduring vitality of controls as a solution technology that industries of recent vintage are recognizing its relevance for solving a new generation of problems. In areas such as communication networks, many of these problems simply did not exist in anything like their present form when the foundations of modern control were being laid some decades ago.

Communication networks pose a number of control problems, all of which can generally be related to ensuring reliable and rapid transmission of heterogeneous digital data over large-scale, geographically distributed channels and processing nodes. Four problems of particular importance are admission control when real-time service is desired, congestion control through regulation of transmission rates, packet routing, and scheduling of available node bandwidth among multiple sources. These problems have different manifestations, depending on the type of network and traffic and on quality of service (QoS) requirements. The chapter notes specific considerations for the Internet and asynchronous transfer mode (ATM) networks. Other types of communication networks, as used in distributed control systems in the building control and process control industries, are briefly discussed in Chapters 16 and 12, respectively.

Communication network control problems are spawning new theoretical developments in control, for example, in discrete-event systems (Chapter 2). At the same time, established control techniques such as linear quadratic gaussian (LQG) and calculus of variations remain directly applicable. This chapter also identifies a number of central implementation issues for the control of communication networks. These include the measurement interval used for calculating queue lengths, the feedback mechanisms employed for network control and "fairness" issues in allocating bandwidth.

R. Srikant is an assistant professor in the Department of Electrical Engineering at the University of Illinois at Urbana-Champaign. He chairs the Working Group on Communication Networks for IEEE Control Systems Society.

19.1 INTRODUCTION

A communication network is a collection of nodes and links, in which each node can potentially serve as an origin or destination for a traffic source. Each link is a transmission medium (optical fiber, satellite link, etc.) whose traffic-carrying capacity is measured in bits-per-second (bps). This capacity is often referred to as bandwidth or data rate. The traffic sources could be one of many types, such as voice, video, and data. Thus, in general, today's networks are a heterogeneous mix of transmission links, nodes and traffic sources, as depicted in Figure 19.1.

The most common example of a heterogeneous network is the ubiquitous Internet, which is really a collection of many networks. Communication in networks is done through the transmission of information in "packets," which can be of fixed or variable size depending on the network. The specific mechanisms for packet transmission in a communication network are referred to as the network protocol. The protocol used to transfer packets in the Internet is called the Transmission Control Protocol/Internet Protocol (TCP/IP). Another widely studied protocol to support a large variety of traffic sources is the set of standards developed for the so-called ATM (Asynchronous Transfer Mode) networks. An ATM network could be one of the many subnetworks that an Internet packet traverses in getting from its origin to its destination. Although an Internet packet can be of variable size, a "packet" in an ATM network is called a *cell*, and it has a fixed size of 53 bytes. Thus, when a TCP packet passes through an ATM network, it is segmented into 53-byte cells.

When packets from a traffic source traverse a network, they are subject to delays and losses. Delays are due to queueing at the nodes, which is a controllable factor, and to the uncontrollable law of physics, which dictates that packets cannot travel from one node to another faster than the speed of light. Packets can be lost because buffers used to store packets at the nodes are of finite size. Thus, two measures of the quality-of-service (QoS) delivered to a traffic source are *delay* and *packet loss*. Traffic sources can be broadly classified into two major classes depending on their QoS requirements: *real-time* sources and *non-real-time* sources. Real-time sources have stringent delay and delay-variation (jitter) requirements but may have either stringent or loose packet loss requirements depending on the application. On the other hand, non-real-time sources have stringent packet loss requirements but can tolerate more delay than real-time sources. Examples of real-time sources include voice, video conferencing, and so on, while e-mail, Web browsing, and the like are examples of non-real-time traffic.

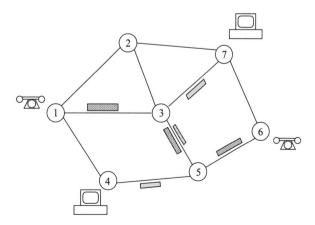

Figure 19.1 High-speed network with packets from many types of traffic sources.

19.2 NETWORK CONTROL AND MANAGEMENT

From a network control and management point of view, real-time sources are admitted into the network only if there are sufficient resources to satisfy their QoS requirements. On the other hand, non-real-time sources are always admitted into the network with the understanding that the resources in the network would be allocated to them on a best-effort basis; that is, real-time sources are given higher priority, and whatever bandwidth is left unused by the real-time sources is allocated to the non-real-time sources. In what follows, we identify four major problems in controlling and managing a network with real-time and non-real-time sources. Some aspects of each of these problems can be studied using control-theoretic tools.

19.2.1 Admission Control for Real-Time Sources

As mentioned earlier, real-time sources are admitted into the network only if the network has sufficient capacity to meet their stated QoS requirements. In the simplest one-node example, suppose that the link connected to the node has a capacity of 50 bps and that all sources are of the ON-OFF type. An ON-OFF source is one that randomly switches between an ON state and an OFF state. In the ON state, let us assume that it generates data at a rate of 1 bps and that it does not generate any data when it is OFF. Let us also suppose that there is no buffer at the node, and therefore, if we admit more than 50 sources into the network, some data would be lost since, occasionally, more than 50 of the sources can be in the ON state at the same time. Thus, if the QoS requirement is in terms of fraction of bits lost, a very conservative admission control scheme would be to admit a maximum of 50 sources into the network at any time. Under this conservative scheme, if a source arrives when there are already 50 sources in the system, then it is blocked, that is, denied admission by the network.

In reality, most real-time sources can tolerate some packet loss. If we know the fraction of time a source is ON, then we can calculate the fraction of packets lost if we admit more than 50 sources into the network at one time. Thus, a better admission control scheme would be to stop admitting sources only when the fraction of lost packets would exceed the QoS requested by the sources. In a real node, there is a nonzero buffer that can store packets when the total arrival rate temporarily exceeds the bandwidth of the link. This buffer space further increases the amount of sources that can be admitted. Thus, one can admit many more sources than the conservative admission control scheme by allowing small amounts of packet loss. This idea is called *statistical multiplexing*. A significant portion of the mathematical research in communication networks over the last decade has been devoted to computing the probability of packet loss for many different types of arrival processes and many different scheduling disciplines at the nodes. Several techniques have been used to solve this problem, including spectral expansion [4], Laplace transform techniques [1], and large deviations [23]. In Section 19.3, we present a connection between this admission control problem and deterministic optimal control using the theory of large deviations.

A different point of view of admission control which has also received widespread interest is one in which the network negotiates a deterministic contract with each source. The contract is in terms of the maximum and mean arrival rates and burstiness (which is a measure of the variability of the traffic about the mean) for the packet arrival process. As long as a source conforms to the contract, the network agrees to transfer its packets with no loss. This type of arrival model has been analyzed in [31].

19.2.2 Congestion Control for Best-Effort Sources

From an application point of view, earlier we categorized sources as real-time and non-real-time sources. From a network point of view, the service provided to sources in high-speed communication networks can be classified as guaranteed service and best-effort service. Guaranteed service refers to a contract between the network service provider and the end user, which requires the network to provide a fixed quality-of-service (QoS) to the traffic. The QoS guarantees could be in the form of upper bounds on packet loss probability, delay, and so on. In contrast, sources that subscribe to the best-effort service are provided minimal or no guarantees *a priori*. In the Internet, there are no guarantees, while in Asynchronous Transfer Mode (ATM) networks, the best-effort traffic (known as Available Bit Rate [ABR] service) may be guaranteed a minimum data rate. Under the best-effort service, instead of guaranteeing a fixed QoS, the idea is to *fairly* allocate the network resources to competing users. The guaranteed-service traffic—referred to as either Constant Bit Rate (CBR) or Variable Bit Rate (VBR) traffic in ATM networks—gets a higher scheduling priority compared to the best-effort traffic. In other words, at a given node, when both best-effort traffic and guaranteed-service traffic are backlogged (i.e., have packets that are waiting to be sent), the packets from the guaranteed-service traffic are processed first, and the best-effort traffic is served only if there are no packets in the guaranteed-service queue. Typically, real-time sources would request guaranteed service, while non-real-time sources would request best-effort service. However, if the congestion in the network is small, real-time sources may also use the best-effort service.

From a control point of view, each best-effort source or user may be thought of as an entity that generates data at a rate specified by the network. The network exercises control over the best-effort traffic by either assigning these rates based on the congestion in the network or providing congestion information to the sources that would individually choose their transmission rates. This is referred to as *congestion control*. In the absence of such a control mechanism, the buffers at each node in the network (which store packets temporarily) may overflow and lead to packet losses.

There are two basic approaches to congestion control. In the Internet, the protocol used for this purpose is called TCP [26]. In the TCP protocol, each source slowly increases the rate at which it transmits data, and upon detection of congestion, the data rate is reduced. Congestion is detected when buffers overflow and packets are lost. It is the responsibility of the destination to inform the source of lost packets, and intermediate nodes in the network do not provide any feedback directly to the source. In contrast, in ATM ABR service, the communication protocol between the source and the network allows for intermediate nodes to suggest a data rate to the ABR sources. In Section 19.4, we adopt this framework. Thus, from the point of view of the current technology, the approach presented in that section can be viewed as a congestion control mechanism for ATM ABR sources.

Congestion control can also be viewed in a game-theoretic context, in which each user minimizes its own performance objective. This would correspond to a situation in which each source makes its decision based on the information provided by the network, somewhat along the lines of TCP. The interested reader can refer to [33] (and references within) for various versions of this problem.

Recent work by Kelly and his co-workers has cast congestion control in a new light. Kelly considers pricing as a mechanism to induce the correct behavior of selfish users during times of network congestion. This point of view is presented in [24].

19.2.3 Routing

Our discussion so far has been confined to a single-node network. A real network consists of a large number of nodes, and therefore, a packet can take many routes to go from an origin to a destination. Routing in communication networks is performed in one of three ways: *circuit-switched* routing, *virtual circuit* routing, and *datagram* routing.

The term *circuit-switched routing* has its origins in the routing techniques used in telephone networks. For a telephone call, a fixed amount of bandwidth is reserved on every link in its route. Thus a call is blocked if there is no route with enough spare capacity connecting the voice source's origin and destination. A call may also be blocked if the only available paths consist of many links since a call that occupies multiple links may potentially block many calls in the future. Circuit-switched routing was extensively analyzed in the 1980s, and an excellent survey of the control issues associated with this type of routing can be found in [25].

Virtual-circuit routing is similar to circuit-switched routing in the sense that all packets that belong to a traffic source take the same route from origin to destination, and this route is predetermined when a source is admitted into the network. However, in contrast to circuit-switched routing, a fixed bandwidth need not be reserved for each source along its route. The amount of bandwidth allocated to a source could be time-varying depending on the type of service. For real-time sources, a common technique is to convert the QoS requirements of a source into an *effective bandwidth* [23] (Section 19.3), in which case, from a network management point of view, the situation becomes similar to circuit-switched routing.

In datagram routing, each packet from a traffic source could take a different path from origin to destination. This is the type of routing used in the Internet. In the simplest form of datagram routing, the routes are chosen to minimize the number of links that a packet traverses. If the links in the network do not fail, then essentially all packets from a traffic source will follow the same route. In more complicated routing schemes, where the route is based on some other criterion such as attempting to minimize average delay in the network, routes will be periodically recomputed, and there is a greater chance that different packets from the same traffic source could choose different routes. For a thorough introduction to different types of routing schemes, see [5].

From a control-theoretic point of view, optimal virtual-circuit routing can be thought of as a stochastic control problem involving a dynamic program over a very large state space (see [30] and references within). Solving this problem exactly is practically impossible because of the dimensionality of the state space. Indeed, given a particular routing scheme, computing its performance analytically is fundamentally difficult, as shown in [21], and simulation can be very expensive even when simulation speed-up techniques such as the ones in [34, 35] are used. Thus, the only way to obtain reasonably good routing schemes is to solve the dynamic program approximately; this approach has been applied successfully to realistic networks in [30]. It would be interesting to compare the results in [30] to the performance bounds given in [20] for the case in which all calls request the same amount of bandwidth and in [12] for the case where different call classes request different amounts of bandwidth.

In a single-link network with many call classes, the routing problem reduces to a simple admission control decision: depending on the state of the network, should a call be accepted or not? When the call classes request different amounts of bandwidth, even this is a difficult problem to solve. This problem has been approached in a novel way in [3]. By considering an appropriate fluid limit, Altman et al. convert the stochastic control problem into a deterministic optimal control problem. The solution to this deterministic optimal control problem can provide useful insight into the structure of the optimal policy for the original stochastic problem.

A different view of routing than the one presented so far is one in which users have a certain amount of flow that they have to route and they select routes to optimize their individual performance objectives. Here the users could be either end users or service providers, depending on the context. The issue of interest here is to understand the interaction of noncooperative, selfish users in a network. A natural approach to such problems is to use game theory, and many variants of these problems have been studied in [29].

19.2.4 Scheduling

Once sources are admitted and routed on specific paths, nodes have to decide how to allocate their bandwidth among various sources. For instance, suppose that we admit five sources, each requiring 0.2 Kbps to meet their QoS requirements, and that all five sources pass through a common node whose capacity is 1 Kbps. A naive scheme would simply allocate 0.2 Kbps to each source. Thus, suppose four of the sources have empty queues for some period of time and the fifth source is backlogged (i.e., its queue is not empty) during this time; then the fifth source would continue to get 0.2 Kbps even though the entire link capacity is available. A better scheme would be to give a *weight* of 0.2 to each source and divide the node capacity only among *backlogged* sources in proportion to their weights. In the above example, if there are only three backlogged sources, each one would get $\frac{1}{3}$ Kbps. This is called Fluid Fair Queueing (FFQ) or Generalized Processor Sharing (GPS) [16, 31]. However, such a scheme cannot be implemented in practice since packets are indivisible entities, not fluids; therefore, it is not possible to instantaneously divide the link capacity among competing sources. Thus, close approximations of this fluid scheduling algorithm, called Weighted Fair Queueing (WFQ) or Packet-by-Packet Generalized Processor Sharing (PGPS) are used in practice [16, 31].

A novel modification of this algorithm has been studied recently in [28] using game-theoretic tools. Suppose that $A_i(s, t)$ is the number of arrivals from source i in the time interval $[s, t)$, and we assume that $A_i(s, t)$ is constrained as follows:

$$A_i(s, t) \leq \rho_i(t - s) + \sigma_i.$$

Such a constraint is called a *leaky bucket* constraint [5]. Note many arrival processes can satisfy the constraint for a given ρ_i and σ_i. Let us also suppose that we are interested in designing a scheduling algorithm to minimize losses given the leaky bucket parameters of the sources. Then, the scheduler can be thought of as playing a zero-sum game where it attempts to minimize losses and the arrival processes attempt to maximize the losses subject to their (ρ_i, σ_i) constraints. The results in [28] show that one can improve upon GPS and PGPS using such a game formulation. Using game theory to address scheduling problems is a very recent development. One can easily think of many variants to the basic problem addressed in [28], such as allowing other performance measures, other types of arrival processes, and other types of servers to model wireless networks. These are interesting open issues that remain to be addressed.

19.3 QOS, ADMISSION CONTROL, AND CALCULUS OF VARIATIONS

In this section, we provide an informal discussion of the relationship between computing the probability of rare QoS violations in queues and the calculus of variations. For the purposes of simplicity, our discussion here is not meant to be mathematically rigorous but rather to provide insight into the queueing behavior that leads to rare QoS violations. For rigorous proofs of the results here, see, for example, [15], and for an informal introduction to variational problems that arise in large deviations, see [8]. Since our goal is to provide a survey of the control issues in communication networks, we narrowly focus our attention on the application of large deviations to estimate rare event probabilities in queues. However, a more fundamental connection exists between optimal control and large deviations, and this is explored in [17, 37].

19.3.1 Large Deviations of the Empirical Mean of a Sequence of Random Variables

Let $A_1, A_2, \ldots$ be a sequence of independent, identically distributed random variables with mean μ, and let the moment-generating function $M(\theta) = E(e^{\theta A_1})$ be well defined for θ in a neighborhood of 0. Let $\tilde{S}_n := A_1 + A_2 + \ldots + A_n$ and $S_n := \frac{\tilde{S}_n}{n}$. By the law of large numbers, the empirical mean S_n will converge to μ in an appropriate sense. However, we are often interested in estimating the probability that this empirical mean will have a large deviation from the mean. For example, consider an experiment in which we toss a fair coin 1000 times. We expect, on average, 500 heads and 500 tails. If we repeat this experiment many times, in some instances of the experiment, the number of heads will be greater than 700. In other words, the empirical mean would sometimes be larger than 0.7. We would like to estimate the probability of this rare large deviation from the mean behavior.

To estimate such probabilities, we first recall Markov's inequality: for any random variable X with mean $E(X)$, $P(X > x) \leq E(X)/x$. Noting that $E(e^{\theta \tilde{S}_n}) = M^n(\theta)$, it follows from Markov's inequality that

$$P(S_n > a) = P(\tilde{S}_n > na) = P(e^{\theta \tilde{S}_n} > e^{na}) \leq M^n(\theta) e^{-n\theta a} = e^{-n(a\theta - \Lambda(\theta))},$$

where $\theta > 0$ and $\Lambda(\theta) := \log M(\theta)$ is known as the log-moment generating function (lmgf), or the cumulant-generating function (cgf). To get the best bound on the probability $P(S_n > a)$, we minimize the righthand side of the above inequality with respect to a. Defining

Section 19.3 QoS, Admission Control, and Calculus of Variations

$$I(a) := \sup_{\theta}\{a\theta - \Lambda(\theta)\},$$

we have

$$P(S_n > a) \le e^{-nI(a)}. \tag{19.1}$$

The above bound on $P(S_n > a)$ is called the Chernoff bound, and $I(a)$ is called the rate function. If we assume that $a > 0$, then we can dispense with the condition that $\theta > 0$ [15, Lemma 2.2.5], and therefore, we have allowed θ to be unrestricted in sign in the definition of $I(a)$. The rate function $I(a)$ is also called the convex transform (or the Cramér-Fenchel-Legendre transform) of $\Lambda(\theta)$. A very useful property of $I(a)$ and $\Lambda(\theta)$ is that they are convex duals of each other. In other words, the convex transform of $I(a)$ is $\Lambda(\theta)$, that is,

$$\Lambda(\theta) = \sup_{a}\{a\theta - I(a)\}.$$

It is easy to verify that $I(a)$ and $\Lambda(\theta)$ are convex functions.

The bound (19.1) is known to be asymptotically tight in the following sense:

$$\lim_{n\to\infty} \frac{1}{n} \log P(S_n > a) = -I(a). \tag{19.2}$$

This result is known as Cramér's theorem [15, Theorem 2.2.3]. Under mild conditions, given by the Gartner-Ellis theorem [15, Theorem 2.3.6], this result remains valid for (possibly) dependent random variables $A_1, A_2, \ldots$ by redefining the lmgf as follows:

$$\Lambda(\theta) = \lim_{n\to\infty} \frac{1}{n} \log E(e^{\theta \sum_{i=1}^{n} A_i}).$$

In large deviations terminology, a sequence of random variables is said to satisfy a large deviations principle (LDP) if there exists a function $I(\cdot)$ such that (19.2) holds. A more precise definition of when a sequence of random variables is said to satisfy an LDP can be found in [15].

19.3.2 Large Deviations of a Random Process from its Fluid Limit

Let $A_1, A_2, \ldots$ be a sequence of random variables with mean μ, and let us assume that they satisfy an LDP with rate function $I(x)$. Define

$$S_n(t) := \frac{1}{n} \sum_{i=1}^{nt} A_i, \qquad \forall t = k/n, \qquad k = 0, 1, 2, \ldots$$

For values of t that are not of the form k/n, define $S_n(t)$ by linear interpolation. As $n \to \infty$, by the law of large numbers, $\frac{1}{nt}\sum_{i=1}^{nt} A_i \to \mu$, and hence $S_n(t) \to \mu t$, which is called the *fluid limit* of the process $S_n(t)$. Since the function $S_n(t)$ behaves like the linear function μt for very large n, a natural *large deviations* question is the following: What is the probability that the function will deviate from this behavior and *look* like some other function $a(t)$?[1] To keep the discussion simple, let us assume that $t \in [0, t_f]$, $t_f < \infty$. We will now informally argue that, for very large n,

$$P(S_n(t) \approx a(t)) \approx e^{-n \int_0^{t_f} I(\dot{a}(s))ds}.$$

A precise version of the above result is in [14].

Let us divide the interval $[0, t_f]$ into M small subintervals, each of length Δt.

$P(S_n(t) \approx a(t))$
$\approx P(S_n(0) = a(0), S_n(\Delta t) = a(\Delta t), S_n(2\Delta t) = a(2\Delta t), \ldots, S_n(M\Delta t) = a(M\Delta t))$
$= P(S_n(M\Delta t) = a(M\Delta t) | S_n((M-1)\Delta t) = a((M-1)\Delta t)) \times$
$\quad P(S_n((M-1)\Delta t) = a((M-1)\Delta t) | S_n((M-2)\Delta t) = a((M-2)\Delta t) \times \ldots \times$
$\quad P(S_n(\Delta t) = a(\Delta t) | S_n(0) = a(0)).$

Note that

$$S_n(M\Delta t) = \frac{1}{n}\sum_{i=1}^{nM\Delta t} A_i$$

$$= S_n((M-1)\Delta t) + \frac{1}{n}\sum_{i=n(M-1)\Delta t+1}^{nM\Delta t} A_i.$$

Thus,

$P(S_n(M\Delta t) = a(M\Delta t) | S_n((M-1)\Delta t) = a((M-1)\Delta t))$
$$= P\left(\sum_{i=n(M-1)\Delta t}^{nM\Delta t} A_i \approx \dot{a}((M-1)\Delta t)n\Delta t\right)$$
$$\approx e^{-n\Delta t I(\dot{a}((M-1)\Delta t))}.$$

Proceeding similarly, we get

$$P(S_n(t) \approx a(t)) \approx \prod_{i=1}^{M-1} e^{-n\Delta t I(\dot{a}(i\Delta t))}$$
$$= e^{-n \sum_{i=1}^{M-1} \Delta t \dot{a}(i\Delta t)}$$
$$\approx e^{-n \int_0^{t_f} I(\dot{a}(s))ds}.$$

Thus $S_n(t)$ satisfies an LDP with rate function $\int_0^{t_f} I(\dot{a}(s))ds$.

---

[1] We are really interested in the probability that $S_n(t)$ is in the neighborhood of $a(t)$. Hence, we use the notation $P(S_n(t) \approx a(t))$ rather than $P(S_n(t) = a(t))$.

19.3.3 Estimating Probabilities of Rare Events in Queues

Consider a discrete-time queue with the capacity to serve c bits in each time unit. Let A_k denote the number of arriving bits generated in time unit k by the source accessing the queue. Let us suppose that the current time is 0, and the arrival process started at time $-\infty$. Thus, at the current time, the system is in steady state. Assuming large buffer sizes, we are interested in the steady-state probability of buffer overflow. If the buffer size is denoted by n, and queue length at time k is denoted by Q_k, we are interested in estimating $P(Q_0 > n)$ for large n.

We first note the following well-known equation called Loynes' formula:

$$P(Q_0 > n) = P\left(\sup_{K>0} \sum_{i=-K}^{1} A_i - cK > n\right). \tag{19.3}$$

To understand this equation, note that

$$Q_0 = \max\{Q_{-1} + A_{-1} - c, 0\}$$
$$Q_{-1} = \max\{Q_{-2} + A_{-2} - c, 0\}.$$

These two equations are called Lindley's equations. Thus

$$Q_0 = \max\{Q_{-2} + \sum_{i=1}^{2} A_{-i} - 2c, A_{-1} - c, 0\}.$$

Continuing as above, and assuming $Q_{-\infty} = 0$, gives the relationship (19.3). From (19.3), we can upper and lower bound the probability $P(Q_0 > n)$ as

$$\sup_{K>0} P\left(\sum_{i=-K}^{1} A_i - cK > n\right) \leq P(Q_0 > n) \leq \sum_{K>0} P\left(\sum_{i=-K}^{1} A_i - cK > n\right). \tag{19.4}$$

If, for each K, $P(\sum_{i=-K}^{1} A_i - cK > n)$ is of the form $e^{-n\gamma_K}$ for some $\gamma_K > 0$, then the largest of these probabilities will dominate as $n \to \infty$; that is, the lower bound will be approximately equal to the upper bound in (19.4). In what follows, we will argue that $P(\sum_{i=-K}^{1} A_i - cK > n)$ goes to zero exponentially fast in n, and thus we have the approximation

$$P(Q_0 > n) \approx \sup_{K>0} P\left(\sum_{i=-K}^{1} A_i - cK > n\right). \tag{19.5}$$

This approximation (19.5) illustrates an important idea in large deviations analysis: Rare events occur in the most likely way. Specifically, in this case, we have replaced the probability of the union of some events by the probability of the most likely event.

Define

$$W_{k-1} := \sum_{i=k}^{K} A_{-i} - c(K-k+1).$$

W_0 can be interpreted as follows: if $-K$ is the last time before $k=0$ that the queue was empty, then W_0 is simply the queue length at time 0. To use the results from the previous subsection, first we have to convert the probabilistic buffer overflow event into a deterministic event using the *fluid limit* scaling. To this end, we let $T = K/n$, which gives

$$W_0 = \sum_{i=1}^{nT} A_{-i} - cnT.$$

Thus,

$$\frac{W_0}{n} = \frac{1}{n}\sum_{i=1}^{nT} A_{-i} - cT.$$

Define $x(t) := \frac{W_{nt}}{n}$, where $t = k/n$. Since we are interested in the event $\{W_0 > n\}$, we impose the condition $x(0) \geq 1$, and since $W_K = 0$, we have $x(-T) = 0$. Let us suppose that $\frac{1}{n}\sum_{i=-nt}^{-1} A_i$ exhibits a large deviation from its fluid limit μt and behaves like some function $a(t)$ in the interval $[-T, 0]$ as $n \to \infty$. Then, $x(t) = a(t) - ct$, $t \in [-T, 0]$.

From our discussion in the previous subsection,

$$P\left(\frac{1}{n}\sum_{i=1}^{nt} A_i \approx a(t)\right) \approx e^{-n \int_{-T}^{0} I(\dot{a}(s))ds}.$$

Since many possible trajectories $a(t)$ can lead to buffer overflow, we approximate the buffer overflow probability by the probability of the most likely trajectory. Thus

$$P(x(0) > 1) \approx \sup_{T, a(\cdot)} \left\{ e^{-n \int_{-T}^{0} I(\dot{a}(s))ds} : x(-T) = 0, x(0) \geq 1, x(t) = a(t) - ct, -T \leq t \leq 0 \right\}.$$

To convert the above problem into a variational form encountered in control theory, we first define $u(t) = \dot{a}(t)$. Since $x(t) = a(t) - ct$, we have $\dot{x}(t) = u(t) - c$. Thus $P(x(0) > 1) \approx e^{-nJ^*}$, where J^* is the optimal cost in the following optimal control problem:

$$\inf_u \int_{-T}^{0} I(u(s))ds,$$

such that

Section 19.3 QoS, Admission Control, and Calculus of Variations

$$\dot{x} = u(t) - c$$
$$x(-T) = 0, \quad x(0) \geq 1, \quad T \text{ is free.}$$

In general, this problem could be difficult to solve. But it turns out that, by exploiting certain properties of the rate function, one can solve the problem without solving complicated Euler–Lagrange equations.

The first observation is that, since I is a convex function, by Jensen's inequality, for any $u(t)$,

$$\frac{1}{T} \int_{-T}^{0} I(u(s))ds \geq I(\bar{u}),$$

where $\bar{u} = \frac{1}{T}\int_{-T}^{0} u(s)ds$. Thus, it is sufficient to consider constant controls of the form $u(t) \equiv \bar{u}$. Since $x(-T) = 0$ and $x(0) \geq 1$, the constant $\bar{u}$ has to satisfy $(\bar{u} - c)T \geq 1$.

We next argue that it is sufficient to only consider constant controls of the form $\bar{u} = c + \frac{1}{T}$. For stable operation of the queue, we assume that the mean service rate is larger than the mean arrival rate, that is, $c > E(A_1)$. Since the probability of the empirical mean being larger than x is approximately equal to $e^{-nI(x)}$, and since it is reasonable to expect the probability of such large deviations to decrease when $(x - E(A_1))$ increases, one would expect $I(x)$ to be a nondecreasing function of x for $x > E(A_1)$. This is indeed true [15, Lemma 2.2.5]. Hence

$$I(\bar{u}) \geq I\left(c + \frac{1}{T}\right), \quad \bar{u} \geq c + \frac{1}{T}.$$

Thus the objective of the optimal control problem simplifies to

$$\inf_{T>0} \int_{-T}^{0} I\left(c + \frac{1}{T}\right) ds = \inf_{T>0} TI\left(c + \frac{1}{T}\right) = \inf_{x>0} \frac{I(c+x)}{x}.$$

Now, suppose that there is a quality-of-service (QoS) requirement which stipulates that the probability of loss should not exceed some L. Then, the admission control problem is to decide whether or not to admit the source, depending on whether the QoS requirement can be met. Defining $\delta := -\frac{1}{n}\log L$, where we recall that n is the buffer size, we have $L = e^{-\delta n}$. Thus we would admit the source if $\inf_{x>0} \frac{I(c+x)}{x} > \delta$. A necessary and sufficient condition to meet the QoS requirement is

$$\frac{\Lambda(\delta)}{\delta} < c. \tag{19.6}$$

The necessity and sufficiency of the above condition can be easily seen by exploiting the relationship between $I(\cdot)$ and $\Lambda(\cdot)$:

$$I(c+x) > \delta x \Rightarrow \delta(c+x) - I(c+x) < \delta c \Rightarrow \Lambda(\delta) < \delta c,$$

and

$$c\delta - \Lambda(\delta) > 0 \Rightarrow (c+x)\delta - \Lambda(\delta) > \delta x \Rightarrow I(c+x) > \delta x, \forall x.$$

Since the minimum bandwidth needed to meet the QoS requirement of the source is $\frac{\Lambda(\delta)}{\delta}$, this quantity is known as the *effective bandwidth* of the source. Even though we have derived the effective bandwidth for discrete-time sources only, the result is valid for continuous-time sources too.

The effective bandwidth has an appealing *risk-sensitive* interpretation. Recall that

$$\frac{\Lambda(\delta)}{\delta} = \lim_{N \to \infty} \frac{1}{\delta} \ln E\left(e^{\delta \sum_{i=1}^{N} A_i}\right).$$

Note that the righthand-side of this equation is similar to the risk-sensitive exponential cost functions, which, in turn, are closely related to the cost functions in H_∞ optimal control problems [37]. Suppose that $A_i \leq P$, almost surely, and let $\bar{A} = E(A_i)$. Thus P is the peak rate, and $\bar{A}$ is the mean rate of the source. Then, it is easy to show that

$$\lim_{\delta \to 0} \frac{\Lambda(\delta)}{\delta} = \bar{A}, \text{ and } \lim_{\delta \to \infty} \frac{\Lambda(\delta)}{\delta} = P.$$

Thus the effective bandwidth is a function of the risk of *buffer overflow* in the network. If the source is willing to accept the risk of frequent occurrence of buffer overflows (which would correspond to $\delta \to 0$), then the capacity required to meet the QoS demand is close to the mean rate $\bar{A}$. On the other hand, if the source wishes to minimize the risk of buffer overflows ($\delta \to \infty$), then the required capacity is close to the peak rate.

19.3.4 Examples

The effective bandwidth formula given in the previous section can be easily computed for many realistic arrival processes. Here, we give one such example for continuous-time ON-OFF models, which are widely used to model traffic sources in communication networks. Consider a source that produces data at μ bps when ON and produces no data when OFF. Let the ON and OFF periods be exponentially distributed with means $\frac{1}{q_d}$ and $\frac{1}{q_u}$ respectively.

We derive the effective bandwidth for this source along the lines of a similar derivation for a discrete-time source in [9]. Define

$$M_t^u(\theta) := E(e^{\theta A(0,t)} | x(0) = ON) \text{ and } M_t^d(\theta) := E(e^{\theta A(0,t)} | x(0) = OFF).$$

Noting that for small δ, $P(x(\delta) = ON | x(0) = OFF) \approx q_u \delta$ and $P(x(\delta) = OFF | x(0) = ON) \approx q_d \delta$, it is easy to see that

$$M_t^d(\theta) \approx E(e^{\theta A(\delta,t)} | x(\delta) = OFF)(1 - q_u \delta) + E(e^{\theta A(\delta,t)} | x(0) = ON) q_u \delta.$$

Note that

$$E(e^{\theta A(\delta,t)} | x(\delta) = OFF) = E(e^{\theta A(0,t-\delta)} | x(0) = OFF) = M_{t-\delta}^d.$$

Using this and letting $\delta \to 0$, we get

Section 19.3 QoS, Admission Control, and Calculus of Variations

$$\frac{dM_t^d}{dt} = -q_u M_t^d(\theta) + q_u M_t^u(\theta). \tag{19.7}$$

Similarly,

$$M_t^u(\theta) = M_{t-\delta}^u(1 - q_d\delta)e^{\theta\mu\delta} + M_{t-\delta}^u q_d\delta e^{\theta\mu\delta}.$$

The term $e^{\theta\mu\delta}$ arises because $A(0, \delta) = \mu\delta$ given that $x(0) = ON$. Now, using $e^{\theta\mu\delta} \approx 1 + \theta\mu\delta$, and letting $\delta \to 0$, we get

$$\frac{dM_t^u}{dt} = (-q_d + \theta\mu)M_t^u(\theta) + q_d M_t^d(\theta). \tag{19.8}$$

Define

$$M_t(\theta) := E(e^{\theta A(0,t)}).$$

Then, $M_t(\theta) = M_t^u(\theta)\pi_u + M_t^d(\theta)\pi_d$, where π_d and π_u are the steady-state probabilities of being in the OFF and ON states, respectively. From (19.7)–(19.8), it is clear that $M_t(\theta)$ is of the form $M_t = a_1 e^{\theta_1 t} + a_2 e^{\theta_2 t}$ for some θ_1 and θ_2. Thus

$$\lambda_A(\theta) = \lim_{t\to\infty} \log M_t(\theta) = \theta_1,$$

where θ_1 denotes the largest eigenvalue of the matrix

$$\begin{pmatrix} -q_u & q_u \\ q_d & -q_d + \theta\mu \end{pmatrix}.$$

It is easy to see that θ_1 is given by

$$\theta_1 = \frac{1}{2}\left[\theta\mu - q_u - q_d + \sqrt{(\theta\mu - q_u - q_d)^2 + 4\theta\mu q_u}\right].$$

To get an idea of the amount of statistical multiplexing gain that is possible due to large buffers, we consider an example with $\mu = 1$ and the mean data rate of the source, $\frac{q_u}{q_u+q_d}\mu$, equal to 1. Thus, effective bandwidth would be some number between 0.5 and 1.0 depending on the value of δ. In Table 19.1, we present the values of the effective bandwidth for various values of q_u and the QoS requirement δ, while keeping the mean rate equal to 0.5. The table also gives the percentage reduction in the capacity required by using the effective bandwidth as opposed to peak-rate provisioning. It is clear from the table that allowing small probabilities of loss would allow for dramatically improved network utilization when compared to the conservative design for zero cell loss.

In this section, we have only dealt with the statistical multiplexing gain associated with large buffers. As mentioned in Section 19.2, significant statistical multiplexing gain is also achieved when multiplexing a lot of bursty sources even with a zero buffer. Large deviations asymptotics obtained by simultaneously scaling both the buffer and bandwidth are provided in [6].

TABLE 19.1 Effective Bandwidth for ON-OFF Sources with Rate 1

q_u	δ	Eff. bandwidth
0.01	0.01	0.6180
0.01	0.02	0.7071
0.01	0.1	0.9099
0.1	0.01	0.5125
0.1	0.02	0.5249
0.1	0.1	0.6180
0.2	0.01	0.5062
0.2	0.02	0.5125
0.2	0.1	0.5616

The results presented here are only valid for a single node. Extensions to certain special types of networks can be found in [10]. Obtaining results for general networks is still an open issue. In [13], a different approach is explored. There, the bandwidth needed to decouple the network into a collection of noninteracting nodes has been studied. A different direction of research is to extend the results to multiple classes of calls in which each node discriminates between the different call types by using some scheduling policy. Large-deviation results using optimal control techniques similar to those described in this chapter have been studied for the GPS policy and for the Generalized Longest Queue First (GLQF) policy (see references in [32]). Although most of these papers deal with the issue of computing rare buffer overflow, the similar problem for delay has been addressed in [32]. When using the effective bandwidth-type results as above, one should also keep in mind that these are approximations. Although the approximations mostly work well in practice, in some situations one could get bad estimates. Examples of such cases are provided in [36] and in the references within.

19.4 CONGESTION CONTROL

To develop a mathematical model of congestion control, we will adopt the point of view that there is a single bottleneck node that plays a dominant role in determining the performance of a given set of sources. In this case, the simplest feedback control mechanism is called *rate matching*. In rate matching, the node measures the average rate available to ABR sources at periodic intervals and simply divides a fraction of this capacity equally among the various users. The main advantage of this scheme is its simplicity, but it is difficult to optimally control queue length to avoid buffer overflows. However, this scheme is stable; that is, the queue length remains bounded in an appropriate stochastic sense. Queue length information is not used in this basic algorithm.

Alternatively, this problem can also be viewed as a feedback control problem in which queue length is used as the explicit feedback. One can study this problem using classical control techniques or a state space approach. As in rate matching, the primary goal is not optimality but simply queue-length stability. In these approaches, the available bandwidth to ABR sources is treated as an unmodeled disturbance. Thus these algorithms ensure stability in the presence of this disturbance. A comprehensive list of references on congestion control is provided in [2].

Here, we review the results in [2] in which both available rate and queue length are used to compute the data rates for the ABR sources. For a control-theoretic formulation, we model the available bandwidth as an autoregressive (AR) process driven by a white noise process. Modeling the available bandwidth allows this algorithm to achieve better performance than other existing algorithms.

There is a fundamental difficulty in obtaining good congestion control performance, namely, *action delay*, which is defined in terms of two components. The first component is the *downstream delay*, that is, the delay between the time that the bottleneck node issues its command to the time it takes for a source to receive this command. The second component is the *upstream delay*, that is, the time it takes for the data packets generated by the source to reach the bottleneck node. The sum of these delays is the action delay. It is well known in the control literature that the presence of delays in the feedback path poses difficulties. In our problem, this problem is further magnified because the action delays are different for different sources. Although the simple rate matching algorithms do not account for delay, the control-theoretic approaches account for feedback delay in their solutions. In the solution presented here, delay is explicitly taken into account.

The congestion control problem is formulated as an LQG stochastic control problem, in which the control actions of all users are actually determined centrally by the node, which, however, has to take into account the fact that these different actions will affect the queue dynamics at different times due to upstream and downstream delays. Even though this is not directly related to the analysis of this chapter, one can actually show that [2] the centralized control problem with action delays is equivalent to a decentralized team problem with information delays, where now the decisions are made by the users. In the parlance of team theory, this fits into the class of LQG teams with *nested information*, and hence the congestion control problem is not as intractable as problems with nonclassical information. The problem posed here does in fact admit an optimal solution, which is characterized in terms of the solution of a discrete-time algebraic riccati equation (DARE) whose dimension is determined by the magnitude of the largest delay and the order of the AR process describing the available capacity [22]. However, there are other solutions, easier to implement (they involve the solution of a scalar DARE), which share a common, appealing feature of *certainty equivalence* in addition to being stabilizing. These simpler solutions are presented here.

19.4.1 Model

We consider here a discrete-time model, in which a time unit corresponds to the length of the minimum measurement interval. Let q_n denote the queue length at a bottleneck link, and μ_n denote the effective service rate available for ABR traffic in that link at the beginning of the nth time slot. Let r_{mn} denote the effective source rate for source m ($m = 1, \ldots, M$) at the input of the bottleneck link during the nth time slot, which is actually the outcome of an action taken by source m several time steps earlier, based on a command signal sent by the node even earlier. We denote the total time it takes for the decision by the node regarding the transmission rate of source m to reach that source and subsequently for the effect of this decision to reach the bottleneck node (i.e., the sum of *downstream* and *upstream* delays—using the terminology introduced earlier) by d_m,[2] and the command decision of the node for source m at time n by v_{mn}, which we will sometimes also write as $v_{m,n}$. Hence we have the relationship:

$$v_{m,n-d_m} = r_{mn}. \tag{19.9}$$

Now, in terms of the notation introduced, the queue length evolves according to

$$q_{n+1} = q_n + \sum_{m=1}^{M} r_{mn} - \mu_n \equiv q_n + \sum_{m=1}^{M} v_{m,n-d_m} - \mu_n. \tag{19.10}$$

This equation corresponds to a linearized version of the actual queue dynamics, since we have ignored the fact that the queue length cannot be negative. Simulations show that this linearization is in fact valid when the controllers are successful in maintaining the queue size around a positive target value Q, sufficiently away from *zero*. The service rate μ_n available to the sources may change over time in an unpredictable way since this is the capacity left over from high-priority traffic. We model this available capacity by a p-dimensional stable AR process:

$$\mu_n = \mu + \xi_n \tag{19.11}$$

$$\xi_n = \sum_{i=1}^{p} \alpha_i \xi_{n-i} + \phi_{n-1}, \tag{19.12}$$

where μ is the known constant nominal service rate, α_i, $i = 1, \ldots, p$, are known parameters, and $\{\phi_n\}_{n \geq 1}$ is a zero-mean i.i.d. sequence with finite variance.

The objective function, to be minimized by the node, involves the transmission rates of all the sources that use the isolated bottleneck node, as well as the length of the queue at that node, and is given by

$$J = \limsup_{N \to \infty} \frac{1}{N} E \left\{ \sum_{n=1}^{N} [(q_n - Q)^2 + \sum_{m=1}^{M} \frac{1}{c_m^2} (r_{mn} - a_m \mu_n)^2] \right\} \tag{19.13}$$

where Q is the target queue length, c_m's are some positive constants, and $\sum_{m=1}^{M} a_m = 1$.

Loosely speaking, the network attempts to operate on the following principle: If the sources obey the control commands issued by the network, then the network attempts to transfer the packets without any loss. Of course, this can be achieved by making the data rates equal to zero for all the sources, which is clearly not desirable. Thus another goal of the network is to maximize utilization; that is, the sum of the data rates of all the sources should be nearly equal to the total capacity available for best-effort sources. Keeping this in mind, the cost function can be interpreted as follows. The *first additive term* above represents a penalty for deviating from a desirable queue length. The *second additive term* is a measure of the quality with which the input rate for each source tracks a given fraction of the available service rate, where the c_m's are weighting terms that serve to prioritize the relative importance of these individual terms

---

[2] Without any loss of generality, we take the d_m's to be orderred in accordance with their indices, that is, $d_1 \leq d_2 \leq \ldots, \leq d_M$.

Section 19.4 Congestion Control

(among different sources as well as collectively with respect to the first additive term). For example, if we desire "fair" sharing of the available bandwidth, we would choose

$$a_1 = a_2 = \cdots = a_M = \frac{1}{M},$$

assuming that everything else is also symmetric for the sources.

The information available to the node at time n is I_n, where

$$I_n = \{\xi_n, \xi_{n-1}, \ldots; q_n, q_{n-1}, \ldots; v_{jk}, j = 1, \ldots, M, k \leq n\}$$

Hence,

$$v_{mn} = \gamma_{mn}(I_n), \quad n = 1, 2, \ldots, ; \quad m = 1, 2, \ldots, M,$$

where γ_{mn} is some measurable function, with respect to which J will be minimized.

From [2], by introducing the new (appropriately shifted) variables

$$x_n := q_n - Q \tag{19.14}$$

$$u_{mn} := v_{mn} - a_m \mu, \tag{19.15}$$

the optimal solution when all users have perfect-state information is given by

$$u_{mn} = -p_m x_n + a_m \xi_n, \quad m = 1 \ldots, M, \tag{19.16}$$

where

$$p_m = c_m^2 - \frac{s c_m^4}{1 + s \sum_{\ell=1}^{M} c_\ell^2} - \frac{s c_m^2 \sum_{\ell=1, \ell \neq m}^{M} c_\ell^2}{1 + s \sum_{\ell=1}^{M} c_\ell^2}$$

$$= c_m^2 \bigg/ \left(1 + s \sum_{\ell=1}^{M} c_\ell^2\right). \tag{19.17}$$

This is therefore the optimum transmission rate for source m if there were no delay (upstream or downstream) in the network. In the presence of delay, however, these rates could lead to an unstable queue system. Hence there is a need to take into account the upstream and downstream delays on various links.

Here we present a specific form of a certainty-equivalent controller, which we call STARC (Stochastic Team Algorithm for Rate Control), where we simply replace x_n and ξ_n by their best estimates in the expression for u_{mn}:

$$u_{m,n}^* = -p_m \hat{x}_{n+d_m|n} + a_m \hat{\xi}_{n+d_m|n}, \quad m = 1, \ldots, M. \tag{19.18}$$

Here p_m is as defined by (19.17), and $\hat{x}_{n+d_m|n}, \hat{\xi}_{n+d_m|n}$ are the predicted values of x_{n+d_m} and ξ_{n+d_m}, respectively, based on the information I_n, and given that all other controllers are also in the form (19.18). These predictors are generated by

$$\hat{x}_{n+j|n} = \hat{x}_{n+j-1|n} + \sum_{i=1}^{M} \hat{u}_{m,n-d_m+j-1|n} - \hat{\xi}_{n+j-1|n}, \quad j \geq 1; \qquad (19.19)$$

$$\hat{x}_{n|n} = x_n$$

$$\hat{\xi}_{n+j|n} = \sum_{i=1}^{p} \alpha_i \hat{\xi}_{n+j-i|n}, \quad j \geq 1; \quad \hat{\xi}_{n-k|n} = \xi_{n-k}, \quad k \geq 0, \qquad (19.20)$$

and

$$\hat{u}_{m,n-d_m+j-1|n} := \begin{cases} -p_m \hat{x}_{n+j-1|n} + a_m \hat{\xi}_{n+j-1|n} & \text{if } j \geq d_m + 1 \\ u_{m,n-d_m+j-1} & \text{if } j < d_m + 1. \end{cases} \qquad (19.21)$$

These are the recursive equations generating the predictors for the queue length and rate information at a future time, where the future time is the current time plus the action delay for the corresponding source. For example, $\hat{\xi}_{n+j|n}$ denotes the predicted value at time n of the value of ξ at some future time $n+j$, based on the information available at time n, which is I_n. A similar interpretation holds for $\hat{x}_{n+j|n}$.

The above algorithm is relatively easy to implement. The estimator algorithms are simple scalar operations and the scalar solution of the Riccati equation has already been obtained explicitly.

19.4.2 Implementation Issues

Measurement Interval: In deriving the LQG-based congestion controller, we assumed a discrete-time control problem. In reality, queue size changes at every arrival or departure event. To convert it into a discrete-time control problem, we use the *measurement interval* over which the available rate for ABR sources is measured at the node as our basic time unit. The queue length is, of course, variable within this time unit. However, for mathematical modeling purposes, we consider the queue length at any time unit to be the queue length at the end of the measurement interval. Extensive simulations confirm that this modeling error is not significant.

The length of the measurement interval also has another impact on the mathematical model. Ignoring queueing delay, a traffic source that is transmitting between an origin and destination of 1000 km has a round-trip delay of roughly 19 ms, assuming that the transmission takes place over an optical fiber with refractive index 0.7. Suppose that we have a bottleneck node with capacity 1 Gbps and that the length of our measurement interval is the time taken to process 5000 cells. Thus, the measurement interval is 2.12 ms. Thus, the round-trip delay is $\frac{19}{2.12} = 8.96$ time units. Since our mathematical model requires the delay to be an integer, this would be rounded to 9 time units. Again, despite this modeling inaccuracy, we have shown through simulations that our controller continues to perform well.

Rate Management (RM) Cells: In our model, we assumed that feedback is available once during every measurement interval. However, the actual feedback mechanism in ATM ABR service is more complicated. Each source generates a cell called an RM cell for purposes of collecting feedback information from the nodes in its route. ATM Forum, the standards organizations that defines protocol standards for ATM networks, has defined a parameter called *NRm*, which is used to decide how often RM

Section 19.4 Congestion Control   481

cells are generated by each source. One RM cell is generated after every ($NRm - 1$) data cells. Typically, NRm is taken to be 32, and thus, one RM cell is generated for every 32 cells. This is done for reasons of *scalability*; that is, as the network size becomes large and the number of sources is correspondingly large, this scheme for RM cell generation will ensure that a maximum of $\frac{1}{32}$ of the network bandwidth is used for congestion feedback. Thus it limits the percentage of overhead associated with ATM ABR service. This would not be true if, for example, one uses some other scheme such as generating an RM cell every x ms for some x. There are some exceptions to the rule described above for generating RM cells. If a traffic source is sending information at a rate that is too slow for a reasonable number of RM cells to be generated, then it is allowed to generate cells more frequently than one every 32 cells.

The RM cell traverses the virtual circuit setup for a traffic source, and once it reaches the destination, it is turned back toward the source. It is during the reverse path that it actually collects the feedback information. The reason stated for this is that, when the RM cell returns to the traffic source, it would have the newer information than if the information was collected in the forward path. In the reverse path, the RM cell has a field called the ER (explicit rate) field, which is altered by each node through which the RM cell passes. If a node decides that the traffic source has to transmit at rate r, then it compares r with the current value of ER. If $r < ER$, then ER is set to r; otherwise, the ER field is not changed. Thus, when the RM cell returns to the traffic source, the smallest of the rates computed by the nodes in the path is reported to the source and the source uses this rate until it receives the next RM cell back.

The above mechanism for feedback implies that feedback may not be available at every time instant. In fact, whether or not feedback is available every time instant depends on the rate at which each traffic source can generate data and the length of the measurement interval. If the measurement interval is long enough so that each traffic source generates at least one RM cell during each such period, then this mechanism for collecting feedback information does not violate our modeling assumptions. Indeed, simulations indicate that this is true.

Variable delay: We have so far assumed that the action delay for each source is fixed. In reality, as mentioned earlier, delay has two components: propagation delay and queueing delay. While propagation delay is fixed, queueing delay is variable. However, in practice, queueing delay is small compared to propagation delay, so that the small variability in delay does not seem to affect the performance of the congestion controllers.

Max-min and Proportional Fairness: In our controller design for a single node, we remarked that a sensible choice for the a_i's in the case of a single bottleneck node would be $a_i = 1/M$, where M is the number of sources currently using the node. By such a choice of the a_i's, we aim to divide the available bandwidth equally between the competing sources. However, this may not be the best choice when there are many nodes.

Consider the network depicted in Figure 19.2. There are three nodes and two links both with capacity 90 units. Sources 1 and 2 share Link 1, and Sources 2, 3, and 4 share Link 2. On Link 2, the bandwidth is divided equally between the three sources and each one gets 30 units. On Link 1, with our current choice of a_i's, Sources 1 and 2 would attempt to get 45 each. However, Source 2 can use only 30 units becuase it is bottlenecked at Link 2. Source 1 can use its full share of 45 units, and thus, the total used bandwidth on Link 1 is only 75 units and 15 units of bandwidth are wasted.

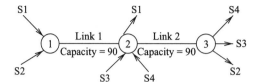

Figure 19.2 Under a max-min fair allocation, Sources 2, 3, and 4 get 30 units of bandwidth, while Source 1 gets 60 units of bandwidth.

Clearly, in this situation, it seems better to allocate the remaining 15 units on Link 1 to Source 1. Thus, on Link 1, a_1 should be $\frac{2}{3}$, and a_2 should be $\frac{1}{3}$, whereas on Link 2, a_2, a_3 and a_4 should be $\frac{1}{3}$. Such an allocation of bandwidths is called *max-min* fair [5]. Under a max-min fair allocation, the bandwidth allocated to a source can be increased only if the bandwidth allocated to the user using the smallest amount of bandwidth is reduced. Thus we protect "small" users with such an allocation. Alternatively, one could use other bandwidth allocation schemes such as proportional fairness [24] to distribute the bandwidth in a multi-node network. From our point of view, however, this simply means that the a_i's have to be adapted to be compatible with the notion of fairness that we want to use.

Peak rate constraints: When we designed the congestion controllers, we assumed that the sources could transmit at the rate suggested by the nodes. However, for various reasons, the maximum rate at which a source can transmit may be limited. For example, the source may access the ATM network through an access link, in which case the maximum rate at which the network receives data from this source is limited by the access speed. Another reason for a peak rate could be simply that the application can only transmit at a certain maximum rate. Thus, when the bottleneck node's rate command is larger than the source's peak rate constraint, then the source will simply transmit at the peak rate. Clearly, this would lead to underutilization of the bottleneck node.

If the sum of the peak rates of all the sources is less than the available capacity at the bottleneck node, there is nothing that one can do to fully utilize the node. Assuming this is not true, if one of the sources sends at a rate lower than its allocated rate, then it makes sense to redistribute the bandwidth unused by this source to other sources. For example, suppose that there are two sources, Source 1 and Source 2, and that we start with $a_1 = a_2 = \frac{1}{2}$. If, on the average, Source 1 uses only one-third of the link bandwidth, then the node should *adapt* the a_i's to be $a_1 = \frac{1}{3}$ and $a_2 = \frac{2}{3}$.

Bursty Sources: In addition to sources being peak-rate constrained, they could also be *bursty*. In other words, sometimes they may have no data to send and, at other times, they may have data to send. Thus, the source would ignore rate commands during inactive periods. If the source's active and inactive periods are on a very fast time scale compared to the measurement interval, the node cannot adapt to the burstiness of the source. However, if the source's active and inactive periods are relatively long compared to the measurement interval, then, by measuring the data transmission from each source, each node can adapt the a_i's to redistribute the bandwidth unused by the inactive sources among those that can use this bandwidth. Note that, in all three cases—to achieve max-min or proportional fairness, or to account for peak rate constraints or bursty sources—the a_i's can be adapted to redistribute the bandwidth allocation between the various sources. This requires that the node measure the utilization due to each source to perform the adaptation. A proof of the convergence of such an adaptive algorithm is still an open issue.

Measuring delay: As mentioned earlier, propagation delay is fixed and is also the main component of the action delay experienced by each source. Thus this can be measured during virtual-circuit setup of the ATM ABR source and will remain constant during the life of the virtual circuit. Indeed, the ATM routing protocol allows one to measure this delay for each link during virtual-circuit setup, and thus the algorithm presented here uses this information effectively to improve performance.

AR process: In addition to measurements needed to adapt the a_i's, measurements are also used to estimate the parameters of the AR process. From our experience, the Yule–Walker algorithm (see, for instance, [7]) works well to perform this estimation. But other techniques may work as well. Again, the joint estimation and control problem is an adaptive control problem that could perhaps benefit from a theoretical analysis.

19.4.3 Simulations

Consider the communication network depicted in Figure 19.3 with high-priority VBR sources only. Let us suppose that the link speed is the speed of light, the service rate offered by every link is 1 Gbit/sec, and the distance between adjacent nodes is constant and equal to 1000 Km. Let us take the time unit to be the time required to serve 5,000 cells ($5000 \times 53 \times 8$ bits). Then the propagation delay for a cell, that is, the time required for a cell to traverse a link, is

$$t = \frac{1 \times 10^6}{3 \times 10^8} \times \frac{10^9}{53 \times 8} \times \frac{1}{5,000} \approx 1.6 \text{ time units}.$$

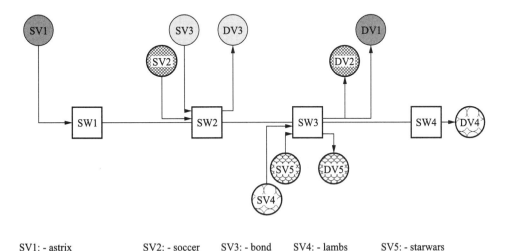

SV1: - astrix   SV2: - soccer   SV3: - bond   SV4: - lambs   SV5: - starwars
      - terminator   - simpsons   - ATP
      - Two On/Off Sources   - Two On/Off Sources

Figure 19.3 Four node configuration: VBR sources.

In the figure, *SVi* represents a set of VBR sources as detailed under the figure, and *DVi* is the destination of *SVi*. Moreover, the VBR sources considered are of two types: video traces obtained from various publicly available Web sites and simulated ON-OFF sources. The latter are bursty sources simulated by us which alternate between ON and OFF states according to a Markov chain, and when in the ON state, cells arrive at a constant rate.

In addition to the VBR sources, we have four ABR sources, as shown in Figure 19.4, which are subject to rate control using STARC. Three of the ABR sources are bottlenecked at the third node. The fourth ABR source will then use the remaining capacity at the second node (according to the max-min criterion [5]), which then becomes bottlenecked. The network eventually ends up with two bottleneck nodes. Table 19.2 specifies the action delays used in the mathematical model.

For the simulation, we used the following parameter values:

- Time unit: Time required to serve 5000 cells.
- Target queue length: $Q = 700$.
- Weights: $c_m = 1 \quad \forall m$.
- The nominal service rate μ and the AR process parameters α_i are estimated on-line using the Yule–Walker algorithm [7], assuming that the order of the AR process is 8.
- The propagation delay from one node to the next is 1.6. Note that the actual delay is variable since the cells go through node buffers, which leads to additional queueing delays.

Following the current ATM standards, the feedback mechanism has been implemented using RM cells that are generated by the sources every 32 data cells.

The VBR sources used in this example are such that the mean available ABR capacities are around 4870, 4720, 4480, and 4860 cells per time unit at Nodes 1, 2, 3, and 4, respectively. The main bottleneck node is the third one where the capacity should be equally distributed between the sources ABR1, ABR2, and ABR4. So those three sources should transmit at around 1500 cells per time unit. Then, since ABR1 does not use its fair share at Node 2 (which is around 2350), ABR2 should use the remaining

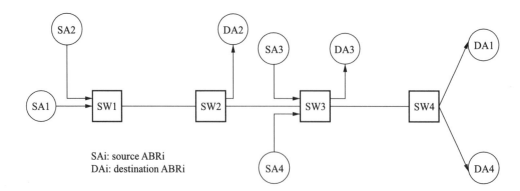

Figure 19.4 Max-min fairness configuration: ABR sources.

Section 19.4 Congestion Control

TABLE 19.2 Action Delays for the Max-Min Fairness Model

	SW1	SW2	SW3	SW4
ABR1	0	3	6	10
ABR2	0	3		
ABR3			0	
ABR4			0	3

capacity and increase its cell rate to 3200. Nevertheless, our basic algorithm sets the weights to the exact fair share ($1/M$ for M sources), and the drop observed in the queue length at Node 2 is not enough to increase the rate of ABR2. Therefore, max-min fairness is not achieved. Thus, as mentioned earlier, the weight a_i's have to be adapted to achieve max-min fairness.

The simulations done with the adaptive weight algorithm indicate that max-min fairness is indeed achieved (Figures 19.5 and 19.6). Initially (with all a_m's set to the fair share), the bottleneck node is the third one, and sources ABR1, ABR3, and ABR4 share fairly the capacity at this node (around 1500 cells per time unit). Meanwhile, source ABR2 only uses half of the capacity of Node 2 with a mean rate of 2400 cells per time unit. Actually, its cell rate is slightly higher than the fair share, the latter being around 2360. Because of the small queue length, the design appears to increase the allowed cell rate slightly. Once the weights a_m's at each node are adapted, source ABR2

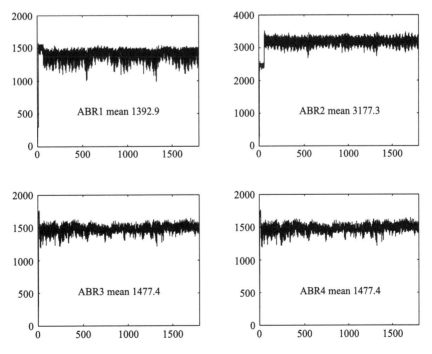

Figure 19.5 Max-min fairness model: source rates.

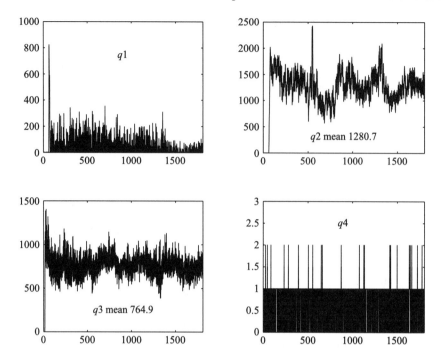

Figure 19.6 Max-min fairness model: queue lengths.

should increase its rate to 3220 cells per time unit while the other sources still transmit at 1500 cells per time unit. The simulations show that source ABR2 effectively catches up with a mean cell rate of 3180 cells per time unit (Figure 19.5). Note that source ABR1 sees its cell rate decrease to 1390 cells per time unit, while sources ABR3 and ABR4 transmit at 1480 cells per time unit (Figure 19.5). Thus there is some deviation from exact fair share between these three sources. We believe that the reason for this is that ABR1 takes the smallest value between the rates computed by nodes 2 and 3, and the mean value of the minimum between two variables with the same mean does not necessarily achieve the mean of those two variables. Figure 19.6 shows that the queue length for the third node is still regulated with a mean value of 765 cells. On the other hand, with a mean value of 1280 cells, the queue length for the second node is not regulated around 700 cells. This is due to the nonlinearities introduced in the system when there is more than one bottleneck node. However, it is important to note that the queue length for the second node remains stable.

19.5 CONCLUSIONS

In this chapter, we have presented a survey of the important control issues encountered in operating a high-speed communication network. We have provided both an introduction to the high-level control issues and examples of mathematical models that are amenable to control-theoretic analysis. Since the subject is vast, when providing mathematical details, we have focused our attention on two of the problems:

admission control and congestion control. For the other problems, we have provided an overview of the results along with several references to recent research in these areas. Furthermore, throughout the chapter, we have identified several open problems that may be of interest to control specialists.

We have focused primarily on wide-area high-speed networks in this survey. For applications of control methods to multiple access communications, see [19]. There is also considerable work on control of queueing networks motivated by applications in manufacturing which could prove useful in understanding the behavior of communication networks as well. For instance, see the work of Kumar, Meyn, and their co-workers on the stability and performance of general queueing networks [27, 11].

Finally, we end the chapter with a note of caution and a note of optimism. Control problems in communication networks are complex and challenging and are becoming extremely important because of the explosive growth of the networking industry. The note of caution is that, while there are exceptions, most problems in this area do not directly fit into the mold of traditional control theory. Indeed, the successful network engineer requires a knowledge of various disciplines, including control theory, operations research, computer science, and information theory. On the other hand, we believe that the general concepts underlying traditional feedback control and optimal state space control can play a central role in building the networks of the future. As in the case of information theory and communication networks [18], however, the realization of this promise awaits a union between control theory and communication networks which is still unconsummated. This could be a source for optimism since, in the future, control theorists may be able to play an important role in facilitating this union.

ACKNOWLEDGMENTS

It is a pleasure to thank Eitan Altman, Tamer Başar and Sonia Compans for earlier collaboration that resulted in the work presented in Section 19.4, Bruce Hajek for discussions on the material in Section 19.3, and P. R. Kumar for bringing [37] to our attention.

Part of the work presented here is a result of research supported by an NSF CAREER Award NCR 9701525, a grant from the Nokia Research Center and a grant from Rome Labs (Air Force Contract F30602-96C0156) through Scientific Systems Company, Inc.

Related Chapters

- Communication networks are an example of discrete event systems, a general introduction to which can be found in Ch. 2.
- See Ch. 16 for a review of communication networks and protocols as used in and proposed for building control system.
- A brief discussion of communication networks for process control systems is included in Ch. 12.

REFERENCES

[1] J. Abate, G. L. Choudhury, D. M. Lucantoni, and W. Whitt, "Asymptotic analysis of tail probabilities based on the computation of moments." *Annals of Applied Probability*, Vol. 5, pp. 983–1007, 1995.

[2] E. Altman, T. Başar, and R. Srikant, "Congestion control as a stochastic control problem with action delays." *Automatica*, December 1999. Special Issue on Control Methods for communication networks, V. Anantharam and J. Walrand, editors.

[3] E. Altman, T. Jimenez, and G. Koole, "On optimal call admission control." *Proceedings of the IEEE Conference on Decision and Control*, Tampa, FL, 1998.

[4] D. Anick, D. Mitra, and M. Sondhi, "Stochastic theory of a data-handling system." *Bell System Technical Journal*, Vol. 61, pp. 1871–1894, 1982.

[5] D. Bertsekas and R. Gallager, *Data Networks*. Englewood Cliffs, NJ: Prentice Hall, 1987.

[6] D. D. Botvich and N. G. Duffield, "Large deviations, economies of scale, and the shape of the loss curve in large multiplexers." *Queueing Systems*, Vol. 20, pp. 293–320, 1995.

[7] P. J. Brockwell and R. A. Davis, *Time Series: Theory and Methods*. 2nd ed. New York: Springer, 1991.

[8] J. A. Bucklew, *Large Deviation Techniques in Decision, Simulation and Estimation*, New York: John Wiley, 1990.

[9] C-S. Chang, "Stability, queue length and delay of deterministic and stochastic queuing networks. *IEEE Transactions on Automatic Control*, Vol. 39, pp. 913–931, 1994.

[10] C. S. Chang, "Sample path large deviations and intree networks." *Queueing Systems*, Vol. 20, pp. 7–36, 1995.

[11] R-R. Chen and S. P. Meyn, "Value iteration and optimization of multiclass queueing networks," 1998. Preprint.

[12] A. Dasylva and R. Srikant, "Bounds on the performance of admission control and routing schemes in general topology networks." *Proceedings of the IEEE INFOCOM*, New York, April 1999.

[13] G. de Veciana, C. Courcoubetis, and J. Walrand, "Decoupling bandwidths: A decompositon approach to resource management in networks." *Proceedings of the IEEE INFOCOM*, pp. 446–474, 1994.

[14] A. Dembo and T. Zajic, "Large deviations: From empirical mean and measure to partial sums process." *Stochastic Processes and Applications*, Vol. 67, pp. 195–211, 1995.

[15] A. Dembo and O. Zeitouni, *Large Deviations Techniques and Applications*, 2nd ed. New York: Springer, 1998.

[16] A. Demers, S. Keshav, and S. Shenker, "Analysis and simulation of a fair queueing algorithm.' *ACM SIGCOMM*, pp. 1–12, 1989.

[17] P. Dupuis and R. S. Ellis, *A Weak Convergence Approach to the Theory of Large Deviations*. New York: John Wiley, 1997.

[18] A. Ephremides and B. Hajek, "Information theory and communication networks: An unconsummated union." *IEEE Transactions on Information Theory*, Vol. 44, pp. 2384–2415, 1998.

[19] A. Ephremides and S. Verdú, "Control and optimization methods in communication network problems." *IEEE Transactions on Automatic Control*, Vol. 34, pp. 930–942, September 1989.

[20] R. J. Gibbens and F. P. Kelly, "'Network programming methods for loss networks," *IEEE Journal on Selected Areas in Communications*. Vol. , pp. 1995.

[21] A. G. Greenberg and R. Srikant, "Computational techniques for accurate performance evaluation in multirate, multihop communication networks." *IEEE/ACM Transactions on Networking*, Vol. 5, pp. 253–277, March 1997.

References

[22] O. C. Imer and T. Başar, "Optimum solution to a team problem with information delays: An application in flow control in communication networks," 1999. *Proc. IEEE Conf. Decision and Contr.*, Phoenix, AZ, December 1999.

[23] F. P. Kelly, "Notes on effective bandwidths." In F. P. Kelly, S. Zachary, and I. B. Ziedins (eds.), *Stochastic Networks: Theory and Applications*, pp. 141–168, 1996.

[24] F. P. Kelly, A. Maulloo, and D. Tan, "Rate control in communication networks: Shadow prices, proportional fairness and stability." *Journal of the Operational Research Society*, Vol. 49, pp. 237–252, 1998.

[25] F. P. Kelly, Loss networks. *The Annals of Applied Probability*, Vol. 1, no. 3, pp. 319–378, August 1991.

[26] S. Keshav, *An Engineering Approach to Computer Networks*. Reading, MA: Addison-Wesley, 1997.

[27] P. R. Kumar, "A tutorial on some new methods for performance evaluation of queueing networks." *IEEE Journal on Selected Areas in Communications*, Vol. 13, pp. 970–980, August 1995.

[28] R. La and V. Anantharam, "Adaptive modification of generalized processor sharing," 1998. Available at http://diva.eecs.berkeley.edn/vananth

[29] R. La and V. Anantharam, "Optimal routing control: Game-theoretic approach." *Proc. IEEE Conf. on Decision and Control*, December 1997, San Diego, CA.

[30] P. Marbach, O. Mihatsch, and J. N. Tsitsiklis, "Call admission control and routing in integrated service networks using reinforcement learning." *Proceedings of the 37th IEEE Conference on Decision and Control*, Tampa, FL, 1998.

[31] A. Parekh and R. Gallager, "A generalized processor sharing approach to flow control in integrated services networks: The single node case." *IEEE/ACM Transactions on Networking*, 1993.

[32] I. Paschalidis, "Performance analysis and admission control in multimedia communication networks." *Automatica*, December 1999. Special Issue on Control Methods for communication networks, V. Anantharam and J. Walrand, editors.

[33] S. Shenker, "Making greed work in networks: A game-theoretic analysis of switch service disciplines." *IEEE/ACM Transactions on Networking*, Vol. 3, pp. 819–831, 1995.

[34] R. Srikant and W. Whitt, "Simulation run lengths to estimate blocking probabilities in multi-server loss models." *ACM Transactions on Modelling and Computer Simulation*, pp. 7–52, January 1996.

[35] R. Srikant and W. Whitt, "Variance reduction in simulation of loss models." *Operations Research*, July-August 1999, pp. 509–523.

[36] V. Subramanian and R. Srikant, "Statistical multiplexing with priorities: Tail probabilities of queue lengths, workloads and waiting times." *Proceedings of the IEEE Conference on Decision and Control*, San Diego, CA, 1997.

[37] P. Whittle, *Optimal Control: Basics and Beyond*. New York: John Wiley & Sons, 1996.

INDEX

A

activation function   111
active control
  bridge towers   423–4
  building/towers   421–2
active mass damper (AMD) system   418–19
actuator nonlinearity   135
A/D (analog-to-digital) converter   2, 6, 14, 23
adaptive bounding   154
adaptive control   121, 126–30, 313, 435, 443, 451
  cruise control   358
  direct   127
  genetic   126–8
  indirect   126
  neural networks   126–8
adaptive law   161
admission control, real-time sources   464–5
advanced cruise control (ACC)   358, 366
aerospace systems, multivehicle   239–58
affine functions   143
agent-based systems   227, 254–5, 343
air charge temperature sensor (ACT)   125
air-handling units (AHUs)   394, 396, 405, 410–12
air traffic management (ATM)   240
  airline operating center (AOC)   248
  airspace   241–2
  constrained airspace   249
  control system design   245–6
  control variables   244
  distributed separation assurance procedures   247
  disturbances   242–6
  flight deck considerations   248
  future controls applications and challenges   241–9
  initial conditions and framework   243–4
  state variables   244
  terminal area operations   249
  traffic flow management (TFM) constraints   247–8
  uncontrolled variables   245
  user-preferred trajectories (UPTs)   246–9
aircraft
  control law design   270–3
  dynamics   260–8
  dynamics modeling   49
  flight control   259–90
  landing approach   55
  linear models   260–8
algorithmic tuning   62
AND gate   170
ANDECS   46
angular velocity sensor system   126
antilock brakes (ABS)   222, 357–8
approximation, on-line   152
approximation-based control   134–64
  components   135–8
  control architectures   140–2
  cost function   148
  neural networks   139–40
  problem statement   138–9
approximation errors   137, 153, 156–7, 162
approximation function
  linearly parameterized   154
  structure   152
approximation problem
  nonlinear system   152–3
  parameter estimation algorithms   152
  parametric models   152
approximator   136–7
  parameter (non)linearity   145–6
  properties   142–52
  structure   137

approximator (*continued*)
 transparency   151–2
 with local influence functions   149–51
arc firing   174
ARCnet   400, 403, 405
area control error (ACE)   333–4
arrival rate   33
artificial neural networks (ANNs)
 see neural networks
asymptotically stable state   199–200
asynchronous sequential circuits
  169–70
Asynchronous Transfer Mode (ATM)
  networks   463, 465, 482
AT-MIO-16F-5 A/D timing board   126
atomic equations   173
Automated Highway System (AHS)
  350–1, 354–7
automatic generation control (AGC)   328,
  333–4, 343
Automatic Synthesis Program (ASP)   43
automatic tuning   60–3
automation, control design   68
automation systems
 building   393–416
 hardware structure   404–6
 objectives   221
 software features   406–7
automaton   26
automotive engine failure estimation   125–6
automotive powertrain controller
  development   370–92
 organizational structure   372–3
 process requirements   374–6
 role of   371–2
 software requirements   374–6
 see also CACSD
autonomous systems   88, 121–2, 224
autonomous vehicles   229–31, 252–3,
  287–90
auto-regressive moving average with
  exogenous input (ARMAX)   83
auto-regressive with exogenous input (ARX)
  83
autotuning   310
available bit rate (ABR)   465, 476–7,
  480–1, 483–6

B

backlash   91
backpropagation algorithm   115, 156

backpropagation through time   156
backward differences   99
BACnet   402–3
balanced reduction   87–8
basis-influence functions   150
Batibus   402
benchmarking   305–6, 419
bifurcation   89
biological systems   104, 121, 226–7, 234–5,
  255
blocking time   16
body axis coordinate system   260
boundary layer   198
Box-Jenkins (BJ)   83
bridge towers, active control   423–4
BridgeVIEW   295
buffer overflow   474
building control and automation
  systems   393–416
 advanced technologies   407–13
 development   393–5
 existing technologies   395–9
bursty sources   482–3
business systems integration   229
busy periods   31

C

CACSD   42–70, 370–92
 computation chain   63
 configuration management   388
 control algorithm   379
 control design life cycle   45–7
 debugging   381
 functional verification   384–5
 further technology   66–8
 generic computation setup   54
 interoperable system   68
 rapid prototyping   381
 requirements   376–7
 software application
  architecture design   377–9
  functional verification   386
  structural verification   385–6
  validation   382
 software design   382
 software engineering project
  management   388–9
 software implementation   383
 software/module integration
  verification   386–7
 software requirements capture   377

Index

structural verification 383–4
toolboxes 61
unit testing 384
user documentation 387–8
validation 380
see also automotive powertrain controller development
calculus of variations 468
capacity allocation problem 38–9
car-following control 358–61
central differences 98
central nervous systems (CNS) 226
central processing unit (CPU) 23
 cycles 4
certainty equivalence principle 140
chaos 90–1, 234–5
chattering 191, 197
 avoidance 203
 hybrid system trajectories 168
 prevention by continuous approximation 197–8
 solutions 168
Chebyshev polynomials 77
Cholesky decomposition 76
chromosomes 116–18
circuit-switched routing 466
circuit verification 170
civil engineering infrastructures 417–41
 control problem 417–20
 hybrid control systems 420–30
 semiactive control systems 430–9
closed-loop inverse kinematics algorithm 444
closed-loop planning system 120
combinational circuit 170
communication networks 462–89
 congestion control 476–86
 control and management 464–8
 examples 474–6
 implementation issues 480–3
 overview 462–3
 simulations 483–6
comparative design exploration, visualization for 59–60
complex adaptive systems (CAS) 227
complex instruction set architectures (CISC) 4
complex systems, emerging control technologies 225–8
complexity management 218–38
 future 221
 "schools" 231–6

compound quality functions 57
computation tree logic (CTL) 178–81
computed torque control 449
computer-aided control engineering (CACE) environment 68
computer-aided control system design see CACSD
computer vision 412–13, 457
condition-based maintenance (CBM) 407
congestion control
 best-effort sources 465–6
 communication networks 476–86
constant bit rate (CBR) 465
continuous approximation 203
continuous-time adaptive algorithms 155
continuous-time Markov process 32
continuous-time systems 79
control action, positive 412
control architecture 135–6, 443, 451, 457
control design automation 68
control hierarchy in manufacturing systems 24
control law
 design life cycle 45
 parameterization 66
 structures 51, 53
control networks 400–2
control performance evaluation setup 53–4
control performance indices 299, 410
control synthesis algorithms 52
controllability grammian 87
controllable fluid dampers 432–4
controller gains 65
controller modeling 51–4
convection equation 97
coordination 240, 250–1
cost function, instantaneous error 155
cost index 303
covariance matrix 157
covariance resetting 158
C*-quality versus control-effort 65
critical sections 11, 174
crossover operation 118
cruise assist systems 358
cruise control 357
CSTAR 59
CTAS 244
curse of dimensionality 151, 226, 344

D

data acquisition system 125

data analysis, intelligent   231
data-based nonlinear estimation   124–5
data communications   296
data dependence   10
data mining   231
datagram routing   466
dc transmission   336
DCS/SCADA   319
deadline monotonic (DM) algorithm   9
deadlock   177
deadlock-free system   179
dead-zone   161, 365
    modification   162
decentralized distributed control   254–5
decentralized models   100
decision making   54–60
decision process   59
declarative compromising   63–5
declarative system dynamics model building   48
deductive verification   182
describing function approximation   93–4
design analysis   54–60
design efficiency   233–4
design modeling   46–53
desktop PCs   17
difference equation   23
differential   23
diffusion equation   97
digital circuit   170
digital communications technology   296
digital control system
    direct   399
    distributed   298
    general form   3
digital magnetic zero-speed sensor   126
digital signal processors (DSPs)   18
direct adaptive control   127
direct control   140, 142
direct control law parameterization   53
direct digital control (DDC)   297–8, 399, 401–2, 406
direct truncation   88
directed graphs   170
discrete-event signals   167
discrete-event systems   20–42, 80, 167, 172, 180, 184
    definition   21, 23–4
    models   25–33
    need for   24
    optimization   33–40
    overview   22–5
    simulations   32
    state model   26–9
    state trajectory   25–6
discrete-time algebraic riccati equation (DARE)   477
discrete-time models   79, 477–80
discretization visualization of flight envelope   53
distributed control system (DCS)   293–4, 319
distributed DDC (DDDC)   399
distributed digital control   298
distributed parameter system   79, 228
    models   95–9
distributed processing   255–6
documentation   387–8
domain knowledge   219–20
drive-by-wire vehicles   357–8
DTC-1 throttle controller   126
Duffing's equation   90
DYN-LOC IV speed/torque controller   126
dynamic backpropagation   156
dynamic inversion   270–3
dynamic matrix control (DMC) algorithm   319
dynamic programming   251, 353, 466
dynamic setpoint maneuvers   315–16
dynamic system   22–3
dynamical models   79–88

E

EASY5x   294
EEC-IV controller   125
effective bandwidth   466, 474
eigenstructure methods   52
electric power system *see* power system control and estimation
electronic control units (ECUs)   357
electronic throttle control   357
electrorheological (ER) fluids   433–4
ELEVRATE   59
energy efficiency   225
energy management systems   394
engine test cell   126
enterprisewide optimization   228–9
environmental safety   222–3, 232–3

Index

EPA IM240 cycle   126
ϵ-modification   162
equilibrium point   182
error state   175
estimation error, small-in-the-mean   162
estimation model   159
estimator construction methodology   124–5
Ethernet   400, 403
Euler's first-order approximation   114
evaluation cases   53
evaluation criteria   55
evaluation model   47
event clocks   30–1
event labels   172
event lifetime   31, 36
existence set   147
expert systems   119–20, 407–13
explicit models   98
extractive distillation process   303–4
extrapolation   148

F

failure detection and identification (FDI) algorithms   121
fault detection indices   299
fault tolerance/safety enhancement   368
feasibility assessment   59
feasible design   58–9
   alternatives   59
feasible events   30
feasible region   33
Federal Energy Regulatory Commission (FERC)   337
feedback control systems   2, 88
feedback linearization   95, 140, 264, 270
feedback loop   2
feedback process   46–7
feedforward multilayer perceptron   110
Fiber Distributed Data Interface (FDDI)   401
finite difference models of PDEs   97–9
finite impulse response (FIR)   83
finite state machines   27, 171
fire protection   406
fitness function   116
fixed point   180
fixed-priority scheduling theory   8–9
Flexible AC Transmission (FACTS)   342, 344
flexible-link robot, vibration damping   123–4

flight control systems   64, 259–90
   analysis tools   281–6
   design   49, 269–80
   development   286–8
   law   51
   simulation tools   268–9
   unmanned aircraft   289–90
   zero shaping   279–80
flight envelope, discretization visualization of   53
flocking   255
flying qualities   270–1
FMRLC   130–1
force control, robotic systems   451–7
forced system   88
formation flying   253–6
forward differences   98
Fourier series   93–4
Fourier series coefficients   94
Fourier series expansion   94
free-floating robotic system   169
free flight   229, 240
frequency-to-voltage converter   126
function approximation   113, 115, 150
function approximation error   139
fuzzy adaptive control, generalization applications   147
fuzzy control   51, 53, 105–10, 119, 419, 420, 435
   adaptive control   126–8
   design   106–10
   ship example   109
   ship steering   128–30
fuzzy estimation   125–1
fuzzy logic   57
fuzzy logic-based control   410–12
fuzzy model reference learning control (FMRLC)   128
fuzzy systems   124, 150
   approximation transparency   152

G

gain scheduling   445
game theory   467–8
GARTEUR design challenge   66
gas turbines   339–40
GEN4 program   286
generalised predictive control (GPC) algorithm   319
generalization   147–9
generalized Lyapunov function   196–7

generalized semi-Markov process   32
genetic adaptive control   126–8
genetic algorithms (GA)   116–19, 234
  design concerns   118–19
genetic operators   117–18
global approximation structure   149–50
global asymptotic stability   200
global positioning system (GPS)   249, 341, 356, 368
global system optimization   314–19
gradient algorithm   34–5, 155–7
  stability   156
gradient estimation   35–6
gradient vector   78
graphical user interfaces (GUIs)   281, 294
grid control   327–37
  structures   338–9
grid frequency regulation   331–4
guard equations   173
gyrators   81

H

Hankel matrix   85
hardware-in-the-loop (HIL)   260, 287, 386–7
Hatley–Pirbhai design method   379
heated exhaust gas oxygen sensors (HEGO)   125
Hessian matrix   78
heuristic construction of nonlinear controllers   122
hierarchical models   100
Hopf bifurcation   90
human safety   222–3, 232–3
human systems, power laws   235
HVAC systems   395–6, 407, 412–14
hybrid automaton   166, 170–7
  definition   171
  dynamics   173
  robotic systems   176
hybrid base isolation   429–30
hybrid mass damper (HMD)   420–9
hybrid specifications   177–9
hybrid state   172
hybrid state space   172
hybrid systems   29, 80, 354, 365–6
  analysis   179–83
  civil structures   420–30
  continuous state   172
  synthesis   183–5
  trajectory   168

hybrid trajectories   172
hysteresis   91

I

IDCOM   316, 319
idle periods   31
impedance control   454
implicit models   99
Independent System Operator (ISO)   340–1, 343–5
indirect adaptive control   126
indirect control systems   140–2
individuals, population of   116–17
industrial process control   291–323
  applications/production processes   297–300
  applications software   294–6
  information issues   299–300
  information technology infrastructure   293–4
  operations hierarchy   298–300
  standard options   296
  state of the art   292–6
  strategy issues   298–9
industrial three term control   306–13
infinitesimal perturbation analysis (IPA)   35–6, 39
influence functions   149–51
information processing power   225
information systems interoperability   67
information technology   399–404
initial state   27
input   22
input-output system   22
Inspection and Maintenance (IM)   240
  cycle   126
instantaneous error, cost function   155
integrated squared error (ISE)   177
intelligent control
  applications   122–30
  current research   131
  in robotics   443
  outlook on   130–1
  overview   104–33
  techniques   105–22
intelligent cruise control   358
intelligent data analysis   231
Intelligent Transportation Systems (ITS)   348–69
  fault tolerance/safety enhancement   368

Index

related problems   366–8
intelligent vehicle and highway systems (IVHS)   121, 350, 357–66
IntelliScout   231
Internet   231, 463
interpolation   148
interrupt latency   4
interrupt service routine (ISR)   4, 7–8, 11
interruptive systems   31
intersections, networks of   352–3
interval quality criteria   56
interval quality functions   56
   damping values   58
invariant equations   173
inverse dynamics control   450
inverse Hessian iteration   78
inverse kinematics   444
iterative modeling process   72

K

Kalman filter   157
kernel   11

L

labeling function   173
Labview   17, 295–6
ladder logic   18, 398
Lagrange multipliers   79
lane change controller   365
lane-keeping controller   363–4
lane tracking   361–4
Laplace equation   97
large deviations principle (LDP)   469–70
large-scale optimization   228–9
lattice-based approximators   151
learning algorithms   155, 160–2
least-mean-square (LMS) algorithm   155
least-squares
   algorithms   157–8
   identification problem   83
   parameter identification scheme   84
   with forgetting factor   158
lighting control   406
likelihood ratio   35
limit cycles   89
limited authority adaptive control   313
linear analysis models   281
linear control systems   134
linear-in-parameter (LIP) approximation   145–6

linear least-squares approximation, linearization for   76–7
linear models   75–7
linear parametric model   154
linear quadratic gaussian (LQG) control synthesis   42, 480
linear quadratic optimal control   273–9
linearization   92–5
   for linear least-squares approximation   76–7
linearly parameterized approximators   159
Linux   12, 17
load management strategy   316
local approximation structure   149
local area networks (LANs)   293, 399–401
local functions   144
local linearized approximation   92–3
logical models   29
Lon Talk   403–4
low level control strategies   315
lumped parameter models   80–1
lumped parameter system   79
Lyapunov stability and design methods   52, 142, 155, 158–60, 182, 196–7, 344, 447, 451
Lynx   6, 17

M

macromodels, scope and future   99–101
magnetorheological dampers   435–9
magnetorheological (MR) fluids   433–4
management information system (MIS)   299
manipulation feedback process   46
manufacturing systems, control hierarchy in   24
manufacturing yield   224–5
market behavior   345
market-oriented programming   255
marketing time and cost   223
Markov chain   28–9
Markov inequality   468
Markov parameters   83, 85
Markov process   22, 32–3, 39
MARX   256
mass airflow sensor (MAF)   125
mathematical models   74
MATRIXx   281, 286–7, 294
maximum overshoot   177

max-min fairness model   485–6
membership function   107–8, 124, 410–12
metrics   388–90
microelectromechanical systems (MEMS)   228
micromodels   99
miniaturization   225
minimizing parameter vector   146
minimum convergence rate line (MCRL)   359–60
missiles
   control law design   273–9
   dynamics   260–8
   flight control   259–90
   linear models   260–8
MIT rule   156
$M/M/1$ queue   32–4, 37
modal truncation   86
model-based predictive control (MPC)   316–19
   architectures   319
   industrial varieties   319
   key advantages   318–19
   tuning parameters   318
model checking iteration   181
model-free control   122–3
model predictive control (MPC)   128
model reduction   86–8
model reference adaptive control (MRAC)   128
modeling errors   160
modification feedback process   46
Monte Carlo   226
motherboard   18
MR fluid damper, full-scale seismic   436
multi-agent systems   254–5
multi-hidden layer networks   143
multi-input multi-output fuzzy systems   108
multilayer perceptrons   110–13
multilink flexible robots   123–4
multimodel compromising   66
multiobjective parameter optimization   62–3
multiple limit cycles   89
multitask scheduling   378
mutexes   175–6
mutual exclusion   174

N

network control units (NCUs)   404
network marking   171
networked DDC systems   401
networks of intersections   352–3
neural networks   110–16, 124, 408–10, 419, 430
   adaptive control   126–8
   approximation-based control   139–40
   control algorithm   410
   design concerns   115–16
   generalization applications   147
   performance index   410
   recognition applications   147
   training   113–15, 156
Newton's law   95
nodal processors   144
nonautonomous system   88
nondeterministic state machines   27–8
nondeterministic trajectories   168
noninterruption   31
nonlinear control law parameterization   53
nonlinear controllers, heuristic construction of   122
nonlinear dynamical systems   88–95, 234–5
nonlinear estimation, data-based   124–5
nonlinear models   77–9
nonlinear programming algorithms   62
nonlinear systems   134–5
   approximation problem   152–3
   stability   158–60
nonlinearities, common effects   89–92
normalized gradient algorithm   155
North American Electric Reliability Council (NERC)   333

O

objective function   33–4
observability grammian   87
on-board diagnostics   371–3
on-line approximation functions   140
on-line optimization   37–40
Open Access Same-time Information System (OASIS)   337
Open Systems Interconnection (OSI)   296
operator control language (OCL)   399
output error   83
overmodeled system   87

Index

P

parallel coordinates   59
parameter drift   161
parameter estimation error   159
parametric models   152–5
Pareto analysis   300, 302
Pareto-Optimal   59, 64
partial classification tree   74
partial differential equations (PDEs)   95–6
   classification   96–7
   elliptic   97
   finite difference models of   97–9
   hyperbolic   97
   parabolic   96–7
partition of unity   151
parts bin   168, 174
PATH group   354
peak rate constraints   482
performance
   analysis   283
   index   303
   monitoring   300–6
   quality indices   302–5
   versus cost   59
perturbation analysis   35
Petri nets   29, 40, 51, 171
phase-plane representation   91
physical systems
   integration   229
   modeling   48–9
PID control   1–7, 17, 51–2, 117, 121–3, 294–7, 306–13, 396–8, 443–8
   ramp response   6
   tuning   309–10
pitch autopilot block diagram   278
pitch loop autopilot   278
pitchfork bifurcation   90
planning systems   120–1
plant performance indices   300
platoon   354
pole placement   295, 449
population of individuals   116–17
positive control action   412
power laws   234–5
power system control and estimation   324–47
   institutional changes impacting control techniques   337
   new technologies   339–42

power system control development and research, future directions   343–6
power system control objectives   327–8
power system dynamics   327–37
power system stability-enhancing controls   334–7
power system stabilizers (PSS)   335–6
predictive model equation   317
primary industries   291
priority ceiling protocol   16–17
priority inversion   16
probabilistic methods   344
process cost function   317–18
process industries *see* industrial process control
Profit Suite   294
programmable logic controllers (PLCs)   2, 18–19, 24, 293, 394, 398
projection modification   161
proportional fairness   481–2
protocols   402–4
Public Utility Regulatory Policies Act of 1978 (PURPA)   326
pulse-width modulation (PWM)
   motor driver   12
   switching frequency   12

Q

QNX kernel   12
QNX operating system   13
QNX process states   13–15
QR decomposition   76
quality functions   54–6, 66, 68
quality levels   56
quality modeling   46, 54–60
quality-of-service (QoS)   465, 468, 473, 475
quantitative feedback theory (QFT)   314
queues   26–7, 30–1, 351, 467, 476–87
   probabilities of rare events   471–4

R

ramp control and merging   354–5
random process   469–70
random variables   468–9
randomized algorithms   225–6
rate management (RM) cells   480–1
rate matching   476
rate monotonic (RM) algorithm   8
reaction control system (RCS)   261
real-time computing and control   1–19
   hardware issues   17–18

real-time low-level programming 6–11
real-time operating systems 11–17
  at run time 13–17
real-time programming languages 11–17
real-time sources, admission control 464–5
rectangular hybrid automata 174
recursive least-squares algorithm, stability 158
redundancy control logic 52
regulatory compliance 223
relay experiment 310–12
response time 9
risk assessment 232–3
roadway transportation 348–69
robotic systems
  control 152, 442–61
  example 174–7
  force control 451–7
  free-floating 169
  historical perspective 442–3
  humanoid robot 230
  hybrid automaton 176
  kinematic control 443–6
  single degree of freedom 203–5
  visual servoing 457–9
robust control 53, 66, 109, 191, 195–6, 201–3, 251, 256, 260, 268, 274, 283–6, 313–14, 344, 361, 430, 443, 451
robust learning algorithms 160–2
robust multivariable predictive control technology (RPMCT)
  algorithm 319
robust PID algorithm (R-PID) 294
Robust Servomechanism Linear Quadratic Regulator (RSLQR) 273–9
robustness 283–6, 313–14
roll-yaw autopilot block diagram 279
routing 353–4, 466–7
RT-ARM 256
rule-base 53, 105–9, 119, 411–12

S

safety analysis, UCAVs 251–2
safety issues 222–3, 232–3
sampled data systems 80, 206–15
sampling frequency 13
satellite clusters 240, 253–6
  emergent behavior 255
scale-model studies 435–6
scenario resolution 365–6

scheduling 8–13, 25, 297–9, 378, 467–8
scheduling level 299
score function 35
script files 281
secondary industries 291
semaphores 15, 16, 175
semiactive control systems, civil engineering infrastructures 430–9
semiactive impact dampers 434
semi-Markov process 32–3
SEPIA software tool 230
sequential circuits 170
service rate 33
servo control system 3
setpoint optimization 316
shared memory 11
ship steering, fuzzy control 128–30
σ-modification 161
sigmoid (logistic) function 111
signalization 351–2
signum function 193
simulation tools, flight control systems 268–9
simulations, communication networks 483–6
SIMULINK 294, 379–84
single-board computers 17–18
single-output network functions 143
single-season model 111
singular, implicit, or differential-algebraic system 79
singular perturbation 86–8
singular value decomposition (SVD) 76
six degree-of-freedom equations of motion (EOM) 260
sliding manifold 190
sliding-mode control 168, 189–217, 364, 419, 429–30, 451
  applications 192
  basic principle 193–8
  general case 198–205
  problem formulation 199
  robust 195–6, 201–3
  sampled data control systems 206–15
sliding surface 190, 194, 200–1
  control law 195
societal connections 235–6
software project management 388–9
software testing 287
sojourn times 33, 36
solar system 72

Index

solution technologies 219–20
squashing functions 144
stability indicator eigenvalue
    damping 58
stable queue 33
stable training algorithm 137–8
standard network variable types
    (SNVTs) 404
star topology 100
STARC (Stochastic Team Algorithm for
    Rate Control) 479, 484
state event systems 79
Stateflow 379–83
state machines 26–7, 29
state models 22, 26–33
state space 26
    models 81, 84–6
state trajectory 26
state transition 26–8
    diagram 26–8
static models 75–9
static nonlinear approximation
    problem 154
statistical multiplexing 464
statistical process control (SPC) 301
steady-state average sojourn time 33, 36
steady-state mean sojourn time 36
steepest descent 78
steering control 361
step sizes 34, 97–8
stochastic approximation 35
stochastic modeling 101
Stone-Weierstrass Theorem 145
superfast algorithms 67
supervisory control 29, 101
    models 100–1
supervisory hybrid systems 165–217
    definition 167
    examples 166–70
    theory 166
supervisory level 299
supervisory system command
    structure 315–16
supply chain 294
sustained oscillation procedure 307–9
SWARM 254
switched systems 167–70
switching surface 190
symbolic model checking (SMC) 170, 180–1
synchronous generator dynamics 328–31
synchronous sequential digital circuits 170
synthesis algorithms 51–4

synthesis model 47
synthesis-specific criteria 55
synthesis tuning 54
synthesize-search feedback process 46
system
    analysis 22
    definition 71
    identification 82–8, 120
system control, difficulties with building
    413–14
system engineering 233–4, 457
    model 374
    process 373–6
system modeling 71–103
    classification 73–4
    common variables 80
    control systems 73
    historical perspective 72–3

T

Takagi-Sugeno fuzzy systems 108
task-specific control requirements 55
Taylor series expansion 78, 92
TCAS 245, 248
TCP 465–6
technoscience 236
threshold function 111
throttle position sensor (TPS) 125
thrust vector control (TVC) 252, 266–8
time- (or frequency-) discretized
    indicator function 57
timed automaton 174
timed-discrete-event system model 30
timed state trajectory 31
timing 4–5, 18
timing jitter 4, 18
tracking errors 4, 138, 203, 205
tracking errors dynamics 193–4
tracking problem 193–4
TRACON 242
traffic control on highways 354–5
traffic related issues 351–7
transducers 80–1
transfer function coefficients 83
transfer function models 81, 83–4
transformers 80
transition rule 27
Transmission Control Protocol/Internet
    Protocol (TCP/IP) 463
transportation capacity 225
tunable functions 124

tuned liquid column dampers (TLCD) 431–2
tuned mass damper (TMD) 420, 425, 432
tuned sloshing dampers (TSD) 432
Tuning & Compromising 46
tuning loop 63

U

UCAVs 240, 249–53
    autonomy 252–3
    conflict resolution 251–2
    inter-fleet and central-command-to-fleet communications 250–1
    safety analysis 251–2
underwater vehicles 230
uniform information model 67
uninhabited combat air vehicles see UCAVs
unique minimum 146
uniqueness set 147
universal approximation 144
    Theorems 147
universal approximator 143–5
Universal Power Flow Controller 342
universe of discourse 107
unmanned aircraft, flight control systems 289–90
unsteady diffusion equation 97
untimed models 29

V

validation tests 179, 182
Vapnik-Chervonenkis (VC) dimension 226
variable bit rate (VBR) 465, 484
variable delay 481
variable-friction dampers 431
variable-orifice dampers 430–1
variable structure 189–217

vector output single hidden layer networks 143
vehicle autonomy 229–31
vehicle dynamics 362
vehicle location systems 356
vehicle model 362–3
verification tests 179
very large scale integrated (VLSI) chips 170, 234
viability kernels 182
vibration damping, flexible-link robot 123–4
video traffic 38–9
virtual circuit routing 466
virtual engineering 44, 47, 233–4
virtual instrumentation 295
visual servoing, robotic systems 457–9
visualization for comparative design exploration 59–60
voltage stability (or voltage collapse) problem 336–7
VxWorks 451

W

WALRAS 256
wiggle system 276
Windows NT 12, 17
work area 174
World Wide Web (WWW) 231

Y

Yule-Walker algorithm 483

Z

z-domain transfer function 83
Zeno systems 168
Ziegler-Nichols tuning 307–9

ABOUT THE EDITOR

Tariq Samad is Chief Fellow at Honeywell Technology Center, where he is involved in a number of projects and initiatives in various areas of systems and control technology. Much of his research has focused on intelligent control systems, including neural networks, genetic algorithms, and agent-based systems, and he has explored applications of these techniques to aerospace, industrial, and other domains. Dr. Samad has also been active in establishing several university/industry collaborations, both within the United States and internationally, and he is currently an adjunct professor in the department of electrical engineering at Georgia Institute of Technology. Dr. Samad received a B.S. degree in engineering and applied science from Yale University, and M.S. and Ph.D. degrees in electrical and computer engineering from Carnegie Mellon University.

Dr. Samad holds eleven patents and is a Honeywell STAR inventor. He has authored or co-authored about 100 publications. He is the author of *A Natural Language Interface for Computer Aided Design* (Kluwer Academic Publishers, 1986) and co-editor of *Automation, Control and Complexity: An Integrated View* (John Wiley and Sons, in press). He is on the editorial board of *Neural Processing Letters* and is the Editor-in-Chief of *IEEE Control Systems Magazine*, the largest circulation of a technical periodical dedicated to all aspects of control systems.

Dr. Samad is currently serving his second term as an elected member of the IEEE Control Systems Society's Board of Governors (1997-2002). His past service to CSS has included chairing the Technical Committee on Industrial Process Control (1996-97) and serving as vice president of Technical Activities (1998); it was in the latter capacity that he conceived and initiated this volume.